Fiberglass and Glass Technology

Frederick T. Wallenberger · Paul A. Bingham
Editors

Fiberglass and Glass Technology

Energy-Friendly Compositions and Applications

 Springer

Editors
Frederick T. Wallenberger
Consultant
708 Duncan Avenue
Apartment 1108
Pittsburgh, Pennsylvania 15237
United States
wallenbergerf@aol.com

Paul A. Bingham
Department of Engineering Materials
University of Sheffield
Sir Robert Hadfield Building
Sheffield S1 3JD
United Kingdom
p.a.bingham@sheffield.ac.uk

ISBN 978-1-4419-0735-6   e-ISBN 978-1-4419-0736-3
DOI 10.1007/978-1-4419-0736-3
Springer New York Dordrecht Heidelberg London

Library of Congress Control Number: 2009938639

Printed on acid-free paper

Springer is part of Springer Science+Business Media (www.springer.com)

Preface

This book offers a comprehensive view of fiberglass and glass technology with emphasis on energy-friendly compositions, manufacturing practices, and applications, which have recently emerged and continue to emerge. Energy-friendly compositions are variants of incumbent fiberglass and glass compositions. They are obtained by reformulation of incumbent glass compositions in order to reduce the melt viscosity and increase the melting rate, thereby saving process energy and reducing environmental emissions. As a result, new energy-friendly compositions are expected to become a key factor in the future for the fiberglass and glass industries. The contributors to the book consist of both academic and industrial scientists. This book is therefore dedicated to those in the academic and industrial community who seek an understanding of the past in order to make progress in the future.

This book consists of three interrelated sections. Part I reviews a wide range of continuous glass fibers, their compositions, and properties. Dr. F. T. Wallenberger authored Chapter 1, which reviews important glass fibers ranging from commercial 100% SiO2 glass fibers and commercial multi-oxide glass fibers to experimental glass fibers containing 81% Al_2O_3. Dr. Wallenberger also authored Chapter 2. This chapter offers a new method (trend line design) for designing environmentally and energy-friendly E-, ECR-, A-, and C-glass compositions to reduce the process energy by compositional reformulation. Few fiberglass applications are based on yarns; most are based on composites. Dr. J. H. A van der Woude and Dr. E. L. Lawton authored Chapter 3, which reviews fiberglass composite engineering with an important sub-chapter on windmill blade construction. Dr. A. V. Longobardo wrote Chapter 4. It reviews the glass fibers which became available as reinforcement for printed circuit boards and analyzes their compositions as well as the needs of the market. Finally Dr. R. L. Hausrath and Dr. A. Longobardo authored Chapter 5, which reviews high-strength glass fibers and analyzes existing and emerging markets for these products.

Part II of the book deals with soda–lime–silica glass technology. The first two chapters (Chapters 6 and 7) parallel the first two chapters in Part I (Chapters 1 and 2). Dr. Ing. A. Smrček wrote Chapter 6. It is devoted to a wide range of industrial flat, container, and technical glass compositions and to an in-depth review of their properties. Dr. P. A. Bingham authored Chapter 7. It deals with the design of new energy-friendly flat, container, and technical glass melts through reformulation of existing compositional variants to reduce the process energy.

Part III of the book deals with emerging glass melting science and technology, and is conceptually applicable to both glass and fiberglass melts. Prof. Dr. H.-J. Hoffmann authored Chapter 8, which offers new insights into the basics of melting and glass formation at the most fundamental level. Prof. Dr. R. Conradt wrote Chapter 9, which deals with the thermodynamics of glass melting, offers a model to predict the thermodynamic properties of industrial multi-component glasses from their chemical compositions, addresses the role of individual raw materials in the melting process of E-glass, and facilitates the calculation of the heat of the batch-to-melt conversion. Prof. Dr. H. A. Schaeffer and Priv. Doz. Dr.-Ing. H. Müller-Simon authored Chapter 10, which reviews the use of in situ sensors for monitoring glass melt properties and monitoring species in the combustion space and also reviews redox control of glass melting with high levels of recycled glass to enhance the environmentally friendly value of the resulting glass. Dr. R. Gonterman and Dr. M. A. Weinstein authored Chapter 11, which deals with the recently emerging plasma melt technology and its potential applications.

Dr. Wallenberger wishes to acknowledge his years at PPG Industries from 1995 to 2008, especially the invitation he received from the late John Horgan, Vice President, Fiberglass, to join PPG, and the support of Dr. Jaap van der Woude, who as Director of Research, encouraged him in 1997 to pursue innovative compositional fiberglass research. Dr. Wallenberger gratefully acknowledges the help of Dr. Bingham, co-editor of the book, and the valuable contributions of the chapter authors. Finally, Dr. Wallenberger wishes to acknowledge the thoughtful support of Jennifer Mirski who, as assistant editor, Springer Publisher, helped in editing the entire book.

Dr. Bingham wishes to thank the following people for their help, insight, comments, and encouragement: Dr. Fred Wallenberger, co-editor of the book; Prof. Michael Cable for many interesting discussions; all of our contributing authors; colleagues present and past at the Society of Glass Technology, the University of Sheffield, and the British Glass Manufacturer's Confederation; and the many other individuals with whom he has held discussions over the years on the fascinating subject of glass. Finally and most importantly, he thanks his family for their endless patience, love, support, and encouragement.

F. T. Wallenberger   Pittsburgh, PA, USA
P. A. Bingham   Sheffield, UK

About the Editors

Dr. Frederick T. Wallenberger was an instructor in Chemistry at Fordham University (1957–1958), a research fellow at Harvard (1958–1959), and a scientist at DuPont Fibers, Pioneering Research Laboratory (1959–1992). He retired in 1992 from DuPont and became a research professor (Materials Science) at the University of Illinois in Urbana-Champaign (1992) and a visiting professor (Textiles) at the University of California in Davis (1994). He joined PPG as a staff scientist in 1995, retired in 2008, and serves as a consultant now. He studied the relationships between structures, properties, and value-in-use of new materials and contributed to the commercialization of new fibers through "intrapreneurial" research, project management, and technology transfer. He is an expert in the fields of advanced glass fibers, ceramic fibers, carbon fibers, natural fibers, polymer fibers, single crystal fibers, and composites.

Dr. Wallenberger has over 150 papers in the refereed scientific literature, including three in the journal science, edited four books (*Advanced Inorganic Fibers*, 1999; *Advanced Fibers*, 2002; *Natural Fibers*, 2004; and *Fiberglass and Glass Technology*, 2009), wrote two recent review articles (*Introduction to Reinforcing Fibers* and *Glass Fibers*, *ASM Handbook on Composites*, 2002), and received 10 US Patents. He chaired three major symposiums ("Chemistry and Environment," American Chemical Society, 1974; "Advanced Fibers, Plastics and Composites," Materials Research Society, 2001; "Behavior of Glass Melts," Gordon Research Conference, 2005), gave three invited Gordon Research Conference lectures ("Aramid Fibers," 1964; "Foamed Polyester Fibers," 1975; and "Glass Fibers," 1992) and wrote the first review article that included Kevlar^{TM} ("The Chemistry of Heat Resistant Polymer Fibers," Angewandte Chemie, 1964)

Dr. Wallenberger received the Environmental Respect Award from DuPont (1992). He is a member of the Association of Harvard Chemists (1958–) and was elected a Fellow of the American Ceramic Society (2005). His biography appears in several standard references including *Who's Who in the World*, *Who's Who in America*, and *Who's Who in Science and Engineering*.

Dr. Paul A. Bingham received his BEng (Hons) degree in Materials Science and Engineering from the University of Sheffield in 1995. He then studied toward a Ph.D. (1995 to 1999, thesis title "The Environment of Iron in Silicate Glasses") at the same institution, which hosts one of the world's premier glass science departments. From 1999 to 2003 he was employed as a technologist by Glass Technology Services Ltd (GTS) a subsidiary of the British Glass manufacturers confederation where he carried out a wide range of glass and ceramic-related R&D, project management, and industrial production problem solving. He was promoted to senior technologist in 2001. Projects ranged from laboratory-scale development of soda–lime–silica glass, E-glass, lead crystal glass, sealing glass, and glasses with novel physical properties to managing industrial-scale glass compositions, melting, recycling, energy, and emissions. Environmentally friendly glass compositions that he developed have reached full-scale production. In 2004, Dr. Bingham took up a position as postdoctoral research associate at the Immobilisation Science Laboratory (ISL), University of Sheffield. Dr. Bingham's chief research interests lie at the boundary between materials science and environmental disciplines. These include composition/structure/property relations in novel glass-forming systems; formulation of glasses for the safe immobilization of legacy and problematic nuclear and toxic wastes; modification of commercial glass and ceramic materials and the re-use of waste materials therein for environmental benefit, energy efficiency, and reduced atmospheric emissions.

Dr. Bingham has published over 30 peer-reviewed scientific papers and he regularly reviews manuscripts for several major international journals. He has presented his research, chaired sessions, and given invited presentations at a number of international conferences on glass science and waste management. He has a strong track record in obtaining funding and access to scientific facilities from both academic and industrial sources and he has built many lasting national and international collaborations. He has been an active member of the Society of Glass Technology for over 10 years and was elected onto the Basic Science and Technology Committee in 2000. In 2004 he was elected committee secretary.

Contents

Part I   Continuous Glass Fibers

1   Commercial and Experimental Glass Fibers   3
Frederick T. Wallenberger
   1.1   Overview: Glass Melt and Fiber Formation   3
      1.1.1   Principles of Glass Melt Formation   3
      1.1.2   Principles of Glass Fiber Formation   9
      1.1.3   Structure of Melts and Fibers   11
      1.1.4   Summary and Conclusions   15
   1.2   Silica Fibers, Sliver, and Fabrics (95–100% SiO_2)   15
      1.2.1   Ultrapure Silica Fibers (99.99–99.999% SiO_2)   15
      1.2.2   Pure Silica Sliver and Fabrics (95.5–99.5% SiO_2) . .   19
      1.2.3   Summary and Conclusions   22
   1.3   Silicate Glass Fibers (50–70% SiO_2, 1–25% Al_2O_3)   23
      1.3.1   Forming Glass Fibers from Strong Viscous Melts . . .   23
      1.3.2   General-Purpose Silicate Glass Fibers   28
      1.3.3   Special-Purpose Silicate Glass Fibers   34
      1.3.4   Non-round, Bicomponent and Hollow Silicate Fibers .   54
      1.3.5   Summary and Conclusions   60
   1.4   Aluminate Glass Fibers (≤81% Al_2O_3, ≤50% SiO_2)   60
      1.4.1   Glass Fibers from Fragile Melts (25–50%
            Al_2O_3, 10–4% SiO_2)   60
      1.4.2   Glass Fibers from Inviscid Melts
            (55–81% Al_2O_3, 4–0% SiO_2)   66
   1.5   Appendix: Single-Crystal Alumina Fibers   77
      1.5.1   Single-Crystal Fibers from Inviscid Melts   77
      1.5.2   The Future of Alumina and Aluminate Fibers   82
   References   84

2   Design of Energy-Friendly Glass Fibers   91
Frederick T. Wallenberger
   2.1   Principles of Designing New Compositions   91
      2.1.1   Compositional, Energy, and Environmental Issues . .   91

        2.1.2   Trend Line Design of New Fiberglass Compositions .   94

2.2   Energy-Friendly Aluminosilicate Glass Fibers   99

        2.2.1   New Energy-Friendly E-Glass Variants with

                 $< 2\% \; B_2O_3$ 99

        2.2.2   New Energy-Friendly E-Glass Variants with

                 $2-10\% \; B_2O_3$ 111

        2.2.3   New Energy- and Environmentally Friendly

                 ECR-Glass Variants   114

2.3   Energy-Friendly Soda–Lime–Silica Glass Fibers   116

        2.3.1   New Energy-Friendly A- and C-Glass Compositions .   117

2.4   Summary, Conclusions, and Path Forward   119

References   121

3   Composite Design and Engineering   125

J.H.A. van der Woude and E.L. Lawton

3.1   Introduction   125

        3.1.1   Continuous Fibers for Reinforcement   125

        3.1.2   E-Glass Fibers   127

        3.1.3   Fiberglass Manufacturing   128

        3.1.4   Fiberglass Size   129

        3.1.5   Composite Mechanical Properties   130

        3.1.6   Products   138

3.2   Thermoset Composite Material   141

        3.2.1   Liquid Resin Processing Techniques   142

        3.2.2   Thermosetting Matrix Resins   148

        3.2.3   Fillers   154

        3.2.4   Release Agents   155

3.3   Reinforced Thermoplastic Materials   156

        3.3.1   Introduction   156

        3.3.2   Semifinished Materials Based on Thermoplastics . . .   158

3.4   Composites for Wind Turbines   168

        3.4.1   Introduction   168

        3.4.2   Raw Materials   169

        3.4.3   Blade-Manufacturing Techniques   169

        3.4.4   Blade Design Methodologies   170

References   172

4   Glass Fibers for Printed Circuit Boards   175

Anthony V. Longobardo

4.1   Introduction   175

        4.1.1   Printed Circuit Board Requirements and Their

                 Implications for Fiberglass   176

        4.1.2   Fiberglass' Role in PCB Construction   177

        4.1.3   Electrical Aspects   179

        4.1.4   Structural Aspects   181

	4.2	Glass Compositional Families	184
		4.2.1 Improvements Initially Based on E-Glass	184
		4.2.2 D-Glass and Its Compositional Improvements	188
	4.3	Future Needs of the PCB Market	191
		4.3.1 The Electronics Manufacturer's Roadmap	191
		4.3.2 What This Means for the Board and Yarn Makers	192
	References		195

5 High-Strength Glass Fibers and Markets   **197**
Robert L. Hausrath and Anthony V. Longobardo
	5.1	Attributes of High-Strength Glass	197
		5.1.1 Strength	198
		5.1.2 Elastic Modulus	203
		5.1.3 Thermal Stability	205
	5.2	Glass Compositional Families	206
		5.2.1 S-Glass	207
		5.2.2 R-Glass	208
		5.2.3 Other High-Strength Glasses	209
	5.3	High-Strength Glass Fibers in Perspective	210
		5.3.1 The Competitive Material Landscape	210
		5.3.2 Inherent Advantages of Continuous Glass Fibers	215
	5.4	Markets and Applications	215
		5.4.1 Defense – Hard Composite Armor	216
		5.4.2 Aerospace – Rotors and Interiors	218
		5.4.3 Automotive – Belts, Hoses, and Mufflers	220
		5.4.4 Industrial Reinforcements – Pressure Vessels	221
	5.5	Concluding Remarks	222
	References		223

Part II   Soda–Lime–Silica Glasses

6 Compositions of Industrial Glasses   **229**
Antonín Smrček
	6.1	Guidelines for Industrial Glass Composition Selection	229
		6.1.1 Economics	230
		6.1.2 Demands on the Glass Melt	230
		6.1.3 Meltability	232
		6.1.4 Workability	233
		6.1.5 Choice of Raw Materials	235
		6.1.6 Cullet Effect – Glass Melt Production Heat	236
		6.1.7 Glass Refining	237
	6.2	Industrial Glass Compositions	240
		6.2.1 Historical Development	240
		6.2.2 Flat Glass	242
		6.2.3 Container Glass	245
		6.2.4 Lead-Free Utility Glass	250

	6.2.5	Technical Glass	253
	6.2.6	Lead Crystal	259
	6.2.7	Colored Glasses	261
6.3		Example Glass Compositions	261
	6.3.1	Perspectives	261
	6.3.2	Practical Examples of Container Glass Batch Charge	262
References			266

7 Design of New Energy-Friendly Compositions 267
Paul A. Bingham

7.1		Introduction	267
7.2		Design Requirements	268
	7.2.1	Commercial Glass Compositions	269
7.3		Environmental Issues	269
	7.3.1	Specific Energy Consumption	269
	7.3.2	Atmospheric Emission Limits	271
	7.3.3	Pollution Prevention and Control	271
7.4		Fundamental Glass Properties	278
	7.4.1	Viscosity–Temperature Relationship	279
	7.4.2	Devitrification and Crystal Growth	281
	7.4.3	Conductivity and Heat Transfer	286
	7.4.4	Interfaces, Surfaces, and Gases	291
	7.4.5	Chemical Durability	297
	7.4.6	Density and Thermo-mechanical Properties	299
7.5		Design of New SLS Glasses	300
	7.5.1	Batch Processing, Preheating, and Melting	300
	7.5.2	Cullet	302
	7.5.3	Silica, SiO_2	304
	7.5.4	Soda, Na_2O	305
	7.5.5	Calcia, CaO	307
	7.5.6	Magnesia, MgO	309
	7.5.7	Alumina, Al_2O_3	310
	7.5.8	Potassia, K_2O	313
	7.5.9	Lithia, Li_2O	315
	7.5.10	Boric Oxide, B_2O_3	316
	7.5.11	Sulfate, SO_3	318
	7.5.12	Water, H_2O	321
	7.5.13	Chlorides and Fluorides	322
	7.5.14	Baria, BaO	323
	7.5.15	Zinc Oxide, ZnO	323
	7.5.16	Strontia, SrO	324
	7.5.17	Multivalent Constituents	324
	7.5.18	Other Compounds	327
	7.5.19	Recycled Filter Dust	329
	7.5.20	Nitrates	329

7.6   Glass Reformulation Methodologies   330
        7.6.1   Worked Examples and Implementation   330
        7.6.2   Reformulation Benefits and Pitfalls   341
        7.6.3   Research Requirements and Closing Remarks   343
References   345

Part III   Glass Melting Technology

8   Basics of Melting and Glass Formation   355
Hans-Jürgen Hoffmann
    8.1   Motivation   355
    8.2   Former Melting Criteria   356
    8.3   Analysis of the Enthalpy Functions of One-Component
            Systems   359
            8.3.1   Theoretical Preliminaries   359
            8.3.2   Pre-melting Range and the Contribution to the
                    Molar Specific Heat Capacity by Electrons   361
    8.4   Melting and the Glass Transformation   365
    8.5   Effects Occurring in the Glass Transformation Range   368
    8.6   What Makes Solids and Melts Expand?   369
    8.7   Modulus of Compression of the Chemical Elements   375
    8.8   Necessary Criteria for Glass Formation   375
    8.9   Possible Extension to Multi-Component Systems   381
    8.10   Discussion   381
    References   382

9   Thermodynamics of Glass Melting   385
Reinhard Conradt
    9.1   Approach to the Thermodynamics of Glasses
            and Glass Melts   385
            9.1.1   Description Frame for the Thermodynamic
                    Properties of Industrial Glass-Forming Systems . . .   386
            9.1.2   Heat Content of Glass Melts   388
            9.1.3   Chemical Potentials and Vapor Pressures of
                    Individual Oxides   391
            9.1.4   Entropy and Viscosity   394
    9.2   The Role of Individual Raw Materials   395
            9.2.1   Sand   395
            9.2.2   Boron Carriers   397
            9.2.3   Dolomite and Limestone   400
    9.3   The Batch-to-Melt Conversion   404
            9.3.1   Stages of Batch Melting   404
            9.3.2   Heat Demand of the Batch-to-Melt Conversion   405
            9.3.3   Modeling of the Batch-to-Melt Conversion
                    Reaction Path   407
    References   409

10 Glass Melt Stability 413
    Helmut A. Schaeffer and Hayo Müller-Simon
    10.1   Introduction 413
    10.2   Target Properties of Glass Melt and Glass Product 414
           10.2.1   Batch-Related Fluctuations 415
           10.2.2   Combustion-Related Fluctuations 416
           10.2.3   Process-Related Fluctuations 416
    10.3   In Situ Sensors 417
           10.3.1   Sensors for Monitoring Glass Melt Properties 418
           10.3.2   Sensors for Monitoring Species in the
                    Combustion Space 422
    10.4   Examples of Glass Melt Stability Control 423
           10.4.1   Redox Control of Glass Melting with High
                    Portions of Recycled Glass 423
           10.4.2   Redox Control of Amber Glass Melting 425
    10.5   Conclusions and Outlook 427
    References 427

11 Plasma Melting Technology and Applications 431
    J. Ronald Gonterman and M.A. Weinstein
    11.1   Concepts of Modular and Skull Melting 431
    11.2   The Technology of High-Intensity DC-Arc Plasmas 433
           11.2.1   Conductive 434
           11.2.2   Radiant 435
           11.2.3   Joule Heating 436
    11.3   Brief History of Plasma Melting of Glass 437
           11.3.1   Johns-Manville 437
           11.3.2   British Glass Institute 438
           11.3.3   Plasmelt Glass Technologies, LLC 438
           11.3.4   Japanese Consortium Project 439
    11.4   DOE Research Project – 2003–2006 440
           11.4.1   Acknowledgments 440
           11.4.2   Experimental Setup of the Plasmelt Melting System . 440
           11.4.3   Technical Challenges of Plasma Glass Melting 442
           11.4.4   Glasses Melted: Results and Broad Implications . . . 444
           11.4.5   Synthetic Minerals Processing Implications 447
           11.4.6   Energy Efficiency vs. Throughput 448
    11.5   Future Applications for Plasma Melting 450
    11.6   Summary and Conclusions 451
    References 451

Index 453

Contributors

Paul A. Bingham Department of Engineering Materials, The University of Sheffield, Sheffield S1 3JD, UK, p.a.bingham@sheffield.ac.uk

Reinhard Conradt Department of Glass and Ceramic Composites, RWTH Aachen University, Institute of Mineral Engineering, 52064 Aachen, Germany, conradt@ghi.rwth-aachen.de

J. Ronald Gonterman Plasmelt Glass Technologies, LLC, Boulder, CO 80301, USA, Ron@plasmelt.com

Robert L. Hausrath AGY World Headquarters, Aiken, SC 29801, USA, robert.hausrath@agy.com

Hans-Jürgen Hoffmann Institute of Materials Science and Technology: Vitreous Materials, University of Technology of Berlin, 10587 Berlin, Germany, hoffmann.glas@tu-berlin.de

E. L. Lawton Fiber Polymer Composite Consulting, LLC, 3432 Kilcash Drive, Clemmons, North Carolina 27012, USA, elawton150@aol.com

Anthony V. Longobardo AGY World Headquarters, Aiken, SC 29801, USA, Anthony.Longobardo@agy.com

Hayo Müller-Simon Research Association of the German Glass Industry (HVG), 63071, Offenbach, Germany, mueller-simon@hvg.dgg.de

Helmut A. Schaeffer Research Association of the German Glass Industry (HVG), 63071, Offenbach, Germany, helmut.schaeffer@gmx.net

Antonin Smrček VÚSU, Teplice, Czech Republic, sklarakeramik@seznam.cz

J. H. A. van der Woude Fiber Glass Science and Technology, Europe, PPG Industries Inc., Hoogezand, The Netherlands 9600AB, vanderwoude@ppg.com

Frederick T. Wallenberger Consultant, 708 Duncan Avenue, Apartment 1108, Pittsburgh PA 15237, USA, wallenbergerf@aol.com

Michael A. Weinstein Plasmelt Glass Technologies, LLC, Boulder, CO 80301, USA, Mike@plasmelt.com

Part I
Continuous Glass Fibers

Chapter 1
Commercial and Experimental Glass Fibers

Frederick T. Wallenberger

Abstract Continuous glass fibers can be formed from melts with a wide range of compositions and viscosities. This chapter reviews pure silica fibers which are formed from highly viscous melts, silicate glass fibers with 50–70% SiO_2 which are formed from moderately viscous melts, aluminate glass fibers with 50–80% Al_2O_3, as well as yttria-alumina-garnet (YAG) glass fibers which are formed from inviscid (literally non-viscous) melts. Commercial glass fibers are made for a variety of applications from pure silica rods and from silicate melts containing 50–70% SiO_2 and 10–25% Al_2O_3. Boron-free, essentially boron-free, and borosilicate E-glass are general-purpose fibers. ERC-glass offers high corrosion resistance, HS-glass offers high-strength composites, D-glass offers a low dielectric constant, and A-glass offers the possibility of using waste container glass for less demanding applications.

Keywords Continuous glass fibers · Viscous and inviscid fiberglass melts · Glass melting and fiber formation · Experimental and commercial glass fibers · Commercial E-glass · ECR-glass · D-glass · HS-glass and A-glass fibers · Structures and properties

1.1 Overview: Glass Melt and Fiber Formation

The viscosity of glass melts depends on the SiO_2 content, the crystallization resistance at the liquidus temperature of these melts depends, at the same Al_2O_3 level, on the SiO_2:RO ratio (RO = CaO + MgO), and the modulus of the resulting glass fibers depends on the Al_2O_3 content at the same SiO_2 content.

1.1.1 Principles of Glass Melt Formation

Viscous melts can be strong melts typically containing 50–100% SiO_2 and 25–0% Al_2O_3 [1, 2] or fragile melts typically containing 4–10% SiO_2 and 50–25% Al_2O_3

F.T. Wallenberger (✉)
Consultant, 708 Duncan Avenue, Apartment 1108, Pittsburgh, PA 15237, USA
e-mail: wallenbergerf@aol.com

F.T. Wallenberger, P.A. Bingham (eds.), *Fiberglass and Glass Technology*,
DOI 10.1007/978-1-4419-0736-3_1, © Springer Science+Business Media, LLC 2010

[3, 5]. Inviscid (literally non-viscous) melts contain 0–4% SiO_2 and 50–80% Al_2O_3 [4, 5]. All melts except those based on 100% SiO_2 may also contain CaO, MgO, B_2O_3, and/or other oxides (Table 1.1).

1.1.1.1 Important Glass Melt Properties

The most important melt properties of fiberglass melts are the fiber-forming viscosity (FV), the temperature at which fibers are formed (FT or T_F), the liquidus temperature at which crystals can form within hours (LT or T_L) and remain in equilibrium with the melt, and the delta temperature (ΔT) [1, 2], or the difference between the fiber-forming and liquidus temperatures.

In the technical and patent literature, the relationship between melt viscosity and melt temperature is expressed by a Fulcher curve that relates the melt viscosity (in log poise) to the linear melt temperature (in °C). At the point of fiber formation, glass melts have viscosities ranging from log 2.5 to log 3.0 poise and the temperature of a melt at a viscosity of log 3.0 poise is generally accepted as the "fiber-forming temperature" (or log 3 FT).

Other melt parameters, which are frequently used to characterize fiberglass melts, include the glass transition temperature (T_g), the liquidus temperature (LT), and the difference between the log 3 forming and liquidus temperatures that is also known as delta temperature (ΔT). The glass transition and log 3 forming temperatures are related to the viscosity. The liquidus temperature is related to the phase diagram and the delta temperature relates the phase diagram (the crystallization behavior of a melt) to the melt viscosity.

The melt viscosity (in log poise) has been correlated with the glass transition temperature (T_g, K) divided by the melt temperature (T_m, K) [3]. This theory distinguishes "strong" melts from "fragile" melts. Strong melts include the silica melt that is seen as the ideal melt because it exhibits a perfect straight-line behavior while fragile melts include all silicate melts, which deviate to various degrees from the ideal straight-line behavior.

According to this theory [3], fragile melts include glass fiber-forming aluminum silicate melts, which are viscous at the fiber-forming temperatures but only in a supercooled state. These melts show notable deviations from the straight-line behavior of an ideal strong melt. However, it would also include glass fiber-forming aluminate melts with <80% alumina as well as yttrium aluminum garnet melts. Both melts deviate dramatically from the ideal straight-line behavior. They have melt viscosities, which are comparable to that of honey at room temperature, and sharp melting points.

Another point of view [5], therefore, considers fiberglass melts with ultralow melt viscosities as "inviscid" (literally non-viscous) and as representing a third class of glass melts with regard to glass fiber formation. The most important reason for this classification is the fact that these melts require dramatically different fiber-forming processes. As a result, this point of view has been accepted throughout this book as the basis for classifying the fiberglass melts as strong, fragile, or inviscid.

The relationship between viscosity and melt temperature is generally represented by a Fulcher curve, wherein the viscosity is expressed in log poise and the

Table 1.1 Fiber formation from viscous and inviscid melts [1] (with kind permission of Springer Science and Business Media)

Melt	Viscosity			Composition	Glass fiber formation		
	$< LT^a$	$> LT^a$	$> MP^a$		By	From	Environment
Strong viscous	Very high	Very high		Pure silica	Downdrawing > LT	Preform	Ambient air
Strong viscous	Very high	Very high		Silicate melts	Downdrawing > LT	Bushing	Ambient air
Fragile viscous	Very high	Very low		Yttria silicate	Downdrawing < LT	Precision bushings	Ambient air
Fragile viscous	Very high	Very low		Fluoride and aluminate	Up- or down- drawing < LT	Supercooled melts	Ambient air
Inviscid melt			Very low	Yttria-aluminate	Downdrawing < MP	Levitated, laser-heated melt	Containerless
Inviscid melt			Very low	Binary aluminate	Melt spinning > MP	Stabilized jet surface	Reactive gas

^{a}Liquidus temperature (LT), melting point (MP), rapid solidification (RS) process

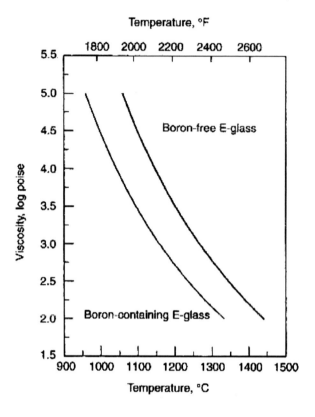

temperature in degree Celsius. Figure 1.1 shows the typical Fulcher curves of a standard borosilicate E-glass containing 6.6% B_2O_3 and 0.7% fluorine and of a boron-free E-glass with <0.2% fluorine.

By replacing the log viscosity with a linear viscosity [1] it becomes evident that the relationship between the viscosity of strong, fragile, and inviscid melts in the fiber-forming range between 317 (log 2.5) poise and 1000 (log 3.0) poise is essentially linear (Fig. 1.2). Minor changes of the melt viscosity in a commercial furnace will therefore not disproportionately affect the fiber formation at the bushing tips.

Figure 1.2 compares six fiberglass melts in the range of their fiber-forming viscosity (FV) between log 2.5 and log 3.0 poise, by comparing their log 3 fiber-forming temperature (FT) and their liquidus temperature (LT). In Fig. 1.2, the strong viscous E-glass melt with ~6% B_2O_3 has the lowest log 3 forming temperature (log 3 FT) and the inviscid yttrium aluminum garnet (or YAG) melt has the highest log 3 forming temperature.

The liquidus temperature (LT), i.e., the highest temperature at which crystals can form within hours and remain in an equilibrium state with the melt is an important factor in commercial use. Commercial experience suggests that the log 3 forming temperature of a strong melt must be at least 50°C higher in a continuous furnace than the liquidus temperature.

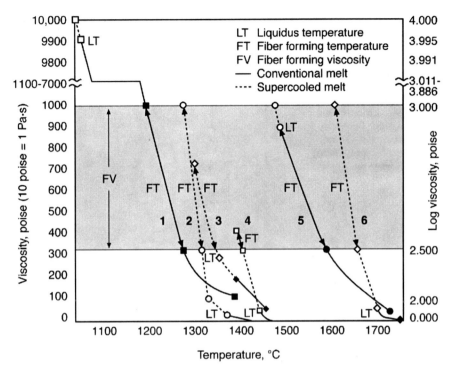

Fig. 1.2 Relationship between linear viscosity (poise) and melt temperature [1] (with kind permission of Springer Science and Business Media)

Commercial E-glass (Fig. 1.2, #1) and S-glass fibers (Fig. 1.2, #5) were formed from strong viscous melts in a continuous process. The yttria silicate (Fig. 1.2, #2) and the quaternary aluminate fiber (Fig. 1.2, #3) were formed from supercooled fragile viscous melts. The binary aluminate (Fig. 1.2, #4) and amorphous yttria-alumina-garnet fiber (Fig. 1.2, #6) were formed from inviscid melts. A viscosity-building fiber-forming process must be used to raise the viscosity of YAG glass fibers from <2 to >log 3 poise.

Two other important factors are the morphology of the crystalline phase and the crystal growth rate, both at and below the liquidus temperature (LT). The crystallization or devitrification at, and below, the liquidus temperature is related to the phase diagram. The crystallization rates of four fiber-forming glass melts [6] are shown in Fig. 1.3. They increase from left to right, i.e., from that of a commercial E-glass melt to that of an experimental yttrium-silicate glass melt.

In summary, in a continuous commercial process, strong viscous melts are operated above their liquidus temperature (LT) and a low crystallization rate at and below the liquidus temperature is preferred. In a stationary process of very short duration, pristine glass fibers can be formed from strong melts at and below the liquidus temperature before crystallization occurs. Fragile viscous melts must be supercooled to a temperature below the liquidus temperature before fibers can be

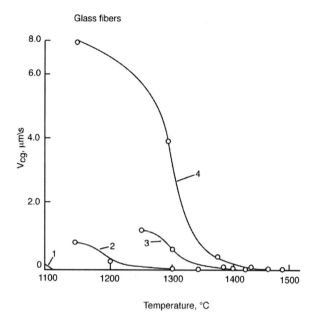

Fig. 1.3 Selected crystallization rates of fiber-forming melts: (1) commercial E-glass, (2) commercial S-glass, (3) experimental silicate glass, (4) yttria-modified silicate glass. Redrawn from Khazanov et al. [1, 6] (with kind permission of Springer Science and Business Media)

formed. Inviscid melts require special fiber-forming processes whereby a liquid jet is rapidly solidified and a solid glass fiber is formed before liquid droplets and shot are formed.

1.1.1.2 Behavior of Strong Viscous Melts

Commercial fiberglass compositions with 50–100% SiO_2, 0–25% Al_2O_3, 0–20% RO ($= CaO + MgO$), and 0–10% B_2O_3 [1] yield strong viscous melts. Three examples will suffice. Ultrapure SiO_2 fibers have melt temperature of >1800°C at a viscosity of log 3 poise. High-strength S-glass with 65% SiO_2 (Fig. 1.2, #5) has a log 3.0 forming temperature of 1560°C, a liquidus temperature of 1500°C, and therefore a delta temperature of 60°C. E-glass with 54% SiO_2 and ~6% B_2O_3 (Fig. 1.2, #1) is also based on strong viscous silicate melts. It has a low log 3.0 poise fiber-forming temperature of 1185°C and a delta temperature of 135°C. Although both silica and the above silicate glass fibers are based on strong viscous melts, they require different fiber-forming processes (Sections 1.2 and 1.3).

1.1.1.3 Behavior of Fragile Viscous Melts

Fragile viscous melt can be made from compositions with 4–10% SiO_2 and 50–25% Al_2O_3, 55–25% CaO, and 5% MgO [1]. Some also contain small amounts of Y_2O_3, ZnO, and/or Li_2O. Two examples may suffice. The melt of the fragile yttria-modified glass fiber (Fig. 1.2, #2) has melt temperature of 1280°C at a log 3 poise viscosity (or log 3 FT), a liquidus temperature (LT) of 1390°C, and therefore a delta temperature (ΔT) that is 110°C *higher* than the log 3 melt temperature. A related melt (Fig. 1.2 #3), which does not contain yttria, has a log 3 fiber-forming temperature (or log 3 FT) of 1300°C, a liquidus temperature (LT) of 1350°C, and therefore a delta temperature (ΔT) that is 50°C *higher* than the log 3 forming temperature.

Both melts not only have relatively high log 3 forming temperatures (1280–1390°C) but also 50–110°C *higher* liquidus temperatures than log 3 forming temperatures [1]. If these melts were held in a continuous commercial furnace at or above their log 3 melt temperature, they would crystallize within hours and could not be fiberized. Nonetheless, glass fibers can be formed from supercooled fragile melts (Section 1.4.1).

1.1.1.4 Behavior of Inviscid Glass Melts

Many common inorganic oxides are crystalline in nature, and therefore have high as well as sharp melting points. These melts are inviscid and may contain 4–0% SiO_2 and 50–80% Al_2O_3 [1], as well as perhaps minor amounts of CaO and/or MgO. Above their melting point, their viscosities are comparable to that of heavy motor oil or honey at room temperature and when they crystallize, they instantly revert to polycrystalline ceramic solids.

Crystallization of inviscid melts can be prevented and glass fibers can be formed by increasing the quench rate of a liquid aluminate jet without increasing the viscosity (Fig. 1.2 #4) or by levitating a laser-heated YAG melt, i.e., by increasing the viscosity without increasing the quench rate (Fig. 1.2, #6) [1]. For details, see Section 1.4.2.

1.1.2 Principles of Glass Fiber Formation

Formation of continuous glass fibers requires the melt to have a viscosity of log 2.5–3.0 poise at the point of fiber formation. Five generic fiber-forming processes [1] are known to yield glass fibers either from viscous or inviscid melts. One process yields glassy metal ribbons and one process yields glass fibers from viscous solutions or sol–gels of organic or inorganic precursors.

1.1.2.1 Generic Fiber-Forming Processes

As illustrated in Fig. 1.4, glass fibers can be made from strong viscous melts by melt spinning, i.e., by downdrawing them from bushing tips (Example A) or by downdrawing them from solid preforms (Example B). Glass fibers can also be made by dry spinning from solutions or sol–gels. Conceptually, dry spinning is a variant of melt spinning (Example A). This process relies on removing the solvent in a spinning tower while the fiber consolidates.

In addition, glass fibers can be made by updrawing from fragile viscous, but supercooled, melts (Example C) and from inviscid melts by rapidly quenching of ribbon-like jets (Example D), by extruding inviscid jets into a chemically reactive environment (Example E), and by downdrawing fibers from acoustically levitated melts in a containerless process (Example F).

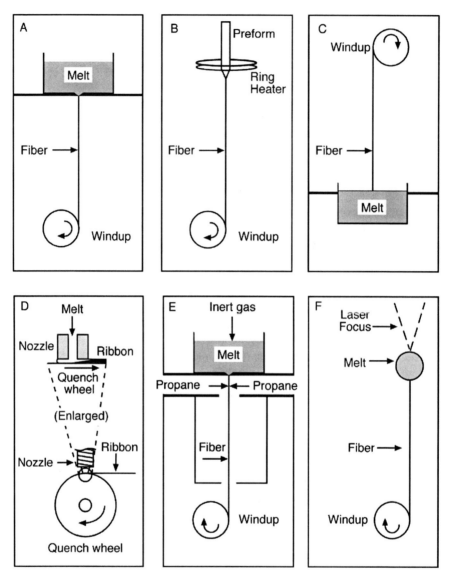

Fig. 1.4 Generic fiber-forming processes. Redrawn from Wallenberger and Brown [35, 36] (reprinted with permission of Elsevier Science Publishers)

1.1.2.2 Fibers from Strong Melts and Solutions

Structural silica fibers with 95–100% SiO_2 have extremely high melt viscosities. The temperature, at which they have a viscosity of log 3 poise (>1800°C), exceeds the capability of even the most costly precious metal bushing alloys. They cannot be melt spun but must be downdrawn from the surface of a solid silica preform rod (Fig. 1.4b). This process exposes only the surface of solid silica rods to such high temperatures, and therefore does not require precious metal bushings (for details, see Section 1.2.1.1).

Silica glass fibers with 99.5–100% SiO_2 can also be made by "dry spinning" [7], i.e., by a variant of the melt spinning process (Fig. 1.4a) whereby the solvent that contains the silica precursor is removed from the as-spun fibers in a hot spinning tower before they are collected (Sections 1.2.1.2 and 1.2.2).

Structural silicate glass fibers with 50–65% SiO_2 and 10–25% Al_2O_3 can be formed from their strong viscous melts in a conventional melt spinning process. They have log 3.0 forming temperatures ranging from 1150 to 1500°C. And they are downdrawn from bushing tips (Fig. 1.4a) in an ambient, rapidly cooling atmosphere and collected on a winder (see Section 1.3).

1.1.2.3 Fibers from Fragile and Inviscid Melts

Continuous fibers can be updrawn (Fig. 1.4c) from fragile calcium aluminate glass melts with 4–10% SiO_2 and 25–50% Al_2O_3, which must be maintained in a super-cooled state at a carefully controlled temperature [8, 5]. For details about fiber formation from fragile melts, see Section 1.4.1.

Continuous aluminate glass fibers with 55–80% Al_2O_3 [9, 10] and metal wires [11] can be melt spun from inviscid melts (Fig. 1.4e) by increasing the lifetime of the liquid jet without increasing the quench rate, i.e., by chemically stabilizing the surface of the inviscid jet; see Section 1.4.2.3. Continuous YAG glass fibers [12, 13] and mullite glass fibers [14] can be formed in a containerless process (Fig. 1.4f) from levitated, laser-heated, inviscid melts (Section 1.4.2.2). Continuous amorphous metal ribbons were made by increasing the quench rates of inviscid melts [15–17] from 10^4 to 10^6 K/s. In this process (Fig. 1.4d), a ribbon is extruded and cast onto the surface of a quench wheel. No ribbons were so far made from inviscid oxide glass melts by this process. For details, see Section 1.4.2.4.

These processes are fast and yield amorphous aluminate and YAG fibers from inviscid melts. Continuous single-crystal sapphire and YAG fibers can likewise be grown from inviscid melts [1] but the respective processes such as laser-assisted [18] or flux-assisted [19] crystal growth are very slow.

1.1.3 Structure of Melts and Fibers

The liquidus temperature and the delta between log 3 forming and liquidus tempera-tures of a glass melt can not only be determined but must actually be designed by the compositional trend line design model [2]. The modulus of a glass fiber increases with its internal structural order and the strength of a fiber depends on the uniformity of its internal structure.

1.1.3.1 From Glass Melts to Fibers

Pure silica melts have a uniform anisotropic network structure. As a result, they have an extremely high viscosity, a log 3 forming temperature of >1800°C, and a lower fiber modulus. The two following examples conceptually highlight the effect of disrupting the anisotropic silica network structure.

Example (1): The anisotropic structure of a conceptually pure SiO_2 melt can be disrupted adding Al_2O_3 and MgO to yield the SiO_2–Al_2O_3–MgO eutectic or S-glass melt. As a result, the viscosity and log 3 forming temperature will drop and the liquidus temperature, crystallization potential, and fiber modulus will increase. Example (2): The ternary SiO_2–Al_2O_3–MgO structure of the eutectic melt can be disrupted by adding CaO and B_2O_3 to yield a borosilicate E-glass melt that is based on the quaternary SiO_2–Al_2O_3–CaO–MgO eutectic. As a result the melt viscosity, the log 3 forming temperature, the liquidus temperature, the crystallization potential, and the fiber modulus drop.

1.1.3.2 Melt Structure vs. Liquidus

The liquidus temperature and the delta between log 3 forming and liquidus temperatures are key properties [2, 20, 21, 22] with regard to both a continuous commercial melt and fiber-forming process and a brief experimental fiber-forming process with a single-tip bushing. In either case, crystallization (devitrification) will occur within a few hours when the log 3 forming temperature is the same as, or lower than, the liquidus temperature.

In a continuous commercial process the resulting crystals would plug the bushing tips and impair fiber formation. An adequate delta between the log 3 forming and liquidus temperatures is therefore required as a safety factor. In a brief experimental single-tip process with small melt volumes, fiber formation is complete before crystallization would otherwise start.

The delta temperature of commercial borosilicate melts is greater than 100°C [1] and that of other commercial fiberglass melts ranges from 55 to 85°C [1]. The magnitude of the delta temperature depends on the individual furnace configuration, but on average a delta temperature of at least 50°C is thought to be required by the fiberglass industry. An excessively high-forming and delta temperatures would translate into a higher energy demand, and therefore also into a higher energy use and cost than really required.

The design of new fiberglass compositions, which are more energy- and environmentally friendly than incumbent compositions [2, 20, 21–25], is based on the discovery [26, 27] that the delta temperature is a function of the SiO_2:RO ratio (RO = CaO + MgO). It is possible to obtain that compositional variant that has the lowest log 3 forming temperature (energy demand) at a pre-selected delta temperature (crystallization potential).

The design of new energy-friendly fiberglass compositions for commercial consideration will be reviewed in Chapter 2. Attempts to model soda-lime-silica glass compositions [28–30] and to reformulate them for environmental benefit [31] will be reported in Chapter 7.

1.1.3.3 Fiber Structure vs. Modulus

The modulus of a fiber reflects its structural or internal order [4]. The fibers in Table 1.2 have compositions that range from 100% silica to 100% alumina.

Table 1.2   Effect of alumina on melt behavior, fiber morphology, and modulus

Fibers from SiO_2 to Al_2O_3	Al_2O_3 (wt%)	Fiber precursor	Fiber morphology	Modulus (GPa)		References
Ultrapure SiO_2	0.0	Strong melt	Amorphous	69		[6]
E-glass (6.6% B_2O_3)	13.0	Strong melt	Amorphous	72		[67]
E-glass (<1.5% B_2O_3)	13.0	Strong melt	Amorphous	81		[21]
HS-glass (S-glass)	25.0	Strong melt	Amorphous	89		[50]
HM-glass (6% Y_2O_3)	35.8	Strong melt	Amorphous	100		[6]
HM-glass (4% SiO_2)	44.5	Fragile melt	Amorphous	109		[80]
IMS glass (4% SiO_2)	54.0	Inviscid melt	Amorphous	50		[9]
RIMS (redrawn IMS)	54.0	Glass fiber	Polycrystalline		135	[9]
Nextel 440	70.4	Sol–gel	Polycrystalline		190	[32]
Mullite glass	71.8	Inviscid melt	Amorphous	45		[9]
Nextel 480	71.8	Sol–gel	Polycrystalline		230	[32]
IMS glass (0% SiO_2)	80.2	Inviscid melt	Amorphous	40		[9]
Safimax	98.8	Sol–gel	Polycrystalline		300	[32]
Fiber FP	100.0	Sol–gel	Polycrystalline		380	[32]
Saphikon	100.0	Inviscid melt	Single crystal		430	[33]

Ultrapure amorphous SiO_2 (Quartzel or Astroquartz) fibers, borosilicate E-glass fibers, boron-free E-glass fibers, high-strength (HS) glass fibers including S-glass, and high-modulus (HM) glass fibers, including that which contains 6% Y_2O_3, are derived from strong viscous melts. These glass fibers have moduli ranging from 69 to 100 GPa. The HM-glass fiber with 4% SiO_2 is derived from a fragile melt. It has 44.4% Al_2O_3 and a modulus of 109 GPa but it cannot be made in a conventional glass melt furnace. The processes, which yield strong and fragile melts, are discussed in Sections 1.3.1 and 1.4.1.

The designation IMS refers to inviscid melt spun aluminate glass fibers which contain 54–80% Al_2O_3. Unlike strong and fragile viscous melts, inviscid (literally non-viscous) melts have a melt viscosity of <2 poise. The IMS glass with 72% Al_2O_3 has a mullite composition. The fiberizing processes, which yield glass fibers from inviscid melts, will be discussed in Sections 1.1.2 and 1.4.2. The extremely low melt viscosities and fiber moduli suggest that the melt as well as the resulting glass fibers lack internal order. The designation RIMS refers to a redrawn IMS glass fiber with 54% Al_2O_3 that became polycrystalline on redrawing.

Nextel 440, Nextel 480, Safimax, and Fiber FP are polycrystalline fibers which are of commercial value. They contain 70–99% Al_2O_3 and have moduli ranging from 190 to 380 GPa [32]. They are made by variants of either the generic sol–gel or the generic slurry spinning process. The amorphous IMS mullite glass fiber in Table 1.2 and the polycrystalline Nextel 480 fiber have nominally the same composition but, due to their different morphology, a vastly different modulus (i.e., 45 GPa vs. 230 GPa). Saphikon is a commercial single-crystal sapphire (100% Al_2O_3)

fiber that is made by variants of the very slow crystal growing process, e.g., by the edge-defined film-fed growth (EDG) process [33].

Except for the inviscid melt spun fibers, the modulus in Table 1.2 increases with increasing Al_2O_3 content and therefore with increasing directional structural order. Specifically, the modulus increases from 69 to 109 GPa for glass fibers from strong and fragile viscous melts, to 135 to 380 GPa for polycrystalline (including redrawn IMS) fibers, and to 410 GPa for single-crystal sapphire fibers. Although the observed modulus trend parallels the percent increase in Al_2O_3, truly amorphous and therefore very low modulus fibers are formed if the formation of any internal structure is prevented.

1.1.3.4 Fiber Structure vs. Strength

Strength is a measure of the internal structure as well as surface uniformity. Table 1.3 shows the average tensile strength of 16 glass fibers. Their strength ranges from 5.57 GPa for a military optical glass fiber (FOG-M) to 0.37 GPa for a highly porous, high silica fiber obtained by leaching E-glass with hydrochloric acid. These fibers have diameters ranging from 4 to 20 μm [6], except FOG-M [34] and the binary calcium aluminate fibers [35], which have a fiber diameter of >100 μm.

Table 1.3 Effect of composition on tensile strength [1] (with kind permission of Springer Science and Business Media)

Generic glass fiber	Strength (GPa)	References
Military fiber optics glass fiber, FOG-M	5.57	[34]
Magnesium aluminosilicate glass fiber, 10% MgO	4.80	[6]
Magnesium aluminosilicate glass fiber, 15% MgO	4.00	[6]
Ultrapure silica fiber, Astroquartz	3.50	[6]
Zn/Ti magnesium aluminosilicate glass fiber	3.20	[6]
Sodium calcium aluminosilicate glass fiber	2.75	[6]
E-type aluminum borosilicate glass fiber	2.70	[6]
Copper aluminum borosilicate glass fiber	2.70	[6]
Borate glass fiber	1.90	[6]
Lead silicate glass fiber	1.55	[6]
Phosphate glass fiber	1.50	[6]
Sodium silicate glass fiber	1.10	[6]
Calcium aluminate glass fiber, 54% CaO	0.95	[9]
Pure silica fiber from water glass	0.85	[45]
Calcium aluminate glass fiber, 80% CaO	0.50	[9]
Porous silica fiber from E-glass	0.37	[6]

Surface flaws or non-uniformities have a tendency to reduce the strength of individual glass fibers. Directionality or spin orientation, as inferred from birefringence measurements, tends to increase the strength of individual fibers. Surface uniformity and spin orientation are important factors but, even if their relative relationships were fully documented, the up to 15-fold differences in strength among the 16 generic fibers shown in Table 1.3 cannot be attributed to these factors alone.

A major reason for the observed differences in fiber-to-fiber strength must therefore be sought in differences in the uniformity of the internal structures. The extremely uniform network structure of FOG-M silica fibers [1, 34] translates into high tensile strength. The highly non-directional (random) arrangement of calcium oxide and aluminum oxide in a rapidly solidified binary calcium aluminate fiber translates into very low tensile strength. The ultralow strength of porous high silica glass is obviously due to its porosity.

1.1.4 Summary and Conclusions

The structure of a melt and fiber [4] determines the properties of the melt and fiber properties. The crystallization potential of a fiberglass melt depends on the liquidus temperature [2]. The modulus of a glass fiber depends on its structural or internal order [4]. The viscosity drops and the modulus increases with increasing Al_2O_3 content. In contrast, the tensile strength of a glass fiber does not only depend on the surface uniformity but also on the uniformity of the fiber structure itself. The principles governing the design of new energy-friendly fiberglass compositions will be discussed in Chapter 2.

1.2 Silica Fibers, Sliver, and Fabrics (95–100% SiO_2)

Temperatures well above 1800°C would be required to achieve silica melts with a viscosity of log 2.5–3.0 poise. Such temperatures would by far exceed the capability of the refractories of most furnace wall ceramics and of most precious metal bushing alloys. Silica glass fibers are therefore downdrawn from solid silica preform rods, dry spun from precursor solutions, or made from borosilicate or soda-lime-silica glass fibers by acid leaching.

1.2.1 Ultrapure Silica Fibers (99.99–99.999% SiO_2)

A commercial process has been developed to downdraw ultrapure silica glass fibers with 99.999% SiO_2 from the surface melt of solid silica preform rods. An experimental process can be used to make ultrapure silica fibers by dry spinning of tetraethylorthosilicate sol–gels [1, 36, 37].

1.2.1.1 Downdrawing from Strong Viscous Melts at the Preform Surface

In principle, the process of downdrawing fibers from preform rods is actually variant of the melt spinning process. A gas flame or an electrical furnace softens the ends of a solid quartz rod and facilitates the formation of continuous silica melt on its surface. The melt is not extruded though a bushing tip, but a fiber is pulled from the melt at the tip of the preform.

(a) Process concept. The process is schematically shown in Fig. 1.5. Thus, a large diameter silica preform rod is heated in a circumferential heater, i.e., a gas burner, an

Fig. 1.5. Pure silica fibers
from preforms. From
Wallenberger et al. [151]
(reproduced with permission
from Materials Research
Society)

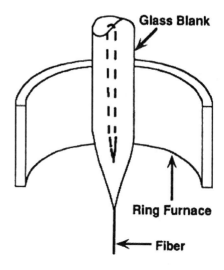

electrical furnace, or a laser source. The molten jet solidifies and the resulting quartz filament is passed over a sizing. The coated filament is collected on a winding drum. A large number of single filaments are then combined to form a multifilament yarn bundle or roving.

(b) Products and properties. Preforms made from natural silica contain 99.99% SiO_2, 20–50 ppm Al, <5 ppm OH^-, and <4 ppm Na [38]. Preforms (blanks) made by oxidation of $SiCl_4$ in a plasma flame [38] afford even purer silica fibers with <1 ppm Al, <0.1 ppm OH^-, <1 ppm Na, and <50 ppm Cl. For details, see [1]. Commercial quartz fibers made by downdrawing can have diameters ranging from 7 to 14 μm (Fig. 1.5). Rovings can be made with up to 4800 filaments.

Two commercial ultrapure silica glass fibers are known, Astroquartz and Quartzel. Despite their name both are amorphous not crystalline. The mechanical and physical properties of a typical commercial silica glass fiber are shown in Table 1.4.

Silica glass fibers have higher glass transition temperatures than general-purpose borosilicate and boron-free E-glass fibers (1150–1200°C vs. 550–600°C) and lower coefficients of thermal expansion (0.5×10^{-6}°C^{-1} vs. 5×10^{-6}°C^{-1}). In addition, they also have lower dielectric constants (3.4 vs. 6.1) and lower loss tangents (2×10^{-4} vs. 30×10^{-4}). These properties translate into significant product advantages but they demand a premium price.

(c) Value-in-use and applications. Structural ultrapure silica glass fibers are known as ultrahigh-temperature fibers. They can be continuously used at temperatures of ≤1090°C. In contrast, S-glass can be continuously used only at ≤815°C [1, 7, 36]. A complete comparison of strength and strength retention of ultrapure silica glass fibers, pure silica glass fibers, and silicate glass fibers (including E-glass and S-glass) is summarized in Section 1.3.3.

Another interesting property of silica glass fibers is their low coefficient of thermal expansion, $\alpha \approx 0.5 \times 10^{-6}$°$C^{-1}$. This value can be lowered and even rendered

Table 1.4 Properties of ultrapure silica glass fibers [36] (with kind permission of Springer Science and Business Media)

Astroquartz and quartzel	
Mechanical properties	
Filament diameter, yarn, and fabric, μm	9.0
Specific gravity (density), g/cc	2.2
Hardness (Mohs scale)	5–6
Pristine filament strength, GPa, RT	6.0
Yarn strand tensile strength, GPa, RT	3.4
Yarn tensile modulus, GPa, RT	69
Physical properties	
Liquidus temperature, °C	1670
Temperature at max. crystallization, °C	1630
Coeff. of thermal exp., 0–1000°C, °C	5.4×10^{-7}
Thermal stability (short term), to °C	2000
Thermal stability (long term), to °C	1200
Thermal conductivity at 20°C, CGS	0.0033
Dielectric constant at 20°C, 1 MHz	3.78
Refractive index at 15°C, n_D	1.4585

negative by adding TiO_2 to silica [36]. Silica glass fibers also display excellent dielectric properties. Finally, they are known for their high resistance to corrosion in neutral or acid chemical environments.

These fibers offer superior heat resistance since they retain useful strength at very high temperatures. They also possess the high ablation resistance and the dielectric, acoustic, optical, and chemical properties of quartz from which they were made. Their pristine modulus is low (69 GPa), their pristine strength is as high as 4.8–6.0 GPa [6], and because of their low density (2.2 g/cc), their specific pristine strength (2.2–2.7 Mm) is the highest of any pristine glass fiber on record.

Quartz fibers offer superior IR and UV transmittance, but contain water, and therefore exhibit a strong hydroxyl group in their IR transmission spectra. The presence of water blocks IR transmission in a critical region of the spectra, but it can be avoided by selecting a synthetic route that yields water-free quartz for the preform fabrication. Mined quartz may contain traces of uranium or thorium, which emit alpha particles and can disturb delicate electronic circuits or signals. Uranium- and thorium-free quartz fibers afford quiet high-temperature circuit boards and/or electronic packaging.

Ultrapure silica or quartz fabrics are used to reinforce radomes, antenna windows for missiles, high-temperature circuit boards, and rocket nose cones. Braided yarns provide high-temperature electrical insulation, e.g., for coaxial cables, thermocouple wires, and space separators. Rovings are used to reinforce polymer matrix composites for ablative and electrical uses, as well as high-performance sporting goods, e.g., tennis racquets and skis, especially when hybridized with carbon fibers. Threads are used to stitch cable tray insulation for nuclear power plants.

1.2.1.2 Ultrapure Silica Fibers from Sol–Gels

An experimental dry spinning process can be used to make ultrapure silica glass fibers from tetraethylorthosilicate (TEOS) solutions in alcohol.

(a) Process concept. In a typical dry spinning process, a viscous solution is spun from, or extruded through, multiple spinneret orifices into a hot column that removes the solvent. The process resembles the typical commercial wet spinning process. In the wet spinning process, nascent filaments are extruded into an ambient environment. In the dry spinning process, they are extruded into a hot column that removes the residual solvent.

(b) Products and properties. Experimental fibers, which have been made from tetraethylorthosilicate (TEOS) sol–gel polymer, have an even higher purity level [39] than ultrapure silica fibers derived from quartz preforms [38]. Since the process relies on reagent grade $Si(OC_2H_5)_4$, the fibers contain 99.999% SiO_2 and absolutely minimal impurities (<1 ppm Al, OH^-, Na, and Cl). The process has been described in detail [1, 7, 36] and will only be reviewed here.

The first step is a polycondensation reaction [40]. TEOS is dissolved in pure ethanol (C_2H_5OH), and a dilute solution of the catalyst, hydrochloric acid (HCl), is added at <25°C. The reaction mixture is heated to 70–80°C, until the polymerization has reached a fiber-forming viscosity (Equations 1.1, 1.2, 1.3, 1.4 and 1.5):

$$(RO)_3Si - OR + HOH \rightarrow (RO)_3Si - OH + ROH(R = C_2H_5) \qquad (1.1)$$

$$(RO)_3Si - OH + OH - Si(OR)_3 \rightarrow (RO)_3Si - O - Si(OR)_3 + H_2O \qquad (1.2)$$

$$(RO)_3Si - OR + OH - Si(OR)_3 \rightarrow (RO)_3Si - O - Si(OR)_3 + ROH \qquad (1.3)$$

for a chain-like polysiloxane polymer:

$$nSi(OR)_4 + H_2O \rightarrow RO- \overset{\displaystyle OR}{\underset{\displaystyle OR}{Si}} - \left[-O - \overset{\displaystyle OR}{\underset{\displaystyle OR}{Si}} - \right]_{n-2} -O- \overset{\displaystyle OR}{\underset{\displaystyle OR}{Si}} -OH + (2n-1)ROH$$

$$(1.4)$$

for complete hydrolysis:

$$nSi(OR)_4 + 4nH_2O \rightarrow nSi(OH)_4 + 4nROH \qquad (1.5)$$

$$nSi(OH)_4 \rightarrow nSiO_2 + 2nH_2O \qquad (1.6)$$

$$nSi(OR)_4 + 2nH_2O \rightarrow nSiO_2 + 4nROH \qquad (1.7)$$

In the second step [40] fibers are dry spun and drawn near a viscosity of log 2.5–3.0 poise. In the third step (Equations 1.6 and 1.7) the gel fiber is converted into a silica fiber by heating it with a low drawing tension in air to 800–900°C [40]. The gel fiber contains solvent and pendant C_2H_5O groups (Equation 1.5). Most of the solvent is removed at <200°C and a porous fiber with a high surface area is obtained [37]. The pendant C_2H_5O groups are removed between 200 and 400°C. Weight loss and fiber shrinkage continue above 400°C. The silica fiber is sintered to full density (2.2 g/cm^3) at 900°C [37, 41].

Sol–gel-derived [42] and preform-derived silica fibers [1, 7, 36] are amorphous. Due to the disordered character of the SiO_4 tetrahedra skeleton, the density of glassy silica and silica glass fibers is lower than that of crystalline silicas (i.e., 2.2 g/cm^3 vs. 2.33 g/cm^3 for cristobalite and 2.65 g/cm^3 for quartz).

(c) Value-in-use and applications. Sol–gel-derived silica glass fibers with a 20 μm diameter have the highest purity of any type of silica fiber that is known. The purity can be selectively changed by adding components which cannot be added to silica preforms. Thus, they are therefore a superior compositional research and design tool.

Sol–gel-derived silica glass fibers have no commercial significance since they have low room temperature strength (800 MPa) [41] relative to that of E-glass (>3000 MPa) and other silica glass fibers (>5900 MPa). A further strength loss occurs when amorphous silica fibers are crystallized [43].

1.2.2 Pure Silica Sliver and Fabrics (95.5–99.5% SiO₂)

Pure silica sliver and fabrics (99.5–99.9% SiO_2) can be made from aqueous sodium silicate solutions [7, 44–46]. High silica fabrics (95.9–99.0% SiO_2) can be made by acid leaching of A- or E-glass fabrics [1, 36].

1.2.2.1 Pure Silica Sliver from Aqueous Solution

The preparation of pure silica sliver from aqueous silicate solutions is an adaptation of the dry spinning process that has been described in Section 1.2.1 for making ultrapure silica fibers from sol–gels. Both processes have their origin in the generic dry spinning process that has been practiced for over 50 years to fabricate acrylic fibers (Table 1.5). A commercial product that is currently being

Table 1.5 Dry spinning and sol-gel processing [7] (with kind permission of Springer Science and Business Media)

Process	Dry spinning	Dry spinning	Sol-gel
Solute	Acrylic polymer	Water glass	Tetraethoxylsilane
Solvent	Dimethylformamide	Water	
Solidification	Remove DMF	Remove water	Remove alcohol
Wet treatment	Remove DMF	Remove Na$^+$	
Drying step	Remove water	Remove water	Remove alcohol
Precursor fiber		Is not isolated	Is wound up
Consolidation		In-line step	Separate step
Process steps	One or two	One	One or two
Final product	Commercial fiber	Commercial fiber	Commercial fiber

made by dry spinning silica sliver from aqueous water glass solutions is known as Silfa yarn [47].

(a) Process concept. The inorganic dry spinning process can be used to make experimental continuous silica fibers but, so far, not commercial continuous multifilament yarns. It is also used to make commercial sliver yarns starting with an aqueous water glass solution. A sliver yarn is a continuous assembly of cohesive, slightly bonded staple fibers in an essentially parallel arrangement.

The commercial process (Fig. 1.6) utilizes a viscous solution of water glass in water and proceeds in 13 stages. A viscous water glass solution is extruded (1) through tiny spinneret holes (orifices) into a drying chimney (2). A spin finish is applied to the resulting sodium silicate filament yarn by a kiss roll (3). The yarn is taken up by a drawing drum and scraped off prior to complete rotation on the drum, intermingled in a conical chamber and drawn laterally (4). The resulting sodium silicate sliver (a cohesive staple yarn) is transported over a godet (5), placed on a conveyer belt and passed through an acid bath (6), a washing station, (7) and a drying zone (8). Silica sliver is formed in the calcining zone (9), and a textile sizing is applied (10) before it is dried (11), further intermingled in air (12) and wound onto a bobbin (13).

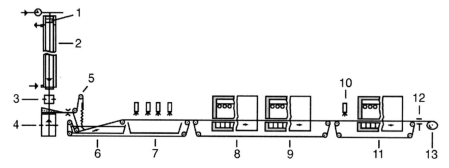

Fig. 1.6 Process for the fabrication of silica sliver from water glass. Redrawn from Achtsnit [152]

The drying step in the continuous process sequence is carried out at ∼150°C. It removes free water and produces sliver yarn containing weak nanoporous silica fibers with a highly hydroxylated surface. Consolidation of the nanoporous fibers within the continuous process sequence at 800°C causes the fibers to dehydroxylate and shrink, and also to increase in strength.

(b) Products and properties. This process facilitates the formation of pure silica Silfa sliver having linear densities of 135–330 g/1000 m of yarn [47]. The individual staple fibers have a length ranging from 50 to 1000 mm and an average diameter of 10 μm. Individual silica fibers have a lower density (1.8–2.0 g/cm^3) than E-glass (2.5 g/cm^3) and crystalline ceramic fibers (3.5 g/cm^3). Unconsolidated fibers are nanoporous and weak and have a density of 1.8 g/cm^3, but consolidated fibers are solid, moderately strong, and have a density of 2.0 g/cm^3. They retain their strength up to 1000°C.

Commercial Silfa silica sliver [46, 47] derived from water glass have a higher impurity level than downdrawn silica fibers. The impurities include 450 ppm Al, 206 ppm Fe, 120 ppm Na, and <0.5% water. And the process resembles the sol–gel process for dry spinning ultrapure silica fibers [1, 7]. At room temperature, their pristine strength is one-half that of ultrapure silica fibers derived from preforms (1.7 GPa vs. 3.4 GPa). At 600°C, the pristine strength of both fibers is the same (1.4 GPa). Between 600 and 1000°C, it is also identical but drops to 0.9 GPa.

In summary, the purity and cost of silica fibers made from water glass is much lower than that of downdrawn silica fibers from pure silica preforms and that of sol–gel-based silica fibers made from pure tetraorthosilicate. Furthermore silica fibers derived from water glass are not available as continuous filaments, only as sliver. Yet in high-temperature applications, the performance of dry spun and downdrawn silica fibers is nearly the same.

(c) Value-in-use and applications. Dry spun silica fibers made from water glass are less costly than downdrawn silica fibers, and may find increasing use in high-temperature markets except those requiring high specific strength at high temperatures, e.g., in aircraft or aerospace applications, or in high-temperature applications requiring high radiation resistance or electromagnetic shielding. Selected products include plied yarns for weaving narrow and broad fabrics, braided and twisted ropes, and sealings and sleevings for high-temperature uses, including turbine insulation, removable flexible insulation, and compensators [46, 47].

1.2.2.2 Acid-Leached E- and A-Glass Fabrics (95.0–95.5% SiO_2)

High silica glass fibers (95–99.5% SiO_2) represent the lowest cost variant of silica fibers. They also represent the weakest form of silica fibers. They are therefore not commercially available in fiber or sliver form. They must be obtained by leaching of E-glass fabrics and are available only in fabric form.

(a) Process concept. In the United States, borosilicate E-glass fabrics are leached with aqueous HCl solutions. A typical commercial product is called Refrasil. In Russia and Japan, sodium silicate A-glass fabrics with SiO_2/Na_2O ratios of 3:1–4:1 are leached with aqueous H_2SO_4 or HNO_3 solutions [1, 6, 7, 36]. Leaching removes nearly all oxides except SiO_2.

Leaching must be carried by leaching woven, braided, or other pre-designed E-glass or A-glass fabrics because the individual yarn strands are too weak after leaching to survive weaving. Almost all alumina, magnesia, calcia, iron oxide, and boron oxide [48] can be extracted from E-glass fibers in 8 h at 95°C in dilute H_2SO_4. Leaching proceeds by ion diffusion. The chemical nature of the acids does not affect leaching kinetics [6].

(b) Products and properties. High silica glass fibers containing 95–98% SiO_2 become porous on leaching [1]. The leached fabrics are sintered to turn the porous fiber structure into a solid silica structure. The resulting high silica fabrics can be used from room temperature to 1000°C. But because of their ultralow strength,

even the resulting high silica fabrics can be used only in non-load-bearing end uses. Pristine fiber strength decreases from room temperature (0.4 GPa) to 1000°C (0.1 GPa). Application of a chromium oxide coating increases the upper in-use temperature to 1200°C for long-term and multiple-cycle heat loads [6]. With increasing temperature, the decrease in fabric strength is accompanied by removal of water from the hydroxylated fiber surfaces and by increasing nanocrystallinity that may cause slight fabric shrinkage.

(c) Value-in-use and applications. High silica fabrics obtained by acid leaching of E-glass were introduced in the 1960s for aircraft and aerospace applications. Their use has declined in these applications by the advent of ultrapure and pure silica fibers which are stronger and can be woven and braided, but they are also more expensive. They are very desirable and cost-effective heat-insulating materials for less demanding uses than those which require pure or ultrapure silica fibers. Among others, they are used as high-temperature filtration media for ferrous, non-ferrous, and corrosive metals, as reinforcement for polymer composites, and as heat-resistant electrical insulation in nuclear reactors [1, 6, 7, 36].

1.2.3 Summary and Conclusions

Silica fibers, irrespective of process, are amorphous, including those called quartz fibers. Ultrapure silica fibers (99.99–99.999% SiO_2) can be formed by downdrawing from preforms or by dry spinning from a viscous tetraethylorthosilicate solutions. Pure silica sliver (95.5–98.9% SiO_2) can be formed by dry spinning from aqueous water glass solutions. High silica fabrics (95.0–95.5% SiO_2) are obtained by acid leaching A- or E-glass fabrics.

At room temperature (Table 1.6), the strongest fiber is S-glass (4.6 GPa), followed by E-glass (3.5 GPa); Nextel, a glass ceramic fiber (2.6 GPa); Astroquartz, an ultrapure silica fiber (2.4 GPa); Silfa, a pure silica sliver (1.7 GPa); a sol–gel-derived experimental ultrapure silica fiber (0.9 GPa); and Refrasil, a high silica fabric (0.4 GPa). Only S-glass, E-glass, Nextel, Astroquartz, and Silfa fabrics can be used in load-bearing applications.

Table 1.6 Tensile strength of high- and ultrahigh-temperature glass [36] (with kind permission of Springer Science and Business Media)

Strength (GPa)	Silicate		Silica		Ultrapure silica		
	E-glass	S-glass	High	Pure	Preform	Sol–gel	Nextel aluminate
RT,°C	3.5	4.6	0.4	1.7	2.4	0.9	2.6
400, °C	1.8	3.8	0.2	1.6	1.8	0.9	2.4
600,°C	0.9	2.4	0.1	1.4	1.4	0.9	2.4
800, °C	–	0.7	0.1	1.2	1.0	0.1	2.1
1000, °C	–	–	0.1	0.9	0.9	0.1	1.9
Cont. use	600°C	815°C	1040°C	1090°C	1090°C	1090°C	1200°C
Rel. cost	1	6	10	30	68	80	150

Borosilicate E-glass fibers support continuous use to 600°C, S-glass to 815°C, high silica glass fibers to 1040°C, pure and ultrapure silica glass fibers to 1090°C, and Nextel to 1200°C. The estimated cost per kilogram of fiber increases by a factor of 150, while the estimated continuous in-use temperature doubles.

1.3 Silicate Glass Fibers (50–70% SiO_2, 1–25% Al_2O_3)

The commercial process for forming continuous fibers from strong viscous silicate melts in a conventional furnace has been authoritatively reviewed in 1986 [49], 1988 [50], 1989 [51], 1993 [52], 1999 [1], and 2001 [53].

1.3.1 Forming Glass Fibers from Strong Viscous Melts

This section offers an overview of the generic process. Two specific advances have, however, been made in the last 10 years and will be discussed. One is the development of a novel plasma glass melting process [54] and the other is modeling of fiber drawing technology [55].

1.3.1.1 Critical Properties of Strong Viscous Melts

A generic strong viscous silicate melt is formed at a viscosity of log 1.7–log 2.0 poise, flows through the melt furnace at a viscosity of log 2.0–log 2.5 poise, and fibers are formed at a viscosity of log 2.5–log 3.0 poise. The fiber-forming temperature at a viscosity of log 3.0 poise must exceed the liquidus temperature by at least 50°C to avoid devitrification in the furnace, forehearth, and/or bushings.

The softening point represents the temperature at which the viscosity is log 7.65. At this temperature a glass fiber deforms under its own weight. The annealing point represents a temperature at which the viscosity is log 13.0. The strain point represents a temperature at which the viscosity is log 14.0. The strain point is near the glass transition temperature and approximately projects the highest tolerable in-use temperature.

1.3.1.2 Commercial Manufacturing Process

E-glass with 5–7% B_2O_3 [49] has been selected as the reference melt for the following overview of the highly automated process shown in Fig. 1.7. The furnace consists of three process sections. The ingredients are selected, weighed and mixed, and then entered into a furnace. In the first section of the furnace, the commercial batch ingredients are melted, gaseous inclusions are removed, and the melt is homogenized at ~1370°C. The melt flows into the refiner section where the temperature of the melt is reduced to ~1260°C [53]. The refined glass melt then enters the forehearth section that is located directly over fiber-forming stations and bushings.

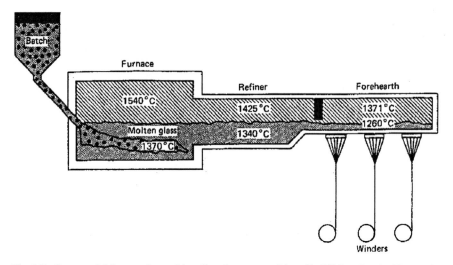

Fig. 1.7 Commercial furnace for melting fiberglass compositions [1, 53] (reprinted with permission of ASM International. All rights reserved www.asminternational.org)

Depending on the fiber diameter, optimum fiber formation is obtained with a melt viscosity ranging from log 2.5 to 3.0 poise as the molten glass passes through the bushing tips. The fibers are drawn down and cool rapidly. After the sizing is applied [53] the filaments are combined in a yarn bundle or strand and delivered to a forming winder, a direct draw winder, or a chopper (Fig. 1.8). A typical multifilament bushing can have 400–8000 tips and the resulting yarn will therefore have a corresponding number of filaments.

The commercial melting and forming process can be used for boron-free Chinese C- (or CC-)glass with >10% Na_2O, E-glass with ~6% B_2O_3, boron-free E-glass, essentially boron-free E-glass with <1.5% B_2O_3, boron-free ECR-glass with >2% ZnO, and boron-free S-glass (Table 1.7), as well as for other strong silicate melts with up to 75% SiO_2, i.e., as long as the bushing alloys and refractory materials can withstand the process temperature. The boron-free CC-glass is compositionally related to the boron-free A-glass that is no longer made in the United States, and it should not be confused with the former C-glass that contained boron and is likewise no longer available in the United States (see Section 1.3.3.5).

As shown in Table 1.7, the log 3 forming temperature (log 3 FT) of E-glass melts with ~6% B_2O_3 [1, 49] is 8°C higher than that of soda-lime-silica CC-glass melts [1, 2, 52]. The log 3 FT of essentially boron-free E-glass melts (<1.5% B_2O_3) is 23°C higher [2, 49, 56]. The log 3 FT of boron-free E-glass melts is 67°C higher [1, 25, 56]. The log 3 FT of boron-free ECR-glass melts (>2.0% ZnO) is 38°C higher [2, 57], and the log 3 FT of boron-free S-glass melts is 368°C higher than that of soda-lime-silica CC-glass melts. All melts have a delta temperature of >50°C, that is required to achieve a crystallization-resistant melt in a continuous commercial furnace.

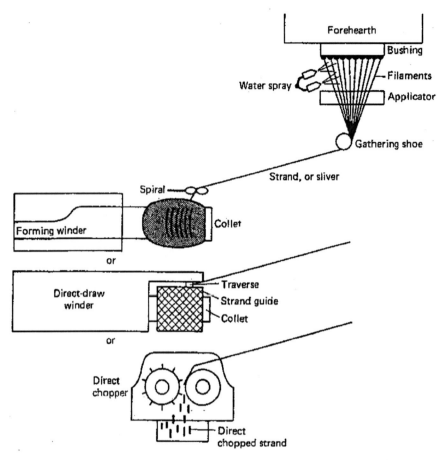

Fig. 1.8 Forming winders, direct-drawing winders, and choppers [1, 53] (reprinted with permission of ASM International. All rights reserved www.asminternational.org)

Energy cost, platinum losses at the bushings, and refractory wear rise with a rising viscosity or log 3 forming temperature. Reinforced bushings, bushings with high Rh content, and more frequent refractory replacement may therefore be needed. All compositions listed in Table 1.7, except S-glass, can be and are made in large conventional furnaces and with large bushings. The viscosity (or log 3 forming temperature) of S-glass melts is so high that a smaller furnace and smaller bushings from special alloys may be required.

In summary, the original E-glass contains ≤1% fluorine and ∼6% B_2O_3 and has a log 3 forming temperature of <1200°C [58]. Borosilicate melts with ∼6% B_2O_3 in Table 1.7 may emit boron and fluorine into the environment and may therefore require the installation of additional and costly emission control equipment. The fluorine- and boron-free E-glass [59, 56] has a log 3 forming temperature of 1260°C. The fluorine-free and essentially boron-free E-glass has a log 3 forming temperature of 1215°C [21, 27].

Table 1.7 Glass fibers by the commercial melt process [1]

Composition	CC-glass	E-glass variants			ECR-glass	HS-glass
Boron content	0.0%	>6%	<1.5%	0.0%	0.0%	0.0%
Weight %						
SiO_2	65.15	54.5	56.50	60.10	58.2	65.0
Al_2O_3	7.70	14.0	13.45	12.99	11.6	25.0
B_2O_3	–	6.6	1.30	–	–	–
CaO	9.23	22.1	24.50	22.13	21.7	–
MgO	4.33	0.6	2.55	3.11	2.0	10.0
ZnO	0.20	–	–	–	2.9	–
TiO_2	–	–	0.55	–	2.5	–
K_2O	–	0.2	–	0.63	–	–
Na_2O	12.92	0.8	0.90	0.14	1.0	–
Fe_2O_3	0.27	0.2	0.25	0.25	0.1	–
F_2	0.96	0.5		–	tr.	–
Log 3 FT, °C	1192	1200	1215	1259	1230	1560
LT, °C	1068	1064	1157	1174	1163	1500
Delta T, °C	123	136	58	85	67	60
Density, g/ccm		2.54	2.62	2.67	2.68	2.49
Strength, GPa	–	3.40	3.5	3.5	3.5	4.6
Modulus, GPa	–	72	81	81	80	88
References	[91]	[58]	[21, 27]	[59, 56]	[52, 57]	[50]

Fluorine-free and either boron-free or essentially boron-free E-glass variants are clearly more environmentally friendly than borosilicate E-glass. And the fluorine-free and essentially boron-free E-glass may be considered to be both energy- and environmentally friendly on the basis of its low log 3 forming temperature. The design of energy- and environmentally friendly E-glass compositions since 2000 will be reviewed in detail in Chapter 2.

1.3.1.3 Experimental Plasma Melt Process

In the typical fiberglass process, that is the backbone of the fiberglass industry, the energy sources are oil, natural gas, or electrical heat, and the throughput is measured in hours [1]. In recent years, extensive research has been conducted into glass melting with high-intensity DC arc plasmas. The still experimental technology differs in several ways from conventional glass furnace technology [54]: It will be described here by way of a brief review. Chapter 11 describes the technology in detail:

- It uses DC electrical arc plasmas
- It has energy densities that are 2.5 times greater than conventional electrical melting, resulting in more rapid melting [60]
- It has a small melter footprint, typically 1–2 m in diameter resulting in significantly lower capital costs
- It is highly flexible in converting from one glass composition to another
- It is capable of melting high-temperature glasses and other materials

Early work into the application of plasma melting was conducted by both Johns-Manville in the mid- to late 1980s [60, 61] and the British glass industry in the mid-1990s [62]. This body of work showed promising results of efficient specialty glass melting systems approaching 4 million BTU/ton, but both projects suffered from the lack of a reliable DC electrical torch design. Most recently, Plasmelt Glass Technologies has completed a DOE-sponsored project in which the torch design and operational systems were specifically addressed and significantly improved [63].

The technology involved in plasma melting uses the skull-melting concept in which the glass-batch forms its own refractory liner, using a high thermal gradient from the hot spot to the outer container. If properly designed and operated, the melter does not require water-jacketing or refractory liners, which greatly improve its energy efficiency. The small glass volume gives the melter a short response time, which has positive and negative aspects.

The benefits include immediate feedback on process adjustments, reduced process development times for new glasses, and rapid turnover for converting from one glass formulation to another. These attributes make the melter an ideal design for small, low-cost, research melting applications in which several melts of various compositions are desired.

Perhaps the main application of the new process is as a pilot melter for the continuous melting of smaller tonnages of newly developed R&D glasses beyond the laboratory scale, prior to the need for high production volumes. For applications requiring low seeds, insufficient refining time must be addressed by add-on refiners.

The experimental plasma melting process can melt a broad range of fiberglass compositions at throughputs at several hundred pounds per hour [54]. Most of the work on plasma melting has focused on making direct melt E-glass patties that were shown to be suitable for fiberizing in marble re-melt bushings. Zero stones, minimal cord, and high seeds were a common feature of these patties. The high seed levels do slightly increase the fine fiber break rates during fiberization, unless mitigated through add-on refining.

Other glasses have also been melted by Plasmelt (but not fiberized) using this technology [54] and they include boron-free E-glass, S-glass, C-glass, AR-glass, high-boron–low DK-glass, frit compositions, and mineral melts (including wollastonite, anorthite, albite, zircon, and kaolinite). Using cold batch to start up, rapid melt pool development and rapid startup of all glasses – even high electrically resistant glass compositions that are difficult to melt via conventional technologies – are routinely possible.

In summary, the benefits of using plasma melting are most directly applicable to those specialty fiberglass applications involving small tonnages, requiring intermittent production, rapid product changes with minimal scrap, high temperatures beyond conventional electrical melting systems, laboratory glass melting of wide-ranging compositions, or as a continuous pilot melter to produce tonnages of new experimental glass candidates before they reach production quantities.

1.3.1.4 Modeling of Glass Fiber Drawing

A computer model has recently been developed [55] for an industrial-scale, single-position fiber drawing tower and a dual package winder to anticipate break levels in the continuous commercial glass fiber drawing process.

In a typical continuous fiber-forming process, a vertical jet is generated [50] as the molten glass, which has a viscosity of log 2.5–3.0 poise, is extruded through a nozzle, called bushing tip. The jet is cooled as it is exposed to the atmosphere that surrounds it and it is simultaneously attenuated (Fig. 1.7) by the mechanical tension that is applied to it by the winder speed.

Any fiber-forming process would be perfectly continuous if (1) the mixed batch ingredients result in a perfectly uniform and melt at uniformly one end of the melt furnace; (2) a perfectly homogeneous melt results and reaches the other end of the furnace, i.e., the forehearth; (3) each jet at each bushing tip has the same diameter and cooling rate; and (4) each resulting fiber receives the same windup tension. Any deviation from this ideal behavior along can and does cause either breaks at the bushings or weak and more fragile fibers.

The recent computer modeling study [55] that has been developed (1) deals with process analysis and computer simulation of homogeneous melts from uniformly mixed ingredients for an experimental pilot plant process with a single position 200-tip bushing and (2) demonstrates how to achieve optimum jet cone shapes, fiber-forming stresses, fiber strength, and breaks at the bushing. The modeling study does not address the issue of how to achieve (1) uniformly mixed batch ingredients and (2) homogeneous melts.

1.3.2 General-Purpose Silicate Glass Fibers

Fiberglass is the world's most important continuous inorganic composite reinforcing fiber [52, 36, 20]. Commodity or general-purpose fibers are called E-glass. They contain 50–60% SiO_2, 10–15% Al_2O_3, and 0–7% B_2O_3 and are characterized by universal applicability, large sales volumes, and low unit cost. They represent ~99% of the commercial fiberglass market. The former A-glass in the United States and the current C-glass in China are alternate low-cost substrates containing 65–70% SiO_2, 1–7% Al_2O_3, and 0% B_2O_3.

1.3.2.1 Borosilicate E-Glass Fibers

Two variants of the commercial general-purpose E-glass fiber are on the market [36, 20]. The original variant is based on a calcium aluminum borosilicate glass composition [49] that generally contains 5–7% B_2O_3 and <1% fluorine. The more recent, and environmentally friendly, variant is based on a fluorine-free calcium magnesium aluminum silicate glass composition that is either boron-free [36] or essentially boron-free [20].

(a) *Industry E-glass specifications.* A fiberglass product that is sold as an E-glass must meet the compositional requirements of ASTM specification D-578-00 [64]. These requirements ensure that a product that is sold as an E-glass has about the same chemical composition, subject only to variations induced by the use and variations in local batch ingredients, and therefore the same in-use performance.

One specification governs a generic E-glass composition that is aimed at printed circuit board applications and the other is aimed at general reinforcement applications (Table 1.8). Since E-glass is an international commodity, other countries have similar specifications, e.g., DIN (Germany) and BS (the United Kingdom).

Table 1.8 ASTM E-glass specification D-578-00 [2, 64]

E-glass for printed circuit boards		E-glass for general applications	
Chemical	% by weight	Chemical	% by weight
B_2O_3	5–10	B_2O_3	0–10
CaO	16–25	CaO	16–25
Al_2O_3	12–16	Al_2O_3	12–16
SiO_2	52–56	SiO_2	52–62
MgO	0–5	MgO	0–5
Na_2O and K_2O	0–2	$Li_2O + Na_2O + K_2O$	0–2
TiO_2	0–0.8	TiO_2	0–1.5
Fe_2O_3	0.05–0.4	Fe_2O_3	0.05–0.8
Fluoride	0–1.0	Fluoride	0–1.0

There are notable differences between the two major E-glass specifications, but the primary difference pertains to the permissible percent boron content. An E-glass for printed circuit board applications must contain 5-10% B_2O_3, while an E-glass for general reinforcement applications may contain 0–10% B_2O_3. The use of Li_2O was approved by ASTM in 2000 [64] for use in general reinforcement applications but not for use in printed circuit boards.

(b) *Commercial borosilicate E-glass fibers.* The age of E-glass started in 1940 (Table 1.9) with a patent claiming new boron-free compositions near the eutectic in the quaternary SiO_2–Al_2O_3–CaO–MgO phase diagram [65] and with a product in 1943 claiming derivatives of the quaternary phase diagram modified with B_2O_3, fluorine, and Na_2O [66].

The batch cost was reduced in 1951 by simplifying the E-glass composition [67]. As a result, the B_2O_3 level was reduced from 10.0 to 5–7%, and the previously required addition of dolomite (4.5% MgO) was eliminated. But $\sim0.5\%$ MgO continues to be present. It is due to an impurity (or tramp) in commercial batch ingredients. The reformulation of the E-glass composition [67] in 1951 represents a basic shift from a modified quaternary eutectic (60% SiO_2 – 9% Al_2O_3 – 27% CaO – 4% MgO) to a modified ternary eutectic (62.2 SiO_2 – 14.5% Al_2O_3 – 23.3% CaO), excluding tramp. This specific E-glass variant [67] has since become the generic borosilicate E-glass standard.

(c) *Boron- and fluorine-free E-glass fibers.* By the 1960s, concerns arose about the effect of boron and fluorine when emitted from a melt in a commercial furnace

Table 1.9 Evolution of commercial general-purpose E-glass fibers [36]

Year	1940	1943	1951	1996	2000
Composition, wt%					
SiO_2	60.0	54.0	54.5	60.01	56.50
B_2O_3	–	10.0	6.6	–	1.30
Al_2O_3	15.0	14.0	14.0	12.99	13.45
CaO	20.0	17.5	22.1	22.13	24.50
MgO	5.0	4.5	0.6	3.11	2.55
TiO_2	–	–	0.5	0.55	0.55
Na_2O	–	–	0.8	0.63	0.90
K_2O	–	–	0.2	0.14	–
Fe_2O_3	–	–	0.2	0.25	0.25
F_2	–	–	0.5	0.04	–
Log 3 forming temp., °C	1288	1200	1200	1259	1215
Liquidus temperature, °C	1233	1120	1064	1174	1157
Delta temperature, °C	55	80	136	85	55
Pristine strength, GPa	3.6	3.2	3.3	3.5	3.5
Added emission control	No	yes	yes	no	no
References	[65]	[66]	[67]	[59, 56]	[21, 27]

into the environment. Boron oxide is a costly batch ingredient and up to 15% of the added boron can volatilize. Fluorine [36] causes a similar problem and can emit fluorides into the atmosphere. The atmospheric temperature above the melt in most E-glass furnaces is between 1400 and 1500°C and, in the exhaust, boron and fluorine can react with potassium, sodium, and sulfur oxides to form particulates which are then emitted from the melt into the environment [36].

The emission of particulates was therefore regulated in North America and Europe. As a result, new and costly pollution control devices were required for the production of fluorine containing high-boron E-glass aimed at printed circuit board applications. And new fluorine- and boron-free, as well as new fluorine-free and essentially boron-free, E-glass compositions were therefore designed and commercialized for use in general reinforcement applications.

The first environmentally friendly boron- and fluorine-free E-glass variant that meets the ASTM specifications for general reinforcement applications was patented in 1985 [68] and the first environmentally friendly boron- and fluorine-free E-glass was commercialized in 1996 [59]. Two fluorine-free and essentially boron-free E-glass variants became known between 1997 and 2000 [2, 20, 150].

(d) New energy-friendly E-glass compositions. The log 3 forming, liquidus, and delta temperatures of a specific compositional variant that is based on the same set of oxides can change significantly with relatively minor percent changes of its oxides, specifically SiO_2, Al_2O_3, CaO, MgO, and B_2O_3. Several compositional design and/or reformulation studies were carried out between 1998 and 2008 to achieve the lowest log 3 forming temperature for a range of fiberglass compositions with a delta temperature >50°C [2, 20, 21–25, 26, 27].

The design of new eco-friendly fiberglass compositions is based on four premises: (1) the desired composition must have the lowest log 3 forming temperature (melt temperature at a viscosity of log 3 poise) and a crystallization-resistant delta between log 3 forming and liquidus temperatures; (2) the most energy- and environmentally friendly variant that meets these requirements must be fluorine-free and boron-free or essentially boron-free; (3) the target composition must be cost-competitive within the industry; and (4) E-glass compositions must meet the ASTM specifications.

Many new fluorine-free compositional variants were designed since 2000. Some were boron free [2, 20, 23], others contain <2% ZnO [2], $\leq$1.5% B_2O_3 [21], $\leq$1% Li_2O [22], as well as $\leq$1% Li_2O and $\leq$1.5% B_2O_3 [24]. The design of both energy- and environmentally friendly compositions and the effect of key oxides on melt viscosity, liquidus temperature, and energy demand [25, 26, 27] will be discussed in detail in Chapter 2.

1.3.2.2 E-Glass Properties and Fiber Structures

E-glass is mostly used to reinforce composite structures and, as such, it is the lowest cost candidate with the widest range of applications. E-glass has therefore become the benchmark for the product designer against which the value-in-use of other organic and inorganic fibers must be evaluated. The commercial composite applications of E-glass are discussed in Chapter 3.

(a) Mechanical properties. The mechanical properties of the general-purpose borosilicate and boron-free E-glass fibers are comparable (Table 1.9). Specifically, single filament strength (3.4–3.5 GPa) and elongation-at-break (4.5–4.9%) are about the same for both fibers, but the modulus of borosilicate E-glass is lower (72 GPa vs. 81 GPa) than that of the boron-free E-glass [59, 56]. Tensile strength and break elongations were determined by ASTM method D2101 at 23°C (2-in. gauge). The modulus was obtained by the sonic method [56].

Fibers are purchased on a cost-per-weight basis but they are generally tested on a function-per-weight basis. Two derived properties that are important to the product designer are specific fiber strength and specific fiber modulus, i.e., strength and modulus corrected for density or weight at equal volume. For composites aimed at weight-sensitive transportation uses, a similar correction is often applied to compare the weight of different reinforcing fibers at an equal fiber volume fraction.

Figure 1.9 compares the specific tensile properties of important reinforcing fibers [36, 56]. Borosilicate E-glass serves as the reference point since it offers a universally acceptable balance of properties at the lowest cost. The other inorganic fibers offer specific mechanical properties, which are not attainable with E-glass. But, S-glass, a generic high-strength fiber, costs $\sim$5× as much as E-glass. Carbon or aramid fibers cost $\sim$10× as much as E-glass and boron or SiC fibers cost >100× as much as E-glass. These inorganic reinforcing fibers offer increasing value-in-use with increasing cost. Still, E-glass continues to serve as the reinforcing fiber of choice composites having the highest sale volumes and lowest cost.

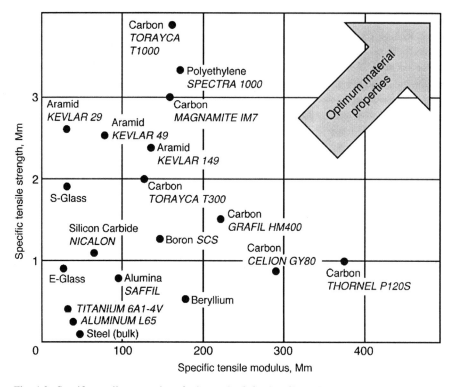

Fig. 1.9 Specific tensile properties of advanced reinforcing fibers. Redrawn from Starr [1, 71] (with kind permission of Springer Science and Business Media)

(b) Other glass fiber properties. Among the general-purpose fibers, the boron-free E-glass fibers exhibit higher acid resistance than borosilicate E-glass fibers (Table 1.10). In this test [56], single filament fibers were weighed, immersed in a 10% sulfuric acid solution for a period of 24 h, and re-weighed. The weight loss for borosilicate E-glass exceeded 40%; that for boron-free E-glass was less than 5%. After 499 days in a 5% sulfuric acid solution, a spray gun laminate made from boron-free E-glass lost half as much flexural strength as a comparable laminate made from a boron-containing E-glass fiber.

Table 1.10 Physical properties of commercial general-purpose E-glass fibers [36] (with kind permission of Springer Science and Business Media)

Fiber properties [59, 56]	Test method	Unit	w/6.6% B_2O_3	w/0.0% B_2O_3
Wt. loss, 10% H_2SO_4	Bare fibers	%/24 h	40	>5
Softening point	ASTM C 338	°C	830–860	916
Refractive index	Oil immersion	Bulk glass	1.547–1.562	1.560–1.562
Thermal expansion	ASTM D 696	ppm/°C	5.4	6.0

Borosilicate and boron-free E-glass exhibit equivalent corrosion resistance and modulus retention in water tests. The acid resistance of boron-free E-glass [68, 59] and of essentially boron-free E-glass with $\leq 1.5\%$ B_2O_3 [2] is higher than that of major borosilicate E-glass variants with 5–7% B_2O_3 [1, 2] and of a minor variant with 4.26% B_2O_3 [69] that is not governed by ASTM E-glass specifications [64].

Boron-free E-glass fibers [56] have a higher softening point and estimated upper service temperature. The softening point of a fiber is the temperature at which a fiber will deform under its own weight. The estimated upper service temperature of a given fiber is the temperature at which it retains useful strength. The softening point of boron-free E-glass is 56–86°C higher than that of the borosilicate E-glass (Table 1.10) and affords a higher useful service temperature (<700°C vs. <600°C). This product advantage is the result of a process penalty, namely its fiber-forming temperature [59].

The electrical properties of both general-purpose glass fibers are comparable (Table 1.11). The refractive index is a property of considerable importance with regard to the appearance of a glass fiber in a laminate or composite. Electrical properties and the coefficient of linear expansion of borosilicate and boron-free E-glass were measured on bulk annealed samples. The differences in test results between both fibers are insignificant [56].

Table 1.11 Electrical bulk properties of general-purpose E-glass fibers [36] (with kind permission of Springer Science and Business Media)

Fiber properties [59, 56]	Test method	Unit	w/6.6% B_2O_3	w/0.0% B_2O_3
Dielectric constant	ASTM D150	23°C/1 MHz	6.9–7.1	7.0
Dissipation factor	ASTM D150	23°C/1 MHz	0.0001	0.0001
Dielectric breakdown	ASTM D149	V/mil	262	258
Volume resistivity	ASTM D257	Log 10/23°C	22.7–28.6	28.1

1.3.2.3 Commercial E-Glass Products and Applications

General-purpose E-glass fibers are the most important composite reinforcing fiber in today's market and they are used to serve a variety of markets: (1) rovings, yarns, chopped strands, and milled fibers; (2) woven rovings, weaver yarns, and braids; and (3) chopped strand, continuous and combination mats.

Rovings are essentially untwisted bundles of fiberglass strands wound up in parallel on cylindrically shaped packages. They are used in open lay-up moldings, woven fabrics, rods, and tubes. Woven rovings are used for open lay and press moldings. Yarns [52] consist of twisted fiber bundles. Weaver yarns are used for electrical and aircraft laminates, insulating tape, window shades, and filtration applications. The major use is in printed circuit boards.

Non-woven mats consist of specific arrays of glass fibers, held together by a binder. A continuous strand mat that consists of fine strands of glass fibers held together by a resin binder or by stitch bonding is used in sheet molding and reinforced thermoplastic sheeting. A chopped strand mat is a non-woven fabric consisting of a random array of chopped glass fiber strands. It is used to reinforce composites for use in boat hulls and decks, car bodies, sheeting, and tanks. A roofing mat is used to manufacture roofing shingles.

A databook [70] that lists all commercial E-glass fibers and a directory and databook [71] that lists all commercial carbon, silica, high-strength glass fibers, and other high-performance fibers including Kevlar are available. Chapter 3 describes the generic fiberglass composite design, engineering and markets since glass fibers are almost exclusively used as composite reinforcing fibers and discusses wind mill blades and the respective market which rely on fiberglass composites. Finally, Chapter 11 is devoted to discussing plasma melting in greater detail.

1.3.3 Special-Purpose Silicate Glass Fibers

Specialty glass fibers are niche products, which may contain up to 65% SiO_2 and 25% Al_2O_3 and a range of other oxides which provide premium properties. Specialty glass fibers are characterized by small sales volumes and high unit cost and support $\sim 1\%$ of the fiberglass market.

1.3.3.1 Designations of Special-Purpose Fibers

Specialty glass fibers have superior functional properties relative to those of general-purpose E-glass fibers. Many other fiberglass variants are known by letter designations [50]. "S" stands for high strength, "A" for high alkali content, "M" for high modulus, "AR" for high alkali resistance, "C" for chemical durability, "CC" for Chinese C-glass, "D" for low dielectric constant, and "ECR" for high acid resistance. "Z" has been used, but not consistently, for Zr-modified compositions. All variants, except the former A- and C-glass and the present E-glass and CC-glass, are premium products.

1.3.3.2 High-Strength–High-Temperature Glass Fibers

This section deals with specialty glass fibers having commercially a desirable balance of high-strength, intermediate modulus (or stiffness), and high-temperature resistance.

(a) Process and products. Higher service temperatures are often a more important design factor than higher strength or higher stiffness. High in-use temperatures require a fiber composition with higher thermal stability than that of E-glass, i.e., higher forming temperature, higher process energy, and higher overall cost. Higher strength or stiffness alone does not necessarily require a more costly fiber.

For a fiber to be usable at higher service temperatures than that of E-glass, it must have a higher glass transition temperature, softening point, forming temperature, and liquidus temperature. With one exception, these fibers can be made in a conventional melt spinning process; special refractory furnace linings and special precious metal bushings are required. The production of Sudaglass basalt fibers requires a different process [72]. Higher materials, process and energy cost translate into a premium price.

HS- and HT-glass fibers possess $\leq 1.3\times$ the strength of E-glass fibers (4.5 GPa vs. 3.4 GPa), $\leq 2\times$ the service or in-use temperature (1250°C vs. 600°C), and $\leq 3.4\times$ the stiffness or modulus (35–37 GPa vs. 72 GPa). In contrast, high-modulus (HM) glass fibers have a modulus of up to 132 GPa and ultrahigh-modulus (UHM) glass fibers can have a modulus up to 248 GPa.

Four typical high-strength high-temperature glass fibers shown in Table 1.12 are commercial products [36, 71]. S-glass [50] and Te-glass [73] are derived from the ternary SiO_2–Al_2O_3–MgO eutectic. R-glass [74] is derived from the quaternary SiO_2–Al_2O_3–CaO–MgO eutectic. Sudaglass [72], a basalt fiber, is based on natural sources and contains high levels of Fe_2O_3. It is not shown in Table 1.12. High-strength glass fibers [71] are suitable for use in typical high-temperature applications. For applications see next section.

Table 1.12 High-strength and high-temperature silicate glass fibers [36]

Glass fiber	S-glass	Te-glass	R-glass	Experimental HS-glass fibers		
Eutectic	Ternary	Ternary	Quaternary	Quaternary	Ternary	Ternary
Wt%						
SiO_2	65.5	65.0	60.0	54.28	64.5	67.5
Al_2O_3	25.0	23.0	25.0	25.53	24.6	15.3
MgO	9.5	11.0	6.0		9.4	15.5
CaO	–	–	9.0	7.57	0.5	–
BaO	–	–	–	12.62	–	–
ZrO_2	–	1.0	–	–	–	–
B_2O_3	–	–	–	–	1.0	–
TiO_2	–	–	–	–	–	1.5
Fe_2O_3	Tr.	0.3	–	–	–	–
Na_2O	Tr.	0.1	–	–	–	–
Properties						
Strength, GPa	4.6	4.7	4.4	–	–	1.2
Elongation, %	5.2	5.5	5.2	–	–	–
Modulus, GPa	88.95	84.3	86.0	–	–	95.3
Density, g/cc	2.53	2.49	2.52	–	–	2.70
Sp. Strength	1.85	1.90	1.74	–	–	0.45
Sp. modulus	35.37	34.0	34.0	–	–	35.3
SP, °C	1056	975	975	–	–	–
Service T, °C	<1000	<900	<900	–	–	<1295
References	[50, 36]	[36, 73]	[36, 74]	[75]	[6, 36]	[36, 78]

Typical HS-glass fibers perform well at high temperatures (900–1000°C) but it may not be the optimum composition for prolonged uses. A novel quaternary glass fiber that is based on 54.28% SiO_2, 25.53% Al_2O_3, 7.57% CaO, and 12.62% BaO has recently been claimed to give superior long-term high-temperature performance as sound-absorbing material in engine muffler uses [75]. It contains 11% less SiO_2

than S-glass, and it contains BaO instead of MgO (Table 1.12). Therefore it has a log 3 forming temperature of only 1478°C and a delta temperature of 83°C [76], while S-glass has a log 3 forming temperature of 1560°C and a delta temperature of 65°C [50].

ZrO_2 and B_2O_3 are fluxes (Table 1.12). The addition of 1% ZrO_2 to the S-glass or ternary SiO_2–Al_2O_3–MgO eutectic composition causes a major drop in melt viscosity [36, 73], and the resulting Te-glass fibers have an 81°C lower softening and a 100°C lower practical use temperature than S-glass fibers. The addition of 1% B_2O_3 has a similar, if not more, pronounced effect. It also reduces both softening point and service temperature.

The addition of 1.5% TiO_2 to a ternary composition (Table 1.12) was found to raise its modulus, density, and service temperature [73, 75, 77, 78]. The resulting experimental fibers have a 295°C higher service temperature than commercial S-glass fibers but the development of a major nanocrystalline phase reduces its overall strength by over 50%.

Importantly, commercial HS-glass fibers, e.g., S-glass, also have higher elongations at break than E-glass [79] and therefore also higher mechanical (not to be confused with fracture) toughness, which translates into higher damage, including ballistic impact, resistance (Table 1.12). To wit, the energy required to break a relatively elastic fiber is proportionate to the area under the stress–strain curve (Fig. 1.10). As a result, the ballistic impact damage resistance of S-glass is also higher than that of E-glass and of nominally stronger, but more brittle, standard modulus (SM) carbon fibers having only 1/10th the elongation at break [70, 71].

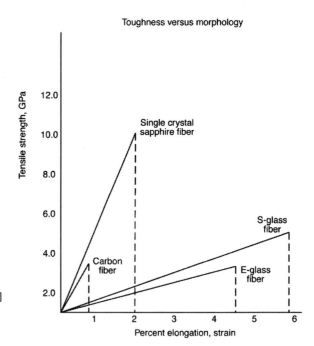

Fig. 1.10 Toughness and impact damage resistance of high-strength glass fibers [36] (with kind permission of Springer Science and Business Media)

Specific properties (properties divided by density) are a key design factor for transportation, especially aircraft, composites, where higher functionality at lower weight translates into higher value-in-use. At room temperature, HS-glass fibers (Table 1.12) have about equal specific strength (174–190 Mm) and specific moduli (34–37 Mm). General-purpose glass E-glass fibers have a specific modulus of 20 Mm at ~1/5th the cost, and standard modulus carbon fibers have a specific modulus of 131 Mm at ~2× the cost.

The utility of HS- and HT-glass fibers is ultimately determined by their softening point, the temperature above which a fiber will deform under its own weight. In practical terms, R- and Te-glass fibers retain usable functionality to about 900°C, B_2O_3–modified S-glass fibers to about 800°C, basalt fibers [72] to about 1000°C, and the TiO_2–modified glass ceramic fiber to 1295°C, while E-glass is limited to 620°C (Table 1.12). In contrast, the upper service temperature limit of unprotected carbon fibers is 350–500°C in an oxidative environment, that of coated carbon fibers is ~1000°C in an oxidative environment, but that of uncoated carbon fibers is (1500°C in an inert environment such as helium.

HS-glass fibers retain useful properties at high temperatures. In the strand tensile test, a bundle of fibers is briefly exposed to a given temperature and broken. The strand tensile strength of R-glass, a typical HS-glass fiber shown in Fig. 1.11,

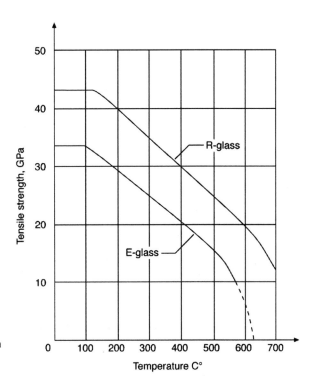

Fig. 1.11 Strength retention of R-glass and E-glass at elevated temperatures. Drawn from data contained in French patent 1,435,739, to Saint Gobain Company [74]

is higher than that of E-glass from room temperature to 750°C (and beyond). E-glass fails at ~600°C. In addition, it should be noted that the modulus of R-glass is also higher than that of E-glass over the entire temperature range tested, from room temperature to 750°C (and beyond).

In another tensile test, R-glass is exposed for prolonged periods of time to a specific elevated temperature before its strength is tested. At 750°C its tensile strength drops within the first 400 h to 1 GPa (or 1/4th its room temperature strength), but then remains nearly constant for at least the next 600 h of exposure. A temperature of 750°C is well beyond the reach of either of the two general-purpose E-glass fibers.

In summary, high-strength fibers have superior high-temperature resistance, superior impact damage resistance, high-strength retention at elevated temperatures, and a high, but lower modulus than high-modulus (HM) glass fibers. The formation of a nanocrystalline structure which facilitates the attainment of HS-glass fibers with the highest service temperatures tends to reduce strength, but in most cases this strength loss is minimal.

(b) Properties and applications. HS-glass fibers occupy an important niche in the composite reinforcement market between E-glass and the lowest cost carbon fibers. They are used in a variety of applications in the aerospace and aircraft industry, automotive industry, electrical and electronics industry, sporting goods industry, and military markets. The higher cost of other inorganic high-strength fibers can be reduced by hybridization, i.e., by designing composites reinforced with mixed HS-glass/carbon, HS-glass/aramid, or HS-glass/boron fibers.

In the aerospace and aircraft market HS-glass fibers are used because of their high specific properties and relatively low cost, e.g., in satellite components, motor cases, nose cones, aircraft flooring, cargo liners, and radome skin sheets. Their use in helicopter rotor blades benefits from the high damage resistance (toughness) of HS-glass fibers. In automotive composites, HS-glass fibers are used because of their high specific strength in pressure vessels and their superior toughness in leaf springs. Their use in compressed natural gas cylinders is one of the fastest growing applications.

In the electrical and electronics market, HS-glass fibers are used as strength members for optical fiber cables, printed circuit boards, and other cables, but mostly when high temperatures are involved. In sporting goods, they are used for product differentiation and because of their toughness in tennis racquets, squash racquets, fishing rods, surfboards, windsurfing masts, skis, archery bows, arrow shafts, and racing yachts.

In high-temperature applications, selected HS-glass fibers have been claimed to be a superior sound-absorbing material in engine exhaust mufflers [76]. In military markets, HS-glass fibers are used in aircraft fuel tanks because of their specific strength, and in rigid armor and helmets because of their superior ballistic impact resistance. They would be an ideal reinforcement for composite windmill blades but because of their high price they are less cost-effective in this application. For products and markets see Chapter 5.

1.3.3.3 High-Modulus–High-Temperature Glass Fibers

Conceptually, HM-glass fibers are achieved by adding BeO [50, 78], La_2O_3 [78], Y_2O_3 [6], or CuO [6, 78] to a typical HS-glass composition. HS-glass fibers have a modulus of 85–95 GPa while HM-glass fibers have a modulus of 100–132 GPa (Table 1.13). The addition of these oxides increases the fiber density, a fact that negatively affects their potential value-in-use [36, 80, 81]. In addition, these oxides are also known to be fluxes, and they should reduce the melt viscosity and therefore the log 3 forming temperature. Unfortunately, a systematic comparison of the properties (log 3 forming temperatures) of an unmodified base composition with compositions having the same levels of key modulus modifying oxides is not available.

Historically, the first HM-glass fiber contained 7.5% BeO. This fiber, YM-31A, had a higher density than a typical HS-glass fiber (2.60 g/cc vs. 2.53 g/cc), higher strength (5.1 GPa vs. 4.6 GPa), higher specific strength (1.96 vs. 1.85), a higher modulus (112 GPa vs. 95 GPa), and a higher specific modulus (43.0 Mm vs. 37.5 Mm). In summary, this composition affords the highest specific fiber modulus and the second highest fiber modulus among silicate glass fibers.

An HM-glass fiber with 5.0% BeO and 30.2% La_2O_3 also has a higher density (3.29 g/cc vs. 2.69 g/cc), higher strength than a typical HS-glass fiber (5.4 GPa vs. 4.6 GPa), lower specific strength (1.64 Mm vs. 1.85 Mm), the highest modulus (132 GPa) among silicate glass fibers, and second highest specific modulus

Table 1.13 High-modulus silicate glass fibers [36] (with kind permission of Springer Science and Business Media)

Modulus modifier	La_2O_3	ZnO	BeO	Y_2O_3	BeO/Y_2O_3
Composition, wt%					
SiO_2	50.0	45.8	50.0	52.6	36.2
Al_2O_3	32.5	–	35.0	35.8	20.5
CaO	–	11.0	–	–	–
MgO	12.5	7.7	7.5	5.4	8.1
La_2O_3	5.0	–	–	–	–
ZnO	–	22.4	–	–	–
BeO	–	–	7.5	–	5.0
Y_2O_3	–	–	–	6.2	30.2
ZrO_2	–	1.8	–	–	–
TiO_2	–	6.8	–	–	–
Li_2O	–	1.7	–	–	–
CeO_2	–	2.5	–	–	–
Fe_2O_3	–	0.3	–	–	–
Pristine properties					
Strength, GPa	4.8	–	5.1	4.5	5.4
Modulus, GPa	100	104	112	100	132
Density, g/cc	2.69	2.77	2.60	3.0	3.29
Sp. Strength, Mm	1.78	–	1.96	1.36	1.64
Sp. modulus, Mm	37.2	37.5	43.0	33.0	40.0
References	[78, 80]	[118]	[78]	[6]	[78, 80]

(40.0 Mm). Values for strength and therefore for specific strength must be accepted with caution for individual fibers for which no statistical support is available. Modulus data are more reliable for such compositions.

The addition of 22.4% ZnO and other oxides raised the density from 2.53 to 2.77 g/cc and the modulus from 95 to 104 GPa. Likewise the specific modulus remained unchanged (37 Mm). The effectiveness of BeO as a modulus builder is the result of the high field strength of the Be^{+2} ion and its ability to coordinate four oxygen ions tightly to it [1]. BeO has a wurtzite structure. The only other wurtzite structure with oxygen is ZnO, and $ZnSiO_4$ is isomorphous with Be_2SiO_4. On a molar basis, the ZnO-modified and the original BeO-modified YM-31A compositions are identical [36]. Because it has a higher density, ZnO is a less effective modulus builder than BeO.

The addition of 5.0% La_2O_3 (Table 1.13) raised the density of a typical HS-glass fiber from 2.53 to 2.69 g/cc and the modulus from 95 to 100 GPa. The specific modulus did not decrease but remained the same ($\sim$37 Mm). The addition of 6.2% Y_2O_3 raised the density of a typical HS-glass fiber from 2.53 to 3.0 g/cc and the modulus from 95 to 100 GPa. The specific modulus dropped from 37 to 33 Mm. The addition of 26.8% Y_2O_3 and 4.3% La_2O_3 raised the density from 2.53 to 4.3 g/cc and the fiber modulus from 95 to 130 GPa. But the specific modulus dropped from 37 to 30 Mm.

These HM-glass fibers could be used only in applications which are not driven by the specific modulus, therefore not in aircraft, automotive, or windmill applications, which require a high specific modulus at the lowest possible cost. In both types of applications, carbon and aramid fiber reinforced composites afford much high stiffness at equal loading, i.e., a much higher modulus and specific modulus because of their low densities.

For example, the specific modulus of commercially available aramid fibers ranges from 40 to 122 Mm and that of commercially available carbon fibers ranges from 131 to 379 Mm [1, 71]. In contrast, the specific modulus of an unmodified HS-glass fiber is 37.5 Mm. When modified with 5.0% La_2O_3 or 7.5% BeO, the specific modulus of the resulting HM-glass fibers ranges from 37.2 to 43.0 Mm (Table 1.13).

BeO- and Y_2O_3/BeO-modified HM-glass fibers would not only be costly but also commercially unattractive because of known hazards involved in the process of handling BeO as an ingredient [36]. With regard to the others, the ingredient cost and availability rises from ZnO to La_2O_3 and Y_2O_3. Aside from availability and ingredient cost, CuO might be a useful modulus modifier but it affords melts with a deep blue color. So would be MnO, but it affords a dark brown color. Because of their dark color, both would increase the energy demand of the melt. For that reason they are not useful.

1.3.3.4 Ultrahigh-Modulus Glass Ceramic Fibers

By modifying the ternary SiO_2–Al_2O_3–MgO or the quaternary SiO_2–Al_2O_3–MgO–CaO eutectic compositions with appropriate oxide modifiers one can raise the fiber strength to 5.4 GPa (vs. 3.4 GPa for E-glass) and the fiber modulus to 132 GPa (vs.

72 GPa for E-glass). The increase in modulus is caused by an increase in internal order as evidenced by a change from an amorphous to a nanocrystalline structure.

An increase in fiber modulus up to 248 GPa can be achieved by inserting nitrogen into the oxide network [36], thus creating a nitride-modified silicate, Si–Al–O–N, or oxynitride glass fibers. This increase is caused by an increase in surface tension as evidenced by microhardness. Thus, one mechanism seems to depend on increasing structural order, the other on crosslinking.

(a) Process and products. Oxynitride fibers are formed by an adaptation of the conventional bushing process. Since nitrides would corrode precious metals in an oxidative environment, the bushings with up to 200 tips are therefore made from boron nitride-coated carbon or from molybdenum. The melts are formed under nitrogen at 1600–1750°C and refined at a lower temperature. Fibers with diameters ranging from 12 to 20 μm are continuously drawn from the melt, and mechanically wound at 1000–2000 m/min [82, 83].

Silicon formation:

$$SiO_2 + Si_3N_4 \longrightarrow 2Si + 2SiO + 2N_2 \tag{1.8}$$

$$SiO_2 \longrightarrow Si + O_2 \tag{1.9}$$

Oxygen formation:

$$Si_3N_4 \cdot SiO_2 + 12Mo \longrightarrow 4Mo_3Si + 2N_2 + O_2 \tag{1.10}$$

$$3Si_3N_4 \cdot SiO_2 + 20Mo \longrightarrow 4Mo_3Si_3 + 6N_2 + 3O_2 \tag{1.11}$$

Silicon oxidation:

$$Si + O_2 \longrightarrow SiO_2 \tag{1.12}$$

The melt process may cause the reduction of silica yielding particulate silicon. Increasing numbers of silicon defects, when formed, impart a blue-gray (hazy) to dark brown (opaque) appearance to the glass (Equations 1.8 and 1.9), and these defects may proportionately reduce the strength of the resulting fibers [83]. The formation of silicon defects can, however, be reversed by using molybdenum (but not boron nitride) as the bushing material (Equations 1.10, 1.11, and 1.12), and by inserting an 8 h refining cycle midway between the melt and fiber-forming temperatures (see second entry in Table 1.14).

A refined Si–Al–O–N glass is colorless and clear [83] and the resulting defect-free glass fibers [36] are stronger (>4.0 GPa) than those obtained from unrefined melts (2.0–3.0 GPa). Si–Al–O–N compositions with <15% nitrogen yield glass fibers with moduli ranging from 100 to 140 GPa, and a microhardness ranging from 660 to 700 kg/mm^2 [36] as shown in Fig. 1.12. Compositions with >15% nitrogen yield glass ceramic fibers with moduli ranging from 140 to 248 GPa and a micro-hardness ranging from 900 to 1200 kg/mm^2 [83]. The substitution of nitride for

Table 1.14 Ultrahigh-modulus Si–Al–O–N glass fibers [36] (with kind permission of Springer Science and Business Media)

Glass fiber	S-glass	Si–Al–O–N glass fiber		Si–Al–O–N glass ceramic	
Composition	Control	Mg⁻	Ca⁻	Y⁻	Ca-Mg
Wt%					
SiO_2	65.50	56.75	32.38	50.3	11.1
Al_2O_3	25.00	25.33	–	7.6	2.2
CaO	–	–	50.78	–	62.6
MgO	9.50	10.42	–	–	0.7
Y_2O_3	–	–	–	25.1	–
Si_3N_4	0.00	7.51	16.84	–	23.4
Al_2N_2	–	–	–	17.0	–
N-content	0.00	2.54	10.00	4.37	13.5
Melting, °C	1650	1720	1750	1600	1790
Refining, °C	–	1600	–	–	–
Log 3 FTT, °C	1565	1500	1380	1560	1590
Bushing tips	200	200	–	1	1
Pristine fibers					
Strength, GPa	4.60	4.50	–	–	–
Modulus, GPa	88–95	115.00	137.0	213.00	248.0
Density, g/cc	2.53	2.80	2.8	3.94	(3.3)
Sp. Strength	1.85	1.61	–	–	–
Sp. Modulus	35–37	41.10	48.9	54.00	(75)
Structure	Amorphous	Amorphous	Amorphous	Nanocryst	Nanocryst
References	[50]	[83]	[83]	[82]	[83]

25% of a given oxide composition under the same conditions of synthesis no longer yields glasses. These melts will foam and crystallize [36].

Selected glass and/or glass ceramic Si–Al–O–N fibers are seen in Table 1.14. The first [83] is S-glass and the second is a Mg–Si–Al–O–N composition that is similar to the S-glass composition but modified to contain 7.5 wt.% Si_3N_4 [83]. Since the latter had been properly melt refined before it was fiberized, it yielded high-strength fibers. Its low nitrogen content produced a significant modulus increase beyond that of the S-glass control. The third example in Table 1.14, a Ca–Si–Al–O–N, has an even higher nitrogen content [36] and therefore a higher absolute and specific modulus. Its low strength suggests that the melt was not or could not be adequately refined.

The remaining examples in Table 1.14 are ultrahigh-modulus (UHM) glass ceramic fibers. The first example, a glass ceramic Y–Si–Al–O–N fiber, exhibits the very predictable effect of yttria that dramatically increases both the measured modulus and the density but offers only a small increase of the specific modulus in return. The second example, a glass ceramic Ca–Mg–Si–Al–O–N fiber, has the highest measured modulus (248.0 GPa) and the highest specific modulus (75 Mm) of any known oxide glass or glass ceramic fiber.

In summary, nitrides offer only a modest crystallization potential but instead seem to act as modulus builders by crosslinking the oxide structure they modify.

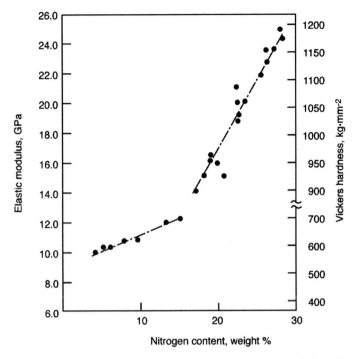

Fig. 1.12 Modulus and microhardness of oxynitride fibers. Redrawn from Kobayashi et al. [153]

And the presence of nitrogen in the network structure, once introduced, may restrict the dimensions of crystals which can be formed.

A selected oxynitride melt that has a high nitrogen content may yield nanocrystalline fibers with an ultrahigh modulus, while a nitrogen-free oxide melt with the same cation ratio may already yield microcrystalline fibers with lower moduli. Thus, the modulus of oxynitride fibers can be increased to much higher levels than that of fibers from oxide melts. Crystallinity of oxynitride glasses can be correlated with their nitrogen content by infrared methods. HM-oxynitride glass fibers are x-ray amorphous and UHM-glass ceramic fibers are nanocrystalline [36].

(b) Properties and applications. Glass and glass ceramic oxynitride fibers can be produced with high strength by properly refining melts while they are formed, and with ultrahigh moduli by inserting nitrogen into a suitable oxide network structure. The highest measured modulus (248 GPa) that has been reported lies between those of standard modulus (SM) and intermediate modulus (IM) carbon fibers (230 and 303 GPa, respectively) and the highest specific modulus (75 Mm) lies midway between those of SM carbon fibers (131 Mm) and E-glass fibers (27 Mm). Two types of applications are being pursued with oxynitride fibers.

Sialons are reinforcing fibers for metal matrix composites. An aluminum alloy 6601 matrix reinforced with a development fiber (Table 1.14) had a strength of 4.0

GPa and a modulus of 180 GPa. The bending strength of the MMC was 25% higher than that of an alumina fiber reinforced control, and almost the same as that of a silicon carbide fiber reinforced control. A high value-in-use may result if this fiber were to cost less than the incumbents did. Oxynitride fibers have high alkali durability and may be useful as a diaphragm material in the electrolytic production of chlorine from aqueous NaOH [36] and as a reinforcement for cementicious composites [36].

1.3.3.5 Glass Fibers with High Chemical Stability

The relationship between fiber composition and chemical stability in water, acids, and bases is complex. It depends on the interaction between (1) the chemical agent [48] to which the glass fiber surface is exposed, (2) the pH of the glass composition [84] in the fiber surface, and (3) the internal microstructure of the fiber [48].

The chemical stability of glass fibers, yarns, or fabrics is complex and depends on whether they have generic finish or not, or even an acid- or alkali-resistant coating. These issues will be discussed here. Due to the presence of the composite matrix, the corrosion resistance of fiberglass reinforced composites is even more complex in sensitive applications.

(a) Chemical resistance of glass fibers. The first step in the attack of water on the bare surface of an alkali-free or near-alkali-free glass fiber is its adsorption. The adsorbed water molecules hydrolyze the siloxane bond by protonation of the oxygen atom and yield a highly hydroxylated fiber surface:

$$\equiv Si-O-Si \equiv -\xrightarrow{\delta-} \equiv Si-O \cdots Si \equiv -\xrightarrow{\delta+}$$
$$\equiv S-OH + HO-S \equiv H^+ + OH^- \tag{1.13}$$

With high alkali-glass fiber surfaces, the reaction of water represents an electrophilic attack by the addition of the proton (H^+) to the negatively charged oxygen atom of the $\equiv Si-O-M$ bond. It proceeds in the same fashion and results in the ion exchange between H^+ (or H_3O^+) and either alkali ions and/or network-modifying alkaline earth ions, and leads to the formation of SiOH groups [48]. The reaction products of water with a highly alkaline fiber surface are NaOH, $Ca(OH)_2$, and hydrated sodium silicate:

$$\equiv S-OM + H^+ \longrightarrow \equiv S-OH + M^+ \tag{1.14}$$

However, as soon as the supply of H^+ ions is exhausted, the corrosive attack of H_2O turns into a nucleophilic attack by alkali ions on the fiber surface. In other words, the reaction of a glass fiber with water turns into a reaction of a glass fiber with alkali. Siloxane bonds are broken and Si–O–Na groups are formed until the glass fiber is completely dissolved in the highly alkaline medium that initially consisted only of water.

The reaction of alkaline media starts with a nucleophilic attack by hydroxyl ions on silicon atoms in the bare surface of a glass fiber ($-OH^- + \equiv Si-O-S\equiv$) where it forms new bonds ($\equiv Si-OH$ and/or $\equiv Si-OM$). Monovalent cations (e.g., sodium) are removed from the glass fiber, leaving behind a hydroxylated surface. Bivalent cations (e.g., calcium) remain attached to the glass surface and form a crystalline sheath growing in thickness. Such a sheath develops when bare silicate glass fibers (>50% SiO_2) including E-glass [48], AR-glass [85, 86], basalt [87, 72] and oxynitride fibers [88], or when bare aluminate glass fibers (>50% Al_2O_3), e.g., calcium aluminate fibers [9] are exposed to alkaline media [89].

The crystalline sheath, mostly $Ca(OH)_2$ [36], increases the alkali resistance, but also limits the practical utility of the resulting fibers since it drastically reduces their strength. ZrO_2, SnO, La_2O_3, TiO_2, Fe_2O_3 [6], Y_2O_3, and Na_2O [36] enhance the alkali resistance perhaps by delaying sheath formation, but rather large amounts of Na_2O (>10%), ZrO (>15%), or Y_2O_3 (>30%) and combinations of Na_2O (11%) and ZrO_2 (16%) are often employed. These fibers are not really alkali resistant, only more alkali resistant than E-glass. In the end, all lose their physical integrity and are destroyed as evidenced from the accelerated leach test shown for E-glass in Table 1.15.

Table 1.15 Borosilicate E-glass solutes after acid leaching 8 h at 95°C [36] (with kind permission of Springer Science and Business Media)

Borosilicate E-glass	SiO_2	Al_2O_3	Fe_2O_3	CaO	MgO	B_2O_3
Oxide content (%)	53.8	14.9	0.3	17.1	4.7	8.7
Solute (%), in H_2O	0.2	0.2	–	0.2	0.1	0.7
In 2 N H_2SO_4	1.7	14.6	0.3	16.6	4.6	8.6
In 2 N NaOH	25.6	7.3	0.2	8.4	1.3	6.3

While alkaline media are known to create a crystalline deposit or sheath on the surface of glass fibers, mineral acids selectively dissolve specific components of the glass, first of all the ions of network modifiers [6]. Silanol bonds, as a rule, are not broken and SiO_2 is not dissolved. If, however, the amount of SiO_2 is not sufficient to create a continuous network structure, cations can selectively dissolve in acid media. The addition of Zr, Ti, and Fe oxides substantially increases the acid resistance [6], but in the end, the entire fiber, whether it is compositionally an E-glass or A-glass, is converted to a porous high silica fiber (see Table 1.15 and Section 1.2.2.2).

In summary, the effect on the pH of the bare fiber surface and the effect of the interaction between a chemical agent and a bare fiber surface are predictable. ZnO (as in ECR-glass) increases the acid resistance and ZrO_2 (as in CemFIL or AcroteX) increases both acid and base resistance. The effect of the internal microstructure [48] of a fiber is highly process dependent and not predictable without a thorough prior investigation of its microstructure. Importantly, however, all fibers, except experimental single fibers, have a primary finish; some have an additional secondary coating. These modifications further reduce the predictability of their chemical resistance from their compositional make-up alone.

(b) Alkali-resistant glass fibers. Commercial AR-glass fibers such as CemFil[TM] [1, 36, 90] and ArcoteX[TM] [85], an experimental and AR-glass fiber [86], and AR basalt fibers [87, 72], which are commercial only in Russia, are shown in Table 1.16.

Table 1.16 Glass fibers with high chemical stability [36] (with kind permission of Springer Science and Business Media)

Glass fiber	Alkali-resistant glass fibers			Acid-resistant glass fibers		
Type	CemFil	Exp. ARG	Basalt	A-glass	C-glass	CC-glass
$Wt\%$						
SiO_2	71.0	68.11	49.06	71.8	65.0	65.15
Al_2O_3	1.0	0.78	15.70	1.0	4.0	6.70
B_2O_3	–	–	–	–	5.0	–
CaO	–	4.86	8.95	8.8	14.0	9.23
MgO	–	3.04	6.17	3.8	3.0	4.33
BaO	–	2.43	–	–	–	–
TiO_2	–	–	1.36	–	–	–
ZnO	–	–	–	–	–	0.20
ZrO_2	16.0	6.92	–	–	–	–
Li_2O	1.0	–	–	–	–	–
Na_2O	11.0	13.85	3.11	13.6	8.5	12.92
K_2O	–	–	1.52	0.6	–	–
FeO	–	–	6.37	–	–	–
Fe_2O_3	–	–	5.38	0.4	0.3	0.27
MnO	–	–	0.31	–	–	–
P_2O_5	–	–	0.45	–	–	–
H_2O	–	–	1.62	–	–	–
F_2	–	Trace	–	–	Trace	0.96
Log 3 FT, °C	1300	1212	1300	1280	1200	1192
LT, °C	1200	1074	1220	1010	1135	1068
ΔT, °C	100	138	100	270	65	124
Structure	Amorphous	Amorphous	Nanocryst	Amorphous	Amorphous	Amorphous
References	[50, 86]	[86]	[87]	[52]	[52]	[91]

CemFil and the experimental AR fiber are highly Na_2O- and ZrO_2–modified glass fibers. Basalt fibers are derived from volcanic rock with high Fe_2O_3 + FeO levels. The alkali resistance of basalt fibers lies between that of E- and AR-glass. All have higher fiber-forming temperatures than E-glass and require an energy-intensive, non-standard process.

In a test that simulates their suitability as a cement reinforcement [90], bare AR-glass and bare E-glass fibers were immersed at 25, 50, and 80°C in a solution (NaOH 0.88 g/l, KOH 3.45 g/l, $Ca(OH)_2$ 0.48 g/l, pH 12.5) that simulates the aqueous phase of Portland cement. Both fibers lost strength between 24 and 96 h, i.e., in the time frame during which the alkalinity of cement reaches its peak as it cures.

At 25°C, the strength of bare E-glass dropped to 2/3 of its original value in 24 h and that of AR-glass in 96 h. In the accelerated test at 80°C, the strength of E-glass dropped to 1/3 of its original value in 24 h and that of AR-glass in 96 h (Fig. 1.13). Thus, neither fiber is usable to structurally reinforce cement without having an effective secondary alkali-resistant coating.

The application of an alkali-resistant finish or secondary coating is known to render even E-glass suitable for continued use as a durable reinforcement of cement structures, whether they are composite wraps for bridge columns or net-like

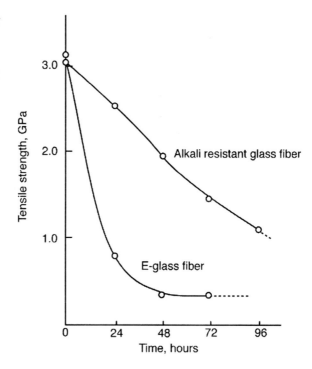

Fig. 1.13 Tensile strength of glass fibers in the aqueous phase of cement. Redrawn from Majumdar [36, 90] (reprinted with permission of Elsevier Science Publishers)

structures aimed at roadbed construction. But since even AR-glass loses strength in <96 h in a solution simulating an aqueous cement phase, it too would require a costly secondary alkali-resistant coating to be acceptable in continuous structural use.

In the Western world and Japan, AR-glass such as CemFIL costs $2\times$ as much as E-glass. AR-coated E-glass rather than AR-coated CemFIL or ArcoteX is therefore the preferred reinforcement of structural, load-bearing cementicious composites. In CIS countries, e.g., Russia, AR-coated basalt fibers are apparently as economically viable as AR-coated E-glass as a replacement for asbestos in the fiber reinforcement of cement pipes.

In addition to specialty cementicious applications such as do-it-yourself driveway repair [85], the cost-considered use of AR-glass (such as ArcoteX or Cemfil) is the reinforcement of non-structural and decorative cement composites ranging from non-structural ready-mix cement and decorative architectural components to non-load bearing, building cladding systems such as balustrades, string courses and cornice elements, corbels and arch units, mullions and window surrounds, and consoles and copings [6].

In these uses [6, 85], AR-glass offers a simple solution to the problem of shrinkage cracking that often occurs during the initial curing phase (<96 h) of cement or concrete mixes, without requiring a costly secondary AR coating. They retain useful strength long enough to survive the alkaline curing cycle and therefore prevent shrinkage cracking. E-glass without a costly AR coating loses its strength and totally disintegrates before the initial, alkaline cement curing cycle is complete.

(c) Acid-resistant glass fibers. Boron is readily removed by acid leaching from borosilicate E-glass fibers with $\sim$6% B_2O_3. Therefore they have a lower acid resistance than boron-free E-glass fibers [36, 20] and essentially boron-free E-glass fibers [2, 24]. An intermediate level of acid resistance is obtained with soda-lime-silica fibers such as the boron-free former A-glass in the United States or the current boron-free CC-glass in China, and the former borosilicate C-glass in the United States.

The typical A- and CC-composition [2, 91] is boron-free but contains 65–72% SiO_2, 1–7% Al_2O_3, and >12% Na_2O. The former C-glass in the United States contained 65% SiO_2, 4% Al_2O_3, 8% Na_2O, and 2–5% B_2O_3 (Table 1.9).

Specifically, the oxides which are preferentially extracted from these fibers on acid leaching, e.g., with dilute H_2SO_4 or HCl [36, 69], are B_2O_3, Na_2O, Al_2O_3, CaO, MgO, Fe_2O_3, and only a very small amounts of SiO_2. Ultimately, a porous silica fiber structure remains after acid leaching of a borosilicate E-glass or a sodium silicate A- (CC-)glass fiber, yarn or fabric that can be consolidated into a near-solid silica fiber, yarn, or fabric structure.

Boron-free A- and CC-glasses have a much higher SiO_2 and Na_2O content and a much lower Al_2O_3 content (Table 1.9) than typical borosilicate E-glass variants. The high SiO_2 content translates into high acid resistance. The low Al_2O_3 content translates into reduced strength and stiffness. The high Na_2O content translates into a lower viscosity and therefore into a lower log 3 forming temperature. The absence of B_2O_3 translates into lower cost.

The cost of A-glass (and presumably also that of CC-glass) would be even lower if a fiberglass forehearth with multiple bushings were added, as has been suggested [52] to a float glass or container glass furnace and/or if the A- or CC-glass product were made from float or container glass waste [92].

The recent technical evaluation of a soda-lime-silica A-glass fiber highlights its value-in-use in automotive composites, especially when made from waste glass [92]. The product performed satisfactorily, but, as expected, the multifilament yarns and the resulting automotive composites had lower tensile strength and a lower modulus than comparable E-glass yarns and composites. To achieve the same strength and stiffness, more A- than E-glass would be required, thus offsetting the initial cost advantage.

Although the boron-free A-glass (or CC-glass) has a better acid resistance than high-boron E-glass, it is not equal to that of boron-free or essentially boron-free E-glass. If a higher acid resistance than that of boron-free E-glass were needed, one might want to consider ECR-, basalt, S-glass, pure silica sliver (Section 1.2.2.1), and ultrapure SiO_2 (quartz) fibers (Section 1.2.1).

Finally, while a commercial process in Russia uses acid leaching of A-glass fabrics to manufacture silica fabrics for low-cost, high-temperature insulation uses, a comparable commercial process in the United States uses acid leaching of high-boron E-glass fabrics (Section 1.2.2.2).

1.3.3.6 Other Special-Purpose Glass Fibers

This group of products includes commercial and experimental glass fibers with low dielectric constants and high dielectric constants, and with superconducting, semiconducting, and bone bioactive properties.

(a) Glass fibers with low dielectric constants. The ASTM specifications which govern E-glass compositions fall into two categories, E-glass for printed circuit board applications and E-glass for general reinforcement applications (Table 1.8). To qualify as an E-glass for printed circuit board applications, the composition may contain 5–10% B_2O_3, 16–25% CaO, 12–16% Al_2O_3, 52–56% SiO_2, 0–5% MgO, 0–2% Na_2O and K_2O, 0–0.8% TiO_2, 0.05–0.4% Fe_2O_3, and 0–1.0% F_2. To qualify as an E-glass for printed circuit board applications, the composition may not contain Li_2O, but an E-glass for general reinforcement uses may contain 0–2% Li_2O.

The electrical properties of glass fibers are characterized by volume resistivity, surface conductivity, dielectric loss, and dielectric constant. Glass fibers which have a lower, dielectric constant than borosilicate E-glass have been designated as D-glass or low Dk-glass. They were originally developed [50, 93] to allow the computer industry to achieve, at a premium cost, a faster response time in reinforced printed wire board composites.

D- and low Dk-glass compositions are not supported by the existing ASTM E-glass specifications and ASTM has not established generic specifications. The potential cost of these compositions to the manufacturer depends on ingredient cost and on melt viscosity, therefore implicitly on energy cost. The potential value-in-use to the computer industry depends on the price relative to specific dielectric properties [36, 94, 95, 149]. This section provides an overview. An in-depth discussion is offered in Chapter 4.

Table 1.17 compares the properties of the original D-glass fiber with those of silica, S-glass, hollow and solid E-glass fibers, and with low Dk-glass fibers. The fibers are arranged by increasing dielectric constants and by decreasing melt viscosities and energy demand. Hollow experimental E-glass fibers [96] have the lowest dielectric constant (2.98) and solid commercial E-glass fibers have the highest dielectric constant (6.86). All fibers in Table 1.17 have a delta temperature of $\geq 55°C$. As the dielectric constant increases, the forming temperature decreases from >1800°C (for silica fibers) to $\leq 1200°C$ (for E-glass), the density increases from 1.80 g/cc (for hollow E-glass) to 6.68 g/cc (for solid E-glass).

Experimental hollow E-glass fibers have the lowest dielectric constant [96], melt viscosity, and log 3 forming temperature and therefore would require the least energy of all eight examples shown in Table 1.17. They can be made in a large conventional fiberglass furnace but would require special bushing technology. But they are not suitable for use in printed circuit boards. After drilling holes into a composite, water can enter into the hollow fiber cores, and the presence of water could uncontrollably raise the dielectric constant.

Twenty-eight energy-friendly compositions with 60–68% SiO_2, 7–13% B_2O_3, 9–15% Al_2O_3, 8–15% MgO, 0–4% CaO, 0.4–2% Li_2O, 0–2% TiO_2, and 0–1% F_2, log 3 forming temperatures ranging from 1372 to 1244°C, and dielectric

Table 1.17 Glass fibers with low dielectric constants [36]

Glass fiber	Hollow E-glass	Commercial D-glass	Silica	Exptl. D-glass	Comm. S-glass	Low-Dk glasses (Chapter 4)		Comm. E-glass
Wt%								
SiO_2	54.3	74.5	99.99	55.7	65.5	52–60	61.03	54.3
Al_2O_3	14.0	0.3	–	13.7	25.0	10–18	12.04	14.0
B_2O_3	6.0	22.0	–	26.5	–	20–30	10.73	6.0
CaO	22.1	0.5	–	2.8	–	4–8	2.98	22.1
MgO	0.6	–	–	1.0	9.5	Trace	9.97	0.6
TiO_2	–	–	–	–	–	Trace	0.50	–
Li_2O	–	–	–	0.1	–		1.05	–
Na_2O	0.8	1.0	–	0.1	–		0.50	0.8
K_2O	0.2	1.3	–	0.1	–		0.40	0.2
Fe_2O_3	0.3	–	–	–	–		0.35	0.3
F_2	0.7	–	–	–	–	<2	0.45	0.7
Properties								
D_k, 1 MHz	2.98	3.56	3.78	4.10	4.35	<5	5.61	6.86
Log 3 FT, °C	1200	1410	>1800	1430	1560	<1340	1244	1200
Density, g/cc	1.80	2.16	2.15	–	2.48	–	2.42	2.54
Special process	Yes	Yes	Yes	Yes	Yes	Yes	No	No
Energy premium	–	+++	+++++	+++	++++	++	+	–
References	[36]	[50]	[36]	[93]	[50]	[95]	[94]	[36]

constants ranging from 5.30 to 5.95 were recently reported [94]. The composition with the lowest forming temperature (1244°C) and a liquidus temperature ≥55°C (Dk = 5.61) is shown in Table 1.17 since it would be the most energy efficient composition among the 28 examples.

The following issues will be discussed in detail in Chapter 4: (1) The scientific principles and technical concepts which support the printed wire board technology; (2) recently emerging compositional D- or low Dk-glass technology; (3) the value-in-use of various combinations of dielectric constant and cost; and (4) the predominant target applications of D-glass fibers such as large mainframe computers, workstations, and general-purpose minicomputers in terms of product requirements and product cost.

(b) Glass fibers with high densities and dielectric constants. Glass fibers with high dielectric constants tend to have high densities also, and their value depends on one or the other property, rarely on both. Three generic applications are known. The oldest application requires radiation resistance, e.g., absorption of gamma rays and relies on a high density [6, 97]. The newest application, which requires a concentration of electro-magnetic energy, e.g., in circuit boards for high-frequency uses, depends also on the electrical properties [98, 99]. Applications, e.g., in capacitors, which require electrical insulation, rely on their electrical properties [99].

Glass fibers with high densities (and high dielectric constants) are known as protective glass fibers [6, 36] since they can absorb gamma rays as well as fast and

Table 1.18 Glass fibers with high densities and/or high dielectric constants [36] (with kind permission of Springer Science and Business Media)

Glass fiber properties	E-glass (6.6% B_2O_3)	Cerium silicate (<30% CeO_2)	Lead silicate (<30% PbO)	Niobium silicate (<15% Ni_2O_5)
Strength, GPa	3.40	4.5	1.7	–
Modulus, GPa	72.00	95.0	51.0	–
Density, g/cc	2.54	3.0–3.8	3.0–4.0	–
Dielectric constant	6.80	7.0–8.0	8.0–10	10–15
References	[50, 52]	[97]	[6, 98]	[98, 99]

slow neutrons. For example, lead-, bismuth-, and/or barium-containing glass fibers have high densities (4.0–4.8 g/cc) and high dielectric constants (8–13) as shown in Table 1.18. They can be used to absorb gamma rays [6] in x-ray equipment and radiation-protective composites. The ability of lead glass fibers to absorb gamma rays increases with increasing density [6], but these fibers have low strengths (<2.0 GPa), low moduli (<55 GPa), and very low resistance to light, humidity, and/or elevated temperatures.

Glass fibers containing boron, cadmium, and/or cerium can be used for protection against neutrons [6, 36]. Their densities (3.0–3.5 g/cc) are not as high as those needed for gamma ray protection (4.0–4.8 g/cc). However, glass fibers in the SiO_2–Al_2O_3–CeO_2 system [36, 93] with 10–28% CeO_2 offer very high strengths (4.5 GPa) and very high moduli (95.0 GPa). They could be considered to be high-strength glass fibers (see Table 1.12), but their thermal stability, unlike that of true high-strength fibers (Section 1.3.3.2), is very low. For example, a major strength loss occurs after heat treatment at 100–200°C due to pre-crystallization and micro-separation of a second solid phase, which is revealed in x-ray diffraction patterns [6, 36].

(c) Glass fibers with very high dielectric constants. Glass fibers with high dielectric constants [98, 99] differ fundamentally in their applications from those requiring glass fibers with low dielectric constants [97]. They are aimed at circuit boards being designed for use in high-frequency applications. Again, using of E-glass as a control, Table 1.19 shows three different approaches to glass fibers with very high dielectric constants, i.e., potentially suitable for the presently emerging market needs [98, 99]. Lead (or L) glass fibers with extremely high levels of PbO (>70%) and silicate fibers with very high combined levels of BaO and TiO_2 offer significantly higher dielectric constants, i.e., 13.0 and 13.5, respectively.

Lead glass compositions possess a dielectric constant of 13.0, but glass fibers from lead glass compositions with such high PbO levels are difficult to form. They are very long melts which have a very high ΔT between log 2.5 forming and liquidus temperatures and are therefore quite crystallization resistant. However, PbO will evaporate violently during the melt process, thus affording highly non-uniform fiber compositions and frequent process discontinuities and fiber breaks [98, 99]. In addition, the use of PbO poses significant environmental concerns.

The silicate glasses based on high levels of BaO and TiO_2 have an equally high dielectric constant of 13.5, but fibers cannot be formed from their melts because

Table 1.19 Glass fibers with very high dielectric constants [36] (with kind permission of Springer Science and Business Media)

Glass fiber composition, wt.%	E-glass (6.6% B_2O_3)	Lead glass (>70% PbO)	$BaO–TiO_2$ silicate	$Nb_2O_5–BaO–TiO_2$ silicates	
SiO_2	54.3	26.0	40.0	55.0	47.16
Al_2O_3	14.0	–	–	2.5	–
B_2O_3	6.6	1.5	–	–	–
PbO	–	72.0	–	–	–
CaO	22.1	–	7.5	9.0	7.03
MgO	0.6	–	–	–	–
SrO	–	–	15.0	6.0	7.03
BaO	–	–	23.0	15.0	14.05
TiO_2	0.5	–	–	7.8	13.95
ZrO_2	–	–	7.0	1.7	3.27
Na_2O	0.8	–	–	–	–
K_2O	0.2	0.5	–	–	–
Nb_2O_5	–	–	–	3.0	7.51
Dk, ε_r, at 1 MHz	6.8	13.0	13.5	10.1	12.3
Log 2.5, °C	1299	850	1077	1199	1136
LT, °C	1063	650	1214	1085	1089
ΔT, °C	+236	+100	−137	+114	+51
Fiber-forming ability	Excellent	Very poor	Infeasible	Very good	Good
References	[1, 79]	[98, 99]	[98, 99]	[98, 99]	[98, 99]

the liquidus temperature is much higher than the preferred forming temperature. Crystallization would occur long before the melt reaches the preferred fiber-forming viscosity of log 2.5 poise. In summary, glass fibers with very high levels of PbO are sublimation prone and glass fibers with high combined levels BaO and TiO_2 are crystallization prone.

In contrast, addition of 0.5–15.0 mol% of Nb_2O_5 to glass compositions having high levels of BaO and TiO_2 slightly reduces the dielectric constant, and dramatically reduces the liquidus temperature, thereby reversing the delta between the log 2.5 forming temperature and liquidus temperature from −137 to +117°C [98, 99]. Customarily the delta temperature (ΔT) quoted in the technical literature refers to the difference between the log 3 forming and liquidus temperatures. The difference does not affect the conclusions.

High-speed and high-frequency information transmission is becoming increasingly important with the recent development of advanced information systems including mobile communication by car telephones and personal radios, as well as satellite broadcasting and cable television. As a result, there is an increasing demand for miniaturizing electronic devices and also microwave circuit elements such as dielectric resonators in conjunction with the electronic devices.

Nb_2O_5-containing glass fibers are embedded for these applications in a resin such as polyethylene oxide that has a low dielectric tangent (tan δ) loss to obtain the desired high-frequency performance of the resulting circuit boards for microwave applications [98, 99]. In summary, microwave circuit elements can be made more compact by using a circuit board having a high dielectric constant. It acts to concentrate the electromagnetic energy within the board and thereby minimizes the leakage of electromagnetic waves.

(d) Glass fibers with super- and semiconducting properties. Superconducting glass fibers are obtained by incorporating a suitable ceramic material in the fiber core, yielding superconducting bicomponent sheath/core glass fibers [36]. Superconducting single-component fibers can be made drawing single superconducting ceramic preform rods by the laser-heated float zone or pedestal growth process [36]

Semiconducting glass fibers have been known for over 30 years [6, 36] but have never attracted commercial interest because significant technical problems have never been solved. Such fibers (e.g., $CuO–CaO–Al_2O_3–SiO_2$) can be made by adding oxides of monovalent metals such as copper or silver to a suitable base glass and by subsequently reducing the glass fibers in various gaseous media, e.g., hydrogen. However these fibers are moisture sensitive, and prolonged storage leads to increased glass conductance [6].

The electrical properties of glass fibers are best modified by (1) applying a permanent chemical coating to the fiber surface, (2) adding a suitable material to the binder or finish formulations, or (3) modifying the composite matrix that is being reinforced with a glass fiber. A semiconducting or conductive coating is applied by vacuum deposition, metallization from metal salts, decomposition of organometallic compounds, or chemical metallization. Also, carbon black can be added to the composite matrix [36].

(e) Glass fibers with bone bioactive oxide compositions. Bioactive glasses have been developed since 1969 [100] and continuous bioactive glass fibers since 1983 [101, 102]. And very recently [103], bone bioactive glass fibers were found to bond to bone tissue and help bone tissue growth when they are preferentially placed on the surface of a thermoplastic composite while carbon fibers are used to stiffen the core of the structure.

A glass composition which was found particularly suitable for these applications [100] contained 52% SiO_2, 30% Na_2O, 15% CaO, and <3% P_2O_5 (in mole %). Fiber bundles or tow of up to 5000 filaments were drawn from a melt of this composition and were interwoven with carbon fiber tow into a cylindrically braided sheath/core textile preform. The carbon fibers formed the core of the braided structure and provided the required stiffness and load support, and the bone bioactive fibers formed the sheath and functionality.

To create a practical reinforcing structure of this nature, two braids are in effect braided simultaneously, one forming the carbon fiber core, the other the bioactive fiber surface layer or sheath, and both are suitably interwoven, overlaid, or otherwise intermingled. The carbon fibers in the core are first co-mingled with a suitable polymer such as a polysulfone and coarse fibers of the same polymer are intermingled with the bone bioactive glass fibers.

The hybrid preform is then processed in a closed die in a hot press. The combined amount of polymer is calculated to give the final total volume fraction; no additional polymer or injection molding step is required. In summary, superior composite materials can now be designed and manufactured for use in practical prosthetic devices, facilitating bone tissue growth that has a structural stiffness matching that of human bone.

1.3.4 Non-round, Bicomponent and Hollow Silicate Fibers

Fibers having a special shape, a non-round fiber cross section or multiple compositions across the fiber composition can offer special properties.

1.3.4.1 Glass Fibers with Non-round Cross Sections

A circular cross section has the smallest circumference for a given area. Any increase in circumference of a fiber while retaining the cross-sectional area will increase the fiber surface. Any increase in the fiber surface offers significant product advantages including higher adhesion and strength in composites, but also creates increased process complexity.

(a) Processes and structures. The effect of the surface tension of a molten jet is the driving force toward the formation of fibers with round cross sections. Thus, a fiber with a round cross section is obtained from a bushing tip with a round cross section. With a bushing tip having a non-round cross section, the effect of surface tension tends to force the melt into a round cross section. This tendency is opposed by the effect of the quench rate which would otherwise stabilize a non-round fiber cross section. Since glass melts have higher surface tensions than polymer organic melts, it is more difficult from glass fibers with non-round cross sections than polymer fibers with non-round cross sections [104].

The basic fiber cross section technology was developed in the early 1950s for nylon yarns. Not surprisingly, the development of glass fibers with non-round cross sections was more demanding and the required technology is still not being practiced on a commercial scale. Two processes were developed for forming continuous single glass ribbons [36], tapered, trilobal fibers [104], and for continuous oval shaped as well as trilobal multifilament glass fibers [105] as shown in Fig. 1.14. But these early processes were slow.

A recent redesign of nozzle tips [106] has been claimed to afford non-round multifilament E-glass at fast, i.e., potentially commercial production rates (Fig. 1.15). For example, ribbon, oval, trilobal triangular, or rectangular cross sections can be produced with 400 tip bushings and windup speeds of 1800–3000 m/min under otherwise conventional E-glass process conditions. This critical technological advance uses an entirely new bushing tip design. An edge, shield, or baffle protrudes from the surface of each tip at critical points to induce preferential cooling of the fibers and thereby to assist the formation of well-defined non-round cross sections.

Non-round fiber cross sections are characterized by their modification or mod ratio. For ribbons, the mod ratio is their width-to-thickness ratio. For trilobal fibers (Fig. 1.14) it is the ratio of the diameter B of the outer circle around the lobes of the fiber cross section to the diameter A of the inner circle within the core of the

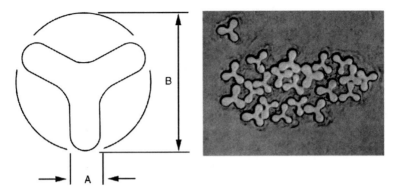

Fig. 1.14 Trilobal glass fibers with high modification ratio. Redrawn from Huey [104]

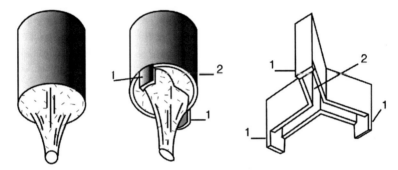

Fig. 1.15 Tip design for the fabrication of non-round glass fibers at high process speeds. Redrawn from Taguchi et al. [106]

fiber cross section. The term mod ratio has been used since the early development of trilobal nylon fibers. The newer term, "deformation ratio" [106], is less intuitive and therefore less desirable.

(b) Products and applications. Several potentially desirable properties increase with increasing mod ratio. Nylon fibers with a high mod ratio offer a high degree of external light reflection and, therefore, the appearance of a silk-like sheen or luster in a yarn or fabric. Nylon ribbons with a high mod ratio offer the appearance of a metallic sparkle in a fabric or garment. They offer large reflective surfaces which constantly shift with the motion of the garment. The visual effect of a trilobal or ribbon-shaped glass fiber does not seem to possess practical value, but the structural effect appears to be useful in limited applications.

The older technology for forming non-round glass fibers afforded a premium application for lead glass ribbons in ribbon wound capacitors [36]. The process, however, is slow and not sufficiently cost-effective to facilitate the growth of large volume applications in the composites market. The newer technology [106] offers higher process speeds and therefore a more cost-effective route to ribbon-shaped and trilobal glass fibers. The potential value of fibers with non-round cross sections has long been understood [36]. When used as composite reinforcing fibers, they have a higher surface area, therefore offer higher matrix adhesion and higher composite strength.

1.3.4.2 Bicomponent Silicate Glass Fibers

Commercially significant inorganic bicomponent fibers have either a concentric sheath/core or a side-by-side fiber structure. Polymer organic fibers with an additional eccentric sheath/core structure are known, but inorganic fibers with this structure are not known.

(a) Sheath/core and side-by-side bicomponent fibers. Structural bicomponent ceramic fibers [36] and optical bicomponent glass fibers [36] have a concentric sheath/core structure (Fig. 1.16). A sheath of one composition surrounds a core of another composition. Either the core or the sheath is responsible for the functionality. For example, the boron sheath of boron/tungsten fibers is responsible for the functionality while the core is sacrificial. The core (or wave guide) of optical fibers is responsible for the functionality while the silica sheath provides strength and load support. Side-by-side bicomponent structures afford non-straight, crimped, or bulky fibers since the components show differential shrinkage during processing.

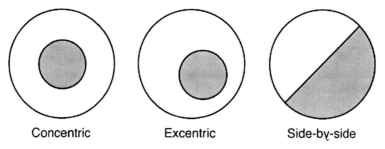

Fig. 1.16 Schematic drawing of concentric, eccentric, and side-by-side bicomponent fiber cross sections [36] (with kind permission of Springer Science and Business Media)

Solid bicomponent glass fibers have different sheath-core (s-c) or side-by-side (s-b-s) compositions. Hollow glass fibers are sheath-core bicomponent fibers having a continuous glass sheath that surrounds a continuous void or hollow core. Hollow porous glass fibers have a microporous, instead of a solid, glass sheath. Hollow superconducting glass fibers have three different compositions, an outer silicate sheath that surrounds a functional ceramic sheath that, in turn surrounds, a hollow core.

(b) Hollow sheath/core silicate glass fibers. Hollow fibers have a concentric bicomponent structure with a solid sheath and a void- or air-filled core. Flame drawing from glass tubes yielded the earliest examples of hollow fibers. Commercial technology relies on introducing air into the core of a molten glass jet, the fiber precursor, and dates to 1966 [36]. In one process, air is injected through feeder tubes

into the core of the melt in each tip [107]. In another process, the bushing tips aspirate air from the forming zone into the core of the forming cone [108]. In the third process, a double-crucible design is used to manufacture three-component superconducting glass fibers, whereby air is aspirated into the core of the forming cone from the inner crucible [109].

Hollow glass fibers have a lower weight at equal volume than solid glass fibers, greater thermal insulation, a lower dielectric constant, better long-term fatigue characteristics, but also lower strength and modulus. Hollow E-glass fibers and hollow S-glass fibers are known. Hollow S-glass fibers have short melts and a much greater crystallization behavior. Hollow E-glass fibers have long melts and a more forgiving crystallization behavior. Because of the additional manufacturing cost, both are specialty fibers, but since hollow E-glass fibers are much easier to fabricate, they have found a wider range of potential applications than hollow S-glass fibers [6].

Hollow borosilicate E-glass fibers with an outside diameter (OD) of 13 μm and an inside diameter (ID) of 8 μm have a coefficient of capillarity ($K =$ ID/OD) of 0.62 and therefore a nearly 40% lower specific gravity than solid E-glass fibers. These fibers are potentially available in the form of yarns, rovings, fabrics, mat, chopped strand, and tape. By definition, hollow E-glass fibers have lower tensile strength (2.5–2.8 GPa) than solid E-glass fibers (3.3 GPa). Their modulus is also lower than that of solid E-glass fibers, and it increases with decreasing capillarity [6]. However, the specific modulus of hollow E-glass fibers is higher than that of solid fibers [6].

Hollow E-glass fibers with a capillarity of $K = 0.5$–0.6 are more sensitive to moisture and/or heat treatment than solid E-glass fibers. Upon exposure to high humidity (>90%), the residual strength drops to about 80–85% of the low original strength. Upon exposure to temperatures between 300 and 500°C, surface crystallization occurs and residual tensile strength drops to about 35–50% of the original strength. Because of their thin walls, hollow fibers are more fragile than solid fibers and need to be handled more carefully. They will not, however, crush under pressure in a laminating press.

Hollow E-glass fibers can produce composite parts with equal thickness and up to 25% lower part weight than E-glass, or composite parts with equal weight and up to 25% higher thickness. In equal weight cylinders, the hydrostatic collapse pressure is increased by 30% and the dielectric constant is reduced from 6.8 to 4.0. In equal thickness laminates, acoustic transmission is increased due to lower mass, dynamic fatigue is significantly increased, and thermal conductivity is reduced by up to 40%. In a hybrid laminate, partial substitution for graphite does not reduce the high specific modulus while removing up to one-half of the more expensive graphite fiber.

Weight reduction is a powerful incentive in aircraft design and construction. Every kilogram of structure that can be eliminated facilitates a major reduction in fuel consumption and cost or a commensurate increase in payload at equal fuel cost. Hollow glass fibers qualify for use in non-load-bearing interior aircraft applications such as sidewall and ceiling panels. The use of hollow glass fibers is more prevalent in Russia and CIS countries [110] than in the United States and the Western world.

(c) Hollow porous sheath/core silicate glass fibers. Hollow porous glass fibers with >95% SiO_2 were produced by acid leaching of hollow glass fibers including borosilicate E-glass [111]. Acid leaching of hollow glass fiber-based compositions having 35–62% SiO_2, 1–11% Al_2O_3, 0–54% B_2O_3, 3–9% ZrO_2, and 1–29% Na_2O (but no CaO) gave silica-rich, porous hollow glass fibers with good alkali resistance. Acid leaching of hollow glass fibers containing 54.0% SiO_2, 22.4% CaO, 14.3% Al_2O_3, 7.2% B_2O_3, and 1.0% Na_2O (but no ZrO_2) gave porous, hollow glass fibers with low alkali resistance [111].

One of the most challenging tasks is to produce porous, hollow glass fibers with controlled pore size and uniform mechanical properties including strength and stiffness, thermal and chemical stability, photochemical and biochemical durability, and superior resistance to compaction under high membrane pressures. The required control over the desired pore size in fibers has been demonstrated [110] with a glass primarily consisting of 57.2 mol%, 22.8% B_2O_3, 9.2% CaO. This study serves as a model for the evaluation of hollow porous glass fibers in reverse osmosis, phase separations, salt extraction, and biochemical research.

(d) Solid side-by-side bicomponent glass fibers. Dual or bicomponent glass fibers [112] can be made as continuous fibers in a standard melt process or as staple or discontinuous fibers in a centrifuge process. Either way, two separate glass melts, each having a different composition and therefore viscosity, are supplied to round cross section tips in a typical multifilament bushing. The melt streams meet under the tips, fuse, and, as they cool, yield side-by-side bicomponent glass fibers with essentially round cross sections. The fibers develop an irregular crimp since a differential stress develops at the interface of the components.

Backscattered electron images (BEI) show that the individual bicomponent fibers have cross sections ranging from round to oval and low, but widely variable, fiber diameters (3–10 μm). Individual fibers can split at their dual glass interface either during manufacture or in subsequent processing and handling, thus producing fine, low diameter chaff considered to be undesirable in use. An individual cross section is shown in Fig. 1.17. Component (A) has a different composition than component (B), and the interface between both components is clearly discernable.

Successive secondary electron image (SEI) profiles can be used to determine the composition in components (A) and (B). This is accomplished by traversing a polished cross section of the fiber at a right angle to the interface between the two components (Fig. 1.18). Accordingly, the Mg (or MgO) level in this sample is higher in component (A) than in component (B), and the Ca (or CaO) level is higher in component (B) than in component (A).

In addition, the SiO_2 and Na_2O levels (not shown in Fig. 1.20) were also higher in component (A) than in component (B) and the Al_2O_3 levels were lower in component (A) than in component (B). Only component (A) contained potassium and only component (B) contained boron.

In summary, the individual components of the dual component fiber differ with regard to their SiO_2, MgO, CaO, Al_2O_3, B_2O_3, K_2O, and Na_2O content. The differential stress between the two components during fiber formation produces a three-dimensional, crimped, non-straight fiber geometry. In thermal insulation batts,

Fig. 1.17 Backscattered electron image (BEI) of a polished cross section of a commercial bicomponent glass fiber [36] (with kind permission of Springer Science and Business Media)

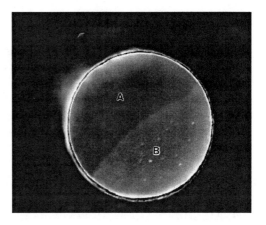

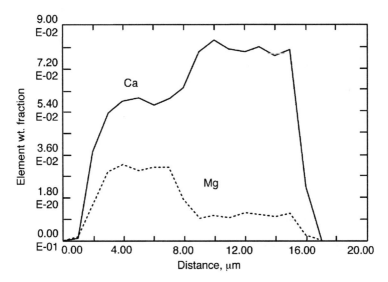

Fig. 1.18 Secondary electron image (SEI) profiles of calcium and magnesium [36] (with kind permission of Springer Science and Business Media)

a main application, arrays of crimped bicomponent glass fibers offer higher bulk and loft than possible with a comparable array of straight fibers. Arrays of crimped fibers can trap more air in their dead spaces and therefore offer higher thermal insulation than comparable arrays of straight fibers having the same basis weight.

Arrays of crimped staple fibers can be processed on conventional textile staple process equipment by carding, needle punching, or air laying [113]. In these conventional textile processes, crimped bicomponent glass staple fibers aim to compete with commodity fibers, e.g., cotton and polyester staple [114].

1.3.5 Summary and Conclusions

Melt spinning glass fibers from strong viscous silicate melts represents the most significant glass fiber-forming process among the multitude processes which have been described in Section 1.3. The most important commercial fibers which are being made by this process include borosilicate E-glass, boron-free, and essentially boron-free E-glass, S-glass, D-glass, ECR-glass, AR-glass, bicomponent insulation glass fibers, and Chinese C-glass (i.e., the former A-glass in the United States and Europe).

1.4 Aluminate Glass Fibers ($\leq$81% Al_2O_3, $\leq$50% SiO_2)

Strong viscous E- and S-glass melts (50–100% SiO_2, 0–25% Al_2O_3) have high viscosities above and below their liquidus temperature. Fragile viscous melts (25–50% Al_2O_3, 4–10% SiO_2) have high viscosities below their liquidus temperature, but low viscosities at and above their liquidus temperature. Inviscid, literally non-viscous, melts (50–81% Al_2O_3, 4–0% SiO_2) have a sharp melting and crystallization point. Thus, a decrease in %SiO_2 and increase in %Al_2O_3 leads from strong to fragile and inviscid melts [4].

1.4.1 Glass Fibers from Fragile Melts (25–50% Al_2O_3, 10–4% SiO_2)

Strong silicate melts are preferred for developing commercial glass fibers. Fragile silicate [7], aluminate [8, 5], tellurite [115], fluoride [116, 117], chalcogenide [128], and fluorophosphate [119] melts have been used to explore theoretical relationships in optical and other specialty fields.

1.4.1.1 Downdrawing from Supercooled Melts

Melt spinning from supercooled fluoride melts affords optical single and bicomponent glass fibers [116]. The latter have a concentric fluoride core and a fluoride sheath or clad with a slightly different composition.

(a) Single- and double-crucible process. The double-crucible apparatus (Fig. 1.19) yields concentric bicomponent fluoride fibers. Two melts are separately maintained under controlled conditions in a supercooled state that is well below the liquidus temperature of the fluoride glasses [116, 117]. The inner and outer crucibles form the concentric tip system through which the glass melts flow.

In this process [116], the inner and outer crucibles are heated to about 750–800°C and held at that temperature for approximately 30 min to allow the melts

Fig. 1.19 Double-crucible melts spinning process depicting the outer and inner crucible. Redrawn from Nice [116]

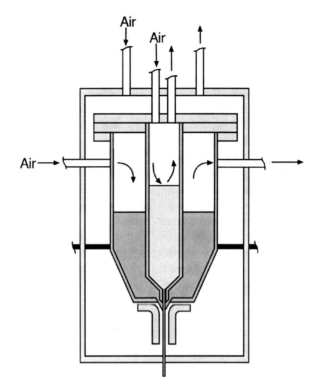

to cure. The melts are then rapidly quenched until the core melt reaches about 350°C. The outer crucible is then rapidly raised to a fiberizing temperature of about 315°C, and fibers are drawn, depending on the pressure in the crucibles, at speeds of 900–3300 m/s. Low pressures are adequate to facilitate high drawing speeds.

This transient operating cycle [116] minimizes devitrification as the melt passes from refining to fiberizing conditions. The low pressures which are independently applied to each crucible facilitate the formation of solid concentric sheath/core fibers and prevent the formation of hollow fibers. The core diameters can range from 8.4 to 75.1 μm and the overall fiber diameter, including sheath or clad, can range from 140 to 149 μm.

(b) Single and bicomponent fluoride fibers. Core/clad bicomponent fibers were obtained with core diameters ranging from 8.4 to 75.1 μm, and single-component fluoride fibers having these diameters were obtained by a single-crucible version of the double-crucible process [116]. Single-component calcia–alumina and alumina–telluria glass fibers could also be made by this process, and the process could be used to fabricate concentric sheath/core fibers from fragile dual calcia–alumina or fragile dual alumina–telluria melts, if there were demand for such fibers.

1.4.1.2 Updrawing from Supercooled Melts

A narrow region exists near the eutectic of alumina–telluria with 5–11% Al_2O_3 and calcia–alumina melts with 25–50% Al_2O_3. They have fiber-forming viscosities (log 2.5–log 3.0 poise) below, but not above the liquidus. These compositions are discussed in this section. Calcia–alumina compositions with higher alumina levels (e.g., 50 and 81%) form inviscid melts, have very low melt viscosities (<5 poise), and they instantly crystallize at a sharp melting point. They will be discussed in Section 1.4.2.

(a) Tellurite glass fibers. The alumina–telluria phase diagram offers a glass-forming region between 5.0 and 11.0 weight (or 7.8–16.8 mol) percent of alumina, where high viscosity compositions can be found surrounded by low viscosity compositions. Within this region, alumina–telluria glass fibers could be manually updrawn from supercooled melts [115] at 800–850°C, yielding the first self-supporting tellurite glass fibers on record. These fibers had tensile strengths ranging from 0.4 to 1.1 GPa, moduli ranging from 69 to 83 GPa, and diameters ranging from 40 to 150 μm. They were x-ray amorphous, clear but pale yellow, and are of interest primarily because of their potentially valuable spectral transmission properties [115].

(b) Aluminate glass fibers. Over 500 quaternary aluminate glass fibers were manually updrawn from fragile melts, which were supercooled to about 50°C below their liquidus while carefully controlling the temperature of the melt [8, 120, 121]. The glass compositions in this narrow region of the phase diagram contained 30–55% calcia, 45–55% alumina, and 0–10% silica. The ratio of oxygen ions to network-forming ions (Al^{+3} and Si^{+4}, and an occasional minor component) was between 2.35 and 2.6 [35], a ratio that is a necessary but not sufficient condition for the formation of fragile fiber-forming aluminate glass melts.

A continuous updrawing process (Fig. 1.20) was designed [35, 122] using a quaternary low silica composition (44.3% CaO–48.7% Al_2O_3–3.5% MgO–3.5% SiO_2). The melt properties of this composition (Table 1.20) confirm the steep drop in melt viscosity from 182 poise at a temperature 50°C above the liquidus temperature to 278 poise at the liquidus temperature and to 720 poise at a log 3 fiber-forming temperature that is 50°C below the liquidus.

This calcium aluminate fiber was evaluated in structural applications but it was not suitable for infrared optical applications because it contained bound water as evidenced by a strong hydroxyl band at 2.9 μm in the IR transmission spectra. Hydroxyl-free compositions were made in carbon crucibles by the Davy process [123], i.e., by a process that is commercially used to fabricate optical calcium aluminate bulk glasses for applications in disposable optical windows [124].

One quaternary calcium aluminate glass fiber was also downdrawn from a solid preforms with a development unit [125]. Downdrawing from preforms is a rare, and costly, method for fiberizing fragile melts. And only one of >500 compositions could be fiberized on a conventional melt fiberizing unit [35]. This process is therefore not generally suitable for fragile melts either.

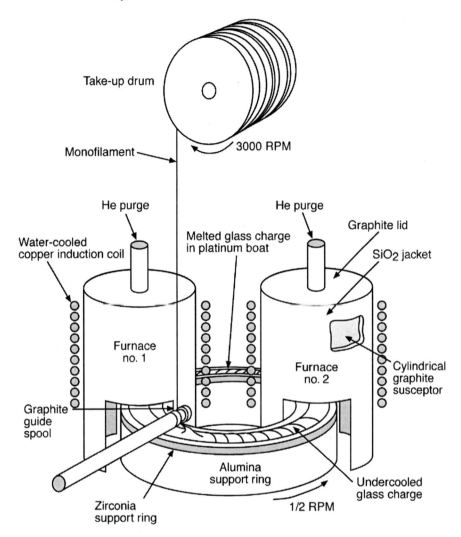

Fig. 1.20 Continuous updrawing process. Redrawn from Schroeder et al. [122]

Table 1.20 Fragile aluminate melt (3.5% SiO_2, 44.3% Al_2O_3, and 48.7% CaO) [1] (with kind permission of Springer Science and Business Media)

Melt temperature (°C)	Melt viscosity, poise
LT + 100	58.0
LT + 50	182.0
LT	278.0
LT – 50	720.0
LT – 200	10000.0
Literature references [8, 5]	

1.4.1.3 Quaternary Calcium Aluminate Fibers

Quaternary calcium aluminate glass fibers made by updrawing from a super-cooled fragile melt offer superior mechanical properties and sapphire-like infrared transmission spectra.

(a) Structural and optical fiber properties. The highest reported pristine strength (8.3 GPa) was obtained (Table 1.21) with low silica (e.g., 44.3% Al_2O_3–48.7% CaO–3.5% MgO–3.5% SiO_2) glass fibers when they were updrawn from super-cooled melts in an induction furnace [121]. When updrawn in an oxyacetylene furnace, they had a somewhat lower strength (4.2 GPa), but it was still higher than that of E-glass (3.5 GPa). The modulus (110.3 GPa) of these fibers was 1.5× that of E-glass. The lowest modulus of any calcium aluminate fiber shown in Table 1.21 was 10% higher than that of E-glass. The stiffest fibers in this table were zinc oxide-modified calcium aluminate glass fibers with moduli of up to 122.7 GPa or 1.7× the stiffness of E-glass.

Table 1.21 Modulus of calcium aluminate fibers updrawn from fragile melts [1] (with kind permission of Springer Science and Business Media)

Examples [80]	Composition, wt%				Other oxides	Modulus (GPa)	References
	SiO_2	Al_2O_3	CaO	MgO			
Example (1)	3.5	44.5	46.6	5.3	–	108.9	[35]
Example (2)	–	46.2	36.0	4.0	13.8% BaO	109.6	[35]
Example (3)	10.0	30.0	30.0	–	30.0% ZnO	109.6	[35]
Example (4)	3.5	44.3	48.7	3.5	–	110.3	[35]
Example (5)	4.1	42.6	47.7	5.6	–	110.3	[35]
Example (6)	2.8	35.2	36.2	–	25.7% PbO	110.3	[35]
Example (7)	10.0	30.0	35.0	–	25.0% ZnO	111.7	[35]
Example (8)	4.0	32.0	44.0	–	20.0% ZnO	115.8	[35]
Example (9)	4.0	32.0	44.0	–	ZnO and Li_2O	122.7	[35]
Controls							
E-glass	54.5	14.0	22.1	0.6	6.6% B_2O_3	72.0	[50]
S-glass	65.5	25.0	–	9.5	–	86.9	[50]
Si–Al–O–N	11.1	2.2	62.6	0.7	23.4% Si_3N_4	248.0	[83]

With two exceptions, all fibers shown in Table 1.21 had strong hydroxyl bands in their infrared transmission spectra. The exceptions are two recent hydroxyl-free compositions made by the Davy process [123]. One is a low silica composition and the other is a non-silica composition (46.2% Al_2O_3 – 36.0% CaO – 4.0% MgO – 13.8% BaO). An in-depth analysis of the physical properties of fibers is shown in Table 1.21 [120, 121].

Their resistance to alkaline media exceeds that of commercially available AR silicate glass fibers (Section 1.3.3) having a zirconia content of up to 15% [52]. Hydroxyl-free quaternary calcium aluminate glass fibers (Fig. 1.21), e.g., non-silica fibers containing 46.2% Al_2O_3–36.0% CaO–4.0% MgO–13.8% BaO, afford sapphire-like infrared transmission properties.

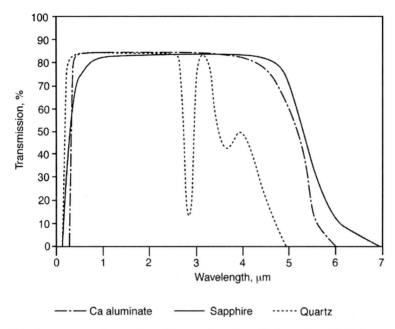

Fig. 1.21 Spectral transmission of calcialumina glass fibers. Redrawn from Wallenberger et al. [126] (Reprinted with permission of Elsevier Science Publishers)

(b) Potential applications. The commercial potential of updrawn quaternary calcium aluminate glass fibers was tested in two stages. In the early 1960s, they were evaluated because they yielded higher moduli than those which could then be achieved with silicate glass fibers [120, 121]. Timing for this development coincided with the onset of the commercial development of carbon fibers and no new aluminate or silicate glass fiber was commercialized until 1995.

In the early 1990s, a renewed evaluation of quaternary calcium aluminate glass fibers was triggered by their sapphire-like optical properties [35, 126, 127]. Coated aluminate fibers might afford sapphire-like sensors at a more affordable cost than single-crystal sapphire fibers, but unlike the latter, they would be limited to ambient and moderately high temperatures [127]. Calcium aluminate fibers are not commercially available, but so far they offer valuable models for important structure–property relationships [80] with regard to strength, modulus, and optical properties.

1.4.1.4 Hybrid Fiber-Forming Processes

The most important glass-forming systems contain elements from the sixth, or chalcogenide, column of the periodic table which includes oxygen, sulfur, selenium, and tellurium. Oxygen-containing (or oxide) glasses are insulators; the others tend to be semiconductors. Some melts are fragile such as the tellurite melts described in the previous section. Others are viscous, and whether they represent strong or fragile melts, they are difficult to fiberize.

A hybrid process for forming chalcogenide glass fibers [128] has been described that uses elements of downdrawing from preforms and fiberizing through bushings. Specifically, a cylindrical chalcogenide preform is vertically inserted into a cylindrical crucible furnished with a nozzle in its bottom plate. The crucible is heated only in the vicinity of the nozzle, and a fiber is continuously drawn from the nozzle at a forming temperature that corresponds to a melt viscosity of log 3 poise.

Heavy metal fluoride fibers require a fiber-forming process that relies on a super-cooled melt, while certain fluorophosphate melts can be fiberized by pulling fibers from the melt in a conventional bushing process [129]. Fluorophosphate glass fibers are difficult to pull from the melt. Strong, 27 μm diameter fibers were made with a tensile strength of 334 MPa for making reinforced visible-IR transparent poly (chloro-trifluoroethylene) composites.

1.4.2 Glass Fibers from Inviscid Melts
(55–81% Al_2O_3, 4–0% SiO_2)

Melts of metals as well as crystalline ceramic oxides have low viscosities. When cooled, they crystallize at a sharp melting point. Above their melting point their viscosity is comparable to that of motor oil at room temperature. Yet, continuous glass and metal fibers can be formed from inviscid melts.

1.4.2.1 Principles of Fiber Formation from Inviscid Melts

In a typical melt spinning process the quench rate is $\sim 10^4$ K/s. When it is extruded through a bushing tip, a viscous melt with a viscosity of log 2.5–log 3.0 poise readily solidifies and forms a continuous fiber at this quench rate. When extruded, an inviscid melt with a viscosity of <0.2 poise will form a liquid jet at this quench rate that readily breaks up into droplets and subsequently forms shot.

(a) Jet formation from inviscid melts. The key principle governing the formation and breakup of a liquid jet is well known [10]. A liquid jet is unstable with respect to viscosity, diameter, and surface tension. A jet has a tendency to break up due to axisymmetric surface pressures, and produces Rayleigh waves or periodic variations of increasing amplitude in the jet diameter (Fig. 1.22). Ultimately, these diameter variations cause the jet to break up into separate droplets, and they, in turn, crystallize and form shot. For any process to yield a continuous fiber from a low viscosity liquid, the transient viscosity of the melt exiting the bushing tip or spinneret orifice must quickly reach the fiber-forming level of log 2.5–log 3.0 poise before jet instability (formation of Rayleigh waves) and potentially disruptive crystal growth can occur.

The straight, cylindrical fiber in Fig. 1.22 represents an inviscid melt spun calcium aluminate glass fiber that had been surface stabilized with particulate carbon. The frozen Rayleigh wave structure represents a calcium aluminate fiber that was not surface stabilized and solidified while it was in the process of breaking up into droplets and shot.

Fig. 1.22 Straight fiber and frozen Rayleigh waves. The *straight, cylindrical* fiber represents an inviscid melt spun calcium aluminate glass fiber that had been surface stabilized with particulate carbon. The frozen Rayleigh wave structure represents a calcium aluminate fiber that was not surface stabilized and solidified while it was in the process of breaking up into droplets and shot [1] (with kind permission of Springer Science and Business Media)

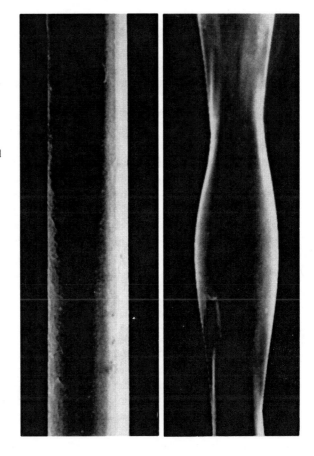

For fiber formation (Equation 1.15), a jet must have a lifetime (t) sufficient for the viscosity to reach log 2.5–log 3.0 poise before the onset of turbulence or breakup [10]. The viscosity (η) of the jet is the key factor since it depends exponentially on the temperature. The diameter (D) of the jet is the next most important factor. Thus, a reduction of the viscosity (η) shortens the jet lifetime (t) and limits the attainment low diameter fibers. The surface tension (γ) and density (ρ) are less important, less sensitive to temperature:

$$t = 14[(\rho D^3/\gamma)^{1/2} + (3\eta D/\gamma)] \tag{1.15}$$

$$d\eta/d\tau \, \Sigma dV_c/d\tau \tag{1.16}$$

$$T_L < T_F \tag{1.17}$$

The conditions (Equation 1.16) under which oxide or metal fibers can be formed from inviscid melts is also defined by the dynamics of the relationship in the forming zone between the rate of solidification or the change of the viscosity (η) and the rate of crystal growth (V_c) with time (t). Any change in melt viscosity, even if only of the surface viscosity, of a molten jet will disproportionately affect its lifetime. If the lifetime is short, the jet will, on cooling, break up into Rayleigh waves, liquid droplets, and shot, and if it is long, the jet will instead solidify into a continuous fiber on cooling.

(b) Fiber formation from inviscid jets. Three processes are known which facilitate the formation of glass and/or metal fibers from inviscid melts. In the containerless laser-heated (CLH) fiber-forming process, a solid aluminate sample is acoustically levitated, laser melted, and instantly supercooled to increase both the viscosity of the inviscid melt and the jet lifetime at a normal quench rate of 10^4 K/s, i.e., to achieve a transient viscous melt from which fibers can be downdrawn.

In the inviscid melt spinning (IMS) process an inviscid aluminate or metal jet is extruded into a chemically reactive environment that raises their surface viscosity and thereby facilitates the formation of continuous fibers. In the rapid jet solidification (RJS) process, an inviscid jet is extruded through an orifice, and the quench rate is increased by casting it onto the surface of a quench wheel where it solidifies.

1.4.2.2 Containerless, Laser Heating (CLH) Process

In this melt process, a solid aluminate sample is acoustically levitated in a containerless environment, laser melted, and instantly supercooled. Both the viscosity of the inviscid melt increases from <2 to >log 2 poise, and the jet lifetime increases at a normal quench rate of 10^4 K/s. As a result, a viscous melt is obtained from which glass fibers can be downdrawn, including yttrium aluminum garnet (YAG or $Y_3Al_3O_{12}$), glass fibers [12], erbium-, neodynium-, and lanthanum-oxide-doped YAG glass fibers [13], mullite ($Al_6Si_2O_{13}$) glass fibers and holmium-, erbium-, and other rare earth-doped mullite glass fibers [14]. Small additions of SiO_2 improve the glass formability but slightly reduce the long wavelength limit for IR transmission.

(a) Process concept. Containerless liquid phase processing has been successfully used to achieve deep undercooling of molten oxides to a temperature where the viscosity is sufficiently high for fiber pulling. And in principle, containerless conditions eliminate heterogeneous nucleation by containers (crucibles, spinning cells or bushings, and precious metal tips) and help deep undercooling [12].

One may assume that the fiber-forming viscosity, although not reported was between log 2.5 and log 3.0 poise. The liquidus temperature and the delta temperature at the fiber-forming viscosity were not reported either. For the following reasons, they might be responsible for spontaneous crystallization that reportedly occurred toward the end of the fiberizing process [13, 14].

In a batch process that lasts only a very short period of time, it would be inconsequential whether the liquidus temperature were higher or lower than the actual fiber-forming viscosity of log 2.5–3.0 poise. By definition [2], the liquidus temperature is the temperature of a melt at which crystals can form over a short period of time and remain in equilibrium with the melt. In a batch process that lasts longer than a very short time, the melt will spontaneously crystallize, if its viscosity at the

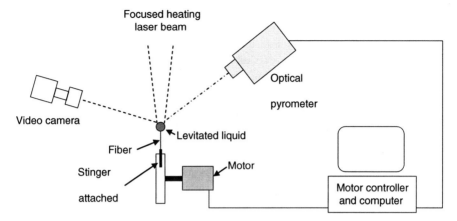

Fig. 1.23 Containerless laser-heated melt process. Redrawn from Weber et al. [12] (supplied by and printed with permission from Richard Weber, April 6, 2009)

fiber-forming temperature were equal to, or lower than, the liquidus temperature (see Section 1.1.3).

(b) YAG glasses and glass fibers. In this process (Fig. 1.23), solid YAG samples, 3 mm in diameter, were levitated in a flow of argon gas. The levitated material was completely melted in a continuous wave CO_2 laser heating beam. The viscosity of the inviscid melt above the melting point was 0.05 Pa s (0.5 poise). The laser beam was then blocked, resulting in a cooling rate of ∼250 K/s.

At a pre-selected temperature between 1600 and 1660°C, i.e., well below the liquidus temperature, fibers were pulled from the levitated droplet by rapidly introducing and withdrawing a 100 μm diameter tungsten wire "stinger." At higher temperatures the stinger pulled out of the melt without forming a fiber, while crystallization was likely to occur at lower temperatures [12].

Glass fibers, up to 0.5 m in length and 5–30 μm in diameter, were pulled from the apparatus at rates of 1.0–1.5 m/s before crystallization terminated the process. The fibers were homogeneous and had smooth surfaces. They were transparent, highly flexible, and x-ray amorphous.

Rare earth oxide–aluminum oxide glasses and glass fibers have properties similar to sapphire and sapphire fibers, respectively. They exhibit infrared transmission up to a wavelength of about 5000 nm. They are hard, strong, stiff, and thermally stable up to about 1000°C. Their homogeneity depends on selected concentrations of optically active dopants. Increasing add-on levels of SiO_2 also increase the homogeneity but reduce the long wavelength limit for infrared transmissions. The average tensile fracture strength of the YAG composition fibers was 5.0±0.3 GPa.

Rare earth oxide–aluminum oxide glasses (REAI^{TM}) are being explored by the Food and Drug Administration as a host glass for mid-IR optical devices in surgical and diagnostic applications, and erbium holmium- and oxide-doped devices in the ∼2100 and 2900 nm wave bands [13]. The economics and scalability of YAG-based glass fibers are promising but remain unexplored. The materials cost is the same for an amorphous YAG sensor fiber made by the containerless laser-heated melt process as for a single-crystal YAG sensor fiber made by laser-heated pedestal growth (Section 1.5.1). But the process for making glass fibers is faster than that for making single-crystal fibers, and it may require less process energy.

(c) Mullite composition glass fibers. Strong and chemically homogeneous mullite-based glass fibers were made with a diameter of 5–40 μm by pulling fibers from supercooled laser-levitated oxide melts. Under transient cooling conditions, the fiber pulling rates ranged from 0.2 to 1.6 m/s. The average tensile fracture strengths of the glass fibers were 5.6±0.7 GPa. Glass fibers with lower strength levels often showed the presence of equiaxial crystallites with diameters ranging from 20 to 50 μm. Transmission electron microscopy confirmed that the crystalline material was mullite [14].

The information that has been presented [14] shows that the crystallization of mullite glass melts begins to occur at approximately 1200 K at a very low heating rate of 2.5 K/min and that bulk crystallization increases when the heating rate is increased. The synthesis of uniform polycrystalline oxide fibers was achieved in a two-step process, i.e., by downdrawing glass fibers from supercooled melts and then crystallizing the resulting glass fibers by an appropriate heat treatment. The mullite-based glass fibers crystallized and produced dense crystalline fibers with grain sizes ranging from 0.5 to 1 μm. A crystalline mullite fiber with a 10 μm radius that had been made by this process had a tensile strength of 1.8 GPa and an elastic modulus of 360 GPa.

(d) Summary and conclusions. In summary, the process of forming glass fibers from laser-heated, acoustically levitated, supercooled, and otherwise inviscid (essentially non-viscous) oxide melts greatly expands the range of materials which have high strength and superior functionality. The process is inherently fast and potentially inexpensive. Glass fibers made by this process may replace costly single-crystal fibers and polycrystalline fibers made by an extension of this process may replace polycrystalline fibers made by other processes.

1.4.2.3 Inviscid Melt Spinning (IMS) Process

In the inviscid melt spinning (IMS) process an inviscid aluminate or metal jet is extruded into a chemically reactive environment that raises their surface viscosity and thereby facilitates the formation of continuous fibers. The theory governing inviscid melt spinning by chemically modifying the jet as well as fiber surface is well established.

(a) Forming metal fibers in a reactive environment. Commercial wire drawing processes produce metal wires with round cross sections but they are highly energy and labor intensive. These processes fall outside the scope of this book. Commercial rapid solidification processes yield amorphous metallic ribbons. Inviscid melt spinning yields metal fibers by a chemically assisted jet stabilization process.

In the inviscid melt spinning process [11, 130], steel fibers are formed by the same mechanism as glass fibers. In this case, the process shown in Fig. 1.24 requires

Fig. 1.24 Inviscid melt
spinning process (schematic
drawing). Redrawn from
Wallenberger et al. [9]
(reprinted with permission of
Elsevier Science Publishers)

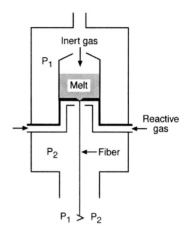

the presence of silicon in the steel formulation and the presence of carbon dioxide
in the process environment.

A commercial pilot production unit based on the schematic process diagram
shown in Fig. 1.24 consisted of a 0.9 m long furnace, a 1.5 m long cooling column, and a windup [9, 11]. In this process, 50 kg of steel was heated and melted
with a 4 kHz power unit supplying 70 KW in a ceramic crucible having an orifice in
its bottom plate. An essential ingredient in the cooling medium was carbon dioxide.
Wires with diameters of 100–200 μm can be made at 1500°C with speeds of 10–20
m/s, respectively. At a rate of 15 m/s, the spinning of a 165 μm diameter wire lasted
4 h. In this pilot process, unbroken wire was obtained for periods in excess of 1 h
which represents a continuous length of about 80 km.

As shown in Equations 1.18, 1.19 and 1.20, carbon dioxide diffuses into the surface of the inviscid molten jet. Its concentration decreases with increasing distance
from the surface, but wherever it finds silicon that is evenly distributed throughout
the melt (and therefore the surface of the liquid jet), it forms silicon dioxide and
carbon which cause a steep increase in the surface viscosity. Carbon may further
react to form carbides. The surface skin is neither a sheath nor a film:

$$Si + 2O \longleftrightarrow SiO_2 \tag{1.18}$$

$$2Si + 2CO \longleftrightarrow SiO_2 + 2C \tag{1.19}$$

$$Si + C \longleftrightarrow SiC \tag{1.20}$$

ESCA analysis confirms the presence of oxidized silicon and oxidized iron in the
fiber surface or skin [11, 130]. The silica peak gradually disappears at a depth of
100 nm, giving way to the peaks of iron and silicon. Analysis with a CAMECA ion
analyzer shows that the intensity of Si^+ peaks decreases to naught between 17 and

55 nm from the surface of a 165 μm diameter steel wire. The results parallel those noted for carbon with aluminate fibers.

Silica is the most viscous inorganic material known, especially at 1500°C. It is the ideal surface viscosity builder to increase the lifetime of a hot inviscid steel jet long enough to prevent formation of Rayleigh waves and shot. Other viscosity builders can be formed in situ at the spinning temperature to stabilize a given molten metal jet [11, 130]. They must have a higher melting point than the metal, and be insoluble in the molten jet [11, 130].

Inviscid melt spinning is considered to be a potentially viable alternative to wire drawing [130] for making steel wires for radial automobile tires, but a prior product development did not reach beyond the pilot plant level. Using silica steels, the complex chemistry (Equations 1.18, 1.19 and 1.20) produces also minute amounts of iron oxides which were detected by ESCA [130] and are a potentially undesirable trace byproduct. The challenge [1, 80] remains to fine-tune the chemistry of this process in a commercial development effort.

(b) Forming oxide glass fibers in a reactive environment. Calcium aluminate glass fibers can also be formed by inviscid melt spinning. In this case, carbon particles which are formed by the decomposition of propane enter into the surface of the molten jet and raise its surface viscosity, a process step that lengthens the lifetime of the jet and prevents its breakup.

In this process variant (Fig. 1.4, Example E, and Fig. 1.24), alumina and calcia are placed in a tungsten crucible having an orifice in its bottom plate [11, 121]. The crucible is placed in a furnace and the surrounding air is replaced with argon. The oxide powder is melted. The melt is maintained 100°C above its melting point. Propane is introduced below the crucible and the argon pressure is increased above the crucible.

A liquid jet of the oxide melt is extruded through the orifice in the bottom of the crucible. Propane decomposes on the hot surface of the molten jet, forms carbon particles which enter into the surface of the jet, and sometimes also deposits an additional secondary carbon sheath on the fiber surface but only after the jet solidifies.

The fibers in the examples contained 51.5–80.2% alumina, 43.5–19.8% calcia, and occasionally also 3.8–4.0% silica and/or 0.1–7.5% magnesia (Table 1.22). Their pristine strength ranged from 0.16 to 1.05 GPa (vs. 3.5 GPa for E-glass) and their moduli ranged from 41.1 to 61.2 GPa (vs. 72.5 GPa for E-glass). Their moduli [9] were much lower than those of updrawn high viscosity calcium aluminate fibers, suggesting much lower internal order. By SEM, the fracture patterns were typical of glass fibers, and fibers with ≤80% alumina were amorphous.

Some fibers had a secondary carbon sheath which was up to 600 nm thick [35]. Not surprisingly, these fibers were black. The secondary overgrowth is not an integral part of the fiber [9, 10]. It does not affect the fiber properties and can be peeled or burned off. A few fibers had no carbon sheath. These fibers were translucent. Carbon that is present in the fiber surface or skin does not affect its transparency,

Table 1.22 Properties of inviscid melt spun calcium aluminate glass fibers [1] (with kind permission of Springer Science and Business Media)

Examples [9]	Composition wt%				FT (°C)	Dia. (µm)	Strength (GPa)	Modulus (GPa)
	Al_2O_3	CaO	MgO	SiO_2				
Example (1)	54.0	39.0	3.0	4.0	1700	170	0.5	50.8
Example (2)	54.6	39.0	2.5	3.9	1500	190	0.7	
Example (3)	54.8	38.9	2.5	3.8	1500	216	0.6	46.6
Example (4)	59.0	40.8	0.2	–	1500	190	0.7	45.9
Example (5)	60.8	39.1	0.1	–	1700	225	0.5	41.0
Example (6)	66.8	33.2	–	–	1800	102	0.9	
Example (7)	67.3	32.7	–	–	1700	167	0.8	
Example (8)	80.2	19.8	–	–	1800	117	1.1	
Controls								
E-glass	14.0	22.1	0.6	54.5	1200	10	3.5	72.0
Silica fiber	–	–	–	100.0	1750	100	3.5	69.0

whether the fiber had originally no carbon sheath as-spun or whether the sheath had been removed from the as-spun fiber [10].

Sputtered neutral mass spectrometry (SNMS) depth profiles document that carbon is present in the skin of all fibers to a depth of about 50 nm (Fig. 1.25), whether a given fiber has a secondary carbon sheath overgrowth or not [9]. X-ray photoelectron spectroscopy (XPS) showed that carbon in the surface or skin [10] consists of carbide (51%), carbon (41%), and carbonate (8%). Infrared depth profiling by diffuse reflectance infrared spectroscopy (DRIFT) provided further important insights [131]. It showed that carbon alters the oxygen environment of the aluminum atoms near the fiber surface from octahedral to tetrahedral coordination and promotes the generation of carbonaceous species such as ethers and esters in addition to carbonates and carbides [80] which have also been found with XPS [10].

In summary, a typical aluminate fiber is spun between 1500 and 1700°C (Table 1.22). Carbon enters into the skin of the still liquid jet and stabilizes it. Some fibers have a carbon sheath. The carbon sheath is obtained after the jet is solidified (<500°C), i.e., when more carbon is present in the reactive propane environment than needed to stabilize the liquid jet and form a fiber.

Inviscid melt spun calcium aluminate glass fibers have low strength (0.5–1.1 GPa) and low moduli (46–58 GPa). Low strength and low stiffness can be attributed to the random structure frozen into the fibers during rapid solidification. As a result, these fibers are not likely to become composite reinforcing fibers, despite their excellent alkali resistance which they share with quaternary calcium aluminate fibers [121].

(c) Mechanism of jet solidification. Continuous aluminate glass fibers are formed in the presence of propane, but not in its absence. A viable mechanism of jet stabilization must therefore explain (1) the function of carbon which enters into the surface or skin of the molten jet, (2) the function of carbides and carbonates which

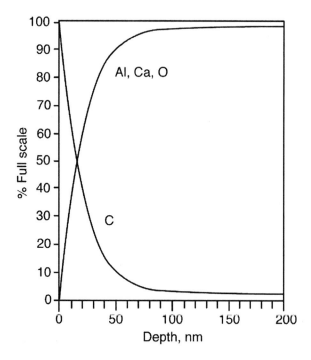

Fig. 1.25 SNMS depth profile of a translucent calcium aluminate fiber. Redrawn from Wallenberger and Brown [1, 35] (reprinted with permission of Elsevier Science Publishers)

are instantly formed in the molten jet surface, and (3) the increase in tetrahedral from octahedral coordination of aluminum atoms in the surface [131] before the fiber solidifies and secondary overgrowth with carbon can occur.

The principle governing jet formation and jet breakup has been discussed in Section 1.4.2.3 [10]. A liquid jet is unstable, degrades into Rayleigh waves (Fig. 1.22, right), and then droplets. The jet stability (Equation 1.15) depends on liquid density, jet diameter, melt viscosity, and surface tension. The lifetime of a jet is the time required for the melt to traverse the continuous length of the jet before the onset of Rayleigh waves [10]. A jet of a silicate glass or an organic polymer melt has a high viscosity ($>10^2$ Pa s) and a lifetime of $>10^0$ s; it can be spun or drawn from the melt by conventional methods, and it solidifies well before it can form Rayleigh waves.

A jet of an aluminate melt with >50% alumina has a low viscosity (<1 Pa s) and a calculated lifetime less than 10^{-2} s (Table 1.23). If ejected into an ambient or neutral environment, it will form Rayleigh waves and droplets (or shot when they freeze) rather than uniform continuous fibers.

Equation 1.15 shows that viscosity is the major factor in determining jet lifetime; surface tension is a secondary factor [10]. Any increase in the surface viscosity of the molten oxide will disproportionately increase the jet lifetime from that calculated for an unassisted jet (2.0×10^{-3} s) to that calculated for an assisted jet (2.0×10^{-1} s) which must have been obtained since continuous fibers were obtained.

Table 1.23 Properties of inviscid calcium aluminate jets [1] (with kind permission of Springer Science and Business Media)

Reference [9] Al$_2$O$_3$(%)	Melt Temp. (°C)	Spin Temp. (°C)	Jet/fiber Diam. (μm)	Melt Density (g/cm^3)	Surface Tension (mN/m)	Melt Viscosity (Pa s)	Unassis- ted jet Life (s)
51.5	1415	1500	375	2.70	680	0.34	1.4×10^{-2}
54.6	1390	1500	190	2.70	680	0.55	8.7×10^{-2}
66.8	1650	1700	105	2.68	625	0.14	2.0×10^{-3}
80.2	1830	1900	118	2.68	575	0.06	1.7×10^{-3}

Particles, especially shaped particles, are known to increase the viscosity of a suspension, following Mooney–Einstein [132]. Thus, carbon particles [10] may enter into the surface or skin of a liquid inviscid jet and increase its viscosity sufficiently and long enough to facilitate its solidification and fiber formation.

For this mechanism to be viable, three conditions must be fulfilled: (1) the increase in the jet surface viscosity must afford a stabilized (assisted) jet lifetime that at least matches the jet cooling time; (2) the assisted lifetime resulting from the viscosity increase must be comparable to the actual (unassisted) lifetime of a typical silicate fiber such as E-glass; and (3) the surface viscosity increase needed to achieve this lifetime must be realistically achievable by carbon insertion in the jet surface.

The pyrolytic production of carbon has been said to create a "snowstorm of large, flat molecules containing the hexagonal ring structure of graphite" [133], i.e., flakes or flat aggregates of smaller particles. They enter into the surface of the molten jet and act as viscosity builders, where they and their instant reaction products such as carbides can be detected by ESCA. Since jet geometry and surface forces tend to constrain particle formation into planar structures parallel to the jet surface, the rheological treatment for flakes is appropriate. The viscosity (η) of a Newtonian fluid containing solid, suspended particles, relative to that (η_0) of the suspending fluid, is governed by the following equation [132]:

$$\ln(\eta/\eta_0) = k_E f_2/[1 - f_2/f_m] \tag{1.21}$$

The Einstein coefficient (k_E) for incorporated particles depends on particle shape, f_2 is the volume fraction of filler, and f_m is the maximum packing fraction for the flakes. The viscosity of the suspending fluid (bulk molten) oxide is 0.14 Pa s, and the required 36.6 Pa s viscosity for the surface layer of the jet can be attained with flake shapes having L/D between 4 and 9, and volume fractions of solids between 0.32 and 0.45. The combined volume fraction of the solids in the oxide skin detected by SNMS was calculated [11] to be 0.508, a value well above that needed for rheological jet stabilization.

E-glass jets with a diameter comparable to that of the 66.8% alumina jet can be melt spun at temperatures ranging from 1100 to 1480°C [9] where they have a viscosity between 6.1 and 816 Pa s and a calculated lifetime between 8.8×10^{-2}

Table 1.24 Chemical stabilization of inviscid calcium aluminate jets [1] (with kind permission of Springer Science and Business Media)

Jet/fiber comp. [10] (105 μm diameter)	Melt behavior	Temp. (°C)	Viscosity (Pa s)		Jet lifetime (s)	
			Bulk	Surface	Unassisted	Assisted
CaO-Al$_2$O$_3$ (66.8%)	Inviscid	1700	0.14	36.6	2.0×10^{-3}	2.6×10^{-1}
E-glass	Strong	1360	31.7	31.7	2.6×10^{-1}	–

and 1.2×10^1 s (Table 1.24). The assisted lifetime of the 66.8% alumina jet (2.6×10^{-1} s) is therefore well within the range of the unassisted E-glass jet lifetimes [10].

In summary, the stabilization of inviscid aluminate jets can be attributed to the increase in surface viscosity due to suspension of solid carbon and carbide particles in the molten oxide surface [10]. The viscosity (rheology)-controlled jet stabilization process is accompanied by a change [131] from an octahedral to a tetrahedral coordination of aluminum atoms near the surface.

1.4.2.4 Rapid Jet Solidification (RJS) Processes

In the rapid jet solidification (RJS) process, the quench rate increases almost instantly from $\sim 10^4$ to $>10^6$ K/s, and the liquid jet quickly reaches a high viscosity range and solidifies before it can crystallize. Unlike the inviscid melt spinning process, a chemically reactive environment that contains, for example, carbon dioxide, is not required to stabilize the metal ribbon [135].

A commercial process has been developed for the production of amorphous metal ribbons [15–17]. The attainment of self-supporting glass ribbons has so net yet been achieved by true rapid solidification.

(a) Amorphous metal ribbons. Glass-forming metal alloys [134] include late transition metal–metalloids (e.g., Fe$_{100-x}$B$_x$), early transition metal–metalloids (e.g., Ti$_{100-x}$Si$_x$), early transition–late transitions metal alloys (e.g., Nb$_{100-x}$Ni$_x$), aluminum-based alloys (e.g., Al$_{83}$Cu$_{16}$V$_7$), lanthanum-based alloys (e.g., La$_{100-x}$Au$_x$), alkaline earth-based alloys (e.g., Mg$_{100-x}$Zn$_x$), and actinide-based alloys (e.g., U$_{100-x}$Co$_x$). The techniques for producing amorphous metal alloys [19] include rapid liquid cooling, supercooling of liquids, physical vapor deposition, chemical methods, irradiation, and mechanical methods [134].

The planar flow casting process facilitates the formation of continuous ribbons of metallic glasses or metal alloys from their inviscid melts without requiring a chemically reactive environment as the inviscid melt spinning process or laser levitation. The molten metal is extruded through a ribbon-shaped orifice onto a cold rotating quench wheel. The ribbon is formed without losing its precision shape and is collected on a continuous windup roll; see [58, 15, 134, 135, 16] for details.

The industrial fabrication of amorphous metal ribbons dates back to the late 1970s and to a process known as planar flow casting [17]. The continuous ribbon that is formed has a width ranging from 1 to 100 mm, a thickness of >10 μm, and a uniform rectangular cross section.

(b) Products and applications. Amorphous magnetic glass ribbons based on alloys of iron, nickel, or cobalt are among the softest magnetic materials known. Their tensile strengths range from 1.0 to 1.7 GPa, their moduli from 100 to 110 GPa, and their service temperatures from 90 to 150°C. Their high degree of softness combined with excellent mechanical properties has benefited applications ranging from microscopic recording heads to large architectural EMI shielding, electronic and power cores, and anti-theft sensors [16].

Tin–lead alloy Metglas ribbons, having moduli as low as 18 GPa, are used as solders for diebonding applications. Their liquidus ranges from 190 to 314°C and their solidus ranges from 182 to 310°C. Metglas brazing foils are made from nickel-, cobalt-, palladium-, and copper-based alloys as melting point depressors, and come in ribbons 10 cm wide and >25 μm in thickness. Solder and brazing ribbons have solidus temperatures from 770 to 1130°C and liquidus temperatures ranging from 925 to 1150°C. Applications include aerospace components, electric motors, and brake pads [16].

(c) Amorphous fiberglass ribbons. No ribbons were so far made from inviscid oxide glass melts. Production of glass ribbons and glass fibers from inviscid melts by a rapid solidification process that does not require a chemically reactive environment or acoustic levitation in a containerless process remains a challenge for the future.

1.5 Appendix: Single-Crystal Alumina Fibers

The following review of single-crystal aluminate and alumina fibers is meant to serve as a reference to the chapters on aluminate and alumina glass fibers – and as a challenge. For example, YAG glass fibers, which are aimed at sensor applications (Section 1.4.2.2), can be made with higher process speeds and are more energy-friendly than single-crystal YAG sensor fibers. As a result, this section provides potential targets for developing the technology to make glass fibers which correspond to a range of other commercial single-crystal fibers, including sapphire fibers.

1.5.1 Single-Crystal Fibers from Inviscid Melts

Continuous single-crystal fibers can be grown from inviscid melts by the very slow edge-defined film-fed growth (EFG) process [18] and by the slow equally slow laser-heated float zone (LHFZ) or laser-heated pedestal growth (LHPG) process with growth rates ranging from 0.3 to 0.7 mm/s [19]. As previously discussed, aluminate glass fibers can be formed at much higher rates from inviscid melts and commercial polycrystalline alumina fibers require different fabrication processes [32].

1.5.1.1 Edge-Defined Film-Fed Growth

Edge-defined film-fed growth (EFG) is a commercial process [18] that facilitates the fabrication of continuous void-free single-crystal oxide fibers from tungsten or other growth orifices.

(a) Growth of sapphire fibers. This process yields commercial single-crystal sapphire fibers. A liquid pool from which the continually growing filamentary crystal is withdrawn is formed on top of a planar surface of the orifice and fed by capillaries which extend down through the orifice into a liquid reservoir. The crystal shaping or edge definition is maintained by the geometry of the top surface of the orifice and the fulfillment of a contact angle of <90° between the liquid and material from which the orifice is fabricated.

Single-crystal alumina (sapphire) fibers grow in this process in the <0001> growth direction. These fibers are void free and have a density of 3.97 g/cm^3, a tensile strength of 2.1–3.4 GPa at room temperature, Young's modulus of 453 GPa, superior electrical and optical properties, and superior chemical stability. Although individual fibers can be obtained with diameters ranging from 64 to 350 μm, their commercial diameters range from 150 to 250 μm.

Sapphire is produced commercially throughout the world and is used in virtually every industry. The optical, electrical, chemical, mechanical, and nuclear properties of sapphire fibers, as described in the literature [18], make them an ideal material for many applications other than their use as sensor or reinforcing fibers for metal and ceramic matrix composites. Frequently, the combination of two or more of its properties renders sapphire as the only material available to solve complex engineering design problems.

(b) Process versatility. The edge-defined film-fed growth method (EFG) is a "near-net-shape" process for growing continuous sapphire fibers and for prototyping a wide variety of other shaped parts. Unlike other crystal growth methods used to make sapphire structures, only the edge-defined film-fed growth method can be used to also produce grown-to-shape tubes, rods, ribbons, and three-dimensional shapes. All other processes form ingots of various sizes, which must be cut to shape by highly skilled workers using costly diamond impregnated tools.

1.5.1.2 Laser-Heated Float Zone Growth

In the laser-heated float zone (LHFZ) or pedestal (LHPG) growth process, a circumferential laser is placed around a preform rod (e.g., polycrystalline alumina) to zone refine a segment of the material while at the same time updrawing a single-crystal fiber (e.g., sapphire).

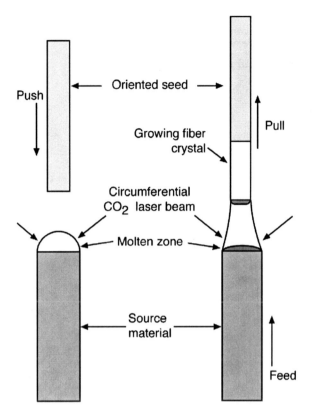

Fig. 1.26 Laser-heated float zone or pedestal growth process (schematic). Redrawn from Feigelson [18, 36] (reprinted with permission of Elsevier Science Publishers)

(a) Growth of single-crystal fibers. Materials may melt congruently or incongruently in the float zone [19, 136, 137]. The singular function of the laser in this process (Fig. 1.26) is to provide uniform circumferential heating. The tip of the preform rod is melted, and a seed rod is introduced into the melt and slowly withdrawn to initiate the growth of a potentially endless filamentary single crystal. Sapphire fibers [19, 33] updrawn from sintered polycrystalline alumina preform rods are the best researched material made by this process. They offer a superior value in end uses requiring prolonged exposure to ultrahigh temperatures.

The properties of these sapphire fibers are comparable to those made by edge-defined film-fed growth, except that potentially lower diameters are possible at about equal growth rates [19]. Growth rates of 0.7 mm/s are possible, but growth rates below 0.3 mm/s are typically used with pure materials [19]. For comparison, sapphire fibers grown by edge-defined film-fed growth were reported to have growth rates up to 0.5 mm/s.

The laser-heated float zone process is not limited to sapphire fibers or even to oxides. Table 1.25 lists fibers other than sapphire which have been grown by

Table 1.25 Single-crystal fibersgrown float zone or pedestal growth [1] (with kind permission of Springer Science and Business Media)

Oxides	MP (°C)	Orientation	Dia. (μm)	Application
Nd:YAG	1940	[124, 126]	6–1000	Laser
YAG	1940	[124, 126]	100–1000	Laser
Al_2O_3	2045	a, c	55–600	Remote sensor
Ti:Al_2O_3	2045	c	200–800	Laser
Cr:Al_2O_3	2045	c	3–170	Laser
$LiNbO_3$	1260	a, c	20–800	Acoustics
Nd:$LiNbO_3$	1260	c	800	Laser
$LiTaO_3$	1650	[124]	600	SAW device
Li_2GeO_3	1170	a, c	100–600	Raman device
$Gd_2(MoO_4)_3$	1157	[124]	200–600	Ferroelastic
$CaSc_2O_4$	2200	a, b, c	100–600	Model
Nd:$CaSc_2O_4$	2200	c	600	Laser
$SrSc_2O_4$	2200	–	600	Model
YIG	1555	[124]	100–600	Insulator
Eu:Al_2O_3	2410	c	500–800	Laser
Ti:$MgAl_2O_4$	2105	–	1000	Laser
Ti:$YAlO_3$	1875	–	1000	Laser
BaB_2O_4	1095	–	500	Nonlinear optics
Nd_2SiO_5	1980	–	750	Laser
$SrTiO_3$	1860	–	600	Model
Nb_2O_5	1495	–	700–1700	Model
$BaTiO_3$	1618	c	300–800	Ferroelectric
$SrBaTi_2O_6$	1700	a, c	200–600	Ferroelectric
$ScTaO_4$	2300	a	200–600	Ferroelectric
$ScNbO_4$	2100	–	200–600	Ferroelectric
$SrBaNb_2O_6$	1500	a, c	600–1700	IR guide
Cr:Y_2O_3	2400	c [124],	600	Laser
Cr:Sc_2O_3	2400	c [124],	600	Laser
Cr:Lu_2O_3	2400	c [124],	600	Laser
$Li_2O–GeO_2$	1106	–	–	Eutectics
($Bi_{1.8}Sr_{1.8}Ca_{1.2}Cu_{2.2}O_8$)	–	–	250–1000	High T_c s/c wires

this process. This includes filamentary single salts, eutectics, oxides, ceramics, and semiconductors, as well as superconductor and metallic fibers.

Sapphire, $LiNbO_3$, $Li_2O–NaF$, Nd:YAG, and Cr:Al_2O_3 fibers have been grown with diameters as low as 55, 40, 20, 6, and 3 μm, respectively. Some of the more important fibers grown by this method are Al_2O_3–$Y_3Al_5O_{12}$ (YAG) eutectic fibers [138] and Y_2O_3–stabilized, cubic ZrO_2 single-crystal fibers [129]. A major motivation for current research is the potential of single-crystal fibers in optoelectronic applications [139].

Single-crystal fibers of potassium lithium niobate are known to be useful materials for electro-optic applications. They can be grown with a growth rate of 11 mm/h with diameters of 0.9 mm by the halogen lamp-assisted laser-heated pedestal growth method, alternatively known as the laser float zone method. The single-crystal fibers, like the ceramic feed rods from which they were grown, consisted of 30% K_2O, %17 Li_2O, and 53% Nb_2O_3 [140].

As core clad waveguide structures of the c-axis, 200 μm diameter single-crystal lithium niobate fibers have potential applications as second harmonic generators, optical modulators, lasers, and optical oscillators. They are obtained by a magnesium ion diffusion process [137].

Filamentary eutectic ZrO/CaO crystals can also be grown by the laser-heated float zone process. They have a high degree of ion conduction and are aimed at advanced applications such as solid electrolytes, where their conductivity at high temperatures may be exploited in heating elements and in situ gas sensors [141]. The starting materials are ZrO_2/CaO powders with a eutectic composition. Single-crystal fibers are prepared by a polymer matrix route or a ceramic route. The phase formation in both routes is similar.

The laser float zone process has also been used successfully to synthesize phosphorescent single crystals of $SrAl_2O_4$:Eu^{2+}, Dy^{3+} and $CaAl_2O_4$:Eu^{2+}, Nd^{3+} [142]. These single-crystal fibers may have important applications in optoelectronic technology. The crystals show bright green and purple phosphorescence, respectively, with persistence times exceeding 16 h. Their brightness is 10 times stronger than that of traditional sulfide phosphors. Sulfide phosphors vaporize near their melting points and single crystals are therefore difficult to grow. Alkaline earth aluminates have low vapor pressures near their melting points and single crystals can be grown directly from their inviscid melts [142].

Typically, fibers having diameters up to 100 μm are readily possible; fibers with diameters above 200 μm are technically speaking no longer fibers but rods. This includes the niobate fibers [140] and the superconducting fibers [19, 143–145] discussed below. In principle, the float zone method is a containerless process. There is a nominal diameter reduction in the float zone from that of the preform rod to that of the final fiber. Zone (or overall process) stability for the growth of oxide and fluoride fibers is usually achieved with diameter reductions in the range of 1/2 to 1/3 [19].

(b) High T_c superconducting fibers. These materials may also melt congruently or incongruently in the float zone [19]. Congruently melting materials do not undergo a phase transformation above room temperature, have a low vapor pressure at their melting, and are easy to grow in single-crystal form by this method. Incongruently melting preform materials first yield a phase that is different from that of the source rod and the melt [136]. As growth proceeds, the melt composition changes, and the liquidus drops until it reaches the peritectic decomposition temperature. At this point, the desired phase crystallizes and steady-state growth will be achieved [19].

The laser float zone process is an effective method for growing oriented $Bi_2Sr_2CaCu_2O_8$ fibers [143]. Single-crystal $Y_3Fe_5O_{15}$ (YIG) and polycrystalline high T_c superconducting $Bi_{1.8}Sr_{1.8}Ca_{1.2}Cu_{2.2}O_8$ fibers [143] (Fig. 1.26) were first grown by the float zone process with diameters of 0.7 and 1 mm, respectively, from incongruently melting compounds. The critical current measurements on the high T_c fiber were 5×10^{-4} A/cm^2 at 68 K with zero resistance near 85 K [144].

More recently, careful processing afforded Pb-substituted Bi–Sr–Cu–Ca–O fibers with critical temperatures (T_c) of 106 K and critical current densities (J_c) of 1015 A/cm^2 (77 K, O T) after annealing [145]. Magnetic susceptibility and critical current density of superconducting fibers of the Bi–Sr–Ca–Cu–O system grown by the laser float zone method are strongly dependent on the pull rates. A favorable grain orientation for current transport and highest critical currents was obtained at high pulling rates, e.g., >4 μm/s [136].

In summary, the laser-heated float zone (LHFZ) method [136], in particular the traveling solvent zone melting (TSZM) configuration [137], is a highly effective technique to grow centimeter long crystals of high T_c and other low-dimensional cuprates. High-temperature superconducting fibers, wires, tapes, and ribbons have also been made by the powder-in-tube method. These are sheath/core bicomponent fibers or ribbons having a protective metal sheath and a functional core consisting of an appropriate multiple oxide. In this fabrication process [146] the superconducting powder is continuously introduced into a metal tube and the filled tube is drawn by conventional wire drawing methods.

Although the powder-in-tube method did yield the highest current density [147] reported so far (> 1,000,000 A/cm^2, 77 K, O T), it is not reviewed here in detail because it is a metal drawing, not a melt-forming process. High-temperature superconducting sheath/core bicomponent fibers have also been made by introducing the superconducting material into the core of hollow glass fibers as they are formed under the bushing (Section 1.3.4.2).

1.5.2 The Future of Alumina and Aluminate Fibers

Continuous single-crystal sapphire fibers offer an impressive performance as reinforcing fibers in ceramic and metal matrix composites. It is therefore appropriate to note some important commonalties and differences between single-crystal and inviscid alumina fibers.

1.5.2.1 Amorphous Alumina vs. Single-Crystal Sapphire Fibers

Sapphire fibers are hard, strong, and scratch resistant to most materials and provide excellent wear surfaces. They can withstand higher pressures than polycrystalline alumina since they lack the grain boundary interface breakdown of the latter. Sapphire fibers transmit ultraviolet, visible, infrared, and microwaves and serve as excellent wave guides between 10.6 and 17 μm, and offer durable and reliable IR transmission. By virtue of their high thermal conductivity they can be rapidly heated and cooled.

EFG sapphire fibers melt sharply at 2050°C and maintain measurable strength at extreme temperatures [18]. Table 1.26 shows tensile strength as a function of test temperature from 25 to 1500°C. The room temperature strength, 3.57 GPa, is low for a single-crystal fiber but typical for sapphire fibers, irrespective of process. The strength of crystal oxide fibers at room temperature should be approaching the

Table 1.26 Strength retention of sapphire fibers at elevated temperature [1] (with kind permission of Springer Science and Business Media)

Fiber testing temperature (°C)	Average tensile strength (GPa)	Standard deviation (GP)
25	3.57	0.66
400	2.08	0.49
800	1.85	0.34
1094	1.03	0.14
1500	0.55	0.09

theoretical value, which is >10 GPa. In fact, room temperature strength of EFG sapphire fibers is only 2.5–3.0 times higher than that of about an equal diameter polycrystalline alumina (1.2–2.6 GPa) such as Fiber FP or PRD-166, respectively [32]. This deficiency of single-crystal EFG sapphire fibers still needs to be corrected.

The tensile strength of EFG sapphire fibers at 1500°C in Table 1.26 is 0.55 GPa. In addition, isolated literature values report strength levels of 0.40 GPa up to 1900°C. These are impressive results since they refer to an oxidative environment. In an oxidative environment, unprotected (uncoated) carbon fibers are stable up to 500°C, especially for longer term exposures.

1.5.2.2 Amorphous YAG vs. Single-Crystal YAG Fibers

Continuous single-crystal oxide fibers, including sapphire, have significant property advantages over the corresponding polycrystalline oxide fibers [32]. They include microstructural stability at high temperatures, retention of high elastic moduli at high temperatures, and high-temperature creep for which even further improvements are being sought [18, 19, 138, 148]. But because of their high fiber diameters, they cannot be woven. They must be filament wound or used as inserts.

1.5.2.3 Summary, Conclusions, and Outlook

Commercial single-crystal sapphire fibers [33, 139], single-crystal yttrium aluminum garnet (YAG) fibers [138], and polycrystalline oxide fibers [32] are commercially available, but they are costly, and commercial opportunities exist for new non-structural continuous single-crystal and polycrystalline oxide fibers for high T_c superconductors, optoelectronics, nonlinear optics and acoustic IR guides, lasers, and remote sensors [1, 33,136–146,148].

The advent of glass fibers, which are made from inviscid oxide melts, serves as a challenge. YAG glass fibers, which are aimed at sensor applications, rely on a more energy-friendly process than the single-crystal YAG sensor fibers and they can be made with higher process speeds. Other single-crystal fibers, including sapphire fibers, offer further potential targets for exploring the development of the corresponding glass fibers and their value-in-use.

References

1. F. T. Wallenberger, Continuous melt spinning processes, in *Advanced inorganic fibers: processes, structures, properties, applications*, Chapter 4, Kluwer Academic Publishers, Dordrecht/Boston/London, pp. 81–122 (1999).
2. F. T. Wallenberger, R. J. Hicks, P. N. Simcic and A. T. Bierhals, New environmentally and energy friendly fiberglass compositions (E-glass, ECR-glass, C-glass and A-glass), Glass Technol. Eur. J. Glass Sci. Technol. A, 48 (6), 305–315 (2007).
3. C. A. Angell, Glass formers and viscous liquid slowdown since David Turnbull: enduring puzzles and new twists, MRS Bull., 33 (5), 544–555 (2008).
4. F. T. Wallenberger, The structure of glasses, Science, 267, 1549 (1995).
5. F. T. Wallenberger, N. E. Weston and S. D. Brown, Calcia-alumina glass fibers: drawing from super-cooled melts versus inviscid melt spinning, Mater. Lett., 11 (89), 229–235 (1991).
6. V. E. Khazanov, Yu. I. Kolesov and N. N. Trofimov, Glass fibers, in *Fibre science and technology*, V. I. Kostikov, ed., Chapman and Hall, London, pp. 15–230 (1995).
7. F. T. Wallenberger, Continuous solvent spinning processes, in *Advanced inorganic fibers: processes, structures, properties, applications*, Chapter 5, Kluwer Academic Publishers, Dordrecht/Boston/London, pp. 123–128 (1999).
8. G. Y. Onoda, Jr. and S. D. Brown, Low silica glasses based on calcia-aluminas, J. Am. Ceram. Soc., 53 (6), 311–316 (1970).
9. F. T. Wallenberger, N. E. Weston and S. A. Dunn, Inviscid melt spinning: as-spun amorphous alumina fibers, Mater. Lett., 2 (4), 121–127 (1990).
10. F. T. Wallenberger, N. E. Weston, K. Motzfeldt, and D. G. Swartzfager, Inviscid melt spinning of alumina fibers: Chemical jet stabilization, J. Am. Ceram. Soc., 75 (3), 629–639 (1992).
11. R. E. Cunningham, L. F. Rakestraw and S. A. Dunn, Inviscid melt spinning of filaments, in *Spinning wire from molten metal*, J. Mottern and W. J. Privott, ed., AIChE Symposium Series, AIChE, New York, 74 (180), 20–32 (1978).
12. J. K. R. Weber, J. J. Felton, B. Cho and P. C. Nordine, Glass fibres of pure and erbium- or neodymium-doped yttria-alumina compositions, Nature, 393, 769–771 (1998).
13. J. K. R. Weber, R. W. Waynant, I. Ilev, T. S. Key and P. C. Nordine, "Rare earth oxide-aluminum oxide glasses for mid-range IR devices," Proc. SPIE, Optical Fibers and Sensors for Medical Applications III, I. Gannot, ed., 4957, 16–22 (2003).
14. J. K. R. Weber, B. Cho, A. D. Hixon, J. G. Abadie, P. C. Nordine, W. M. Kriven, B. R. Johnson and D. Zhu, Growth and crystallization of YAG- and mullite-composition glass fibers, J. Eur. Ceram. Soc., 19, 2543–2550 (1999).
15. H. H. Liebermann, *Rapidly solidified alloys*, Marcel Decker, New York. (1993).
16. H. H. Liebermann, Metglas®, Allied Signal, Parsippany, NJ (1993).
17. D. M. C. Narashima, Planar flow casting of alloys, US Patent 4,142,571 (1979).
18. R. S. Feigelson, Growth of fiber crystals, in *Crystal growth of electronic materials*, E. Kaddis, ed., Elsevier Publishers, London, pp. 127–145 (1985).
19. J. Monbleau, *Single crystal technology, product bulletin*, Saphikon Inc., Milford (1994).
20. F. T. Wallenberger and R. J. Hicks, The effect of boron on the properties of fiberglass melts, Glass Technol. Eur. J. Glass Sci. Technol. A, 47 (5), 148–152 (2006).
21. F. T. Wallenberger, R. J. Hicks and A. T. Bierhals, Design of energy and environmentally friendly fiberglass compositions derived from the quaternary SiO_2–Al_2O_3–CaO–MgO phase diagram – Part I: structures, properties and crystallization potential of selected multi-oxide E-glass compositions, Ceramic Transactions, Volume 170, H. Li et al., eds., 181–199 (2004).
22. F. T. Wallenberger, R. J. Hicks and A. T. Bierhals, Effect of key oxides, including Li_2O, on the melt viscosity and energy demand of E-glass compositions, 66th Conference on Glass Problems, University of Illinois at Urbana-Champaign, Collection of Papers, W. M. Kriven, American Ceramic Society, John Wiley & Sons, 155–165 (2006).

23. F. T. Wallenberger, R. J. Hicks and A. T. Bierhals, Design of environmentally friendly fiber-glass compositions: ternary eutectic SiO_2–Al_2O_3–CaO and related compositions, structures and properties, J. Non-Cryst. Solids, 349, 377–387 (2004).

24. F T. Wallenberger, R. J. Hicks and A. T. Bierhals, Design of energy and environmentally friendly fiberglass compositions derived from the quaternary SiO_2–Al_2O_3–CaO–MgO phase diagram – Part II: Fluorine-free E-Glass compositions containing low levels of B_2O3 and Li_2O, in Proceedings of the Norbert Kreidl Memorial Conference, Glastechnische Berichte – Glass Science and Technology, Vol. 77C, 170–183 (2004).

25. F. T. Wallenberger, R. J. Hicks and A. T. Bierhals, Effect of oxides on decreasing melt viscosity and energy demand of E-glass, The Glass Researcher, Vol. 15, No. 1, Am. Ceram. Soc. Bull., 85 (2), 38–41 (2006).

26. F. T. Wallenberger, Glass fiber forming compositions, US Patent 7,153,799 B2, December 26, 2006.

27. F. T. Wallenberger, Glass fiber forming compositions, US Patent 6,962,886 B2, November 8, 2005.

28. T. P. Seward and T. Vascott, eds., High temperature glass melt property database for process modeling, The American Ceramic Society, Publisher, Westerville, p. 258 (2005).

29. K. H. Karlson and R. Backman, Thermodynamic properties, in Properties of glass-forming melts, L. D. Pye, A. Montenero and I. Joseph, eds., Taylor and Francis, CRC Press, Boca Raton, pp. 11–23 (2005).

30. P. Hrma, D. E. Smith, J. Matyas, J. D. Yeager, J. V. Jones and E. N. Boulos, Effect of float glass composition on liquidus temperature and devitrification behavior, Glass Technol. Eur. J. Glass Sci. Technol. A, 47 (3), 78–90 (2006).

31. P. A. Bingham and M. Marshall, Reformulation of container glasses for environmental benefit through lower melt temperatures, Glass Technol., 46 (1), 11–29 (2005).

32. R. Naslain, Ceramic oxide fibers from sol-gels and slurries, in Advanced inorganic fibers: processes, structures, properties, applications, F. T. Wallenberger, ed., Chapter 8, Kluwer Academic Publishers, Dordrecht/Boston/London, pp. 216–225 (1999).

33. J. T. A. Pollock, Filamentary sapphire – the growth of void-free sapphire filament at rates up to 3.0 cm/min, J. Mater. Sci., 7, 786–792 (1972).

34. J. E. Ritter and J. D. Helfinstine, A tougher fiber for the FOG-M, Photonics Spectra, 8, 90–93 (1967).

35. F. T. Wallenberger and S. D. Brown, High modulus glass fibers for new transportation and infrastructure composites and for new infrared uses, Compos. Sci. Technol., 51, 243–263 (1994).

36. F. T. Wallenberger, Continuous melt spinning processes, in Advanced inorganic fibers: processes, structures, properties, applications, Chapter 6, Kluwer Academic Publishers, Dordrecht/Boston/London, pp. 129–168 (1999).

37. E. M. Rabinovitch, Sol-gel processing: general principles, in Sol-gel optics, processing and applications, L. C. Klein, ed., Kluwer Academic Publishers, Boston, pp. 1–37 (1994).

38. R. Brückner, Silicon Dioxide, in Encyclopedia of Applied Physics, VCH Publishers, Inc., 18, 102–131 (1997).

39. Asahi, Product Bulletin (1986).

40. K. Matsuzaki, D. Arai, N. Taneda, T. Mukaiyama and M. Ikemura, Continuous silica glass fiber produced by sol-gel process, J. Non-Cryst. Solids, 112, 437–441 (1989).

41. K. Kamiya and T. Yoko, Synthesis of SiO_2 glass fibres from Si $(OC_2H_5)_4$ – H_2O–C_2H_5OH–HCl solutions through sol-gel method, J. Mater. Sci., 21, 842–848 (1986).

42. K. Kamiya, R. Uemura, J. Matsuoka and H. Nasu, Effect of preheat treatment on the tensile strength of sol-gel derived SiO_2 glass fibers, J. Ceram. Soc. Jpn, 103 (3), 245 (1995).

43. W. Zhou, Y. Xu, L. Zhang, X. Sun, J. Ma and S. She, Crystallization of silica fibers made from metal alkoxide, Mater. Lett., 11 (10–12), 352–354 (1991).

44. A. Wegerhoff and H. D. Achtsnit, High temperature resistant fibrous silicon dioxide material, US Patent, 4,786,017, November 22, 1988.

45. G. H. Vitzhum, H. U. Herwig, A. Wegerhoff and H. D. Achtsnit, Silica fiber for high temperature applications, Chemiefasern/Textilindustrie, 36/88, E-126-127 (1986).

46. H. D. Achtsnit, Textile silica sliver, its manufacture and use, US Patent 5,567,516, October 23, 1996.
47. Product Bulletin, Silfa silica yarns, Ametek/Haveg, Wilmington, DE (1996).
48. G. Wiedermann and N. Frenzel, Untersuchungen zur chemischen Beständigkeit der Glasseide, Faserforschung und Textiltechnik, 24 (9), 335–340 (1973).
49. P. F. Aubourg and W. W. Wolf, Glass fibers, in *Advances in ceramics, Vol. 18, commercial glasses*, D. C. Boyd and J. F. MacDowell, eds., American Ceramic Society, Westerville, pp. 51–63 (1986).
50. P. K. Gupta, Glass fibers for composite materials, in *Fibre reinforcements for composites materials*, Chapter 2, A. R. Bunsell, ed., Composite materials series 2, Elsevier, Amsterdam, pp. 19–71 (1988).
51. B. A. Proctor, Continuous filament glass fibers, in *Concise encyclopedia of composite materials*, A. Kelly, ed., Pergamon Press, Oxford/New York/Beijing/Frankfurt, pp. 62–69 (1989).
52. K. L. Loewenstein, *The manufacturing technology of continuous glass fibres*, Edition 3, Elsevier, Amsterdam (1993).
53. F. T. Wallenberger, H. Li and J. Watson, Glass fibers, in *ASM Handbook, Vol. 21, composites*, ASM International, Metals Park, pp. 27–35 (2001).
54. J. R.Gonterman and M. A. Weinstein, High Intensity Plasma Glass Melter Project, Final DOE Report, October 27, 2006. The report is available at www1.eere.energy.gov/industry/glass/pdfs/894643_plasmelt.pdf
55. S. Rehkson, J. Leonard and P. Sanger, Continuous glass fiber drawing, Am. Ceram. Soc. Bull., 6, 9401–9407 (2004).
56. F. Rossi and G. Williams, A new era in glass fiber composites, 28th AVK Conference, Baden-Baden, Germany, 1–10, October 1–2, 1997.
57. T. D. Erickson and W. W. Wolf, Glass composition, fibers, and methods for making same, US Patent 4,026,715, May 31, 1977.
58. F. T. Wallenberger, Design factors affecting the fabrication of fiber reinforced infrastructure composites, Annual Wilson Forum, Santa Ana, CA, March 20–21, 1995; in Applications of Composite Materials in the Infrastructure, 1–10 (1995).
59. W. L. Eastes, D. A. Hofmann and J. W. Wingert, Boron-free glass fibers, US Patent 5,789,325, August 4, 1998.
60. M. J. Cusick, M. A. Weinstein and L. Olds, Method for the melting, combustion or incineration of materials and apparatus thereof, US Patent No. 5,548,611, August 29, 1996.
61. J. K Williams, C. P. Heanley and L. E. Olds, Method of melting materials and apparatus thereof, US Patent 5,028,248, July 2, 1991.
62. D. A. Dalton, Plasma and electrical systems in glass manufacturing, IEE Colloquium [Digest], 229, p3/1-2 (1994).
63. C. Ross, P. Tincher and G. Tincher, *Glass melting technology: A technical and economic assessment*, GMIC Publication, October, 2004.
64. ASTM Specification D 578-00, Standard for E-glass fiber strands and stating the composition limits for E-glass, Annual Book of Standards, American Society for Testing and Materials, Conshohocken, PA, March 10, 2000.
65. Naamlooze Vennootschap Maatschapij tot Beheer en Exploitatie van Octoorien, New and improved glass compositions for the production of glass fibers, Brit. Patent, GB 520,247, April 18, 1939.
66. R. A. Schoenlaub, Glass compositions, US Patent 2,334,961, November 23, 1943.
67. R. L. Tiede and F. V. Tooley, Glass composition, US Patent 2,334,961, November 2, 1951.
68. J. F. Sproull, Fiber glass composition, US Patent 4,542,106, September 17, 1985.
69. R. L. Jones, The role of boron in the corrosion of E-glass fibres, Glass Technol. Eur. J. Glass Sci. Technol. A, 47 (6), 167–171 (2006).
70. T. F. Starr, *Glass-fibre databook*, Edition 1, Chapman & Hall, London/Glasgow/New York/Tokyo/Melbourne/Madras (1993).

71. T. F. Starr, *Carbon and high performance fibers, directory and databook*, Edition 6, Chapman & Hall, London/Glasgow/New York/Tokyo/Melbourne/Madras (1995).
72. Product Bulletin, Comperative technical characteristics of filament made from E-glass, basalt and silica, Sudaglass Fiber Technology, Inc., 14714 Perthshire, Suite A, Houston, TX, 77079 USA (2003).
73. S. Tamura, M. Mori and S. Saito, Compositions for the production of high-strength glass fiber, Japanese Patent, 8[1996]-231-240, September 10, 1996.
74. French Patent 1,435,739 to St. Gobain Company, Chambrey, France (1963).
75. P. B. McGinnes, High temperature glass fibers, International Patent Application WO 02/42233 A2, May 30, 2002.
76. P. B. McGinnis, High temperature glass fibers, International Publication WO 02/42233 A2 under the Patent Cooperation Treaty (PCT) on May 30, 2003.
77. Product Bulletin, High strength glass fibers, Advanced Glass Fiber Yarns, LLC, 2558 Wagener Road, Aiken, SC 29801 (2003).
78. J. F. Bacon, High modulus, high temperature glass fibers, Appl. Polym. Symp., 21, 179–200 (1973).
79. P. K. Gupta, Strength of glass fibers, in *Fiber fracture*, M. Elices and J. Llorca, eds., Elsevier, Amsterdam, pp. 127–153 (2002).
80. F. T. Wallenberger, Melt viscosity and modulus of bulk glasses and fibers – challenges for the next decade, in Present State and Future Prospects of Glass Science and Technology, Kreidl Symposium, Triesenberg, Liechtenstein, July 3–8, 1994, Glasstech. Ber. Glass Sci. Technol. 70 C, 63–78 (1997).
81. F. T. Wallenberger, Introduction to reinforcing fibers, in *ASM handbook, Vol. 21, Composites*, ASM International, Metals Park, pp. 23–26 (2001).
82. H. Kaplan-Diedrich and G. H. Frischat, Properties of some oxynitride fibers, J. Non-Cryst. Solids, 184, 133–136 (1995).
83. M. Oota, T. Kanamori, S. Kitamura, H. Fujii, T. Kawasaki, K. Sekine and C. Manabe, Decrease of silicon defects in oxynitride glass, J. Non-Cryst. Solids, 209, 69–75 (1997).
84. A. Carre, F. Roger and C. Variot, Study of acid/base properties of oxide, oxide glass, and glass-ceramic surfaces, J. Colloid Interface Sci., 154 (1), 31–40 (1992).
85. P. C. Almenera and P. Thornburrow, A new glass fiber reinforcement for anti-corrosion composites, Advanced Polymer Composites for Structural Applications in Construction, Proceedings 69IFS8 (2004).
86. P. Simurka, M. Liska, A. Plsko and K. Forkel, Development of a composition suitable for the production of alkali-resistant glass fibres with a low fiberising temperature, Glass Technol., 33 (4), 130–135 (1992).
87. V. I. Kostikov, M. F. Makhova, V. P. Sergeev and V. I. Trefilov, Ceramic fibres, in *Fibre science and technology*, V. I. Kostikov, ed., Chapman and Hall, London, pp. 581–606 (1995).
88. K. Suganuma, H. Minakuchi, K. Kada, H. Osafune and H. Fujii, Properties and microstructure of continuous oxynitride glass fiber and its application to aluminum matrix composite, J. Mater. Res., 8 (1), 178–186 (1993).
89. S. Loud, Composites News, Solana Beach, CA, Infrastructure Newsletter Number 11, page 3, September 1994, and Number 28, page 5, June 30, 1995.
90. A. J. Majumdar, Alkali-resistant glass fibres, in *Strong fibers*, W. Watt and B. V. Perov, eds., Elsevier Publishers, North-Holland, pp. 61–85 (1985).
91. Chinese C-glass, Government Specification (1997).
92. D. A. Steenhammer and J. L. Sullivan, Recycled content of polymer matrix composites through the use of A-glass fibers, Polym. Comp., 18 (3), 300–312 (1997).
93. Nitto Boseki, Glass fibers having low dielectric loss tangent – composed of silica, alumina, boria, calcia, opt. magnesia etc., Japanese Patent 9002839, January 7, 1997.
94. H. Li, Low dielectric glass and fiber glass for electronic applications, US Patent Application 2008/0146430 A1, June 19, 2008.

95. D. S. Boessneck et al., Low Dielectric Glass Fiber, US Patent Application 2008/0103036 A1, May 1 2008.
96. G. Demidov, Hollow fibres make light and strong reinforcements, Reinf. Plas., 9, 19 (1995).
97. J. F. Bacon, Composition of glasses with high modulus of elasticity, US Patent 3,573,078, March 30, 1971.
98. K. Komori, S. Yamakawa, S. Yamamoto, J. Naka and T. Kokubo, Substrate for circuit board including the glass fibers as reinforcing material, US Patent 5,334,645, August 2, 1994.
99. K. Komori, S. Yamakawa, S. Yamamoto, J. Naka and T. Kokubo, Glass fiber forming composition, glass fibers obtained from the composition and substrate for circuit board including the glass fibers as reinforcing material, US Patent 5,407,872, April 18, 1995.
100. L. L. Hench, Bioactive glasses and glass ceramics, in *Handbook of bioactive ceramics, Vol. I*, T. Yamamuro, L. L. Hench and J. Wilson, eds., CRC Press, Boca Raton, pp. 7–23 (1990).
101. H. Tagai et al., Preparation of apatite glass fiber for applications as biomaterials, in *Ceramics in surgery*, P. Vincenzini, ed., Elsevier Sci. Pub. Co., Amsterdam, p. 387 (1983).
102. U. Pazzaglia et al., Study of the osteoconductive properties of bioactive glass fibers, J. Biomed. Mater. Res., 23, 1289–1297 (1989).
103. M. S. Marcolongo, P. Ducheyne, F. Ko and W. La Course, Composite materials using bone bioactive glass and ceramic fibers, US Patent 5,721,049, February 24, 1998.
104. L. J. Huey, Method and apparatus for making tapered mineral and organic fibers, US Patent 4,666,485, May 19, 1987.
105. K. Shioura, S. Yamazaki and H. Shono, Method for producing glass fibers having non-circular cross sections, US Patent 4,698,083, October 6, 1987.
106. H. Taguchi, K. Shioura and M. Sugeno, Nozzle tip for spinning glass fiber having deformed cross section and a plurality of projections, US Patent 5,462,571, October 31, 1995.
107. T. H. Jensen, Hollow glass fiber bushing, method of making hollow fibers and the hollow glass fibers made by that method, US Patent 4,758,259, July 19, 1988.
108. L. J. Huey, Method and apparatus for producing hollow glass filaments, US Patent 4,846,864, July 11, 1989.
109. J. Huang, Hollow high temperature ceramic superconducting fibers, International Patent Application WO 97/22128, June 19, 1997.
110. T. Yazawa, H. Tanaka and K. Eguchi, Preparation of porous hollow fibre from glass based on SiO_2–B_2O_3–RO–ZrO_2 (R = Ca, Zn) system, J. Mater. Sci. Lett., 13, 494–495 (1994).
111. R. P. Beaver, Method for producing porous hollow silica rich fibers, US Patent 4,778,499, October 18, 1988.
112. J. E. Loftus, C. R. Strauss and R. L. Houston, Method for making dual-glass fibers by causing one glass to flow around another as they are spun from a rotating spinner, US Patent 5,529,596, June 25, 1996.
113. N. T. Huff, Innovative technology can create products, Glass Res., 5 (1), 1–9 (1995).
114. M. C. Kenny, S. K. Barlow and S. L. Eikleberry, New glass-fiber geometry – a study of non-woven processability, TAPPI J., 30, 169–177 (1997).
115. F. T. Wallenberger, N. E. Weston and S. D. Brown, Infrared optical tellurite glass fibers, J. Non-Cryst. Solids, 144 (1), 107–110 (1992).
116. M. L. Nice, Apparatus and process for fiberizing fluoride glasses using a double crucible and the compositions produced thereby, US Patent 4,897,100, January 20, 1990.
117. H. Tokiwa, Y. Mimura, T. Nakai and O. Shinbori, Fabrication of long single-mode and multi-mode fluoride glass fibers by the double crucible technique, Electron. Lett., 21 (24), 1130–1131 (1985).
118. F. T. Wallenberger, S. D. Brown and G. Y. Onoda, ZnO-modified high modulus glass fibers, J. Non-Cryst. Solids, 152, 279–283 (1993).
119. H. Lin, W. L. Dechent, D. E. Day and J. O. Stoffer, Preparation and properties of mid-infrared glass fibers and poly(chlorotrifluoroethylene) composites, J. Mater. Sci., 32, 6573–6578 (1997).

120. S. D. Brown and G. Y. Onoda, Jr., High modulus glasses based on ceramic oxides, Report R-6692, Contract NOw-65-0426-d, US Department of the Navy, October 1966.
121. G. Y. Onoda, Jr. and S. D. Brown, High modulus glasses based on ceramic oxides, Report R-7363, Contract N00019-67-C-301, US Department of the Navy, February 1968.
122. T. F. Schroeder, H. W. Carpenter and S. C. Carniglia, High modulus glasses based on ceramic oxides, Technical Report R-8079, Contract N00019-69-C-0150, US Navy Dept., Naval Air Systems Command, Washington, DC, December 1969.
123. J. R. Davy, Development of calcia-alumina glasses for use in the infrared spectrum, US Patent No. 3,338,694 (1967), Glass Technol., 19 (2), 32–36 (1978).
124. R. Maddison, Calcia-aluminas, Product Bulletins WB37A and WB39B, Sassoon Advanced Materials LTD, Dumbarton, UK (1994).
125. P. R. Foy, T. Stockert, J. Bonja, G. H. Sigel, Jr., R. McCauley, E. Snitzer and G. Merberg, Meeting Abstracts, American Ceramic Society, 94th Annual Meeting, Presentation 7-JXV-92, Minneapolis, MN, April 12–16, 1992.
126. F. T. Wallenberger, N. E. Weston and S. A. Dunn, Melt spun calcia-alumina fibers: infrared transmission, J. Non-Cryst. Solids, 12 (1), 116–119 (1990).
127. F. T. Wallenberger, N. E. Weston and S. D. Brown, Melt processed calcia-alumina fibers: optical and structural properties, in *Growth of materials for infrared detectors*, R. E. Longshore and J. Baars, eds., Proceedings of the SPIE, Society of Photo-Optical Instrumentation Engineers, Bellington, Vol. 1484, 116–124 (1991).
128. J. Nishii, I. Inagawa, T. Yamagishi, S. Morimoto and R. Iizuka, Process for producing chalcogenide glass fiber, US Patent, 4,908,053, March 13, 1990.
129. F. T. Wallenberger, New melt spun glass and glass-ceramic fibers for polymer and metal matrix composites, in *High performance composites: Commonalty of phenomena*, K. K. Chawla, P. K. Law and S. G. Fishman, eds., The Minerals, Metals and Materials Soc., Warrendale, PA, pp. 85–92 (1994).
130. J. M. Massoubre and B. F. Pflieger, Small diameter wire making through solidification of silicon steel jet, in *Spinning wire from molten metal*, J. Mottern and W. J. Privott, eds.; AIChE Symp. Ser., (180), Vol. 74, pp. 48–57 (1978).
131. F. Fodeur and B. S. Mitchell, Infrared studies of calcia-alumina fibers, J. Am. Ceram. Soc., 79 (9), 2469–2473 (1996).
132. M. Mooney, The viscosity of a concentrated suspension of spherical particles, J. Colloid Sci., 6 (2), 162–170 (1951).
133. R. J. Diefendorf and E. R. Stover, Pyrolytic graphites: how structure affects properties, Metal Prog., 81 (5), 103–108 (1962).
134. A. L. Greer, Metallic glasses, Science, 267, 1947 (1995).
135. D. E. Polk and B. C. Giessen, Amorphous or glassy materials, in *Rapid solidification technology source book*, R. L. Ashbrook, ed., American Society for Metals, Metals Park, pp. 213–247 (1983).
136. A. Revcoleschi and J. Jegoudez, Growth of large high-T_c single crystals by the floating-zone method: a review, Prog. Mater. Sci., 42, 321–339 (1997).
137. P. H. Keck and M. J. E. Golay, Traveling solvent zone melting, Phys. Rev., 39, 1297 (1953).
138. T. Mah, T. A. Parthasarathy, M. D. Petry and L. E. Matson, Processing, micro-structure, and properties of Al_2O_3–$Y_3Al_5O_{12}$ (YAG) eutectic fibers, Ceramic Engineering and Science Proceedings, 622638, 17th Ann. Conference on Composites and Advanced Ceramic Materials, Am. Ceram. Soc., Westerville OH (1993).
139. W. M. Yen, Preparation of single-crystal fibers, in *Insulating materials for opto-electronics*, F. Agulló-Lopez, ed., World Sci., Singapore (1995).
140. M. Matsukura, Z. Chen, M. Adachi and A. Kawabata, Growth of potassium lithium niobate single-crystal fibers by the laser-heated pedestal growth method, Jpn. J. Appl. Phys. 1, 36 (9B), 5947–5949 (1997).
141. J. I. Peña, H. Miao, R. I. Merino, G. F. de la Fuente and V. M. Orera, Polymer matrix synthesis of zirconia eutectics for directional solidification into single-crystal fibers, Solid State Ionics, 101–103, 143–147 (1997).

142. W. Jia, H. Yuan, L. Lu, H. Liu and W. M. Yen, Phosphorescent dynamics in $SrAl_2O_4:Eu^{2+},Dy^{3+}$ single-crystal fibers, J. Lumin., 76–77, 424 (1998).

143. R. S. Feigelson, D. Gazit, D. K. Fork and T. H. Geballe, Superconducting Si–Ca–Sr–Cu–O fibers grown by the laser-heated pedestal growth method, Science, 240, 1642–1645 (1988).

144. F. M. Costa, R. F. Silva and J. M. Vieira, Influence on epitaxial growth of superconducting properties of LFZ Bi–Sr–Ca–Cu–O fibres, Part I. Physica C, 289, 161–170 (1997) and Part II., Physica C, 289, 171–176 (1997).

145. H. Miao, J. C. Dietz, L. A. Angurel, J. I. Peña and G. F. de la Fuente, Phase formation and micro-structure of laser floating-zone grown BSCCO fibers: reactivity aspects, Solid State Ionics, 101–103, 1025–1032 (1997).

146. U. Balchandran, A. N. Iyer, P. Haldar and L. R. Motowidlo, The powder-in-tube processing and properties of Bi-223, J. Metals, 45 (9), 54–67 (1993).

147. G. Geiger, New record for super-conducting wire, Am. Ceram. Soc. Bull., 74 (12), 19 (1995).

148. K. J. McClellan, H. Sayir, A. H. Heuer, A. Sayir, J, S. Haggerty and J. Sigalovsky, High strength, creep resistant Y_2O_3– stabilized cubic ZrO_2 single-crystal fibers, Ceramic Engineering and Science Proceedings, 651–659, 17th Ann. Conf. on Composites and Advanced Ceramic Materials, Am. Ceram. Soc., Westerville, OH (1993).

149. E. L. Lawton, F. T. Wallenberger and H. Li, Recent advances in oxide glass fiber science – low dielectric constant fibers, in *Advanced fibers, plastics and composites*, F. T. Wallenberger and N. E. Weston et al., eds., Materials Research Society, Symposium Proceedings, Vol. 702, MRS, Warrendale, 165–172 (2002).

150. M. H. Gallo, J. van Genechten, J. P. Bazin, S. Creux and P. Fournier, Glass fibers for reinforcing organic and/or inorganic materials, French Patent Application, 2,768,144 A1, September 10, 1997.

151. F. T. Wallenberger, N. E. Weston and S. A. Dunn, Melt spun calcium IR aluminate fibers: Product value, conference proceedings, Second International Conference on Electronic Materials, R. P. H. Chang, T. Sugano and V. T. Nguyen, eds., Material Research Society, pages 295–300, 1990.

152. H. D. Achtsnit, Textile silica sliver, its manufacture and use, US Patent 5,567,516, October 23, 1996.

153. J. Kobayashi, M. Oota, K. Kada and H. Minakuchi, US Patent 4,957,883, September 18, 1990.

Chapter 2
Design of Energy-Friendly Glass Fibers

Frederick T. Wallenberger

Abstract Incumbent fiberglass compositions rely on decades of commercial experience. From a compositional point of view, many of these melts require more energy than needed in their production, or emit toxic effluents into the environment. This chapter reviews the design of energy- and/or environmentally friendly E-glass, HT-glass, ECR-glass, A-glass, and C-glass compositions, which have lower viscosities or fiber-forming temperatures and therefore require less energy in a commercial furnace than the respective incumbent compositions and/or do not contain ingredients which are of environmental concern.

Keywords Energy- and environmentally friendly compositions · Compositional design principles · Compositional reformulation for reduced energy use and cost · Trend line design · Design of new energy-friendly E-glass · ECR-glass · A-glass and C-glass variants

2.1 Principles of Designing New Compositions

This chapter reviews the principles of designing environmentally and/or energy-friendly fiberglass compositions for commercial use and it discusses compositional, energy, and environmental issues which must be considered in the design of new energy- and/or environmentally friendly compositions.

2.1.1 Compositional, Energy, and Environmental Issues

Energy-friendly fiberglass compositions yield crystallization-resistant melts in a commercial furnace with low fiber-forming viscosities. Environmentally friendly compositions yield crystallization-resistant melts, which will not emit particulate or toxic matter into the environment. Design of new fiberglass compositions is aimed

F.T. Wallenberger (✉)
Consultant, 708 Duncan Avenue, Apartment 1108, Pittsburgh, PA 15237, USA
e-mail: wallenbergerf@aol.com

F.T. Wallenberger, P.A. Bingham (eds.), *Fiberglass and Glass Technology*,
DOI 10.1007/978-1-4419-0736-3_2, © Springer Science+Business Media, LLC 2010

at defining new energy- and/or environmentally friendly compositions which yield crystallization-resistant melt with lower fiber-forming temperatures than incumbent compositions. Compositional design must therefore reflect environmental regulations, industry standards, batch cost, and energy cost [1–12].

2.1.1.1 Environmental Regulations and Emission Control

Environmental regulations, which became effective in 1997, govern all commercial fiberglass furnaces [13]. With incumbent emission control systems, borosilicate E-glass melts, which contain ~6.6% B_2O_3 and <1% fluorine, may emit as much as 15% of their total B_2O_3 content as particulate salts into the environment plus much of their fluorine as toxic fluorides [13].

In order to effectively scrub boron and fluorine emissions, these regulations require a costly additional emission control system for furnaces, which produce borosilicate E-glass with 5–10% B_2O_3. Many incumbent borosilicate E-glass compositions were therefore replaced by fluorine- and boron-free E-glass compositions, which were designed between 1985 [14] and 1997 [15] and do not require an additional environmental control system.

The new boron-free or essentially boron-free E-glass compositions [7, 14, 15] are considered to be environmentally friendly. They are less costly, but they are also less energy friendly than incumbent borosilicate E-glass with ~6% B_2O_3 since their fiber (log_3)-forming temperature is 55–60°C higher. The melt viscosity of boron-free [2, 3, 15] E-glass has already been reduced, as will be shown in this chapter, and the viscosity of incumbent E-glass melts (~6% B_2O_3) can still be reduced further by compositional reformulation [7, 12]. For energy- *and* environmentally friendly E-glass compositions [2, 3, 7], which have about the same fiber-forming temperature as borosilicate E-glass see Section 2.2.3.

2.1.1.2 Industry Standards and Specifications

In addition to governmental regulations, there are other restrictions, which need to be considered in the design of new fiberglass compositions. They include industry standards for the compositions of glass fibers, which can be sold as E-glass, product specifications by individual companies for glass fibers, such as S-glass, which are sold to the government under a contract, and product specifications by individual companies for specialty fibers, such as ECR- and D-glass, which are sold to individual direct customers.

Because of the extraordinary large sales volume of E-glass fibers, the most important restriction to compositional design and/or reformulation is the ASTM E-glass standard D-578-00 [16] shown in Table 2.1. No industry standards have been developed for ECR-glass, S-glass, D-glass, A-glass, and C-glass [7, 13]. In order for a fiberglass product to be sold as an E-glass, its chemical composition must meet the

Table 2.1 ASTM E-glass standard D-570-00 [16]

E-glass for printed circuit boards		E-glass for general applications	
Chemical	% by weight	Chemical	% by weight
B_2O_3	5–10	B_2O_3	0–10
CaO	16–25	CaO	16–25
Al_2O_3	12–16	Al_2O_3	12–16
SiO_2	52–56	SiO_2	52–62
MgO	0–5	MgO	0–5
Na_2O and K_2O	0–2	$Na_2O + K_2O + Li_2O$	0–2
TiO_2	0–0.8	TiO_2	0–1.5
Fe_2O_3	0.05–0.4	Fe_2O_3	0.05–0.8
Fluoride	0–1.0	Fluoride	0–1.0

compositional requirements of ASTM standard D-578-00. Specifically, the composition must therefore fall within the percent ranges that limit the use of each specified oxide and for fluorine.

The ASTM E-glass standard (Table 2.1) recognizes two separate E-glass variants, one for printed circuit boards and one for general applications, and there are notable differences between these standards. The standard for printed circuit boards requires the use of 5–10% B_2O_3; 52–56% SiO_2, 0–2% $Na_2O + K_2O$, 0–0.8% TiO_2, and 0.05–0.4% Fe_2O_3. The standard for general applications requires the use of 0–10% B_2O_3, 52–62% SiO_2, 0–2% $Li_2O + Na_2O + K_2O$, 0–1.5% TiO_2, and 0.05–0.8% Fe_2O_3. Both standards are identical with regard to CaO (16–25%), MgO (0–5%), and fluoride (0–1.0%). And they differ with regard to %B_2O_3, %SiO_2, % alkali oxides, and %TiO_2.

The E-glass standard for printed circuit board applications requires compositions with a high but limited range of B_2O_3 (5–10%). Fluorine-free and boron-free E-glass as well as fluorine-free and essentially boron-free E-glass cannot be sold as E-glass for circuit board (PCB) applications. The E-glass standard for general reinforcement applications permits the use of compositions with a broad range of B_2O_3 (0–10%) that includes (a) borosilicate E-glass with, for example, 6.6% B_2O_3, (b) fluorine- and boron-free E-glass [14, 15], and (c) fluorine-free and essentially boron-free E-glass [2, 3].

For general reinforcement applications the addition of 0–2% Li_2O as a flux would be acceptable but not for printed circuit board applications. TiO_2 is also a flux that reduces the fiber-forming temperature. The ASTM standard for printed circuit boards would allow only ≤0.8% TiO_2, but that for general applications would allow nearly twice as much or ≤1.5% TiO_2 [14].

The ASTM E-glass standard governs the production of E-glass variants in the United States. Similar standards govern the production of E-glass in other countries, for example, DIN in Germany and BS in the UK. These standards provide a focus for the design of new E-glass compositions which are more environmentally and/or energy friendly than the incumbent E-glass compositions. For details see Sections 2.2.2 and 2.2.3.

2.1.2 Trend Line Design of New Fiberglass Compositions

The design of new fiberglass compositions, which are more energy- and/or environmentally friendly than the incumbent fiberglass compositions, requires a compositional protocol that is based on recent structure–property insights into function of key oxides and their effect on the melt properties.

2.1.2.1 Glass Databases and Compositional Models

A recent book provides a new glass melt property database but notes that the liquidus temperature has not been addressed [17]. Another recent book offers a chapter on the viscosity of glass melts and a chapter on the liquidus temperature [18]. Neither chapter addresses the commercially important relationship between liquidus temperature and viscosity. Both chapters deal with Na_2O-CaO-SiO_2 or soda-lime silica glasses, but not with fiberglass compositions, which are based on the quaternary SiO_2-Al_2O_3-CaO-MgO-phase diagram.

A recent paper shows the effect of float glass compositions on the liquidus temperature and crystallization behavior [19]. But this paper neither discusses the relationship between viscosity and liquidus temperature, nor reviews fiberglass compositions. Notably, even major differences were observed between calculated and measured liquidus temperatures. In a related study (Chapter 7, Table 7.20), similarly unreliable predictions were also obtained for the melt viscosities of a set of soda-lime-silica A-fiberglass compositions.

Encouraging attempts were, however, recently reported with regard to modeling and reformulating of selected soda–lime–silica (S–L–S) glass compositions for environmental benefit [20]. This pioneering study does deal with the relationship between viscosity and liquidus temperature and has become the basis of a full-length chapter, Chapter 7.

Seven papers have been reported in the recent literature [1–7], which offer a new compositional fiberglass database and address the issue of designing crystallization-resistant melts which yield more environmentally and/or energy-friendly fiberglass compositions than the incumbent melts. The specific design method has become known as the trend line model. One tool for this model is the designed reformulation of a given composition to attain the desired target properties, and the other tool for this model is a carefully designed use of fluxes such as B_2O_3, Li_2O, TiO_2, and ZnO.

2.1.2.2 Principles of Trend Line Design

Trend line design of experiments is a new compositional design tool that relies on a conceptual understanding of the function and contributions of each oxide toward critical melt properties, especially of %SiO_2 and %RO ($= CaO + MgO$) at a constant level of Al_2O_3 [1–7] and of the other oxides. The SiO_2 level of a melt determines the viscosity, fiber-forming temperature, and energy demand. The RO level governs the liquidus temperature at which crystals can form in a glass melt, and the delta (or difference) between the fiber-forming and the liquidus temperatures governs the crystallization resistance of a glass melt in a commercial furnace.

Trend line design therefore aims to reformulate a specific composition that has a given set and ratio of oxides, to achieve the most energy-friendly compositional variant with the same oxides but with different ratios. Thus, the goal is a melt that has a lower viscosity (fiber-forming temperature, energy demand) and higher liquidus temperature, but one that still affords a crystallization resistant in a commercial furnace. The method will be described in detail in conjunction with the design of new environmentally and energy-friendly fiberglass variants in Sections 2.2.1, 2.2.2, 2.2.3 and 2.3.1.

2.1.2.3 Design Required Melt Properties

The most significant melt properties, which are required for the design of new fiberglass compositions, are the log_3 fiber-forming temperature, the liquidus temperature, and the delta between the fiber-forming and the liquidus temperatures. It should be noted that the fiber-forming temperature is derived from the viscosity–temperature behavior of a given melt, while the liquidus temperature is derived from the phase diagram to which the melt belongs.

The SiO_2-Al_2O_3-CaO phase diagram was selected for an overview [1]. It is a simpler and therefore more intuitive system than the quaternary SiO_2-Al_2O_3-MgO-CaO system [2–4] or the SiO_2-Al_2O_3-CaO-B_2O_3 system [5–7], i.e., systems which will also be described in this chapter with regard to specific compositional design efforts.

(a) Fiber-forming and liquidus temperatures. The fiber-forming temperature is the temperature of the melt at the furnace exit. Specifically, it is the temperature of the melt as it is pulled from the bushing to be drawn and collected on a winder. Typically [21], a fiberglass melt has a viscosity of $log_{2.5}$ poise to $log_{3.0}$ poise at the bushing, depending on fiber diameter, wind-up tension, and other factors. The relationship between melt viscosity and melt temperature is shown in Fig. 2.1 for a typical borosilicate composition with 6.6% B_2O_3 and a typical boron-free E-glass composition.

The melt viscosity changes with compositional changes. For example, if %SiO_2 rises slightly in a commercial furnace, the viscosity (and fiber-forming temperature) will also rise slightly, continuously, and predictably. Or, as shown in Fig. 2.1, if B_2O_3 is entirely removed from a borosilicate E-glass melt, the resulting boron-free E-glass melt has a much higher viscosity. Specifically, the increase of the melt viscosity corresponds to a 90°C higher fiber-forming temperature at a melt viscosity of log_3 poise.

By an industry-wide agreement, the temperature (°C) of a melt that has a viscosity of 1000 (log_3) poise has become a common standard for the fiber-forming temperature (or log_3 forming temperature). In practice, the viscosity of a glass melt is determined by ASTM standard C 965-81 [22]. The liquidus temperature is an equilibrium temperature that is derived from the phase diagram. It is the temperature at which solids (crystals) form within a few hours and remain in equilibrium

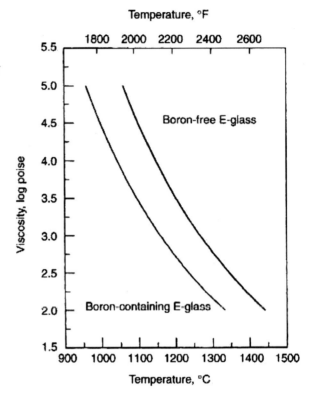

Fig. 2.1 The relationship between melt viscosity and temperature, F. T. Wallenberger et al. [21]. (Reprinted with permission of ASM International. All rights reserved www.asminternational.org)

with the liquid (melt). The liquidus temperature is determined by ASTM standard C 829-81 [23].

Specifically, the liquidus temperature is derived from the phase diagram that represents the phases of a composition as a function of temperature [24]. The ternary SiO_2-Al_2O_3-CaO eutectic is located, as shown in Fig. 2.2, in the narrow area circumscribed by the following oxide ranges: 62.1–62.0% SiO_2, 14.6–14.5% Al_2O_3, and 23.3–23.5% CaO. Figure 2.3 shows the topography of the ternary SiO_2-Al_2O_3-CaO liquidus surface around the eutectic temperature or better around the lowest liquidus temperature [24].

The liquidus temperature at the ternary SiO_2–Al_2O_3-CaO eutectic has been determined to be 1170°C [25]. Recently, and as shown in Table 2.2, the lowest liquidus temperature of nominally the same composition was found to be 1188°C [1]. Considering the rapidly changing topography of liquidus surface (Fig. 2.3), a slight difference in ingredient losses during melting can produce a slight difference in composition and, in turn, a notable difference of the liquidus temperature. Either way, Table 2.2 confirms a rapid drop of the liquidus temperature over a narrow compositional range from 1217°C to a deep valley at 1188°C and a rapid rise again to 1203°C.

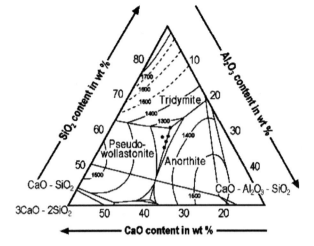

Fig. 2.2 Phase diagram of the ternary SiO_2-Al_2O_3-CaO system, F. T. Wallenberger et al. [1]. (With permission of Elsevier Limited)

Fig. 2.3 Topography of the ternary SiO_2-Al_2O_3-CaO liquidus surface, F. T. Wallenberger et al. [1]. (With permission of Elsevier Limited)

In summary, the $\log_3$ forming temperature (or the melt viscosity) changes gradually from composition to composition (Fig. 2.1), but a slight change of the SiO_2 and CaO may cause a sudden change of the liquidus temperature (crystallization

Table 2.2 Ternary compositions around the ternary SiO_2-Al_2O_3-CaO eutectic

Example	1	2	3	4	5	6	7
				Eutectic			
SiO_2, wt%	62.20	62.10	62.10	62.00	62.00	62.00	61.90
Al_2O_3, wt%	14.60	14.60	14.70	14.50	14.60	14.70	14.80
CaO, wt%	23.20	23.30	23.20	23.50	23.40	23.30	23.30
SiO_2/CaO ratio	2.68	2.67	2.68	2.64	2.65	2.66	2.65
Log_3 forming T, °C	1338	1335	1340	1327	1327	1329	1342
Liquidus T, °C	**1217**	**1217**	**1201**	**1188**	**1191**	**1195**	**1203**
Delta (Δ) T, °C	121	118	139	139	136	134	139
ASTM E-glass	–	–	–	+	+	+	+

potential) of a melt due to a sudden change from one crystal moiety to another, notably near a eutectic (Figs. 2.2 and 2.3 and Table 2.2). New fiberglass compositions cannot effectively be designed without taking this fundamental principle into consideration.

(b) Delta temperature. The delta (or difference) between the fiber-forming and the liquidus temperatures states how much higher or lower the forming temperature is than the liquidus temperature. Importantly, the delta temperature relates a property that is based on the viscosity of a melt to a property that is based on the phase diagram of the melt. This is another principle that must be taken into consideration when new compositions are designed or reformulated.

In a stationary process, pristine fibers can be formed from melts, which have delta temperatures ranging from $\geq 150°C$ to $\leq 150°C$ [1]. In this laboratory process, a single tip bushing is used and a ~ 50 g melt. Fiber formation is complete before crystals can form and freeze up the melt or the bushing tip. In a commercial fiber-forming process the batch ingredients are continuously replenished. Thus, the melt and multifilament yarns are continuously formed.

For a commercial process it is therefore important that the melt has a higher log_3 forming temperature than liquidus temperature in order to avoid crystal formation in dead corners of the furnace or in the bushing tips. Generally, it is believed that a delta temperature of ≥ 50 is required. Because of this requirement, the delta temperature becomes a key parameter for the design of those melts, which have been classified as strong viscous melts (Chapter 1). When, however, a composition is in commercial use that yields a melt with a delta temperature of $\geq 100°C$, it has by definition a log_3 fiber-forming temperature that can be reduced. In other words, its viscosity can be reduced and it can be made more energy friendly by compositional reformulation.

2.1.2.4 Compositions, Energy Use, and Emissions

Two trend line design tools are available, compositional reformulation by trend line design and the addition or removal of a flux such as B_2O_3, Li_2O, TiO_2, or ZnO by trend line design.

(a) Compositional reformulation. A reduction of %SiO_2 and a corresponding increase of %RO (= CaO + MgO) while keeping Al_2O_3 and the reminder of the composition constant, reduces the $\log_3$ forming temperature, but also increases the liquidus temperature or crystallization potential [1–12]. In summary, increasingly energy-friendly compositions can be obtained by stepwise reducing %SiO_2 and by correspondingly increasing %RO of an incumbent composition by designed compositional reformulation.

As few as four melts, but often no more than ten designed experiments, are required to establish the trend lines for the forming and liquidus temperatures. This discovery led to new energy-friendly fiberglass variants not only in the ternary SiO_2-Al_2O_3-CaO phase diagram but also in the quaternary SiO_2-Al_2O_3-CaO-MgO and SiO_2-Al_2O_3-CaO-B_2O_3 phase diagrams [1–12]. Further results will be described in Sections 2.2.2, 2.2.3 , and 2.3.1.

(b) Addition or removal of a flux. The viscosity and energy demand of a fiberglass (or glass) melt can also be reduced by adding a flux, i.e., an oxide that reduces the melt viscosity ($\log_3$ forming temperature). Since the Latin word "flux" means "flow," a flux is therefore an ingredient that increases the flow of a melt at the same temperature. Thus, a carefully designed addition of B_2O_3, Li_2O, TiO_2, and ZnO (or even of a fluoride) to a trend line-designed composition will further reduce the viscosity and $\log_3$ forming temperature. Conversely, partial or complete removal of a flux, if present, will increase the melt viscosity.

B_2O_3 and fluorine are inexpensive fluxes. They are used in commercial borosilicate E-glass compositions for printed circuit board applications (Table 2.1). They are energy friendly but they are not environmentally friendly. Variants with 5–10% B_2O_3 and 1% fluorine have a low forming temperature but require a costly environmental control system to prevent boron and fluorine emissions from melts to reach the environment.

In contrast, Li_2O is one of the most effective fluxes. A relatively small addition of Li_2O yields a relatively large reduction of the viscosity and $\log_3$ forming temperature [2, 3, 7]. No additional environmental control system is required. However, Li_2O is costly. Its rapidly rising ingredient cost is, among others, caused by the rapidly rising demand for lithium ion batteries.

2.2 Energy-Friendly Aluminosilicate Glass Fibers

This chapter deals with the trend line design of (a) boron-free and essentially boron-free E-glass compositions, (b) high-temperature or HT-glass compositions, and (c) corrosion-resistant or ECR-glass compositions which are more energy friendly than the respective first-generation compositions.

2.2.1 New Energy-Friendly E-Glass Variants with < 2% B_2O_3

Fiberglass compositions, which meet one, the other, or both ASTM E-glass standards, can essentially be derived from the ternary SiO_2-Al_2O_3-CaO phase diagram, the quaternary SiO_2-Al_2O_3-CaO-MgO phase diagram, and the quaternary SiO_2-Al_2O_3-CaO-B_2O_3 phase diagram – but not from the soda–lime–silica (S–L–S) phase diagram. E-glass for general uses may also contain L_2O according to ASTM [16]. When made from commercial ingredients all compositions may also contain Na_2O, K_2O, Fe_2O_3, and TiO_2.

2.2.1.1 Ternary SiO_2-Al_2O_3-CaO Phase Diagram

A reduction of %SiO_2 and a corresponding increase of %RO (=CaO) at a constant level of Al_2O_3 reduces the log_3 forming temperature but also increases the liquidus temperature or crystallization potential [1–12].

For example, the SiO_2 level of the ternary eutectic composition was reduced in eight steps [1] from 62.10 to 55.0%, the CaO (RO) level was increased from 23.20 to 28.75%, the Al_2O_3 level was kept about constant, and the SiO_2/RO ratio was thereby reduced from 2.68 to 1.91. The resulting trend lines for the forming and liquidus temperatures are found in reference [1]. Four of the eight compositions are shown in Table 2.3.

By reducing the SiO_2 level, the log_3 forming temperature decreased from 1340 to 1246°C, and by increasing the CaO level, the liquidus temperature increased from 1188 to 1246°C, while the delta temperature decreased from +139 to –23°C. The actual forming and liquidus temperature trend lines [1] cross at 1256°C. At this point, which can also be interpolated from Table 2.3, the log_3 forming temperature is the same as the liquidus temperature.

Thus, if a delta temperature of 139°C were desired for a melt in a continuous commercial process, it could be obtained with the composition that contains 62% SiO_2 and has a log_3 forming temperature of 1327°C (Table 2.3). If a delta temperature of 59°C were judged to be satisfactory for a commercial process, it could be obtained with a far more energy-friendly composition that contains only 60% SiO_2 and therefore has a 33°C lower log_3 forming temperature. In a stationary single tip bushing process, glass fibers can be obtained from an experimental melt that has a delta temperature below 55°C (i.e., +25, 0, or –23°C) but not from a continuous commercial melt.

Table 2.3 Trend line design of ternary compositions

Example	1	2	3	4
SiO_2, wt%	62.00	60.00	57.50	55.00
Al_2O_3, wt%	14.50	15.00	15.00	16.25
CaO, wt%	23.50	25.00	27.50	28.75
SiO_2/CaO ratio	**2.64**	**2.40**	**2.09**	**1.91**
Log_3 forming T, °C	1327	1294	1266	1246
Liquidus T, °C	1188	1235	1241	1269
Delta (Δ) T, °C	**139**	**59**	**25**	**–23**
ASTM E-glass	+	+	–	–

A different series of ternary SiO_2-Al_2O_3-CaO glass fibers, some of which qualify as ASTM E-glass compositions, has been reported for use in high-temperature applications such as sound absorbing materials in engine exhaust mufflers [26]. For these applications a very high glass transition temperature (>850°C) was desired, i.e., one that should be at least 20°C higher than that of a typical S-glass composition. The new compositions were designed [26] in a two-phase trend line design study as shown in Table 2.4.

Table 2.4 Design of high-temperature glass fibers for mufflers to absorb engine sound. P. B. McGinnis, High Temperature Glass Fibers, International Patent Application, PCT, WO 02/42233 A-2/A-3, May 30, 2002, to Owens Corning

Example [26]	1-1	1-2	1-8	1-3 and 2-1	2-10	2-9	*Est.*	2-55
SiO_2, wt%	62.00	65.00	66.50	68.00	66.50	65.00	*62.00*	59.00
Al_2O_3, wt%	15.00	16.87	17.81	18.75	20.25	21.75	*24.75*	27.75
CaO, wt%	23.00	18.13	15.69	13.25	13.25	13.25	*13.25*	13.25
SiO_2/RO ratio	2.69	3.59	4.23	5.13	5.02	4.91	*4.68*	4.45
Log_3 FT, °C	1332	1419	1494	1533	1518	1501	*1479*	1458
LT, °C	1222	1314	1355	1386	1411	1431	*1440*	1449
ΔT, °C	110	105	139	147	107	70	*39*	9
Tg, °C	794	813	837	**859**	**861**	**859**	*858*	**858**
ASTM E-glass	+	–	–	–	–	–	+	–

In the first phase of the trend line study, the starting composition was an ASTM E-glass that consisted of 62% SiO_2 and 23% CaO. The %SiO_2 level of this composition was stepwise increased and %CaO was stepwise decreased at about a constant level of Al_2O_3 (Table 2.4, Examples 1-1, 1-2, 1-8, and 1-3) until the interim target composition was achieved (Examples 1-3 and 2-1) that consisted of 68% SiO_2 and 13.25% CaO. The SiO_2/CaO ratio increased from 2.69 to 4.23, and the glass transition temperature increased from 794°C to a target level of 859°C.

However, the log_3 forming temperature rose so dramatically that it had to be reduced in the second phase of the trend line study (Table 2.4, Examples 2-1, 2-10. 2-9, and 2-55) by stepwise reducing %SiO_2 and by stepwise increasing %Al_2O_3 at about a constant level of CaO. In this design phase, the glass transition temperature remained constant and on target (858–861°C), the log_3 forming temperature dropped to 1458°C, the liquidus temperature rose to 1449°C, and the delta temperature dropped to 9°C.

Although not mentioned in reference [26], a potential target composition can be interpolated from the trends which can be reconstructed from Table 2.4. It would have a log_3 forming temperature of 1479°C, a liquidus temperature of 1440°C, and a delta temperature of 39°C. This ternary SiO_2-Al_2O_3-CaO composition would be more energy friendly and might have a higher value-in-use than a comparable ternary composition (Example 1-8 in Table 2.4) that was derived in the first phase of the trend line design study. It would have a 15°C lower log_3 forming temperature and would require less process energy. It would also have a 21°C higher glass transition temperature and might be a better sound absorbing material in engine exhaust mufflers.

Finally a designed addition of BaO to a ternary SiO_2-Al_2O_3-CaO composition was found to raise the glass transition temperature of these high-temperature or HT-glass fibers from 861 to 875°C [26].

2.2.1.2 Quaternary SiO_2-Al_2O_3-CaO-MgO Phase Diagram

The log_3 forming temperature of the quaternary fluorine- and boron-free eutectic composition in the ternary SiO_2-Al_2O_3-CaO phase diagram was found to be 1327°C (Table 2.5), while eutectic in the quaternary SiO_2-Al_2O_3-CaO-MgO phase diagram was found to be 1299°C or 28°C lower [1]. This means that at least a small amount of MgO reduces the viscosity or log_3 forming temperature of the two original eutectic compositions.

(a) From eutectic to commercial compositions. The SiO_2 level of the ternary eutectic composition was reduced from 62 to 60%, the CaO level was raised from 23.5 to 25% at about a constant level of Al_2O_3 (Table 2.5, Example 1), and the SiO_2/RO was reduced from 2.64 to 2.40. As a result, the log_3 forming temperature was reduced by 33°C to 1294°C, and a crystallization-resistant melt was achieved with a delta temperature of 59°C. But even these forming temperatures were too high for a high volume production of the resulting E-glass composition. Two commercial first-generation fluorine-free as well as boron-free E-glass compositions were introduced in 1985 and 1997, respectively. Conceptually, they may be considered to be derivatives of the eutectic composition in the quaternary SiO_2-Al_2O_3-CaO-MgO phase diagram (Table 2.5).

Table 2.5 First-generation fluorine- and B_2O_3-free E-glass (1997)

Phase diagram	Ternary		Quaternary		
Examples	*Eutectic*	1	*Eutectic*	2	3
SiO_2, wt%	*62.00*	60.00	*60.00*	60.10	59.30
Al_2O_3, wt%	*14.50*	15.00	*15.00*	12.99	12.10
CaO, wt%	*23.50*	25.00	*20.00*	22.13	22.60
MgO, wt%	–	–	*5.00*	3.11	3.40
TiO_2, wt%	–	–	–	0.55	1.50
Fe_2O_3, wt%	–	–	–	0.25	0.20
K_2O, wt%	–	–	–	0.14	Trace
Na_2O, wt%	–	–	–	0.63	0.90
F, wt%	–	–	–	0.04	–
SiO_2/RO ratio	*2.64*	2.40	*2.40*	2.37	2.28
Log_3 forming T, °C	*1327*	1294	*1299*	1259	1258
Liquidus T, °C	*1188*	1235	*1225*	11	1185
Delta (Δ) T, °C	*139*	59	*74*	85	73
ASTM E-glass	+	+	+	+	+
References	[1]	[1]	[27, 28]	[15]	[14]

Among the quaternary compositions which are based on the SiO_2–Al_2O_3–CaO–MgO phase diagram, Examples 2 [15] and 3 [14] have the same or a slightly lower SiO_2 level (60.10 and 59.30%) than the eutectic composition (60.00%) [27], a lower Al_2O_3 level (12.99 and 12.10%) than the eutectic composition (15.00%), but a higher CaO level (22.13 and 22.60 than the eutectic composition (20.00%). Examples 2 and 3 were used with commercial ingredients and therefore contain small amounts of TiO_2, Fe_2O_3, K_2O, and Na_2O. As a result, the SiO_2/RO ratio of Examples 2 and 3 is lower than that of the eutectic composition, and their viscosity and log3-forming temperature is 40–41°C lower than that of the eutectic composition. All three compositions have about the same delta (difference) between the log3-forming and the delta temperature (75–85°C) and therefore about the same crystallization resistance at the bushings.

In summary, both first-generation, quaternary, fluorine- and boron-free E-glass variants [14, 15] (Table 2.5, Examples 2 and 3) were more energy friendly (had lower log_3 forming temperatures) than the ternary and quaternary eutectic E-glass compositions as well as the reformulated ternary E-glass (Example 1). They had delta temperatures of 73 and 85°C, respectively, and have since 2000 been reformulated to achieve an even more energy-friendly log_3 forming temperature at delta temperatures of 50–60°C [1–12].

(b) The quaternary eutectic with regard to MgO. The ASTM E-glass standard for general applications [16] permits the use of 0–5% MgO and 0–1.5% TiO_2. This standard covers all five compositions in Table 2.5, i.e., the ternary SiO_2-Al_2O_3-CaO and the quaternary SiO_2-Al_2O_3-CaO-MgO eutectic compositions, the reformulated ternary E-glass (Example 1), and two reformulated quaternary E-glass variants. Of these, Example 2 was reported in 1997 and contains 3.1% MgO and 0.5% TiO_2 [15] and Example 3 was reported in 1985 and contains 3.4% MgO and 1.5% TiO_2 [14].

MgO and TiO_2 are known to have the opposite effect on the log_3 forming temperature (viscosity), liquidus temperature (crystallization potential), and therefore energy demand of a given melt. An increase of MgO reduces the delta temperature and increases the crystallization potential of the melt, while an increase of TiO_2 (a flux) reduces the melt viscosity and log_3 forming temperature and therefore the energy demand of a given melt. The effect of MgO will be discussed next and that of TiO_2 in the following sections.

The MgO level affects the eutectic of the multi-oxide composition in the quaternary SiO_2-Al_2O_3-CaO-MgO phase diagram. Before beginning to reformulate the original, first-generation fluorine- and boron-free multi-oxide E-glass compositions [14, 15] by trend line design [2], the eutectic of the quaternary compositions with regard to MgO was determined. The MgO level of Example 3 in Table 2.5 was therefore stepwise reduced from 3.4 to 1.5% [2] and the results are shown in Fig. 2.4.

According to Fig. 2.4, the liquidus temperature dropped from 1185°C (at 3.4% MgO) to 1162–1168°C (at 2.5–2.6% MgO) and rose again to 1185°C (at 1.5% MgO). In summary, second-generation fluorine- and boron-free E-glass compositions [9] were found to have their lowest liquidus temperature at ~2.55% MgO (Fig. 2.4). To establish the optimum level of MgO, or the eutectic with regard to

Fig. 2.4 Eutectic of quaternary fluorine- and boron-free SiO_2-Al_2O_3-CaO-MgO E-glass variants, melt chemistry, relaxation, and solidification kinetics of glasses, F. T. Wallenberger et al. [2]. (Reprinted with permission of The American Ceramic Society, www.ceramics.org [2004]. All rights reserved)

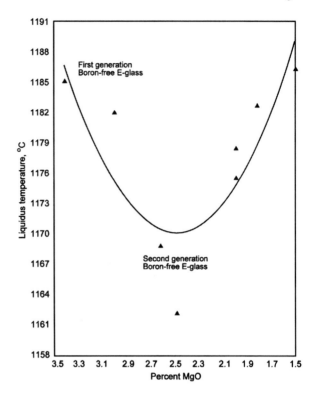

MgO, for third-generation fluorine-free E-glass compositions with <1.5% B_2O_3 and <1% Li_2O, the MgO level of a typical composition was varied from 3.4 to 1.5% [8]. As documented in reference [3], the lowest liquidus temperature, like that of second-generation fluorine- and boron-free E-glass variants, was again found near 2.50–2.55% MgO [3].

(c) *Fluorine- and boron-free E-glass with* ≤*1.5% TiO₂*. Two first-generation E-glass variants with commercial potential were known by 1997, one contained 1.1–1.5% TiO_2 [14], the other 0.1–0.5% TiO_2 [15]. Since 0.1–0.5% TiO_2 is obtained as tramp (impurity) from other commercial ingredients, only the first of these two compositions requires the addition of virgin TiO_2 in commercial use. Both variants had higher delta temperatures than 50–60°C and were therefore separately subjected to trend line design to reduce their energy demand in commercial use [1–3, 7].

The $\log_3$ forming temperature of the original first-generation fluorine- and boron-free E-glass composition that contained 1.5% TiO_2 and 3.40% MgO [2, 14] was first reformulated by trend line design. The incumbent composition was [2] modified to contain 2.55% MgO as required by the eutectic (Fig. 2.4) and 1.1% TiO_2. The trend lines for the $\log_3$ forming and liquidus temperatures were obtained by stepwise reformulation of the starting composition. Eighteen consecutive melts are shown in Fig. 2.5. The trend lines include the original first-generation composition as a control [14].

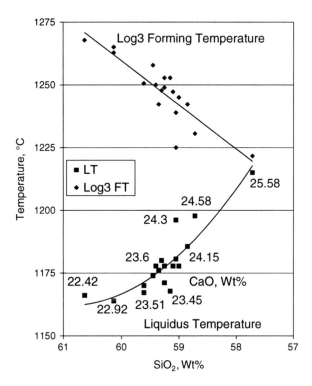

Fig. 2.5 Trend lines for the forming and liquidus temperatures of boron-free E-glass, F. T. Wallenberger et al. [7]

Specifically, SiO_2 was stepwise reduced from >60 to <58%, CaO was stepwise increased from <23% to >25%, Al_2O_3 was maintained between 12.3 and 13.2%, MgO was kept constant at 2.55%, and TiO_2 was kept constant at 1.1%. The values for %CaO (at a constant level of 2.55% MgO) were superimposed on the liquidus temperature trend line in Fig. 2.5. The value for RO (CaO + MgO) can be calculated by adding 2.55% of MgO to the %CaO value.

The log_3 forming temperatures of four compositions with 1.1% TiO_2 were selected from Fig. 2.5 and are shown in Table 2.6. They ranged from 1268 to 1232°C, the SiO_2/RO ratios ranged from 2.43 to 2.15, and the delta temperatures ranged from 102 to 66°C. The control is the corresponding first-generation boron-free E-glass. The goal composition has a log_3 forming temperature of 1232°C, or a 26°C lower log_3 forming temperature than the original first-generation fluorine- and boron-free E-glass [2, 7, 14]. This is the most energy-friendly melt with 1.1% TiO_2 that has a crystallization-resistant delta temperature of 66°C.

The next question was how far the log_3 forming temperature (or the energy demand) of the first-generation fluorine- and boron-free E-glass that contained only 0.55% TiO_2 [2, 7, 8, 15] can be reduced by compositional reformulation by trend

Table 2.6 Design of energy-friendly fluorine- and boron-free E-glass compositions

Compositions	1.10% TiO$_2$				0.55% TiO$_2$	
Example	1	*Control*	2	3	*Control*	4
SiO$_2$, wt%	60.63	*59.30*	57.95	57.75	*60.10*	58.18
Al$_2$O$_3$, wt%	12.27	*12.10*	13.20	13.20	*12.99*	13.68
CaO, wt%	22.42	*22.60*	24.05	24.25	*22.13*	23.75
MgO, wt%	2.50	*3.40*	2.55	2.55	*3.11*	2.55
TiO$_2$, wt%	1.10	*1.50*	1.10	1.10	*0.55*	0.55
Fe$_2$O$_3$, wt%	0.20	*0.20*	0.25	0.25	*0.25*	0.35
Na$_2$O, wt%	0.90	*0.90*	0.90	0.90	*0.63*	0.91
B$_2$O$_3$, wt%	–	–	–	–	–	–
F, wt%	–	–	–	–	*0.04*	–
SiO$_2$/RO	2.43	*2.28*	2.18	2.15	*2.37*	2.20
Log$_3$ FT, °C	1268	*1258*	1235	1232	*1259*	1238
LT, °C	1166	*1185*	1164	1166	*1174*	1183
ΔT, °C	102	*73*	71	66	*85*	55
References	[2, 10]	*[13, 14]*	[2, 10]	[2, 10]	*[13, 15]*	[2, 10]

line design while still offering a crystallization-resistant melt in commercial use with a delta temperature of 50–60°C.

The trend lines for the log$_3$ forming and liquidus temperatures were obtained by a stepwise reformulation of the modified first-generation fluorine- and boron-free E-glass composition that contained 0.55% TiO$_2$ and 2.55% MgO [2, 7, 8, 15]. SiO$_2$ was therefore reduced from 60.10 to 58.18%, CaO was increased from 22.13 to 23.75%, and MgO was reduced from 3.11% to the eutectic value of 2.55%. The actual trend lines can be found in reference [2, 3]. As a result (Table 2.6), the log$_3$ forming temperature dropped from 1259 to 1238°C, the liquidus temperature increased from 1174 to 1183°C, and the delta temperature decreased from 85 to 55°C [2, 10].

In summary, the lowest log$_3$ forming temperature of the fluorine- and boron-free E-glass variant with 2.55% MgO and 0.55% TiO$_2$ was 1238°C at a crystallization-resistant delta temperature of 55°C [7], while that of the fluorine and boron-free E-glass with 2.55% MgO and 1.1% TiO$_2$ was 1232°C at a crystallization-resistant delta temperature of 66°C [7]. The 0.55% increase of TiO$_2$, acting as a flux, translates into a 6°C ($\sim$10°C/%) reduction of the log$_3$ forming temperature or into a notable decrease of the melt viscosity and a corresponding change of the required melt energy.

(d) Fluorine- and boron-free E-glass with <1% Li$_2$O. The addition of Li$_2$O to a fluorine- and boron-free E-glass yields an E-glass that is not only environmentally friendly (potentially free of toxic emissions) but also more energy friendly (potentially affording a lower melt viscosity). But even small amounts of Li$_2$O will substantially increase the batch cost of the resulting E-glass (see Section 2.2.3) since it is not a common ingredient in E-glass melts. But, considering the factors that limit

the use of B_2O_3, it could at least become a partial substitute for B_2O_3 in an age of growing energy cost and environmental concern.

Li_2O, Na_2O, and K_2O are alkali metal oxides and the use of a combined content of 0–2 wt% alkali metal oxide is allowed by ASTM E-glass standards. The relative compositional effect of Li_2O and Na_2O was quantified by separately adding increasing amounts 0.2–1.2% of each oxide to the same base composition, a fluorine-, boron-, and alkali oxide-free E-glass, at a constant SiO_2/RO ratio (RO = CaO + MgO). The other oxides were reduced proportionately to facilitate the stepwise increase in alkali oxide.

The results shown in Fig. 2.6 confirm that Li_2O is a far more powerful flux than Na_2O. The addition of 0.6 wt% Li_2O to the alkali oxide-free base composition produced a 36°C lower log_3 forming temperature than the addition of 0.6 wt% Na_2O and a 14°C lower liquidus temperature. Similarly, the addition of 1.2 wt% Li_2O to the alkali oxide-free base composition produced a 57°C lower log_3 forming temperature than the addition of 1.2 wt% Na_2O and a 23°C lower liquidus temperature.

As shown in Table 2.7, two first-generation fluorine- and boron-free E-glass variants, one with 0.9% Na_2O and 0.55% TiO_2 [15] and the other with 0.9% Na_2O and 1.1% TiO_2 [14], were reformulated. In each variant, 0.9% Na_2O was first replaced with 0.9% Li_2O and then subjected to compositional reformulation by trend line design [2, 3] to find and identify the lowest log_3 forming temperature that can be

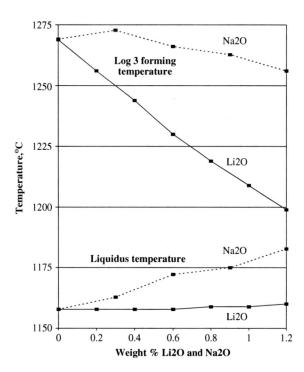

Fig. 2.6 Effect of Li_2O as replacement for Na_2O

Table 2.7 Energy-friendly E-glass compositions with 0.9% Li_2O

Examples	1	2	3	4	5	6
SiO_2, wt%	60.21	58.35	57.65	59.71	58.70	58.96
Al_2O_3, wt%	13.02	13.40	13.40	13.24	13.44	13.24
CaO, wt%	22.18	23.55	24.15	22.90	23.55	23.65
MgO, wt%	2.30	2.55	2.55	2.50	2.50	2.50
TiO_2, wt%	1.10	1.10	1.10	0.50	0.50	0.50
Fe_2O_3, wt%	0.25	0.25	0.25	0.25	0.25	0.25
K_2O, wt%	–	–	–	–	–	–
Na_2O, wt%	–	–	–	–	–	–
Li_2O. wt%	0.90	0.90	0.90	0.90	0.90	0.90
B_2O_3, wt%	–	–	–			
SiO_2/RO	2.46	2.23	2.16	2.35	2.26	2.25
Log$_3$ FT, °C	1233	1215	1206	1227	1216	1212
LT, °C	1142	1146	1154	1142	1153	1158
ΔT, °C	91	65	52	85	63	54
References	[2, 8]	[2, 8]	[2, 8]	[7, 8]	[7, 8]	[7, 8]

achieved with each composition at a delta temperature of ~55°C. The actual trend lines are not shown but can be found in references [2] and [3]. Three examples each are summarized in Table 2.7.

The lowest log$_3$ forming temperature of the compositional variant with 2.55% MgO, 0.9% Li_2O, and 1.1% TiO_2 (Examples 1–3) was 1206°C at a delta temperature of 52°C and that of the compositional variant with 2.55% MgO, 0.9% Li_2O, and 0.55% TiO_2 was 1212°C at a crystallization-resistant delta temperature of 54°C (Examples 4–6). In summary, with and without Li_2O (as a flux), an increase of 0.55% TiO_2 from 0.55 to 1.1% translates into a 6°C (~10°C/%) decrease of the forming temperature or into a notable reduction of the melt viscosity and projected melt energy demand.

(e) Fluorine-free E-glass with <1.5% B_2O_3. The effect of 1.0–1.5% B_2O_3 when added to a second-generation fluorine- and boron-free E-glass composition containing 0.55% TiO_2 was determined by trend line design [2, 7]. Compositions containing 5–10% B_2O_3 require a costly additional environmental control system to remove B_2O_3 emissions from commercial melts, but essentially boron-free compositions with <1.5% B_2O_3 may not require additional emission controls.

A fluorine-free E-glass that contained 1.8% B_2O_3, 60.82% SiO_2, 11.70% Al_2O_3, 21.2% CaO, 2.8% MgO, and 1.8% B_2O_3 that had been reported in 1997 served as a reference [2, 7, 8, 29]. It was modified to contain only 1.3% B_2O_3 and 56.5% SiO_2, but 13.4% Al_2O_3, 24.6% CaO, and 2.5% MgO, and was then subjected to a stepwise reformulation by trend line design. The trend lines for this E-glass variant can be found in references [2, 3].

The compositions and melt properties of three selected melts are shown in Table 2.8 to document the designed trends (Examples 1–3). SiO_2 was reduced from 60.82 to 56.55%, CaO was raised from 21.20 to 24.60%, %MgO was kept constant,

Table 2.8 Energy-friendly E-glass compositions with $\leq 1.3\%$ B_2O_3

Examples	Control	1	2	3	4	5
SiO_2, wt%	60.82	56.55	56.55	58.40	58.30	58.00
Al_2O_3, wt%	11.70	13.35	13.40	13.03	13.03	13.03
CaO, wt%	21.20	24.55	24.60	23.44	23.54	23.84
MgO, wt%	2.80	2.55	2.55	2.50	2.50	2.50
TiO_2, wt%	0.10	0.55	0.55	0.50	0.50	0.50
Fe_2O_3, wt%	0.16	0.25	0.25	0.23		0.23
K_2O, wt%	0.30	–	–	–	–	–
Na_2O, wt%	1.10	0.90	0.90	–	–	–
Li_2O. wt%	–	–	–	0.90	0.90	0.90
B_2O_3, wt%	1.80	1.30	1.30	1.00	1.00	1.00
SiO_2/RO	2.53	2.09	2.08	2.25	2.24	2.20
Log_3 FT, °C	1262	1220	1211	1208	1200	1192
LT, °C	1180	1155	1153	1137	1132	1137
ΔT, °C	82	65	58	71	68	55
References	[29]	[2, 9]	[2, 9]	[3, 9]	[3, 9]	[3, 9]

and the SiO_2/RO ratio dropped from 2.50 to 2.08. As a result, the log_3 forming temperature dropped from 1262 to 1211°C, the liquidus temperature reached 1153°C, and the delta temperature dropped from 82 to 58°C, thus creating an energy-friendly and crystallization-resistant melt [2, 9].

(f) Fluorine-free E-glass with <1.5% B_2O_3 *and* <1% Li_2O [3]. The lowest log_3 forming temperature of the compositional variant with 1.0% B_2O_3, 0.9% Li_2O, and 0.5% TiO_2 was likewise developed by trend line design. The original trend lines are not shown here but can be found in references [2] and [3]. Three examples are shown in Table 2.8 (Examples 4–6). The SiO_2 level was stepwise reduced from 58.40 to 58.00%, the CaO level was increased from 23.44 to 23.84% at a constant MgO level of 2.50%, and the SiO_2/RO ratio decreased from 2.25 to 2.20. The forming temperature dropped from 1208°C at a delta temperature of 71°C to 1192°C at a delta temperature of 55°C [3, 9]. In summary, 1192°C is the lowest possible log_3 forming temperature of this compositional variant for a commercial melt that is crystallization resistant in a commercial furnace.

As a flux 1.3% B_2O_3 appears to be as effective as 0.9% Li_2O at the same TiO_2 level (0.55%) and at a delta temperature (54–58°C) that is required for a crystallization-resistant melt. For melts with <1.5% B_2O_3, an added environmental control system, that is required for melts with >2% B_2O_3, is not needed unless dictated by the configuration of a specific furnace.

(g) Summary and conclusions. The log_3 forming temperatures of a 621 borosilicate E-glass composition [7] and of a ternary SiO_2-Al_2O_3-CaO eutectic control [28] are the controls in Table 2.9, followed by three generic and increasingly energy-friendly E-glass compositions (I, II, and III). The various compositional variants are also schematically shown in Fig. 2.7. The first-generation fluorine- and boron-free E-glass variants became available before 1998 [1, 14, 15]; the others variants were developed since then by trend line design.

Table 2.9 Environmentally and increasingly energy-friendly E-glass compositions

Composition	Std.	*Ternary*	First generation			Second generation	Third generation		
Label	621	*Control*	I-1	I-2	I-3	II	III-1	III-2	III-3
SiO_2, wt%	53.27	*62.10*	60.00	60.00	60.10	57.75	58.35	56.60	58.00
Al_2O_3, wt%	13.98	*14.60*	15.00	15.00	12.99	13.20	13.40	13.25	13.03
CaO, wt%	23.53	*23.30*	25.00	20.00	22.13	24.25	23.55	24.60	23.84
MgO, wt%	0.61	–	–	5.00	3.11	2.55	2.55	2.55	2.50
TiO_2, wt%	0.51	–	–	–	0.55	1.10	1.10	0.55	0.50
Fe_2O_3, wt%	0.36	–	–	–	0.25	0.25	0.25	0.25	0.23
K_2O, wt. %	0.10	–	–	–	0.14	–	–	–	–
Na_2O, wt%	0.93	–	–	–	0.63	0.90	–	0.90	
Li_2O, wt. %	–	–	–	–	–	–	0.90	–	0.90
B_2O_3, wt%	6.00	–	–	–	–	–	–	1.30	1.00
F, wt%	0.48	–	–	–	0.04	–	–	–	
SiO_2/RO	2.21	*2.67*	2.40	2.40	2.37	2.15	2.23	2.08	2.20
Log_3 FT, °C	**1174**	*1335*	**1294**	**1299**	**1259**	**1232**	**1211**	**1211**	**1192**
LT, °C	1068	*1217*	1235	1225	11	1166	1146	1153	1137
ΔT, °C	106	*118*	**59**	**74**	85	**66**	**65**	**61**	**55**
E-glass	+	–	+	+	+	+	+	+	+
References	[13]	[28]	[1]	[13]	[1]	[3, 8]	[3, 8]	[3, 8]	[3, 8]

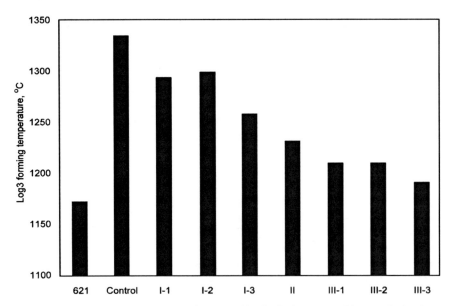

Fig. 2.7 Environmentally and increasingly energy-friendly E-glass compositions, melt chemistry, relaxation, and solidification kinetics of glasses, F. T. Wallenberger et al. [2]. (Reprinted with permission of The American Ceramic Society, www.ceramics.org [2004]. All rights reserved)

The second-generation E-glass variant (II-1) was derived from a first-generation E-glass variant [2]. The third-generation E-glass compositions contain <1% Li_2O (III-1), <2% B_2O_3 (III-2) as well as <1% Li_2O and <2% B_2O_3 (III-3), respectively [3]. The compositions in both the table and the figure contain 0.5–0.55% TiO_2 that is obtained from tramp (impurities) in commercial batch ingredients, or 1.0–1.5% TiO_2 that, although permitted by ASTM E-glass standards, would require the deliberate addition of MgO.

The lowest log_3 forming temperature of second- (II-1) and third-generation E-glass variants (III-1, III-2, and III-3) has been achieved with a goal delta temperature of 55–65°C. Consequently, the third-generation fluorine-free E-glass variant (III-3) that contains 1% B_2O_3 and 0.9% Li_2O has a log_3 forming temperature that is 67°C lower than that of the first-generation fluorine- and boron-free E-glass (I-3) and only 18°C higher than that of the original incumbent E-glass (621) that contains 6% B_2O_3 and 0.5% fluorine.

2.2.2 New Energy-Friendly E-Glass Variants with 2–10% B_2O_3

When glass fiber is manufactured it must meet the stringent requirements of environmental regulations. Glass fibers with <2% B_2O_3 may not require a costly, additional environmental control system or an efficient wet scrubbing system but with >2% B_2O_3 they almost certainly require an additional and costly environmental control system. And to be sold as an E-glass the composition in question must meet the ASTM E-glass standards.

2.2.2.1 Quaternary SiO_2-Al_2O_3-CaO-B_2O_3 Phase Diagram

The addition of 1.0–1.3% B_2O_3 to a composition in the ternary SiO_2-Al_2O_3-CaO phase diagram does, in practical terms, not make it a derivative of the quaternary SiO_2-Al_2O_3-CaO-B_2O_3 phase diagram. Likewise, the addition of 0.55% MgO tramp when received as impurity from commercial ingredients does not, in practical terms, move the resulting composition into a SiO_2-Al_2O_3-CaO-MgO-B_2O_3 system.

However, when >2% B_2O_3 is added to a commercial composition in the ternary phase diagram that contains 0.55% MgO tramp, the resulting composition should, in practical terms, be considered to belong to the quaternary SiO_2-Al_2O_3-CaO-B_2O_3 phase diagram. This distinction is important because the topographical changes of the liquidus surface in the true quaternary system are unpredictable by first-order considerations.

2.2.2.2 Trend Line Design of Energy-Friendly Variants

The eutectic composition of borosilicate E-glass variants with 5–7% B_2O_3 that has the lowest liquidus temperature among neighboring compositions has never been identified, and that is the reason recent publications, while pioneering, must be considered to be diagnostic rather than definitive [4–6, 30, 31] with regard to the effect of boron on the melt properties. In principle, to reduce the energy demand, %SiO_2 and the SiO_2/RO ratio have been reduced to reduce the forming temperature, but the path was difficult because the SiO_2/B_2O_3 ratio also seemed to affect log_3 forming temperature and delta temperature [3, 4].

Specifically, the trend lines for E-glass with 2% and 3% B_2O_3 are found in reference [3]. The results of diverse investigations about the effect of B_2O_3 at levels ranging from 4 to 10% B_2O_3 are reported in references [4–6, 30] and [31]. The effect of 1% B_2O_3 added to, or removed from, a composition at the same delta temperature was estimated to result in a ~10–12°C decrease or increase of the log_3 forming temperature [4, 5]. The literature should be consulted for details on specific effect of the SiO_2/RO ratio on the melt properties of E-glass with 4–10% B_2O_3 [6] and for compositional claims, which may imply commercial significance [7, 31].

The log_3 forming temperatures of commercial 621 borosilicate E-glass melts vary. In Asia, they tend to be higher (1191–1200°C) than in the United States (1178–1180°C). As shown in Table 2.10, the published literature [1–3, 5–7, 30, 31] suggests that the log_3 forming temperature (and melt viscosity) of typical borosilicate E-glass compositions can be reduced by trend line design in laboratory experiments from 1178–1200 to 1154°C when the delta temperature is reduced from >120 to ~65°C. A 29–55°C reduction of the log_3 forming temperature (implicitly melt viscosity) would result in a major reduction of the energy that is required to melt the batch in a commercial furnace.

Table 2.10 Reformulation of 621 E-glass with 6.5% B_2O_3 in laboratory experiments

		Demonstrated	Predicted	Demonstrated	Predicted
References [4–6]	Change of Log_3 FT	For operations in Asia		For operations in the United States	
Log_3 FT, °C	From	1191	1191	1178	1178
	To	1154	1145	1154	1145
		−37	−46	−24	−33
ΔT, °C	From	120	120	111	111
	To	65	55	65	55
		−55	−65	−46	−56

While it is possible to reduce melt energy significantly by trend line design and compositional reformulation, it is also possible to reduce the batch cost of a commercial borosilicate composition by reducing the boron level, for example, by 1% from 6.5 to 5.5%, while still remaining within the limits of the ASTM E-glass standard for printed circuit board applications. That result would be accomplished by designing a composition by trend line design that has a 1% lower B_2O_3 level at the same forming temperature as before, but a higher liquidus temperature and lower delta temperature.

That move, however, may affect other properties of the resulting product. For example, the electronic industry requires an E-glass with a dielectric constant that does not change over time or from supplier to supplier. It is more likely that fluorine can be reduced (even removed) without affecting important product properties but such a conversion would have to be carefully executed.

2.2.2.3 Effect of B_2O_3 at the Same Delta Temperature

When the data from the various trend line design studies were evaluated [4–6], another new model emerged that can be used to project the effect of %B_2O_3 on the $\log_3$ forming temperature of E-glass melts at the same delta temperature, both at 100–120°C and at 55–65°C as shown in Table 2.11.

This model confirms that a 1% change (increase or decrease) of B_2O_3 at the same delta between $\log_3$ forming and liquidus temperatures (e.g., at ∼65 or at ∼ 100°C) would result in a 12°C change (decrease or increase, respectively) of the $\log_3$ forming temperature.

Table 2.11 Effect of boron at equal delta temperatures [4, 5, 6]

B_2O_3	$\log_3$ FT, °C	Lowest $\log_3$ FT (°C)	
%	Change	$\Delta T \sim 100°C$	$\Delta T \sim 65°C$
0.0		1272	1232
1.0	−12	1260	1220
2.0	−24	1248	1208
3.0	−36	1236	1196
4.0	−48	1224	1184
5.0	−60	1212	1172
6.0	−72	1200	1160
7.0	−84	1188	1148
8.0	−96	1176	1136
9.0	−108	1164	1124
10.0	−120	1152	1112

2.2.2.4 Summary and Conclusions

By reducing SiO_2 level of borosilicate E-glass melts with 6.0–6.6% B_2O_3 and therefore by increasing %RO and decreasing the SiO_2/RO ratio it was possible to reduce the $\log_3$ forming temperature from 1174 –1200°C to 1154°C at a delta temperature of 65°C. When the effect of the SiO_2/RO ratio and of the SiO_2/B_2O_3 will have been taken into consideration, a $\log_3$ forming temperature of 1145°C is predicted for a delta temperature of 50–55°C. A significant energy reduction would result from such a reformulation and the design of the ultimate goal composition, therefore, remains a challenge.

2.2.3 New Energy- and Environmentally Friendly ECR-Glass Variants

The addition of 1.2–5.3% ZnO, 2.2–3.7% TiO_2, and/or 0.2–2.4% Li_2O to a boron-free E-glass composition has been reported in 1974 [32] and 1975 [33].

2.2.3.1 Commercial Corrosion-Resistant ECR-Glass

A commercial product is known in the trade as ECR-glass (or a corrosion-resistant E-glass) that contains 2.5% TiO_2 and 2.9% ZnO but no Li_2O. But the commercial and related ECR-glass compositions (Table 2.12) are not E-glass variants because ZnO is not specified in the ASTM E-glass standards and because the TiO_2 level of the commercial ECR-glass variant exceeds 1.5% [16]. Industry has not requested and ASTM has not issued ECR standards. Note: TiO_2 is a flux that reduces the melt viscosity (log_3 forming temperature). ZnO is a flux that is known to increase the corrosion resistance of glass fibers [34].

The boron-free ECR-glass that was commercialized contained 2.5% TiO_2 and 2.9% ZnO [13, 32]. As a premium product it offers higher acid resistance and stress corrosion resistance than E-glass but at a higher price. ERC-glass-reinforced composites can be used in diverse applications, such as sewer pipes and insulator rods [35]. The log_3 forming temperature (Table 2.12) of a commercial ECR-glass (1235°C) is 23°C lower than that of a typical boron-free E-glass [14]. The 1.9% higher level of TiO_2 and the presence of ZnO reduce the log_3 forming temperature and thereby the energy demand.

2.2.3.2 Fluorine- and B_2O_3-Free E-Glass with ZnO, TiO_2, and/or Li_2O

The trend line-designed addition of 1.0% ZnO and 1.0% TiO_2 (Fig. 2.8) was initiated with a composition near the quaternary eutectic having a very high SiO_2 level and a very high log_3 forming temperature or viscosity, a very low liquidus temperature, and therefore a very high delta between log_3 forming and liquidus temperatures. In 14 consecutive experiments, %SiO_2 was stepwise reduced and, at a constant MgO level, %CaO (i.e., %RO) was increased until the compositions reached the lowest forming temperatures, 1234 and 1231°C, respectively, with crystallization-resistant delta temperatures of 59 and 50°C, respectively

The log_3 forming temperatures, liquidus temperatures, and delta temperatures of an E-glass control, an ECR-glass control, and two new ECR-glass variants without Li_2O are shown in Table 2.12. The log_3 forming temperatures of the two target compositions are 1 and 4°C lower, respectively, than that of the commercial ECR-glass control despite the fact that they contain much less TiO_2 and ZnO. As a result, they are considerably more cost-effective than the commercial ECR-glass composition. The results attest to the effectives of trend line design as a compositional design and reformulation tool.

The trend line-designed addition of 0.45% ZnO and 0.45% Li_2O to a typical boron-free E-glass composition is also discussed in reference [7] and is also shown

Table 2.12 New ECR-glass variants with ZnO, TiO₂, and 0.0–0.5% Li₂O

Examples	E-glass Control	ECR-glass Control	New ECR-glass variants		With 0.45% Li₂O	
			Without Li₂O			
SiO_2, wt%	59.45	58.1	59.00	58.70	59.61	59.47
Al_2O_3, wt%	12.29	11.5	12.00	12.90	12.16	12.16
CaO, wt%	23.55	21.7	22.50	22.40	23.50	24.22
MgO, wt%	2.55	2.0	3.40	3.40	2.50	1.90
TiO₂, wt%	**1.10**	**2.8**	**1.00**	**1.00**	**1.10**	**1.10**
Na_2O, wt%	0.90	1.0	0.90	0.90	–	–
ZnO, wt%	–	**2.9**	**1.00**	**1.00**	**0.45**	**0.45**
Li₂O, wt%	–	–	–	–	**0.45**	**0.45**
Fe_2O_3, wt%	0.25	0.1	0.20	0.20	–	–
RO (CaO + MgO)	26.10	23.7	25.90	25.80	26.00	26.12
SiO_2/RO	2.28	2.45	2.28	2.28	2.29	2.28
Log₃ FT, °C	**1258**	**1235**	**1234**	**1231**	**1229**	**1218**
Liquidus, °C	1173	1166	1175	1161	1154	1159
Delta T, °C	**85**	**69**	**59**	**50**	**75**	**59**
Reference/Ex.	[13, 14]	[32, 33]	[7, 10]	[7, 10]	[7, 10]	[7, 10]

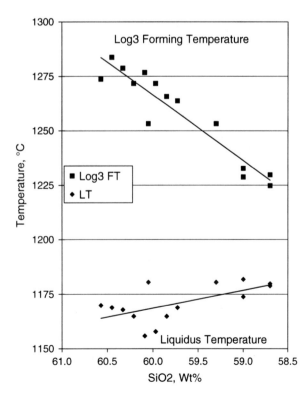

Fig. 2.8 Trend lines design of energy-friendly ECR-glass with 1% TiO₂ and 1% ZnO, F. T. Wallenberger et al. [7]

in Table 2.12. The lowest $\log_3$ forming temperature that was obtained for this compositional variant at a delta temperature of 59°C was 1218°C. The effect of 0.45% Li_2O is dramatic. Thus, the lowest $\log_3$ forming temperature in this compositional variant was 17°C lower than that of the commercial ECR-glass control.

Finally the trend lines were designed for an experimental ECR-glass having 1% ZnO, 0.9% Li_2O, and 0.5% TiO_2. They are based on 20 consecutive experiments and can be found in reference [7]. To obtain the trend lines, the SiO_2 level was stepwise reduced and, at a constant MgO level, the CaO level was correspondingly increased until the compositions had the lowest $\log_3$ forming temperature at a delta temperature of 57°C. As shown in Table 2.13, the target $\log_3$ forming temperature (1204°C) was found to be 55°C lower than that of the original, first-generation fluorine- and boron-free E-glass [15] and 32°C lower than that of the commercial ECR-glass [13, 32].

Table 2.13 New ECR-glass variants with ZnO, TiO_2, and 0.9% Li_2O

Examples	E-glass Control	ECR-glass Control	New ECR-glass variants 1	3	5
SiO_2, wt%	60.10	58.72	58.25	58.10	58.30
Al_2O_3, wt%	12.99	15.85	13.33	13.03	13.03
CaO, wt%	22.13	19.95	23.29	23	23.54
MgO, wt%	3.11	0.18	2.50	2.50	2.50
TiO_2, wt%	**0.55**	**0.29**	**0.50**	**0.50**	**0.50**
Na_2O, wt%	0.63	0.11	–	–	–
K_2O, wt%	0.14	0.13	–	–	–
Li_2O, wt%	–	**2.44**	**0.90**	**0.90**	**0.90**
ZnO, wt%	–	**1.96**	**1.00**	**1.00**	**1.00**
Fe_2O_3, wt%	0.25	0.20	0.23	0.23	0.23
RO, wt%	25.24	20.13	25.79	26.24	26.04
SiO_2/RO	2.38	2.92	2.26	2.21	2.24
Log_3FT, °C	**1259**	**2336**	**1213**	**1206**	**1204**
Liquidus, °C	1174	1191	1146	1144	1147
Delta T, °C	**85**	**45**	**67**	**62**	**57**
Ref./Ex.	[13, 15]	[32, 33]	[7, 10]	[7, 10]	[7, 10]

In summary, the lowest $\log_3$ forming temperature of an E-glass variant with 1.0% B_2O_3, 0.9% Li_2O, and 0.5% TiO_2 (1192°C) was 12°C lower than that of an ECR-glass variant with 1.0% ZnO, 0.9% Li_2O, and 0.5% MgO at the same delta temperature (55–57°C). As a flux, ZnO is therefore not quite as effective as B_2O_3, but it increases the corrosion resistance of a given glass fiber. ZnO is environmentally friendly, B_2O_3 is only conditionally so.

2.3 Energy-Friendly Soda–Lime–Silica Glass Fibers

Commercial glass fibers, which are derived from the ternary Na_2O-CaO-SiO_2 or soda–lime–silica (S–L–S) phase diagram, are compositionally related to container and float glass, and therefore represent an entirely different compositional system along with different properties than commercial glass fibers which are derived from quaternary SiO_2-Al_2O_3-CaO-MgO system.

2.3.1 New Energy-Friendly A- and C-Glass Compositions

Three glass fibers will be discussed, which are derived from the soda–lime–silica phase diagram. They are the former A-glass and the former C-glass in the United States as well as the current C-glass (or CC-glass) in China.

2.3.1.1 Fluorine and Boron-Free A-Glass

The former A-glass in the United States and current C-glass in China are based on boron-free soda–lime–silica compositions. A-glass has not been produced commercially during the past few decades. A suggestion was made in 1996 that it could be formed by attaching a forehearth and multiple bushings to a commercial float glass furnace "Loewenstein, private communication (1997)" and a development effort that was completed in 1997 showed that A-glass fibers could be formed from recycled container glass and could be effectively used to reinforce structural composites in selected automotive applications [36].

The original A-glass composition had a log_3 forming temperature of 1199°C and a delta temperature of 181°C [7]. It is erroneously quoted to have a log_3 forming temperature of 1280°C [34, 37] and a delta temperature of 270°C [37]. A stepwise decrease of SiO_2 (Fig. 2.9) by a trend line-designed compositional reformulation study [7] reduced the log_3 forming temperature (melt viscosity) and a corresponding increase of RO increased the liquidus temperature (thus crystallization potential) until the liquidus temperature was higher than the log_3 forming temperature.

A delta temperature of any desired value can be readily interpolated from the trend lines. At the crossover point, the log_3 forming temperature and the liquidus temperature were of course the same. If such a melt were formed from such a composition in a commercial furnace, it would within hours cause crystallization in dead corners of the furnace and in the bushing tips.

Specifically, when $\%SiO_2$ was reduced by trend line design from 71.80 to 68.75%, the SiO_2/RO ratio dropped from 5.69 to 4.39, the log_3 forming temperature was reduced by 59°C from 1199 to 1140°C, and the delta temperature was reduced by 116°C from 181 to 65°C as shown in Fig. 2.9 and Table 2.14. Comparable trend lines could be developed for the Chinese C-glass and equally significant energy savings could be achieved, if desirable, by compositional reformulation.

2.3.1.2 Fluorine-Free C-Glass with 5% B_2O_3

Finally, diagnostic trend lines were also developed for the reformulation of a C-glass composition that was formerly used in the United States [34, 37]. It has a soda–lime–silica composition with 5% B_2O_3 that, because of its high boron level,

Fig. 2.9 Trend line design of energy-friendly boron-free A-glass, F. T. Wallenberger et al. [7]

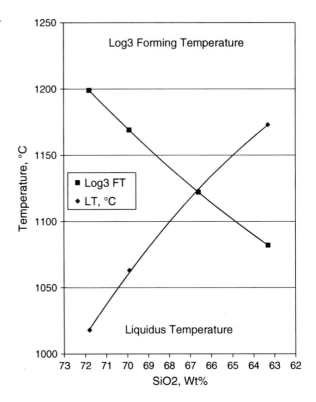

Table 2.14 Environmentally and energy-friendly A- and CC-glass

Examples	1	2	*Target*	3	4
SiO_2, wt%	**71.80**	**69.90**	*68.75*	66.60	63.30
Al_2O_3, wt%	1.00	1.00	*1.00*	1.00	1.00
CaO, wt%	**8.80**	**10.70**	*11.85*	14.00	17.30
MgO, wt%	3.80	3.80	*3.80*	3.80	3.80
Na_2O, wt%	13.60	13.60	*13.60*	13.60	13.60
K_2O, wt%	0.60	0.60	*0.60*	0.60	0.60
Fe_2O_3, wt%	0.40	0.40	*0.40*	0.40	0.40
RO	12.60	14.50	*15.65*	17.80	21.10
SiO_2/RO	**5.69**	**4.82**	*4.39*	3	3.00
Log_3 FT	**1199**	**1169**	*1140*	1122	1082
LT, °C	1018	1063	*1075*	1123	1173
ΔT, °C	**181**	**106**	*65*	−1	−91
References	[34]	[7]	[7]	[7]	[7]

is however no longer of commercial interest. Its forming temperature was reduced by trend line design from 1240 to 1200°C and its delta temperature was reduced from 157 to 65°C. Thus, a 40°C reduction of the $\log_3$ forming temperature (melt viscosity and implicitly energy demand) was achieved. The actual trend lines can be found in reference [7].

2.3.1.3 Future Soda–Lime–Silica Glass Fibers and Glasses

The potential of A-glass fibers is real but limited. One limitation is their relatively low Al_2O_3 level, Thus it came as no surprise in 1997 when it was reported that A-glass-reinforced composites had lower strength and stiffness than corresponding E-glass-reinforced composites, but that "A-glass could be used as reinforcement in composite applications by simply increasing the fiber fraction relative to their E-glass counterpart" [36]. If the energy use for the production of A-glass fibers were reduced by trend line reformulation, their value might increase when derived from container or float glass waste.

The other current limitation is their high $\log_3$ forming and delta temperatures. While it has been shown (Fig. 2.9 and Table 2.14) that the energy demand of the incumbent A-glass composition can be dramatically reduced by trend line-designed compositional reformulation, the energy demand of the very similar soda–lime–silica float and container glass compositions could most likely also be reduced by the application of trend line-designed compositional reformulation. Pioneering research is already under way with regard to the reformulation of commercial soda–lime–silica glasses and initial results have been reported in the literature [20] and in Chapter 6 and Chapter 7 of this book.

2.4 Summary, Conclusions, and Path Forward

Compositional reformulation by trend line design is an exploratory method that is aimed at creating the trend lines for the $\log_3$ forming and the liquidus temperatures (Figs. 2.4, 2.7 and 2.8) and to interpolate and identify the lowest forming temperature (melt viscosity and energy demand) of a compositional variant with a desired delta between $\log_3$ forming and liquidus temperatures that characterize a crystallization-resistant melt in a commercial furnace.

The method is based on a simple discovery [1–12]: The SiO_2 level of a composition governs the melt viscosity, $\log_3$ forming temperature, and energy demand. The RO level ($= CaO + MgO$) governs the liquidus temperature at which crystals form within a few hours, and the SiO_2/RO ratio correlates directly with the delta (or difference) between the $\log_3$ forming and liquidus temperatures and represents a safety zone between forming temperature and crystal formation in a commercial furnace.

Thus, $\%SiO_2$ is stepwise reduced in successive experiments (while keeping the other oxides constant) while $\%RO$ is correspondingly increased. As a result, the viscosity ($\log_3$ forming temperature or implicitly the energy demand) is stepwise reduced, the liquidus temperature is stepwise increased, and the delta temperature is reduced. With a limited set of experiments it is therefore possible to determine the lowest log3 fiber-forming temperature of a compositional variant (i.e., the most energy-friendly composition) with any desired delta temperature (e.g., 55°C).

Trend line design is not a statistical method, but it can be augmented at each step along the design process by making and evaluating near neighbor compositions preferably near the goal composition because the liquidus temperature can change rapidly over short distances in a given phase diagram (see Figs. 2.2 and 2.3).

The log3 forming temperature is derived from the viscosity–temperature behavior of a melt and changes gradually, almost linearly, with temperature and composition. The liquidus temperature is derived from the phase diagram of a compositional system and the liquidus surface may change abruptly with slight compositional changes. The delta temperature is derived from both systems. Since the key melt properties are derived from different systems, the oxides of a composition and their ratios are not independent but codependent variables. The next step would be the development of a mathematical system that takes these realities into consideration.

In summary, many new and energy-friendly compositions were designed by trend line design. The lowest log3 forming temperatures, which were obtained with delta temperatures of 75–120°C and 55–65°C, respectively, are shown in Table 2.15 for one ternary HT-glass, six fluorine-free E-glass variants with 0–1.3% B_2O_3, four boron-free ECR-glass variants, one C-glass variant, one borosilicate E-glass variant with 6.5% B_2O_3, and one boron-free A-glass variant. A delta temperature of 50–60°C is believed to be required to afford a crystallization-resistant melt in a commercial furnace.

Table 2.15 Commercial and experimental fiberglass compositions (2009)

Fiberglass types and compositions[a]						Log3 FT, °C	
Key oxides and fiber type	B_2O_3 %	TiO_2 %	ZnO %	Li_2O %	Na_2O + K_2O, %	@ ΔT = 75 – 120°C	@ ΔT = 55 – 65°C
HT-glass [26]	–	–	–	–	–	1518	1479
E-glass [7, 10]	–	0.1–0.5	–	–	0.90	1259	1238
E-glass [10, 14]	–	1.1–1.5	–	–	0.90	1258	1232
E-glass [7, 8]	–	0.50	–	0.90	–	1227	1212
E-glass [3, 10]	1.30	0.50	–	–	–	1225	1211
E-glass [3, 10]	–	1.10	–	0.90	–	1220	1206
E-glass [5, 10]	1.30	0.50	–	0.90	–	1208	1192
ECR-glass [7, 32]	–	2.80	2.90	–	0.90	1235	–
ECR-glass [7, 10]	–	1.00	1.00	–	0.90	1234	1231
ECR-glass [7, 10]	–	1.10	0.45	0.45	–	1229	1218
ECR-glass [7, 10]	–	0.50	1.00	0.90	–	1213	1204
C-glass [7, 34]	5.00	–	–	–	8.50	1242	1200
E-glass/PCB [7]	6.50	0.1–0.5	–	–	0.90	1191	1154
A- or CC-glass [7]	–	–	–	–	14.00	1199	1140

[a]The fiber types are listed by decreasing log3 FT, melt viscosity, and energy demand.

As a result, the lowest $\log_3$ forming temperature combined with a delta temperature of 55–65°C is 1479°C for the ternary HT-glass. Among the E-glass compositions it is 1238°C and 1232°C, respectively, for the boron-free variants, 1211°C for the variant with 1.3% B_2O_3, and 1192°C for the variant with 1.0% B_2O_3 and 0.9% Li_2O. Among the ECR-glass variants the lowest forming temperature with a delta temperature of 55–65°C ranges from 1204 to 1231°C depending upon the specific composition. For C-glass it is 1200°C, for the borosilicate E-glass it is 1154°C, and for A-glass or the Chinese C-glass it is 1140°C.

Major energy savings have in some cases already being realized commercially, and they can still be realized for the commercial 621 borosilicate E-glass in the United States and in Asia, for the Chinese C-glass, and for the commercial ECR-glass in the United States.

References

1. F. T. Wallenberger, R. J. Hicks, and A. T. Bierhals, Design of environmentally friendly fiberglass compositions: ternary eutectic SiO_2-Al_2O_3-CaO and related compositions, structures and properties, J. Non-Cryst. Solids, 349, 377–387 (2004).
2. F. T. Wallenberger, R. J. Hicks and A. T. Bierhals, Design of energy and environmentally friendly fiberglass compositions derived from the quaternary SiO_2-Al_2O_3-CaO-MgO phase diagram – Part I: Structures, properties and crystallization potential of eutectic and selected multi-oxide E-glass compositions, Ceram. Trans., 170, 181–199 (2004) H. Li et al., eds.
3. F. T. Wallenberger, R. J. Hicks and A. T. Bierhals, "Design of Energy and Environmentally Friendly Fiberglass Compositions Derived from the Quaternary SiO_2-Al_2O_3-CaO-MgO Phase Diagram – Part II: Fluorine-Free E-Glass Compositions Containing Low Levels of B_2O_3 and Li_2O", Proceedings of the Norbert Kreidl Memorial Conference, Glastechnische Berichte – Glass Science and Technology, Vol. 77C, pp. 170–183 (2004).
4. F. T. Wallenberger, R. J. Hicks and A. T. Bierhals, Effect of oxides on decreasing melt viscosity and energy demand of E-glass, Glass Res, 15 (1), Am Ceram Soc Bullet, 85 (2), 38–41 (2006).
5. F. T. Wallenberger, R. J. Hicks and A. T. Bierhals, Effect of key oxides, including Li_2O, on the melt viscosity and energy demand of E-glass compositions, in *A collection of papers presented at the 66th conference on glass problems: Ceramic engineering and science proceedings, Vol. 27, Issue 1*, W. M. Kriven, ed., American Ceramic Society, John Wiley & Sons, Hoboken, pp. 155–165 (2006).
6. F. T. Wallenberger and R. J. Hicks, The effect of boron on the properties of fiberglass melts, Eur. J. Glass Sci. Technol., A, Glass Technol., 47 (5), 148–152 (2006).
7. F. T. Wallenberger, R. J. Hicks, P. N. Simcic and A. T. Bierhals New environmentally and energy friendly fiberglass compositions (E-glass, ECR-glass, C-glass and A-glass). Glass Technol., Eur. J. Glass Sci. Technol. A, 48 (6), 305–315 (2007).
8. F. T. Wallenberger, Glass fiber composition, US Patent 6,686,304 B1 to PPG Industries, February 3, 2004.
9. F. T. Wallenberger, Glass fiber forming compositions, US Patent 6, 818, 575 B2, to PPG Industries, Incorporated, November 16, 2004.
10. F. T. Wallenberger, Glass fiber forming compositions, US Patent 6, 962886 B2, to PPG Industries, Incorporated, November 8, 2005.
11. F. T. Wallenberger, Glass fiber forming compositions, US Patent 7, 144, 836 B2, to PPG Industries, Incorporated, December 5, 2006.

12. F. T. Wallenberger, Glass fiber forming compositions, US Patent 7, 153, 799 B2, to PPG Industries, Incorporated, December 26, 2006.
13. F. T. Wallenberger, *Advanced inorganic fibers: Processes, structures, properties, applications*, Kluwer Academic Publishers, Dordrecht/Boston/London, 346 pages (1999).
14. J. F. Sproull, Fiber glass composition, US Patent 4,542,106, September 17, 1985.
15. W. L. Eastes, D. A. Hofmann and J. W. Wingert, Boron-free glass fibers, US Patent 5,789,325, August 4, 1998.
16. ASTM Standard D 578-00, Standard for E-glass fiber strands and stating the composition limits for E-glass, Annual Book of Standards, American Society for Testing and Materials, Conshohocken, PA, March 10 (2000).
17. T. P. Seward and T. Vascott, eds., *High temperature glass melt property database for process modeling*, The American Ceramic Society, Publisher, Westerville, p. 258 (2005).
18. K. H. Karlson and R. Backman, Thermodynamic properties, in *Properties of glass-forming melts*, L. D. Pye, A. Montenero and I. Joseph, eds., Taylor and Francis, CRC Press, Boca Raton, pp. 11–23 (2005).
19. P. Hrma, D. E. Smith, J. Matyas, J. D. Yeager, J. V. Jones and E. N. Boulos, Effect of float glass composition on liquidus temperature and devitrification behavior, Glass Technol., Eur. J. Glass Technol. A, 47 (3), 78–90 (2006).
20. P. A Bingham and M. Marshall, Reformulation of container glasses for environmental benefit through lower melting temperatures, Glass Technol. 46 (1), 11–19 (2005).
21. F. T. Wallenberger, H. Li and J. Watson, Glass fibers in *ASM Handbook, Vol. 21, Composites*, S. L. Donaldson and D. B. Miracle, eds., ASM International Park, Novelty, pp. 27–35 (2001).
22. ASTM Standard C 965-81, Standard practice for measurement of viscosity of glass above the softening point, Annual Book of ASTM Standards, American Society for Testing and Materials Philadelphia, PA (1990).
23. ASTM Standard C 829-81, Standard practices for measurement of liquidus temperature of glass by gradient furnace method, Annual Book of ASTM Standards, American Society for Testing and Materials Philadelphia (1990).
24. R. Flinn and P. K. Trojan, *Engineering materials and their applications*, Fourth Edition, Houghton Mifflin Company, Boston (1990).
25. E. M. Lewin, C. R. Robbins and H. F. McMurdie, System $CaO-Al_2O_3-SiO_2$ in *Phase equilibrium diagrams for ceramists*, M. K. Reser, ed., The American Ceramic Society, Columbus, Figure 630, p. 219 (1964).
26. P. B. McGinnis, High temperature glass fibers, US Patent 6.809,050 B1, to Owens Corning Fiberglass Technology, October 26, 2004.
27. Y. K. Kim, R. Brückner and J. Murach, Properties of textile glass fibers based on alkali- and boron oxide- free aluminosilicate glasses, Glastechnische Berichte, Glass Sci. Technol., 71 (3), 67– (1998).
28. J. Murach, A. Makat and R. Brückner, Structure-sensitive investigation on glass fibers from the system $SiO_2-Al_2O_3-CaO$, Glastechnische Berichte, Glass Sci. Technol., 71 (11), 327–331 (1998).
29. M. H. Gallo, J. Van Genechten, J. P. Bazin, S. Creux and P. Fournier, Glass fibers for reinforcing organic and/or inorganic materials, French Patent Application, No. 2.768,144 A1. September 10, 1997.
30. H. Li, C. Richards and J. E. Cox, WO2004/020355 A1, March 11, 2004.
31. J. P. Hamilton and H. H. Russell III, European Patent Application, EP 1496026 A1, January 12, 2005.
32. T. D. Erickson and W. W. Wolf, "Glass Compositions and Methods of Making Same", US Patent 3,847,627, to Owens Corning Fiberglass, November 12, 1974.
33. British Patent Specification No. 1,391,384, "Glass compositions, fibers and methods of making same", to Owens Corning Fiberglass Corporation, April 23, 1975.
34. P. Gupta, Glass fibers for composite materials in *Fibre reinforcements for composite materials*, A. R. Bunsell, ed., Elsevier, Amsterdam/Oxford/New York/Tokyo, pp. 19–71 (1988).

35. S. Sundaram, Boron-free glass fibres – the trend for the future? Reinfor. Plast., June issue, 47 (6), 36–40 (2003).
36. D. A. Steenkamer and J. L. Sullivan, Recycled content in polymer matrix composites through the use of A-glass fibers, Polym Compos., 18 (3), 300–312 (1997).
37. K. L. Loewenstein, *The manufacturing technology of continuous glass fibers, complexly revised*, Third Edition, Elsevier, Amsterdam (1993).

Chapter 3
Composite Design and Engineering

J.H.A. van der Woude and E.L. Lawton

Abstract Fiberglass is a versatile and cost-effective reinforcement for composites. Many processes, resins, and forms of fiberglass facilitate this versatility. The design, engineering, manufacture, and properties of fiberglass-reinforced composite products from diverse thermoset and thermoplastic resins are described. The attributes of fiberglass-reinforced composites include its mechanical and chemical properties, lightweight, corrosion resistance, longevity, low total system cost, and Class A surface properties. Specific examples illustrate the importance of the form of the fiberglass reinforcement and of the interfacial bond between the glass fibers and the matrix resin in optimizing composite properties. In addition, recent advances are described with regard to the fabrication of fiberglass-reinforced wind turbine blades.

Keywords: Fiberglass · Composite · Thermoset · Thermoplastic · Resin · Reinforcement · Polymer

3.1 Introduction

3.1.1 Continuous Fibers for Reinforcement

Although the former denomination "reinforced plastics" is now generally replaced by the phrase "reinforced composites," the definition for this group of materials remains unchanged from a combination of at least two physically distinct materials, acting in concert by virtue of the interfacial bond between them. The largest segment of the composite industry is based on the combination of organic polymers and reinforcing glass fibers. The process for combining the polymer and reinforcing fibers

J.H.A. van der Woude (✉)
Fiber Glass Science and Technology, Europe, PPG Industries Inc.,
Hoogezand 9600AB, The Netherlands
e-mail: vanderwoude@ppg.com

F.T. Wallenberger, P.A. Bingham (eds.), *Fiberglass and Glass Technology*,
DOI 10.1007/978-1-4419-0736-3_3, © Springer Science+Business Media, LLC 2010

is termed compounding, and the resulting material is called compound. The terms polymer, resin, and plastic are used almost interchangeably in the composite field to denote an organic macromolecular matrix. The term resin is probably the most frequently used term for the matrix component. The term resin is also sometimes used to denote the reactants, including low molecular weight polymers or oligomers, that are precursors of the final matrix polymer. The reinforcing glass fibers used in composite manufacturing are both continuous filaments and short length filaments generated by cutting continuous filaments into short lengths.

To understand how a material can be made stronger through reinforcement, the basic physical and mechanical characteristics such as stress, strain, modulus of elasticity, tensile strength, elongation at rupture must be considered. Without this knowledge it may be difficult to understand why glass fibers can reinforce organic polymers, and why one combination gives far better mechanical properties than another. The term reinforced generally means that some (or most) of the mechanical properties of a homogeneous material are enhanced by mixing with a reinforcing material. Typical properties are tensile and flexural strength.

To be able to reinforce an organic polymer, the reinforcing fibers must have the following properties:

- A significantly greater modulus of elasticity than the polymer.
- A greater tensile (yield) strength.
- Be in a suitable form to be combined with the polymer.
- Render the best possible adhesion with the polymer.
- Be chemically and physically resistant to the polymer and other additives, such as plasticizers and antioxidants.

The most widely used fiber for reinforcement is the E-glass-type fiber; however, both carbon and aramid fibers are used in special applications. The cost effectiveness of fiberglass accounts for its dominance among the reinforcing fibers. Other synthetic fibers such as high molecular weight polyethylene, ceramic fibers, metal fibers, and inorganic whiskers are used to achieve very specific property enhancement.

Composites find their way into many automotive, transportation, marine, construction, electrical, leisure, and other applications. A large variety of application technologies have been developed to meet specific performance, price, and level of investment requirements. The versatility of fiberglass as a reinforcement aids an industry which is continuously looking for new materials that impart enhanced properties in a cost-effective way. Drivers for the growth of commercial applications of composites center around advantages in mechanical properties, corrosion resistance, weight reduction, recyclability, cost, and scale of operation. The composite industry is built on the lowest system cost.

This chapter focuses on general descriptions of the various technologies that are available for engineering composites that meet the performance and cost targets of a wide variety of applications. This chapter is largely based on PPG's Manual "Introduction to Glass Fibre Composites" [1].

3.1.2 E-Glass Fibers

Glass refers to a group of materials which are basically under-cooled liquids. Glass consists of various oxides, which melt to form eutectics. When the melt is quickly cooled to room temperature, a clear rigid solid forms. This glassy state, unlike the solid state and the fluid state, is not thermodynamically stable, but transition rates are so slow that glass is, for all practical purposes, a stable solid material. Not every glass composition can be fiberized. The viscosity–temperature relationship of the glass predominately determines whether the glass can be drawn into small diameter continuous filaments. The glass composition must provide a high resistance to environmental attack from water because of the extremely large surface area exposed in fiberized form.

Glass fibers are predominately produced from a composition known as E-glass, an alumina–borosilicate glass. E-glass is a glass composition that imparts among others strength, stiffness, corrosion resistance, low electrical conductivity, and essentially isotropic properties. The designation E-glass is specified by a compositional range for each inorganic oxide and element in the glass [1]. Therefore, physical and chemical properties vary to a limited extent depending on the exact composition. Also physical properties of E-glass fibers are dependent on the processing history of the fibers. Illustrative properties of E-glass within the specified compositional range are summarized in Table 3.1.

Table 3.1 Typical physical properties of E-glass [1]

Property	Range of published values[a]
Density – bulk (g/cm^3)	2.54–2.62
Coefficient of thermal expansion – bulk (m/m/K)	4.9–6.0 × 10–6
Specific heat – bulk (kJ/kg K)	0.8
Softening temperature (°C)	830–916
Thermal conductivity – bulk (W/m K)	1.0
Refractive index – bulk	1.547–1.560
Tensile strength – filament (GPa)	3.1–3.8
Tensile modulus – filament (GPa)	76–81
Elongation at break – filament (%)	4.5–4.9
Poisson's ratio	0.18
Dielectric strength – bulk (kV/mm)	10.3
Volume resistivity – bulk (ohm cm)	10^{15}

[a]Includes E-glass compositions with and without boron oxide.

The original E-glass designation included only formulations with boron oxide content in the 5–10 wt% range. Commercial interest in and use of low or no boron oxide variants of E-glass resulted in a revision of the American Society for Testing and Materials specification of the E-glass composition [1]. The range of boron oxide content was widened to 0–10%. The broader range was designated for general applications. The 5–10% range was designated for printed circuit board and aerospace applications (ASTM D578).

3.1.3 Fiberglass Manufacturing

Why fiberglass is such a useful reinforcement fiber for so many different applications can be explained through the nature of the raw materials and the versatile continuous manufacturing technology. The raw materials are relatively inexpensive minerals such as kaolin, quartz, lime, and boric acid. However, capital intensive installations are required to form the glass and fibers. The mineral ingredients are reacted and melted in a furnace where gaseous inclusions are removed and the oxide network of the glass is formed. From the melter the glass flows through channels into a forehearth, while the temperature is lowered to allow fiberization at forming.

A schematic of the formation of continuous glass filaments is presented in Fig. 3.1.

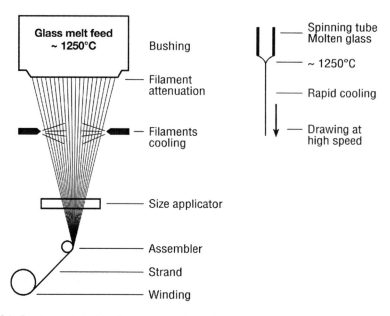

Fig. 3.1 Representation of continuous filament-forming process

Filaments are drawn from the glass, which behaves as a Newtonian fluid, flowing under gravity through orifices in a platinum alloy "bushing." This bushing is fabricated by inserting tubes into a platinum alloy plate with the tips of the tubes extending below the face of the plate. A delicate balance between diameter of the tips, viscosity of the glass (acutely temperature dependent), glass flow, and drawing speed controls the attenuation into solid filaments. The design of the bushing, specific for the final product, will determine throughput, filament diameter, and number of filaments. The cooling of the glass is aided by cooling fins positioned close to the tips and by a controlled forming environment [2]. The versatile process can produce products ranging in individual filament diameter from 4 to 40 μm and in number of filaments per bundle from 200 to 4,000.

The filaments are drawn from the bushing by a cylindrical take-up winder to form a spool of filaments referred to as forming cake, spin cake, or forming package. Between the bushing and the winder, the bundle of filaments is coated with a size that protects and lubricates the brittle glass filaments. The size provides the abrasion protection required for mechanical processing of the fiber bundle. Without the coating, inter-filament abrasion and abrasion at contact surfaces result in breakage of the filaments and in a practical sense would prevent the forming of continuous filament bundles. The composition of the size is critical to both the processing and the use of fiberglass. Characteristics imparted to the filament bundle by the size affect a host of behaviors, such as friction against surfaces and inter-filament tack bonding, that are critical in processing. The size is the agent that determines the properties of the interphase between the glass filaments and the polymer matrix in composites.

After the application of the size, the fiberglass continuous filament bundles can either be directly drawn into a continuous processing step, such as chopping, or be wound as a package which is processed in secondary fabrication steps.

3.1.4 Fiberglass Size

Sizes are waterborne mixtures of dissolved and dispersed ingredients. Terms such as finish, binder, and coating are used interchangeably with size. The most important factor in the performance of glass fibers in composites is the nature of the size that is applied to the strand during the forming process. This size provides lubrication during filament forming and determines to a large extent the processing characteristics of the glass fiber products. The interface interaction between the glass filament surface and the surface of the polymeric matrix is critical in determining the properties of the composite. The exact composition of the size is usually considered a manufacturer's secret and may be protected by patents. Much of the proprietary character of fiberglass products is imparted by the characteristics of the size. The typical amount of size on the filaments can be as low as 0.1 wt% but generally is in the range from 0.3 to 1.5 wt%. The average thickness of the size layer varies from a few to several tens of nanometers depending on the particular application of the product.

The size is a complex mixture of components, each of which is carefully chosen for its specific contribution to the performance of the glass fibers during processing and in the composite. Water as a carrier is the predominate component of a size formulation. In general, the active components in sizes are in three functional groups, as represented in Fig. 3.2.

The film-forming agent is a controlling factor in the degree of tack bonding between the filaments. This inter-filament bonding is important in providing the degree of strand stiffness required for various processing steps. The polymer is also functional in protecting the glass surface and in inducing wetting and compatibility with matrix resins. The predominate polymers in the form of emulsions and dispersions in sizes are polyvinyl acetates, polyesters, epoxies, and polyurethanes, sometimes in combination. Water-soluble polymers, such as polyvinyl pyrrolidone

- **1% to 20% Organofunctional silane**

  $(C_2H_5O)_3SiCH_2CH_2CH_2NH_2$ gamma-aminopropyltriethoxysilane

                          O
                          ‖
  $(C_2H_5O)_3SiCH_2CH_2CH_2OCC=CH_2$ 3-glycidoxypropyltriethoxysilane
                          |
                          CH₃

                                    O
  $(C_2H_5O)_3SiCH_2CH_2CH_2OC$ H₂CHCH₂   gamma-methacryloxytriethoxysilane

- **1% to 40% Process Aids**

  Lubricants, antistats, antioxidants, wetting agents

- **50% to 90% Polymeric Film Former**

  Polyurethane, polyvinylacetate, epoxy, polyester

Fig. 3.2 Functional groupings for typical ingredients in sizes

and polyvinyl alcohol, are also used. The very high coefficients of friction of fiber-glass, filament-to-filament and filament-to-contact surfaces, require that lubricants be a component of the size. Additives, such as antistatic agents and antioxidants, are present to alter the response of the fiberglass size to specific processing conditions. For instance, an antioxidant is added to prevent thermal degradation and color formation when the size is exposed to a damaging temperature during processing.

A third group in the size functions as coupling agents between the surface of the glass filaments and the resin matrix. High mechanical strength and wet strength retention in fiberglass composites can be obtained only when a high-quality, durable glass–matrix bond exists. This bonding can be greatly enhanced by applying a coupling agent to the glass fiber surface [3]. Most coupling agents are organo-functional silane compounds. The success of fiberglass/resin composites is largely dependent on the effectiveness and versatility of organo-functional silanes.

3.1.5 Composite Mechanical Properties

Composites differ from homogeneous materials in that they consist of at least two components that have widely different properties. Glass fiber composites may contain short (chopped) or long (continuous) fibers which can be aligned to coincide with the direction of stress The glass fibers may also be transversely oriented or may be randomly arranged. The interface between the fiber and the continuous phase or matrix polymer is a key contributor to the overall mechanical properties of the composite. The interphase transfers the load from the matrix to the fiber.

The behavior and properties of fiber-reinforced composites are studied at three levels [4]:

1. At the interphase level, where molecular scale parameters, such as van der Waals forces, acid–base interactions, and covalent chemical bonding, determine the interaction between the phases.
2. At a mechanics level, the interfacial interaction is described by parameters which characterize the transfer of stress and load, with the distribution of reinforcing fibers being a crucial parameter.
3. From a macro perspective, the characterization of the properties of the composite as a bulk material.

This chapter focuses on the later two levels, which are guidelines for engineers.

3.1.5.1 Unidirectional Continuous Fibers

Consider the case of a specimen with original length L, consisting of a homogeneous matrix that adheres tenaciously to the imbedded continuous fibers. When load P is applied to this specimen with section F, a stress σ is developed throughout the specimen, which equals

$$\sigma = P/F$$

There are three main kinds of stress: tensile, compressive, and shear stress. Under tensile stress the specimen elongates and the increase in length (ΔL) is referred to as elongation.

The change in length per unit original length ($\Delta L/L$) is the strain (ε). The ratio of stress to corresponding strain is the modulus of elasticity (E) at a specified strain

$$E = \sigma/\varepsilon$$

The modulus of a material is a measure of its stiffness. Depending on the type of stress applied to the specimen, the moduli are referred to as tensile modulus, flexural modulus, etc.

In the case of glass fibers in a homogenous polymer matrix, the moduli of the glass fiber are orders of magnitude greater than the moduli of the organic polymer. The failure stress or the strength of the composite is determined by gradually increasing the stress until rupture occurs. When the continuous fiber or the aligned bundle of fibers is oriented in the lengthwise direction as illustrated in Fig. 3.3, the capability of the fibers to bear the applied stress is maximized.

Strength at rupture can be represented by a simple rule of mixtures where the composite strength is a function of the strength and the volume fraction of the constituent fiber and the matrix.

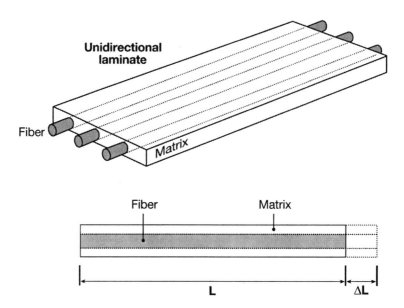

Fig. 3.3 Illustration of strain effect in a unidirectional laminate loaded in the fiber direction

$$\sigma = (V_f)(\sigma_f) + (V_m)(\sigma_m)$$

In this rule of mixtures, σ_c = composite tensile strength, V_f = fiber volume fraction, σ_f = fiber tensile strength, V_m = matrix volume fraction, and σ_m = matrix tensile strength.

This simple rule of mixtures applies only in the theoretical case where there is perfect adhesion between the matrix and the fiber. Also this rule of mixtures applies to composite materials only when the matrix is capable of stress transfer up to the point of rupture. To fulfill this requirement, the matrix resin should have an elongation at break that is at least equal to that of the reinforcing fiber. The hypothetical stress–strain diagram in Fig. 3.4 illustrates this requirement.

In Fig. 3.4 at a specimen strain (a), both matrix resins A and B are still capable of stress transfer, and the relative contribution of resin and fiber to specimen stress can be read from the vertical axis *if there is perfect adhesion*. At strain (b), resin A will start breaking and will progressively fragmentize as strain increases. Beyond strain (b), resin A loses its stress transfer function, and the glass fiber is no longer protected against abrasive influences and will rupture before the theoretical maximum stress is reached. The rule of mixtures applies to resin B if there is perfect stress transfer between the resin and the fiber up to the point of rupture. In the case of resin A, rupture of the resin matrix will occur at a strain prior to the rupture strain of the fiber; thus even with perfect load transfer, the maximum potential reinforcing effect of the fiber is not achieved. Resin B with an elongation at break that matches the fiber's elongation represents the ideal situation with the proviso that strong bonding (stress transfer) exists between the resin and the fiber up to the strain of rupture.

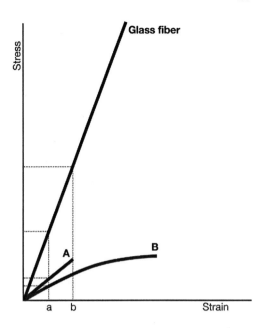

Fig. 3.4 Stress–strain diagram of fiberglass and two resin types

3.1.5.2 Bidirectional (Orthotropic) Reinforcement

Unidirectionally reinforced composites belong to a specialty group of materials; it is more common to design composites to have load-bearing properties in two or more directions. Figure 3.5 is a representation of two stacked layers with unidirectional fibers, crossing at a 90° angle.

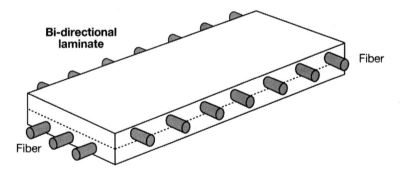

Fig. 3.5 Bidirectional reinforced laminate

When such a composite is loaded in the direction of one of the layers or *laminae*, the response to strain is illustrated in Fig. 3.6.

The layer that is loaded in the fiber direction will respond as described previously for a unidirectional laminate. The layer with the fibers oriented perpendicular to the direction of strain consists of a discontinuous array of alternating areas of resins and

Fig. 3.6 Induced
unidirectional strain in
bidirectional laminate

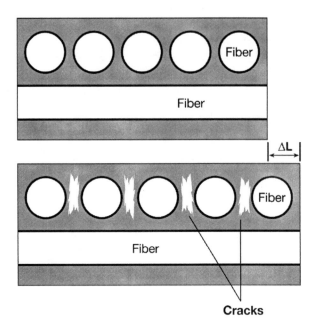

Cracks

fibers. The fiber has a much higher resistance to deformation than the matrix resin
due to its significantly greater modulus. The strain in the layer with perpendicular
orientation of the fiber axis to the direction of strain will mainly take place in the
resin areas between the fibers. The actual strain in the resin component is therefore
much greater than the observed strain in the overall composite. This phenomenon is
known as strain magnification.

Visual and audio evidence of strain magnification occurs when a cross-piled laminate is tested for tensile strength. In clear test specimens, the onset of traverse cracks
at relatively low strains can be observed, while simultaneously the specimen starts
to emit cracking sounds. The stress–strain curve recorded during such a tensile test
may show a deviation to a different angle at the onset of cracking. This point of
deviation is designated as a "knee."

3.1.5.3 Random Short Fibers

A large percentage of continuous filament fiberglass is cut (chopped) into short
lengths (tens of millimeters) for use as reinforcement in a variety of applications.
In these applications the short fibers are usually distributed randomly along two or
three principal axis. Fig. 3.7 depicts the response of a composite reinforced with
random short fibers to strain.

In the unstrained condition a discontinuous fiber with length l is embedded in a
matrix resin. When the matrix is loaded in a tensile mode in the direction of the fiber
axis, a complex deformation pattern is created around the fiber. As the tensile load

Fig. 3.7 Response of random
short fiber composite to strain

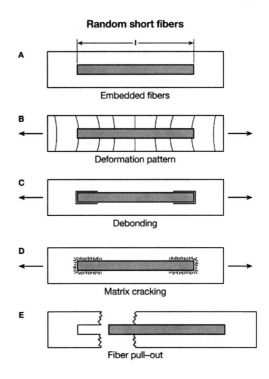

Random short fibers

A

Embedded fibers

B

Deformation pattern

C

Debonding

D

Matrix cracking

E

Fiber pull–out

is gradually increased until the specimen ruptures, the response to strain progresses
through several stages, some of which are visible and some are not.

Provided that the resin matrix shows a ductile behavior, characteristic of most
thermoplastics resins, the stages of fracture that can be identified are the following:

- *cracking* in the resin matrix around the fiber, usually at some distance from the
  interface
- *de-bonding*, starting at the fiber ends and gradually progressing along the fiber
  length toward the fiber's middle part
- *matrix failure*, mostly as a crack-induced fracture
- *fiber rupture*, which occurs only in fibers that are sufficiently long to be stressed
  to failure (see the discussion of "critical length")
- *fiber pullout*, characterized by lengths of fibers protruding from the face of a ruptured specimen and the empty cavities from which they were extracted (Stage E
  in Fig. 3.7 and fracture surface in Fig. 3.8)

Whether de-bonding or matrix cracking is the first phenomenon depends on the
quality of the bond between the resin matrix and the glass fiber and the toughness
of the matrix.

Fig. 3.8 Face of a ruptured
specimen

It is likely and also substantiated by microscopic analysis that failure starts from
the fiber ends. The faces of the fiber ends are created in a chopping process that follows the process of forming continuous fibers and do not have a size applied to their
surface. Thus weak adhesion is expected between the glass surface at the fiber ends
and the resin. In addition, both the fiber and the end faces are stress concentration
points during tensile deformation. As stress increases, either the matrix may start
cracking or there may be de-bonding at the fiber–matrix interface. Both phenomena
will most likely have their initiation point near the fiber ends and gradually proceed
toward the middle part of the fiber. Further stress increase will then lead to rupture
of the specimen.

Figure 3.8 presents the microscopic examination of the face of a typical rupture of
a specimen. Electron micrographs typically reveal matrix rupture and fiber breakage
and also a certain degree of fiber pullout. Fiber pullout is witnessed by the presence
of fiber ends protruding from the face of rupture and corresponding empty fiber
sockets.

Whether a fiber breaks or is pulled out depends on its length. The dependence
on length can be explained by an over-simplified analogy. Assume that a length of
1 m of sewing yarn is firmly gripped in both hands. By exerting a pulling force the
thread will break. Next, one section of the broken thread is broken and the procedure
is repeated many times. There will come a moment when the broken pieces are so
small that it proves impossible to get the necessary grip on the thread to break it.
The thread slips through the fingers.

In composite theory, the shortest length of fiber that can be broken is designated
as the *critical length*. The critical length can be derived from the following equation:

$$l_c = (\sigma d)/2\tau$$

where l_c = critical fiber length, σ = fiber tensile strength, τ = interfacial shear
strength between fiber and matrix resin, and d = fiber diameter.

If the actual fiber length is less than the critical length, the stress on the fiber will not exceed the stress required to break the fiber. In this case, failure of the composite is initiated by failure at the interface or within the matrix resin. In the composite, the interfacial shear stress is greatest near the fiber ends and decreases toward the midpoint of the length. The normal stress in the fiber is least at the fiber ends and increases toward the midpoint of the length. The preceding equation predicts that the critical fiber length will decrease as the strength of the interfacial bond increases.

For fiber lengths greater than the critical length, the reinforcement value effectively increases with fiber length according to the relationship

$$E_f = [l - (l_c/2\,l)]$$

where E_f = fiber efficiency factor, l_c = fiber critical length, and l = actual fiber length.

A graphical representation of the dependence of reinforcing efficiency on fiber length is presented in Fig. 3.9. The equation used for the calculation of the fiber efficiency factor is from a generalized theoretical approach and may or may not be valid depending on the nature of the composite material and the specific property investigated.

Fig. 3.9 Reinforcing efficiency versus fiber length

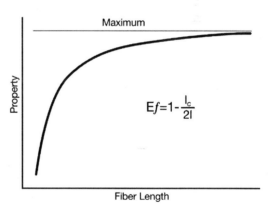

3.1.5.4 Test Methods

Comparison of property data on composite materials is possible only when sample preparation and test procedures are standardized. Standards that are internationally agreed upon are issued by the International Organization for Standardization (ISO), Geneva, Switzerland. Once established, ISO standards are adopted by member countries as national standards. A brief discussion of the tests most often used for the characterization of composite materials follows.

Tensile strength and modulus. The tensile test provides an insight into the stress–strain behavior of a material under uniaxial tensile loading. Tensile test easily distinguishes between brittle and ductile materials. These tests are very useful tools for quality control and general property comparison. Tensile testing is not representative of applications with load–time scales widely different from those of the standard test.

Tensile modulus (initial modulus) is derived from the tangent to the initial linear part of the stress–strain diagram.

Flexural strength and modulus. Flexural strength is determined on bar-shaped specimens with rectangular or circular cross section. The specimen is supported horizontally and a load is applied vertically. Flexural strength is an important parameter for property comparison because flex is a stress-deformation mode that is often encountered under service conditions.

Compressive strength. Compressive strength is the maximum stress borne during a compressive test. The failure in the compressive testing is highly dependent on the interfacial bonding between resin and glass fiber. Although not generally applied for routine composite property characterization, the compressive test is valuable for evaluating the effect of environmental conditions on the resin-to-fiber bonding in composites. For instance, response to exposure to moisture and/or elevated temperature is evaluated by comparing compressive strength before and after exposure.

Shear strength. Shear strength is the maximum load required for complete shear of the specimen divided by the shearing area. The most widespread method for testing interlaminar shear strength (ILSS) is the short span flexural test. If, despite the short span to thickness ratio, the specimen fails in a flexural rather than a shear mode, the result of the test should not be reported as ILSS. Short span flexural testing is applicable to composites with unidirectional and bidirectional reinforcement but does not give satisfactory results with planar random and three-dimensionally random short fiber composites.

ILSS data are often used to specify composite quality. The ILSS value is considered to be a direct function of interfacial adhesion.

Impact strength. Impact strength is probably the most widely tested, but the least understood, of all composite properties. The difficulty in understanding and applying impact testing to predict functional behavior is illustrated by the large number of methods that are in use. A well-recognized variable in impact testing is the finding that specimen shape and size, as well as molding conditions, significantly affect the outcome of impact testing. Lists of ISO standards relevant to glass fibers and glass fiber composites are given in Tables 3.2 and 3.3.

3.1.6 Products

Glass fiber is available in a variety of forms. These include the following product groups:

Glass mats. Glass mats are available in two distinct forms: those known as chopped strand mat (CSM) used primarily for the hand lay-up molding process and continuous strand mat generally used in press molding and pultrusion applications.

Table 3.2 Important standards for mechanical testing of fiberglass-reinforced composites

ISO 62	Plastics – Determination of water absorption
ISO 75	Plastics – Determination of temperature of deflection under load
ISO 178	Plastics – Determination of flexural properties
ISO 179	Plastics – Determination of Charpy impact properties
ISO 180	Plastics – Determination of Izod impact strength of rigid materials
ISO 291	Plastics – Standard atmospheres for conditioning and testing
ISO 472	Plastics – Vocabulary
ISO 527	Plastics – Determination of tensile properties
ISO 604	Plastics – Determination of compressive properties
ISO 1043	Plastics – Symbols and abbreviated terms – Part 2: Fillers and reinforcing materials
ISO 1172	Textile–glass-reinforced plastics – Prepregs, molding compounds, and laminates – Determination of the textile–glass and mineral–filler content – Calcination methods
ISO 1268	Fiber-reinforced plastics – Methods of producing test plates
ISO 3597	Textile–glass-reinforced plastics – Determination of mechanical properties on rods made of roving-reinforced resin
ISO 4899	Textile glass-reinforced thermosetting plastics – Properties and test methods
ISO 7822	Textile glass-reinforced plastics – Determination of void content – Loss on ignition, mechanical disintegration, and statistical counting methods
ISO 8604	Plastics – Prepregs – Definitions of terms and symbols for designations
ISO 8605	Textile–glass-reinforced plastics – Sheet molding compound (SMC) – Basis for a specification
ISO 8606	Plastics – Prepregs – Bulk molding compound (BMC) and dough molding compound (DMC) – Basis for a specification
ISO 9353	Glass-reinforced plastics – Preparation of plates with unidirectional reinforcements by bag molding
ISO 9782	Plastics – Reinforced molding compounds and prepregs – Determination of apparent volatile matter content
ISO 10352	Fiber-reinforced plastics – Molding compounds and prepregs – Determination of mass per unit area
ISO 11667	Fiber-reinforced plastics – Molding compounds and prepregs – Determination of resin, reinforced fiber and mineral–filler content – Dissolution methods
ISO 12114	Fiber-reinforced plastics – Thermosetting molding compounds and prepregs – Determination of cure characteristics
ISO 12115	Fiber-reinforced plastics – Thermosetting molding compounds and prepregs – Determination of flowability, maturation, and shelf life
ISO 14125	Fiber-reinforced plastic composites – Determination of flexural properties
ISO 14126	Fiber-reinforced plastic composites – Determination of compressive properties in the in-plane direction
ISO 14130	Fiber-reinforced plastic composites – Determination of apparent interlaminar shear strength by short-beam method
ISO 15034	Composites – Prepregs – Determination of resin flow
ISO 15040	Composites – Prepregs – Determination of gel time
ISO 15310	Fiber-reinforced plastic composites – Determination of the in-plane shear modulus by the plate twist method

Chopped strand mats are made by chopping glass strand into uniform lengths (typically 50 mm) onto a moving conveyor. The chopped strands fall randomly and evenly onto a mesh belt forming a mat or a fleece. The individual strands are then

Table 3.3 Important standards for dynamic testing of mechanical properties of fiberglass-reinforced composites

ISO 6721	Plastics – Determination of dynamic mechanical properties
ISO 13003	Fiber-reinforced plastics – Determination of fatigue properties under cyclic loading conditions

bound together by the controlled application of a powder, typically a polyester, or an emulsion, typically poly(vinyl acetate). The mat is then passed through a series of ovens, a process which softens the powder causing it to flow or alternatively dries the emulsion to form a bond between strands. The mat cools on leaving the oven and is wound onto cardboard tubes.

Rovings. Rovings are supplied in two district types: those known as direct draw rovings and those commonly referred to as assembled rovings.

Direct draw rovings – This roving is produced by forming a large number of glass continuous filaments and drawing this multitude of filaments into one single strand. The bundle of filaments is wound onto a cardboard tube and then dried. This process gives a product which is totally free of loops (catenary) and is commonly used in weaving, filament winding, and pultrusion applications.

Assembled rovings – These rovings are produced by combining several strands to produce a roving with a number of filaments in the bundle that is a multiple of those in the individual strands. The process consist of winding the individual strands together into a cylindrical package (hence the term assembled).

A major application of assembled roving is in the area of spray deposition where the rovings are required to chop and break up into separate integral strands free from clumps. Shower stall and small boat fabrication utilize spray deposition of assembled rovings. The same requirements are demanded of rovings used in sheet molding compound. Assembled rovings are also used in weaving, filament winding, and pultrusion where special characteristics are derived from the assembly of parallel strands.

Chopped strands. Chopped strands are produced by continuously feeding bundles of continuous glass filaments through a rotating glass cutter. The cut lengths are varied to suit the end use requirements, typically in the range of 3–50 mm. Chopped strands are used for reinforcing both thermosetting and thermoplastic resins. There are three major areas of application for chopped strands. In injection molding, chopped strands reinforce a wide range of thermoplastic polymers. Polymers commonly reinforced include polyamides, polyesters, and polypropylene. Chopped strands reinforce dough molding compound (DMC) which is used in pressure molding applications where intricate shapes of varying wall thickness sections are required. Chopped strands are dispersed in water in the manufacture of glass felts or tissues used to reinforce building materials such as plaster and gypsum

Milled fibers. Milled fibers are manufactured by the continuous milling of glass fibers and are typically available in strand lengths between 50 and 250 μm. Milled glass fiber gives higher strength reinforcements than do most mineral fillers such as chalk or talc, but less than the longer glass strands.

Fabrics woven from rovings. Fabrics woven from rovings are used as reinforcement of thermosetting polymers in a wide variety of applications

Non-woven fabrics. Non-woven fabrics can be unidirectional, bidirectional, and multi-axial in construction. These fabrics are designed to overcome the "crimp effect" of woven fabrics where the warp (machine direction) and the filling (across machine direction) cross over forming "knuckles." The non-woven fabrics are designed with the rovings in a given direction being completely parallel. The completely parallel orientation in a given direction results in optimum use of the unidirectional tensile strength of the rovings. A variety of stitching techniques are used to tie the rovings into the configuration.

Multi-axial fabrics can be made up of several layers of roving each with its own specific direction, typically 30° or 45° to the warp. This allows precise engineering placement of the fabric with the glass roving being placed in the direction of maximum stress or design loading.

Glass yarn. In the fiberglass industry, the term yarn is conventionally applied to bundles of continuous filaments, with an individual filament diameter of 13 μm or less and with less than 1600 filaments in the bundle.

3.2 Thermoset Composite Material

Composites based on thermosetting resins are formed when the liquid monomer or pre-polymer is transformed through a chemical reaction into a cross-linked solid polymer. The reaction produces a three-dimensional network through covalent bonding of macromolecules. The term resin is often applied both to the precursor chemicals and to the resulting polymer. An example is the use of "polyester resin" referring to liquid reactants which are polymerized by free radical initiation to form a cross-linked solid polymer which is also referred to as an "polyester resin."

Thermosetting resin composites are infusible and in general, hard and brittle compared to thermoplastic resin composites. A large variety of resins and reaction systems are available for applications in automotive, transportation, marine, construction, electrical, leisure, defense, and many other applications. Reinforced thermoset composites have highly diverse applications.

Composites from thermosetting polymers with glass fiber reinforcement are typically used for large parts produced at relatively low numbers. Figure 3.10 illustrates applications where the thermoset composites are large and produced in relatively low quantity. Two categories of thermoset composites are exceptions to the preceding generalization. Parts fabricated from sheet molding compound (SMC) and bulk molding compound (BMC) are produced in a mass production mode.

In all forms of fabricating composites from resin and glass fiber, an initial process step is to achieve proper wetting of the fibers by the matrix resin. With liquid monomer or pre-polymer, the ease and speed of wetting is a function of viscosity and the match of surface energies of the liquid and the fiber. Temperature increase or physical factors, such as compressive and shear forces, can be applied to

Fig. 3.10 Reinforced
thermoset composite
applications involving the
molding of large parts

enhance wetting. The chemical compatibility between the matrix polymer and the
size on the surface of the glass filaments is crucial. Composite quality is dependent
on the creation of a perfect interface. An equally important factor is the distribution
throughout the resin matrix of the glass fibers. The degree of fiber dispersion and
the volume fraction are parameters controlled by the fabrication process as well as
the characteristics of the individual components.

Processing of thermosetting resins whose precursor reactants are liquids was the
foundation of the first large-scale production of glass fiber-reinforced composites.
Techniques such as hand lay-up (HLU), spray-up, filament winding, pultrusion,
continuous laminating, centrifugal molding, and resin transfer molding (RTM) are
employed for fabricating thermoset composites. Compression molding with thermoset resins whose precursor reactants are liquids and dry reinforcement bridges
the gap between low series production characteristic of HLU, spray-up, or RTM
and high series production characteristic of SMC and BMC.

3.2.1 Liquid Resin Processing Techniques

Glass fiber composite products of various shapes and sizes can be made by impregnating reinforcing fibers with liquid precursor reactants in or on a suitable mold,
followed by curing at ambient or elevated temperature. The reinforcing glass fibers

can be in continuous filament form or chopped rovings, chopped strand mat, or fabrics.

3.2.1.1 Hand Lay-Up (HLU)

The hand lay-up technique was the first method employed for producing glass-reinforced plastics (GRP) and is still widely used. The molds are relatively inexpensive and can be quickly made from wood, plaster, steel, or more commonly glass-reinforced plastics. The technique results in a part with only one perfectly smooth surface. The smoothness of the back surface can be improved by the application of a surfacing tissue or flow coat. The mold surface is prepared to the desired finish and release agent is applied. The get coat, a fast curing thixotropic liquid, is then sprayed or brushed onto the mold surface and allowed to gel. The purpose of the gel coat is to provide a resin-rich surface on the GRP molding to achieve a good finish and resistance to chemical attack. The gel coat usually carries the color of the laminate in the form of a pigment paste.

The next stage is to evenly apply a coating of reactants followed by the glass fiber reinforcement which is impregnated with brush, roller, or spray. Air bubbles are removed and consolidation effected by the use of special rollers. Curing is accomplished at room temperature by the addition of the necessary curing system. Unsaturated polyester resins are predominantly used.

3.2.1.2 Spray Deposition

Simultaneous deposition of chopped fibers and reactants onto the mold can be considered as a mechanized version of the hand lay-up technique. The process is commonly referred to as spray-up. This technique enables a faster production rate than does hand lay-up particularly where large areas are involved. The requirements of a spray depositor device are that glass fiber roving be chopped to a predetermined length and delivered into the center of the spray of the reactants prior to landing on the mold surface. Gel coats are applied to the mold in the same manner as for hand lay-up.

Combinations of reactants with significant styrene content are common in spray deposition. Styrene emission is a potential problem because of the large surface area of resin exposed to the air. Polyester resins are available which are designed to minimize the styrene emission problem.

3.2.1.3 Resin Transfer Molding (RTM)

Resin transfer molding, also known as resin injection molding or liquid injection molding, involves the placement of engineered reinforcement layers in the cavity of a mold, which is then closed. Low-viscosity liquid reactants are injected into the cavity, assisted by either vacuum or pressure. The reinforcement can be pre-formed from mats or from chopped rovings to match the cavity form. The fibers are tack bonded with an insoluble binder to prevent displacement of the reinforcement by the resin flow. Special fabrics or braids are used to achieve specific property enhancements. The development of reactant systems with low viscosity and high reactivity with automation of the preforming and injection process has lead to significant decreases in cycle time.

3.2.1.4 Reinforced Reaction Injection Molding (RRIM)

The principle of the RRIM process centers on bringing together two fast reacting liquids into a mixing chamber and then into a mold cavity. Relatively low clamping force pressure is used, and external heat is not normally used on the surface of the mold. The rate of the polymerization reaction between the two liquids gives a cycle time of typically less than 60 s in the mold. The chemical reaction starts immediately when the two liquids are combined through a mixing chamber and progresses as material flows into the mold cavity.

In general, RRIM has been developed using polyurethane formulations, with one reactant stream being a polyol and the other being a di-isocyanate. The glass reinforcement normally used in the RRIM process is milled fiber added to the polyol stream. An alternate processing method involves placing a glass fiber pre-form into the mold cavity and then injecting the two liquids into the mold cavity. While polyurethanes are the dominant resin matrix used in this process, other matrix materials such as phenolics are used to achieve specific property requirements.

3.2.1.5 Filament Winding

Continuous glass fiber rovings are impregnated with liquid reactants and wound onto a rotating mandrel in predetermined patterns. The rovings are positioned at various angles to the axis of the mandrel yielding cylindrical objects with designed directional properties. At a near 90° angle, called circumferential or hoop winding, the radial strength is high but the axial strength is very low. The reverse holds for an axial fiber orientation. An example of the axial fiber orientation situation is the polar winding of closed-end pressure vessels. Additional hoop windings provide the necessary radial strength in the closed-end pressure vessel application. When the continuous filament bundles are wound in a helical pattern, both axial and radial strength are a function of the winding angle. Cylindrical shells for silos and containers are often produced by a combined spray-up and winding technique (chop-hoop winding) where layers of chopped roving are alternated with circumferentially wound rovings. A continuous version of the filament winding process has been developed. Unsaturated polyester, vinyl ester, and epoxy resins are most frequently used as the matrix for filament-wound structures.

3.2.1.6 Centrifugal Molding

Cylindrical objects can be made by using centrifugal forces generated when a mass is rotating. Reactants and reinforcing fibers, mostly in the form of chopped rovings, are introduced into a rotating hollow drum by a resin/glass spray gun traversing along the axis of the drum. The combination is pressed against the inside wall of the drum by centrifugal forces and cures at elevated temperature. The parts that are obtained have a smooth outer surface and a resin-rich inner layer.

3.2.1.7 Pultrusion

Shapes with a uniform cross section such as rods, angles, hollow tubes, and squares can be produced by the pultrusion process. Continuous reinforcement is impregnated in a bath containing reactants and then drawn through a heated forming die where the resin cures. Using only continuous glass fiber rovings, the resultant profile has highly directional properties. To obtain isotropic character, fabrics or combinations of mat/fabric are used in the pultrusion process, sometimes in combination with rovings. A variation of pultrusion is known as pull forming which is a combination of pultrusion and compression molding. This process has been used for the manufacture of vehicle leaf springs.

3.2.1.8 Continuous Laminating

Flat and corrugated sheets are produced continuously by depositing chopped rovings or chopped strand mats onto a thin film of liquid unsaturated polyester reactants that have been applied to a moving sheet of cellophane or poly(ethylene terephthalate). The upper surface is covered by a second layer of protecting film, air is removed, and the sandwich passes through an oven where forming members provide the required corrugation while the resin cures.

Tables 3.4 and 3.5 contain summaries of the characteristics of these liquid resin processing techniques.

Table 3.4 Non-continuous liquid resin processing techniques

Technique	HLU + gel coat	Spray-up + gel coat	RTM	Centrifugal molding
Reinforcements	Mats, fabrics	Rovings (chopped)	Mats, fabrics, preforms	Rovings (chopped)
Resins[a]	UP, VE, EP, PF	UP, VE, PF	UP, EP, VE	UP, VE
Weight percentage of glass	25–35	25–35	20–65	25–40
Cure temperature (°C)	Ambient	Ambient	20–50	Ambient
Cycle time (h)	> 8	> 8	0.1–8	0.5–5
Typical products	Boats, industrial moldings	Boats, industrial moldings	Industrial moldings	Tanks, silos, pipes

[a]Unsaturated polyester (UP), vinyl ester (VE), epoxy (EP), phenol–formaldehyde (PF)

Table 3.5 Continuous liquid resin processing techniques

Technique	Winding	Pultrusion	Continuous lamination
Reinforcements	Rovings (continuous)	Fabrics from continuous rovings	Mats, rovings (chopped)
Resins[a]	UP, VE, EP, PUR	UP, VE ,EP	UP
Weight percentage of glass	50–70	65–75	20–30
Cure temperature (°C)	50–100	100–180	50–140
Production rate (kg/h)	20–150	5–60	60–120
Typical products	Pipes, tanks	Rods, beams	Corrugated sheets, panels

[a]Unsaturated polyester (UP), vinyl ester (VE), epoxy (EP), polyurethane (PUR)

3.2.1.9 Pre-combined Materials

The production of glass fiber composites on an industrial scale requires the availability of mechanized or automated processing methods. Continuous laminating and pultrusion techniques are examples of industrial processing where liquid reactants are metered to the machine. The production of large series of individual parts is facilitated when the precursors of matrix and reinforcing material are precombined in the required ratio and only need to be converted to the final product by a single processing step without the addition of further components. Semifinished products play an ever increasing role in glass fiber composite production. These precombined materials may be based either on thermosetting resins or on thermoplastic resins.

Pre-impregnated Fabrics (Prepregs)

Reinforcing fabrics made from glass fiber are impregnated with the precursor of a thermosetting resin, which is in most cases an epoxy resin. The fabric is led through a bath containing the precursor (reactants) solution of relative low viscosity. The excess solution is removed by passing between one or more pairs of rollers, the last pair being calibrated to exactly control the pickup of the solution.

The impregnated fabric then travels through an oven where the solvent is evaporated and the resin is advanced to a B stage, an intermediate state of cure. After leaving the oven, the prepreg is cooled to prevent further advancement of cure and then wound onto rolls with release paper or film between the layers to avoid sticking. In this stage, the resin is dry to the touch but melts on application of heat. A typical resin content for prepreg is 30–40 wt%.

Pre-impregnated continuous rovings and tapes are sometimes used for filament winding. These applications are based on epoxy resins and on special grades of unsaturated polyester resins.

Sheet Molding Compounds (SMC)

Sheet molding compound is a pre-combined material in sheet form containing predominantly chopped rovings as a reinforcement. The capability to flow in all directions enables the molding articles of intricate shape. The process can produce parts with ribs, bosses, and varying wall thickness. The majority of SMC is formulated with unsaturated polyester resins. The resin is mixed with initiator, release agent, fillers, pigments, and optionally with polymer additives as shrinkage reducers. These latter ingredients are also referred to as low-profile additives. Toughened grades of SMC can be formulated by using special styrene–butadiene rubbers as low-profile additives.

Most formulations contain magnesium oxide or hydroxide as a thickening agent, which reacts with terminal carboxy groups of the polyester. Thickening can also be accomplished through formulation of an interpenetrating polymer network by introducing a polyol, followed by the addition of a di-isocyanate.

The resin paste is usually prepared with high shear mixing equipment to attain homogeneous dispersion of all the components. In batch-type processes, reactive thickeners are added as the last component during preparation of the paste. From this moment on, the batch has a limited working life, since viscosity immediately starts rising at a rate that is dependent on temperature.

A representation of the continuous SMC process is presented in Fig. 3.11. The SMC is wound into rolls which are stored under controlled conditions for further thickening. Three commonly distinguished classes of SMC are characterized by the differences in curing shrinkage. Standard SMC has a shrinkage in the range of 0.5–0.2%, low-shrink SMC values lie within the 0.2–0.05% range, and low-profile SMC has a shrinkage of less than 0.05% including zero-shrinkage grades.

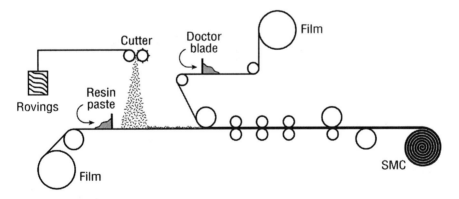

Fig. 3.11 Flow of materials in SMC process

SMC properties depend on the type of glass fiber roving used. Proper formulation of the SMC and stringent quality control present the possibility of automated production of automotive body panels with class A surface. A recent development involves direct SMC, where the time consuming thickening process is overcome by microwave maturation making in-line production concept possible for lower cost and improved performance [5].

Bulk Molding Compounds (BMC)

The formulation of BMC resembles that of SMC except that the material is supplied in bulk form rather than in sheet form. The material can be made in a batch process or in a continuous process. In batch process, the resin paste is mixed with chopped strands with fiber lengths of 6–25 mm in a sigma blade kneader or in a high-speed plough-share mixer. Excessive mixing leads to fiber degradation. Continuous production methods start from raw materials which are metered to the machine and mixed to prepare the resin paste, or a resin paste may be pre-mixed separately. In both cases the paste is combined with glass fibers in the nip of rollers where the impregnation takes place. The reinforcing fibers can be introduced as chopped strands or as rovings which are chopped by a cutter that is positioned above the nip.

Continuous compounding methods usually produce less damage to the fibers and thus produce BMC with longer length fibers than produced by the batch methods.

The majority of BMC applications are processed on automated production lines through injection molding, injection/compression molding, or transfer molding.

3.2.2 Thermosetting Matrix Resins

The choice of resin is determined by the functional requirements of the composite. The choice is based on requirements such as mechanical properties, corrosion resistance, weight, recyclability, cost, and scale of operation. Typical resins are unsaturated polyesters, epoxys, vinyl esters, phenolics, polyurethanes, and silicones. Within each resin classification, a large array of options exist. An overview of characteristics of the resin classes follows.

3.2.2.1 Unsaturated Polyester (UP) Resins

The reaction system of unsaturated polyester resins consists of low molecular weight condensation products of unsaturated and saturated biacids with diols dissolved in styrene monomer or other suitable reactive diluents. The unsaturation in these resin systems provides vinyl sites for cross-linking to occur. The cross-linking forms the thermoset resin. The cross-linking can be initiated by various activators at ambient or elevated temperature. Although most unsaturated polyester resins will gel under the influence of heat and/or ultraviolet light, the accomplishment of cure in a reasonable period of time calls for the use of an initiator for the free radical cross-linking reaction.

Organic peroxides are predominantly used as initiators. On heating the peroxide decomposes, forming free radicals which initiate the cross-linking reaction between the unsaturated sites in the polyester molecule and the styrene monomer. Cobalt complexes are used as accelerators. These complexes catalyze the decomposition of the organic peroxide initiators. Each peroxide has a critical temperature, which is the lowest temperature at which the decomposition rate is high enough for the curing to take place in a reasonably short time. This temperature is lowered in the presence of the cobalt-containing accelerator. Table 3.6 has examples of typical combinations of initiators and accelerators chosen to match the curing conditions used in several thermosetting resin processing techniques.

Table 3.6 Illustrations of initiator systems applicable to various processes

Peroxide	Dibenzoyl peroxide		Acetylacetone peroxide	Perketals	Dicumyl peroxide
Accelerator	None	Diethyl amine	Cobalt	None	None
Hand lay-up (room temperature cure)	–	+	+	–	–
Spray-up (room temperature cure)	–	+	+	–	–
Winding (40–100°C cure)	–	0	+	–	–
Pultrusion (100–180°C cure)	+	–	–	+	0
Continuous laminating (50–140°C cure)	+	–	+	–	–
Cold press molding/RTM (30–70°C cure)	–	+	+	–	–
SMC (135–170°C cure)	–	–	–	+	0
BMC injection molding (150–190°C cure)	–	–	–	+	+

+ = favorable, 0 = possible/fair, – = unfavorable

The curing reaction may be retarded by the addition of a number of chemicals, which, as a group, are referred to as inhibitors (or retarders), to prevent premature cross-linking. Inhibitors may also be added to lengthen the usable life (pot life) of reactants. Inhibitors are chemicals that trap free radicals. Examples are hydroquinone, *p*-benzoquinone, and *p-tert*-butylcatechol. Polyester resins are the most widely used matrix material in thermosetting applications, and they can be engineered to give a wide variety of end product parameters. Illustrative mechanical properties with various forms of glass fiber reinforcement are shown in Table 3.7.

Table 3.7 Typical properties of polyester laminates

Reinforcement form	None	Chopped strand mat	Fabric from roving	Fabric from yarn	Unidirectional roving
Weight percentage of glass	0	30.0	50.00	65.00	75.0
Tensile strength (MPa)	70.0	100.0	270.00	400.00	1150.0
Tensile modulus (GPa)	3.5	7.7	16.00	22.00	42.0
Flexural strength (MPa)	100.0	180.0	350.00	450.00	1200.0
Flexural modulus (GPa)	3.2	7.0	12.00	21.00	40.0
Compressive strength (MPa)	140.0	200.0	160.00	280.00	450.0
Coefficient thermal expansion ($10-6 \ K^{-1}$)	100.0	30.0	20.00	15.00	8.0
Density (kg/dm^3)	1.20	1.40	1.65	1.75	2.00

The following list summarizes characteristics of the polyester resin systems:

Advantages	Limiting factors
Wide choice, easy to use	Styrene emission
Cure at room or elevated temperature	Cure shrinkage
Good composite properties	Flammability
Good chemical resistance	B stage not possible
Good electrical resistance	

3.2.2.2 Epoxy (EP) Resins

Epoxy resins are based on reactions of the three-membered ring known as the epoxy, epoxide, ethoxyline, or oxirane group. Cured epoxy resins result from chemical reactions involving a curing agent. The curing agents are in two categories: catalytic and coreactive. Both Lewis acids and bases compose the catalytic category. The curing agents produce insoluble, intractable, cross-linked thermoset polymers. Tertiary amines are used in small quantities for room temperature cure or as accelerators for acid anhydride cures.

Properties of the cured resin depend to a large extent on type of hardener, cure temperature, and cure schedule. The following list summarizes characteristics of epoxy composite systems:

Advantages	Limiting factors
Excellent composite properties	Long cure cycles
Very good chemical resistance	Low cure shrinkage
Good thermal properties	Best properties achieved with elevated
Very good electrical properties	temperature cure
Can be B staged (prepreg)	

A distinction from polyester resins is that the final properties of the epoxy composite are largely governed by the choice of the hardener and cure conditions, whereas selection of the proper unsaturated polyester resin is essential in engineering-specific properties with polyester resin systems. Illustrative characteristics of laminates produced from a common fiberglass fabric reinforcement, a common bisphenol-A epoxy resin, and three typical hardeners are presented in Table 3.8. The incorporation of the hardener moiety into the polymer structure has a significant effect on properties of the cured resin. When epoxy resins are cured at room temperature, the reaction will slow down as molecular weight and viscosity increase and eventually come to a stop before full cure is attained. The reaction may be completed by heating for several hours at temperatures in the 60–100°C range. Such an after-cure is indispensable for achieving the best properties.

3.2.2.3 Vinyl Ester (VE) Resins

The reaction system for vinyl ester resins consists of an unsaturated low molecular weight base resin dissolved in a co-polymerizable vinyl monomer, which is commonly styrene. A typical vinyl ester reactant is the product resulting from reacting glycidated bisphenol-A with methylacrylic acid. Thus the bisphenol-A moiety is capped with terminal methylacrylate groups. These systems are sometimes referred to as epoxy acrylics. Vinyl esters respond to the initiators normally used with unsaturated polyester resins. The cured resin exhibits excellent chemical resistance. Also the elongation at rupture can be greater than that of polyurethane resins, resulting in mechanical properties that are relatively tough (elastic character). The following list summarizes characteristics of vinyl ester composite systems:

Advantages	Limiting factors
Fatigue resistant	Styrene emission
Excellent composite properties	Cure shrinkage
Very good chemical resistance	Flammability
Good toughness	No B stage

Table 3.8 Examples of adjusting characteristics of bisphenol-A epoxy composites based on the choice of hardener (65 wt% fiberglass fabric)

Hardner	Aliphatic amine	Aromatic amine	Acid anhydride
Pot life	30 min	2 h	> 24 h
Cure schedule	24 h 23°C, 1 h 100°C	1 h 80°C, 1 h 110°C	4 h 80°C, 6 h 150°C
Hand lay-up (room temperature cure)	+	0	−
Winding (70–150°C cure)	0	+	+
Prepegs (140–180°C cure)	−	−	−
Hot press molding (80–160°C cure)	−	+	+
Laminate tensile strength (MPa)	490	435	435
Laminate tensile modulus (GPa)	22.3	24.2	23.5
Laminate flexural strength (MPa)	560	585	655
Flexural strength after 72 h in boiling water (MPa)	430	425	570
Electrical properties	+	+	++
Chemical resistance	0	++	++
Heat resistance	0	+	+

+ = favorable, 0 = fair, − = unfavorable

3.2.2.4 Phenolic (PF) Resins

Liquid phenolic resin systems are used in open mold applications. These phenolic resins are synthesized by a condensation reaction between phenol and formaldehyde catalyzed by alkali. The reactants are heated to achieve the desired molecular weight. Variants can be synthesized by including alkylphenols such as cresol and xylenol in the reaction system. Thus a range of products with widely varying properties are available. Modified phenolic resins have been developed for the manufacture of prepregs and for use in SMC formulations. The dominant reasons for using phenolic resin systems are excellent flammability resistance and low smoke emission in fire situations. Characteristics of phenolic resins are given in the following list:

Characteristics	Limiting factors
Very good thermal properties	Color
Good flammability resistance	Alkali resistance
B stage possible	Inappropriate for food contact
Good electrical properties	

3.2.2.5 Polyurethanes (PUR)

Polyurethanes are based on the reaction of di-isocyanates with hydroxy-terminated polyesters or polyethers. Depending on the functionality of the reactants, linear elastomeric products or branched, stiff materials can be obtained. Additional variation in properties can be designed into the polymer by the use of chain extenders, such as short-chain diols or diamines. Polyurethane chemistry provides one of the broadest range of physical properties available from a single category of polymer types. The versatility of polyurethane chemistry is evidenced by its widespread applications in paints, flooring compounds, elastomers, coatings, casting resins, and foams. Typical composite applications of glass fiber-reinforced polyurethanes are in reaction injection molding parts, SMC-type moldings, foamed panels, and filament-wound structures. General characteristics of thermosetting polyurethane composites are given in the following list:

Advantages	Limiting factors
Good composite properties	Isocyanates as curing agents
Very good chemical resistance	Color
Very high toughness (impact)	Anhydrous cure conditions
Good abrasion resistance	No B-stage

3.2.2.6 Silicone (SI) Resins

Methylsilicone resins or methylphenysilicone resins, usually as solutions in organic solvents such as toluene or xylene, are used as a matrix to achieve rather special sets of properties. The silicone resins may be pre-impregnated into glass fabrics, braids, and tapes. Final cure is accomplished by condensation through silanol groups catalyzed by metal-organic salts or by amines. The silicone resins can also form interpenetrating polymer networks with other polymer types to obtain a composite possessing some of the special properties of the silicone resin, such as very low surface tension, while retaining most of the characteristics of the other resin. Characteristics of glass fiber/silicone resin composites are given in the following list:

Advantages	Limiting factors
Very good thermal properties	Adhesion
Excellent chemical resistance	Long cure cycles
Very good electrical properties	Elevated temperature cure
Resistant to hydrolysis and oxidation	
Flammability resistance (self-extinguishing)	
Non-toxic	

The attributes of thermosetting resin composites are compared in Table 3.9. The applicability of the processing techniques is governed by the inherent characteristics of the resin.

Table 3.9 Property trends for thermosetting resin composites reinforced with glass fibers

Resin	Processing	Mechanical	Electrical	Heat Resistance	Chemical Resistance	Fire Resistance
Polyester	a, b, c, d, e					
Epoxy	a, b, c, e					
Vinyl ester	a, b, c, e					
Phenolic	a, b, c, d, e					
Silicone	e					
Polyurethane	b, c					
Polyimide	e					

a = Hand lay-up/spray-up   d = Continuous laminating
b = RTM/RIM   e = Molding compounds/prepreg
c = Winding/pultrusion
The length of a bar reflects the relative extent of the property

3.2.3 Fillers

The term "fillers" is often associated with extenders added to a matrix to reduce the total cost of the combination. However, fillers in glass fiber-reinforced composites are predominately added to engineering-specific properties with relatively inexpensive materials. The predominant use of fillers is with thermosetting resins, although some filler types are used in thermoplastic compounding. Fillers are used to impart special characteristics in processing or in the composite application. A wide range of both inorganic and organic fillers are available. Inorganic fillers are most widely used in conjunction with unsaturated polyester resins.

The ideal filler material for thermosetting resins should meet requirements such as good dispersibility, suspension stability, low density, light color (when mixed with resin), low moisture content, negligible influence on cure properties, controlled particle size distribution, heat resistance, chemical resistance, and low cost. In addition to these properties, the filler should impart one or more of the following characteristics to the final composite: opaqueness, surface smoothness, improved abrasion resistance, hardness, stiffness, impact resistance, improved electrical, and fire-retardant properties.

Dispersibility, suspension stability, and viscosity increase depend on particle size and particle shape of the filler along with the interaction of the filler surface with the matrix resin. The oil absorption value is a very indirect indicator of the wetting and interaction of the surface of the filler with organic polymer matrices. Water absorption values are indicators of hydrophilic character of fillers. The surface of the filler is often modified to influence these properties, as discussed in the next section. Particle size, in general, is between 1 and 5 μm. Smaller particles tend to cause a sharp viscosity increase due to their higher surface area, whereas particles of 10 μm and larger may be filtered out by the glass fiber reinforcement. The lower level of density of suitable inorganic fillers is approximately 2.5 kg/dm^3, while the higher level is approximate 4.5 kg/dm^3. Generally, filler addition will lead to increased density. An exception to this rule is the addition of hollow micro-spheres or micro-balloons, made from glass, quartz, or silicates, which have densities ranging from 0.2 to 0.8 kg/dm^3.

The use of magnesium oxide as a filler in SMC is common. Careful control of moisture content is important because the thickening reaction of polyesters with magnesium oxide is influenced by traces of water.

Perfect bonding between the matrix resin and the filler requires that the resin in a liquid state wet the filler surface and that the adhesion between the filler and the solidified resin be optimized. To achieve the best possible adhesion, the surface of the solid filler must be completely wetted by the matrix resin. However, good wetting is not a guarantee of good adhesion. Surface treatments of fillers are designed to provide good wet-out and adhesion with specific types of resins.

Organo-functional silanes are common agents used to enhance bonding of the matrix resin to the surface of a filler. Silanes also function as wetting agents in certain systems. Silane coupling agents, when applied to siliceous materials, improve both wetting and adhesion. Table 3.10 contains a summary of typical properties and uses of fillers in the SMC application.

3.2.4 Release Agents

Release agents are used to prevent the molded product from adhering to the mold surface. A variety of release agents have been developed to encompass a wide range of processing temperatures and molding situations. Most commonly, release agents are applied to the mold surface in the form of fluids or waxes, but in some cases, the release agent is actually mixed with the resin to enhance processing and release characteristics.

Release agents are in a variety of physical forms depending on the resin and the molding situation. For instance, silicones are applied as fluid, solution, emulsion, and aerosol spray. Film of poly(ethylene terephthalate) serves as a release agent in

Table 3.10 Characteristics of fillers

Filler	Density (kg/dm^3)	Particle shape	Typical particle size (μm)	SMC/BMC-related properties
Calcium carbonates	2.7	Granular	2–5	Improves surface appearance, good color, good mold flow, good dispersion stability, permits high filler loading
Dolomite	2.8	Granular	2–5	Same as calcium carbonate
China clay	2.6	Plate	5	High viscosity limits loading. Very good chemical resistance. Very good electrical properties. Good surface appearance
Talc	2.9	Plate	5	Only used as partial replacement for other fillers to improve flow and surface appearance
Quartz/silica	2.7	Granular	10	Very good chemical resistance; very good electrical properties
Wollastonite	2.8	Fibrous needles	–	High viscosity. Improved impact properties
Aluminum trihydrate	2.4	Granular	10	Flame retardant. Good electrical properties esp. tracking and arc resistance
Antimony oxide	5.5	Granular	2	Flame-retardant additive, used in conjunction with chlorinated or brominated resins or resin additives
Barites (BaSO$_4$)	4.4	Granular	2	Good chemical resistance, high whiteness, sound deadening
Hollow glass spheres	0.2–0.8	Spheres	50	As partial replacement of other fillers for low density compounds

certain processes. The common feature of the release agents is a low surface energy that hinders adhesion of matrix resin to the surface provided by the release agent.

3.3 Reinforced Thermoplastic Materials

3.3.1 Introduction

Reinforced thermoplastic composites are formed by combining a fusible polymer with fibers. The thermoplastic matrix polymer consists of linear macromolecules with no chemical linkage (covalent bond) between the individual macromolecules.

A thermoplastic polymer flows at temperatures exceeding a characteristic temperature, except for the situation where the decomposition temperature of the polymer is less than the flow temperature. Injection molding technology consists basically of raising the temperature of the thermoplastic polymer to induce flow into a mold where a shape is formed and retained when the temperature of the polymer is decreased below the flow temperature. Injection molding techniques enable production of large numbers of parts at short cycle times. Injection molding technology is attractive for mass production of parts, for instance, in automotive applications. Figure 3.12 shows an automotive air intake manifold. This injection-molded part is an example of a complex design molded from glass fiber-reinforced polyamide resin.

Fig. 3.12 Illustration of applications where glass fiber-reinforced parts are produced in large numbers

In general, thermoplastic polymers are more ductile and impact resistant than are thermosetting polymers. The characteristic of flow at elevated temperatures limits the upper-use temperature of the polymer. For structural applications, new process technologies have enabled mass production of larger, more complex parts with enhanced stiffness, strength, and impact properties. Glass mat thermoplastic (GMT) and long fiber thermoplastic (LFT) processes are at the frontier of thermoplastic composite substitution for metal. GMT and LFT processes use compression as well as injection molding techniques. High-impact properties are obtained while maintaining stiffness and strength by preserving fiber length in the composite. These processes enhance the freedom of part design and the consolidation of parts beyond the weight reduction and recyclability.

3.3.2 Semifinished Materials Based on Thermoplastics

In contrast to thermosetting resins, thermoplastic matrix resins undergo physical changes during the process of compounding with glass fibers and the process of molding. The thermoplastic polymer may undergo a chemical change such as molecular weight decrease but the polymer retains the property of flow (thermal formability) when a characteristic temperature is exceeded. The process of mixing thermoplastic resin and glass fiber as a precursor for subsequent thermal molding is referred to as *compounding*. The shelf life of compounds is virtually unlimited.

3.3.2.1 Reinforced Thermoplastic Compounds (RTP)

Compounds are solid pellets formed by dispersing glass fibers in molten thermoplastic polymer. The compound is used in subsequent molding processes. The majority of RTP is compounded on twin-screw extruders, which most often have co-rotating, intermeshing screws. Thermoplastic resin powder or pelletized resin is metered into the first feeding port of the extruder as illustrated in Fig. 3.13. The polymer is melted and then conveyed by the screws. In the second feed port, chopped glass fibers are introduced. The glass fiber-to-resin weight ratio is controlled by automatic feeding equipment. Wear of the extruder surfaces is minimized because the fibers are quickly wetted by the polymer melt.

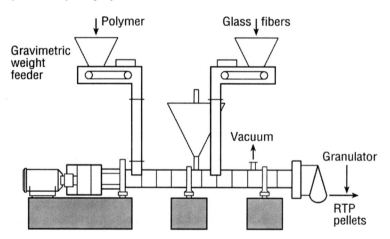

Fig. 3.13 Extruder for fiberglass-reinforced thermoplastic compounding

Instead of chopped strands, continuous rovings can be introduced into the second feed port. The rovings are cut by the shearing action of the screws and much the same ultimate fiber length distribution is obtained as with chopped strands.

Metering of the glass fiber weight fraction is accomplished by applying the correct combination of screw speed and roving linear density. The high shear forces exerted by the screws cause fiber length degradation. The final fiber length is principally a function of screw rotational speed, material throughput, and glass content.

This function takes a different form for each polymer type and is also dependent on the extruder type and the screw geometry. An example of fiber length degradation is presented in Fig. 3.14.

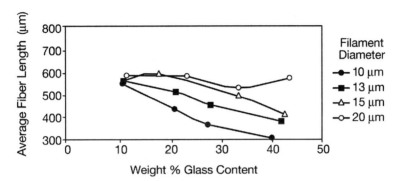

Fig. 3.14 Influence of filament diameter on fiber length degradation for compounding roving with chemically coupled polypropylene

The extrusion experiments in Fig. 3.14 were conducted under conditions of constant total output at a given screw configuration and temperature setting. Clearly fiber length is reduced as glass content increases indicative of more intense fiber–fiber interaction and abrasion in the extrusion process.

The reduction in length is less acute as the diameter increases due to the lower concentration of individual fibers (filaments).

3.3.2.2 Glass Mat Thermoplastic (GMT) and Long Fiber Thermoplastic (LFT)

Long fiber thermoplastics are a rapidly growing sector of the reinforced thermoplastic industry. The most commonly reinforced thermoplastic polymers are polyamides, polyesters, and polypropylene. The attractiveness of longer reinforcing fibers is a response to the improved reinforcement effectiveness discussed in Section 3.1.5.3.

Thomason [6] has experimentally made an estimate of the response of physical properties for 30% fiber glass-reinforced polypropylene like modulus, strength, and impact for different fiber lengths. These properties show an increase to limiting value as the fiber length increases to large lengths.

GMT composites are consolidated sheets containing thermoplastic matrices reinforced with unidirectional, randomly oriented, long chopped, and/or continuous fiberglass mats. The most common matrix polymer is polypropylene. Prior to molding, the sheets are heated in an infrared oven above the melting point of the resin matrix and then transferred hot into a cold compression mold to form the finished part. The melt impregnation process consists of combining glass mat along with a polymer sheet under pressure and temperature through a double belt system. The material is cooled and slit and sheared into specific sizes for subsequent molding. The final parts are produced by compression molding.

Two types of LFTs are generically recognized: granulated (G-LFT) and direct (D-LFT). G-LFT compounds are obtained by impregnating continuous glass fiber rovings with thermoplastic resin, cooling, and then cutting the impregnated roving into rod-like pellets (generally 6–25 mm). These pellets are then injection molded into parts.

D-LFT eliminates the stage of pellet production. The roving is pre-impregnated in combination with compression or injection molding as the part is formed in virtually a one-step process. The process is also referred to as in-line compounding. Typically longer residual fiber lengths are obtained with the D-LFT process. In this field, developments are rapid and new and improved direct-LFT as well as GMT technologies continue to provide new material solutions introduced like injection-molded D-LFT, compression-molded D-LFT, E-LFT, and low weight-reinforced thermoplastic (LWRT) mats [7].

3.3.2.3 Mechanical Properties of Compounds

Through reinforcing with glass fibers, the mechanical and physical properties of thermoplastic resins are saliently modified. Figure 3.15 illustrates the dramatic changes in stiffness, strength, and impact behavior achieved with glass fiber reinforcement. Reinforcing polyamide (PA), poly(butylene terephthalate) (PBT), and polypropylene (PP) with 30 wt% glass fiber result in increases in tensile strength greater than 200%. The increases in notched impact strength vary from 70% for PBT to 300% for PP. Modulus, an essential engineering property, is increased by a factor of 10 for PP and by a factor of 4 for PA and PBT.

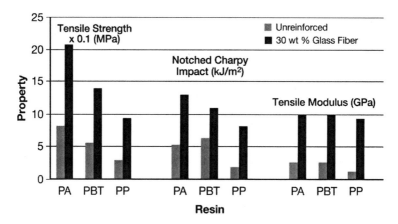

Fig. 3.15 Property improvement of thermoplastic resins with 30 wt% glass fiber reinforcement

The size on the surface of the glass fibers plays a crucial role in transfer of stress at the glass/resin interface. The criticality of the size is clearly demonstrated in

Table 3.11. Much of the proprietary nature of fiberglass products resides in the formulation of sizes. The size must function effectively as a processing aid in fiberglass production and in composite production. In addition, the size must provide bonding (interfacial adhesion) between the fiberglass and the resin used in a particular composite.

Table 3.11 Property enhancement of poly(butylene terephthalate) compound through favorable interaction with size on glass fibers

Property	Size designed for poly(butylene terephthalate) resin	Size designed for polyamide resin
Tensile strength (MPa)	150	117
Tensile elongation (%)	3.2	1.9
Unnotched Charpy impact (kJ/m²)	50	27
Weight percentage of glass	30	30

Composite properties are also influenced by the effect that glass fibers have on the morphology of semicrystalline resins [7]. The temperature stability of semicrystalline resins reinforced with glass fibers is markedly affected by reinforcement. In Fig. 3.16, polycaprolactam (PA 6), poly(hexamethylene adipamide) (PA 66), poly(butylenes terephthalate) (PBT), and polypropylene (PP) are semicrystalline resins. Polycarbonate (PC) is an amorphous resin. Figure 3.16 illustrates the improvement in terms of the heat distortion temperature for the semicrystalline resins. Polycarbonate is not significantly improved in thermal stability by fiberglass reinforcement.

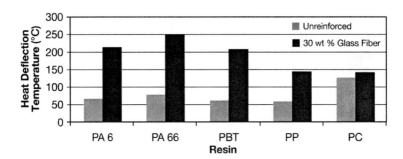

Fig. 3.16 Effect of glass fiber on temperature of deflection of thermoplastic resins

Considering mechanical properties from a macro perspective, Fig. 3.17 demonstrates that reinforcement of chemically coupled polypropylene with a suitable size on the glass fibers does not show significant dependence on filament diameter. The increase in tensile modulus follows the trend predicted by a rule of mixture function:

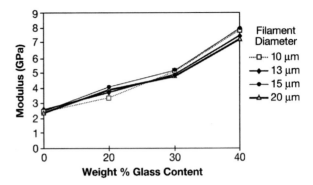

Fig. 3.17 Illustration of compounding chemically coupled polypropylene and glass fibers with no degradation of fiber length

$$E_c = \alpha V_f E_f + V_m E_m$$

where E_c = modulus of composite, E_f = modulus of glass fiber, E_m = modulus of matrix resin, V_f = volume fraction of glass fiber, V_m = volume fraction of matrix, and α = reinforcement geometry-dependent factor. The modulus is not dependent on filament diameter and increases with glass content in this rule of mixture relationship.

The situation depicted in Fig. 3.17 exists only when the compounding process is engineered to avoid damage to the glass fibers. As shown in Fig. 3.14, fiber–fiber interaction can strongly affect residual fiber length. If conditions are not adjusted (screw dimension, throughput, temperature) at increasing glass content, tensile strength will exhibit reduced improvement. Another important property, impact strength, may also suffer as demonstrated in Fig. 3.18. Breakage of fibers during compounding not only decreases fiber length but also increases the concentration of un-sized ends which will have poor bonding with the resin. The challenge for the process engineer is to optimize temperature and thus viscosity of the system, throughput, and screw design so as to achieve a tailoring of composite properties for specific applications.

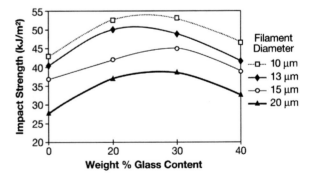

Fig. 3.18 Negative effect of compounding chemically coupled polypropylene with glass fibers under extruder conditions that degrade fiber length

An additional example that demonstrates the role of optimizing the interaction of the size with the resin to obtain optimal properties is given in Fig. 3.19. Injection-molded parts are used as components in automotive radiators. These composites consist of PA 66 resin with fiberglass chopped strand reinforcement. The size on the chopped strand provides adhesion to the resin system and resistance to degradation of adhesion when the composite is exposed to radiator fluid such as solutions of ethylene glycol and water. Fig. 3.19 depicts the situation where a commercial "standard" size was modified to achieve improved resistance to exposure to the radiator fluid. The "special" size significantly improved the retention of tensile strength and impact resistance. Note that the properties prior to exposure were not improved. The modification of the size specifically improved the resistance of the glass–resin bonding to degradation during exposure.

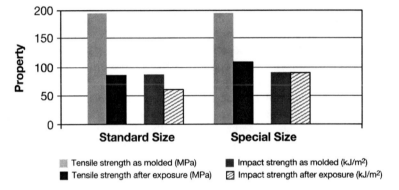

Fig. 3.19 Design of size to enhance the chemical resistance of a composite

The evolution of sizes has resulted in chopped strand products that are designed to obtain optimal properties in a variety of polyamide matrix resins. For instance, enhanced chopped strand products have been developed for PA 66 and PA 6 resins modified for hydrolysis resistance and impact resistance, with and without calcium stearate lubrication. A requirement for these chopped strand products is that extruder throughput or other economic parameters not be impacted negatively.

Mathematical models have been developed to describe mechanical properties of random short fiber composites [8–10]. The complexity and wide range of fiber orientation states result in difficulty in adequately mathematically representing the composite. These orientation states originate from the shape of the part and alignments induced from the flow [11] complicated by the crystallization of the resin induced by the fiber [12]. The enthalpy of crystallization influences flow and subsequent cooling behavior of the part in the mold. The existing models are used on a semiempirical basis.

Work continues to expand existing flow models to account for these effects and to allow better descriptions of critical properties for larger and more complicated parts [11, 13]. The accuracy and the applicability of models for long fiber systems are areas particularly in need of improvement.

3.3.2.4 Semicrystalline Resins

Semicrystalline polymers exist as viscous liquids at temperatures above the melt-ing point of the crystalline domains. Upon cooling, crystals nucleate and grow. Domains remain uncrystallized or amorphous during the crystallization process, hence the term semicrystalline. The amorphous portions of the polymer are dis-persed between the crystalline portions. Mechanical properties such as strength, modulus, and heat distortion resistance are enhanced by the "reinforcing effect" of the crystalline portions of the polymer morphology. Compared with amorphous polymers, the semicrystalline polymers generally offer superior performance in fatigue, creep, friction, and wear, along with good mechanical properties such as stiffness and strength.

An overview of the most common semicrystalline thermoplastic resins used in composites follows.

Polyamide (PA). Polyamide was the first class of polymer to be used as an engineering thermoplastic and is still a major polymer class used in glass fiber-reinforced applications. Polyamides have the amide group as part of the main polymer chain. Polyamide use is dominated by poly(hexamethylene adipamide) (PA 66 or Nylon 66) and polycaprolactam (PA 6 or Nylon 6). Physical properties of the two polyamides are similar, but there are some differences in the chemical prop-erties. As a class, polyamides can be characterized as strong and tough with good resistance to abrasion, fatigue, and impact. Excellent performance when exposed to elevated temperatures or solvents has allowed the development of a wide range of automotive applications. Polyamides are also extensively used in many other appli-cation areas. The use of polyamides has grown parallel to the overall growth of injection molding for mass producing glass fiber-reinforced composites.

Poly(ethylene terephthalate) (PET) and poly(butylene terephthalate) (PBT). Poly(ethylene terephthalate) is synthesized through polycondensation of tereph-thalic acid with ethylene glycol at elevated temperature to form a linear saturated polyester of high molecular weight. PET was predominately used as a fiber and a film until the 1970s. The crystallization rate of PET during cooling from the melt in molds was too slow for the degree of crystallization to occur needed to achieve the desired physical properties. The discovery of effective nucleating additives, such as sodium benzoate, initiated the large-scale use of PBT in injection molding. The nucleating agents provided fast cycles in injection molding processes. Such a mod-ification is not required for PBT which shows high crystallization speed and hence fast molding cycles even at mold temperatures of 65–85°C. PET and PBT are hard, wear resistant, and dimensionally stable. In general, impact properties are inferior to those of polyamides.

Polypropylene (PP). Isotactic polypropylene is synthesized by the stereospecific polymerization of propylene. PP has better mechanical and thermal properties than polyethylene, exhibits high rigidity, excellent flex life, and good surface hardness while having the lowest density of commercially available thermoplastics. Like other polyolefins with low surface energies, PP is a difficult material to adhesively bond to reinforcing fibers. The adhesion of homopolymer PP to a glass fiber surface is improved by modification of the polymer and by design of special sizes for the glass fibers. The polymer is modified by grafting a vinyl compound, such as maleic anhydride, to the PP macromolecules during extrusion. The modified PP is identified as "chemically coupled polypropylene." The resulting carboxylic acid functionality provides sites for chemical and/or physical interaction with the size on the glass fibers. Special glass fiber sizes are required to bring about optimal improvements in adhesion. The adhesion improvement results in improved reinforcement as witnessed by increased tensile strength.

Polyacetal (POM). Polyacetal, also referred to as polyformaldehyde and poly-oxymethylene, is synthesized by the polycondensation of formaldehyde. The resulting polymer has a high degree of crystallinity and is very hard with good strength and stiffness. Incorporation of co-monomers, such as cyclic ether trioxane, yields co-polymers with improved thermal stability.

Typical properties of semicrystalline matrix resins reinforced with glass fibers are summarized in Table 3.12.

Table 3.12 Typical mechanical properties of 30 wt% glass fiber-reinforced compounds

Resin	PA 66[a]	PA 66[b]	PBT	PET	PC	POM	PP[c]
Tensile strength (MPa)	210	160	150	180	140	135	90
Tensile modulus (GPa)	10	8.4	9.9	11.5	9.1	10	7.5
Flexural strength (MPa)	302	215	225	225	220	180	140
Charpy impact un-notched (kJ/m^2)	57	75	50	40	45	30	35
Charpy impact notched (kJ/m^2)	14	24	12	11	14	8	12
Continuous use temperature (°C)	120	120	130	130	130	115	105

[a] As molded.
[b] At equilibrium moisture regain.
[c] Chemically coupled.

3.3.2.5 Amorphous Resins

Amorphous thermoplastic polymers lack the positional order on a molecular scale for crystallization to occur. At temperatures less than the softening range (glass transition temperature), these polymers are glassy in nature. At temperatures in excess of the glass transition temperature, the segmental mobility of the macromolecules results in a condition often described as limp and flexible. At temperatures less than the glass transition temperature, the amorphous polymer becomes relatively stiff and hard. In general, the amorphous polymers are used in composite applications to take advantage of specific properties, such as chemical and temperature resistance.

Polycarbonate (PC). Polycarbonates consist of bisphenol units joined together by carbonate linkages. In general terms, PCs are polyesters from aromatic dihydroxy compounds and carbonic acid. Typical raw materials are bisphenol-A and phosgene. Polycarbonates are especially recognized as an excellent impact-resistant thermoplastics.

Poly(phenylene oxide) (PPO). Poly(phenylene oxide) is synthesized by oxidative coupling of 2.6-xylenol using a copper-catalyzed process. For injection molding, PPO is almost exclusively used in modified versions, consisting of a blend with polystyrene (PPO/PS) or a blend with polyamide (PPO/PA). The PPE/PS combination is one of the rare examples of a fully compatible blend of two polymers. PPO has exceptional moisture resistance, high dimensional and thermal stability, and good flammability resistance.

Polysulfone (PSU) and polyethersulfone (PESU). The polysulfone family of thermoplastics is characterized by its high thermal and oxidative resistance and high mechanical strength. Polysulfone is a linear condensation product from bisphenol-A and dichlorodiphenyl sulfone. Polyethersulfone is synthesized by self-condensation of diphenylene oxide sulfochloride. Both polymers are yellowish transparent, tough materials.

3.3.2.6 Heat-Resistant Polymers (HT)

A group of polymers with the distinctive property of being resistant to high temperatures is referred to as "high-temperature" polymers. The criteria that define heat resistance are not clearly established. Temperature limits, specific for each polymer, are melting point and glass transition temperature. The often-quoted heat deflection temperature (HDT) is not a relevant measure for heat resistance since it represents but one point from a curve, the slope of which is unknown. A more practical approach is the determination of an important property, such as tensile or flexural strength, after ageing for various periods of time at different temperatures. The time it takes for the strength to drop to 50% of the original value is plotted as a function of temperature and through extrapolation the temperature index (TI) is found that indicates the temperature at which 50% of strength is retained over a predetermined period of, for instance, 20,000 or 50,000 h. A similar approach is used in the test method UL 746B (Underwriters Laboratories, Northbrook, IL, USA) to establish a continuous use temperature. The decrease in mechanical property at an elevated temperature is an indicator of basic chemical changes such as polymer chain scissions due to thermal degradation in air.

As a generality, high-temperature polymers are defined as polymers which possess a continuous use temperature in excess of 180°C. Reinforcement with glass fibers generally increases the continuous use temperature.

Table 3.13 presents properties of injection-moldable thermoplastic HT polymers, un-reinforced as well as reinforced with 30 wt% glass fibers. Fluoropolymers are

Table 3.13 Effect of fiberglass reinforcement on properties of high-temperature polymers

Resin	Polyphenylene sulfide		Polyetherimide		Polyether ether ketone	
Melting temperature (°C)	277		340		334	
Glass transition temperature (°C)	90		215		143	
Weight percentage of glass	0	30	0	30	0	30
Heat distortion temperature[a] (°C)	135	260	200	210	160	> 300
UL rating (°C)	220	180	170	180	> 240	> 250
Tensile strength (MPa)	75	135	105	160	100	165
Tensile modulus (GPa)	4.2	11	3.0	10	3.7	10
Notched Izod impact (J/m)	16	22	50	100	85	96

[a] At 1.8 MPa.

not included because the members that meet the HT polymer definition are not melt processable.

Table 3.13 illustrates that glass fiber reinforcement can significantly enhance mechanical properties of high-temperature polymers.

3.3.2.7 Liquid Crystal Polymers (LCPs)

Linear macromolecules with a very rigid, rod-like molecular structure can associate inter-molecularly forming highly ordered domains. These polymers have certain characteristics that are analogous to non-polymeric organic molecules that form liquid crystals. Polymers capable of liquid crystalline behavior in solution are lyotropic, while polymers exhibiting liquid crystalline behavior in the melt are thermotropic. Aromatic polyamides used for the production of aramid fibers are examples of lyotropic LCPs. Thermotropic LCPs can be found among the polyarylate family.

The thermotropic LCPS exhibit a strong tendency toward molecular orientation under flow. Injection-molded LCPs are highly anisotropic, with very high strength in the direction of orientation. Their fracture surface has a fibrous appearance. Although LCPs are sometimes referred to as "self-reinforcing polymers," reinforcement with glass fibers is advantageous in decreasing the anisotropy of properties. The chemical structures that result in the rod-like molecular structure are also very resistant to thermal degradation. Therefore, LCPs can be included in the high-temperature classification.

3.4 Composites for Wind Turbines

3.4.1 Introduction

Wind is currently the largest and fastest growing source of renewable energy [14]. This trend is almost exclusively supported by the high value in use of fiber glass-reinforced windmill blades. Fiberglass offers high performance at low cost.

Figure 3.20 illustrates the recent development of large wind energy parks that are and will be connected to the power generation grid.

Fig. 3.20 Wind energy park

Large wind turbines are intrinsically more efficient than smaller ones, explaining the emphasis on increasingly larger turbines. For offshore projects and low-wind onshore areas, turbines as large as 5–6 MW are in prospect. Composites have been enabling and will continue to play a key role in the development of wind turbines. Composites combining thermosetting resin and fiberglass reinforcement are the material of choice for making the rotor blades.

Four factors that make composite technology an essential enabling component can be enumerated.

1. Composite materials offer lightweight for construction compared to metals or aluminum. This implies less wear and tear and a longer service life for mechanical components. Typical blade weights today are approximately 4–5 t for a 30-m blade and as much as 13 t for a 54-m blade. The weight factor results in fiberglass composite being the material of choice.
2. Average blades with today's length of 30–40 m have complex shapes. The cross section of blades varies along the length axis from root to tip of the blade.

A typical blade consists of a large number of airfoil shapes. Such large and complex shapes can only be made cost effectively and easily using composite materials. Manufacturing technology has moved from labor-intensive open mold techniques to much more sophisticated, labor-effective, vacuum-assisted closed mold technologies using specially designed glass fiber fabrics.

3. Composite fatigue strength is superior to metals. The blade is subjected to complex loading cycles including both tensile and compression modes during rotation. Composite materials can be designed with directed reinforcements to allow a very long fatigue life.

4. Composites have attractive wear and corrosion resistance when installed in an open environment, such as the sea coasts. Composites have weathering properties, resulting in a service life exceeding 20 years and reducing maintenance cost. Maintenance cost is increasingly important for windmills being installed in less easily accessible areas like offshore.

In summary fiberglass-reinforced blades are light enough to require only a small portion of the energy extracted from the wind for their own turning and can tolerate continuous cyclically varying loads without suffering fatigue failure.

3.4.2 Raw Materials

Blade producers predominantly use fiberglass-reinforced laminates with unsaturated polyester resin or epoxy resin as the matrix. The use of wood has progressively diminished. The use of thermoplastic materials is very limited today. Carbon fiber has not made significant inroads yet but is expected to become more important as blade sizes continue to increase. The weight factor results in interest in carbon fiber. The use of carbon fibers may occur when certain limits to the stiffness of the total construction are required without increasing weight significantly. Balsa wood and foam are used as core materials. The most common types of glass fiber reinforcement are unidirectional and $\pm$ 45° woven fabrics, knitted and stitched fabrics, roving, chopped strand mats, and continuous strand mat. These forms of fiberglass reinforcement are used in a variety of combinations to design optimal stiffness and directed reinforcement. New developments include three-dimensional woven constructions for preforms. The choice of the resin is mainly determined by the blade-manufacturing technology used. The technology for blade manufacturing is dynamic with industrial rationalization establishing requirements on quality and economies of scale.

3.4.3 Blade-Manufacturing Techniques

The blade size of wind turbines is too large for technologies such as SMC and injection molding. Therefore the typical techniques used resemble those in boat building.

The two major components in a turbine blade are the outer shell and the longitudinal web or spar. These components are made by lay-up process, resin infusion process, vacuum-assisted resin transfer molding techniques, or prepreg lay-up process. Blades are molded in two halves, which are then bonded together. As blades become larger, laminating by hand becomes increasingly unworkable. Consequently manufacturers are moving to use vacuum infusion (VARTM) or closed mold techniques such as resin transfer molding (RTM). These techniques use designed fabrics to distribute strength and stiffness precisely where needed.

The migration from hand lay-up to RTM and VARTM, albeit at increasing tooling cost, also facilitates higher dimensional consistencies and allows in-mold finishing of the sides. The use of prepregs is increasing due to a need for reduced fiber waviness. Carbon fiber is expected to play a role as reinforcement because of its high specific modulus at a specific strength comparable to fiberglass. The use of carbon fiber is at a significant cost increase, keeping therefore a strong focus on optimizing the design of the increasingly larger blades. Hybrid reinforcement combining glass fiber and carbon fiber is very likely.

3.4.4 Blade Design Methodologies

Drivers for blade design are among others maximum length for a given load envelope, minimum weight, lifetime, manufacturability, reliability, and performance. Several designs are in place in the industry. The blade design illustrated in Fig. 3.21 has a rectangular spar. The spar provides strength and stiffness as the load-bearing member, the construction is tubular, and the fibers are longitudinally aligned. The forward and aft shear webs transfer shear loads and provide buckling resistance. The blade surfaces provide aerodynamic performance and allow optimal thrust. The whole system will be under variable torque forces with tensile and compression modes.

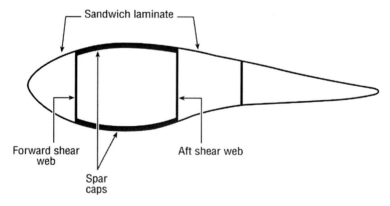

Fig. 3.21 Example of blade design

Past designs were based on safe life with little data and little basic fatigue testing. Currently more elaborate data on materials, architectures, loadings, and structural configurations are available. Fatigue testing in itself is complex, and the choice and execution of the tests is highly dependent on the design of the blade.

ASTM D3479 (The American Society for Testing and Materials, West Conshohocken, PA, USA) is a test standard that provides guidance for the fatigue testing of composite materials. A typical fatigue test uses specimens of glass fiber-reinforced epoxy or vinyl ester laminate fabricated using the VARTM process under well-defined curing conditions. The dimensions of a tensile specimen are defined in ASTM D3039. The tension–tension fatigue test is a commonly used test method to generate fatigue life data for the glass fiber-reinforced composite materials. The amplitude of cyclic loading is expressed as the stress ratio (R), which is the minimum peak stress divided by the maximum peak stress. The constant amplitude tension–tension fatigue test is frequently conducted with a stress ratio of $R = 0.1$ at multiple stress levels. Fatigue data at multiple stress levels are then used to determine the fatigue stress versus fatigue life (S–N fatigue data).

The S–N fatigue data trends for stress amplitude ($S_{amplitude}$) versus cycles to failure (N) are frequently represented by a log–log model. The log–log model has the form of

$$S_a = S_0[N^{[-1/k]}]$$

or

$$\log(S_a) = a + b[\log(N)]$$

where N = cycles to failure, S_a = amplitude of the constant–amplitude cyclic stress, S_0 = constant of the function, k = exponent parameter of the function and sometimes called the "fatigue exponent," a = intercept of the linear function and b = slope of the function, $a = \log(S_0)$ and $b = -1/k$. The quantities, k and S_0 or a and b, are fitting parameters determined by least-squares regression technique. A fitted S–N relationship is shown in Fig. 3.22.

The values of the regression parameters are shown in Table 3.14. The S and k values are important parameters to characterize the fatigue behavior of the test specimen. The slope of the S–N relationship is an indicator of the resistance of the composite to fatigue.

Fatigue below the elastic limit in metals is generally associated with local plastic strains at a defect site or stress riser in the metal. The mechanism of fatigue in metals is described in terms of crack initiation and propagation. The fatigue mechanism in composites can be much more complex. The failure can result from matrix cracking, delamination between plies, and loss of interfacial bonding between the matrix and the reinforcement. Research into the fatigue of composites is intense due to the expanding use of composites in applications that are structural and involve dynamic loading, for instance, wind turbine blades. The combination of good fatigue

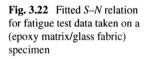

Fig. 3.22 Fitted *S–N* relation for fatigue test data taken on a (epoxy matrix/glass fabric) specimen

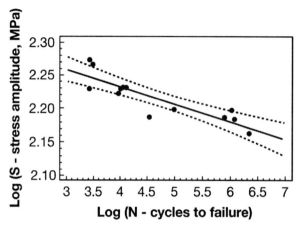

Table 3.14 *S–N* regression parameter estimates – fiberglass fabric/epoxy matrix

Term	Estimate	Lower 95% confidence limit	Upper 95% confidence limit
S_0	216.3	194.9	239.8
k	39.04	28.52	61.84

properties and attractive specific strength and stiffness is desired for structural applications.

Fatigue testing such as that illustrated in Fig. 3.22 is very time consuming and needs to be complemented by compression and more complex modes of loading under applicable environmental conditions. Currently, industry uses these data combined with phenomenological models to improve predictions. For future designs, further improvement of these methodologies are foreseen. Use will increasingly be made of composite design experts from the aerospace community to benefit from well-entrenched engineering of damage-tolerant designs. Expanded use is indicated for design tools from the aerospace industry that limit manufacturing variability in order to limit probability of failure.

Blade producers have capabilities to test the full-size blade as part of their overall design effort. The future will see a further sophistication of materials, manufacturing techniques, tests, and models to maximally use composites. Composites certainly are a material of choice in society's drive for effectively using wind, the most promising cost-competitive renewable energy source.

References

1. J. Klunder and R. Goodwill, "Introduction to Glass Fibre Composites," PPG Manual.
2. K. L. Loewenstein, *The manufacturing technology of continuous glass fibers*, Elsevier, Amsterdam (1983).

3. E. P. Pleuddemann, *Silane coupling agents*, Plenum Press, New York (1982).
4. S. Zhandarov and E. Maeder, Characterization of fiber/matrix interface strength: applicability of different tests, approaches and parameters, Compos. Sci. Technol., 65, 149–160 (2005).
5. T. Portyra, D. Schmidt, F. Henning and P. Elsner, Proceedings ACCE 2008, Troy, MI.
6. J. L. Thomason , The influence of fibre length and concentration on the properties of glass fibre reinforced polypropylene: 5. Injection moulded long and short fibre PP. Composites A 33, 1641–1652 (2002).
7. Proceedings Fachtagung LFT, Wuerzburg, Germany, 2008.
8. J. C. Halpin and J. L. Kardos, The Halpin-Tsai equations: A review, Polym. Eng. Sci., 16, 344–352 (1976).
9. F. W. J. van Hattum and C. A. Bernardo, A model to predict the strength of short fiber composites, Polym. Compos., 20, 524–533 (1999).
10. T. D. Papthanasiou and D. C. Guell, *Flow-induced alignment in composite materials*, Woodhead Publishing Ltd, Cambridge (1997).
11. W. Michaeli and M. Kratz, 3D-FE simulation of injection molding – calculation of fiber orientation and crystallization. Antec 2002 Plastics: Annual Technical Conference, Vol. I: Processing, 556 (2002).
12. W. Feng, A. Ait-Kadi, J. Brisson and B. Riedl , Polymerization compounding composites of nylon-6,6/short glass fiber, Polym. Compos., 24, 512–524 (2003).
13. W. H. Yang, D. C. Hsu, V. Yang and R. Y. Chang, Computer simulation of 3D short fiber orientation in injection molding. Antec 2003 Plastics: Annual Technical Conference, Vol. I: Processing, 651 (2003).
14. Composites – prime enabler for wind energy. Reinforced Plastics 47 , 29–34 (2003).

Chapter 4
Glass Fibers for Printed Circuit Boards

Anthony V. Longobardo

Abstract Fiberglass imparts numerous positive benefits to modern printed circuit boards. Reinforced laminate composites have an excellent cost–performance relationship that makes sense for most applications. At the leading edge of the technology, new glass fibers with improved properties, in combination with the best resin systems available, are able to meet very challenging performance, cost, and regulatory demands while remaining manufacturable.

Keywords Aramid · Borosilicate · Composite · Coefficient of thermal expansion (CTE) · D-glass · Dielectric loss (Df) · Dielectric constant (Dk) · E-glass · Elastic modulus (E) · Electronics · Epoxy · Fabric · Fiberglass · Glass · Hollow filament · Laminate · L-glass · NE-glass · Printed circuit board (PCB) · Permittivity · Prepreg · Resin · Restriction of hazardous substances (RoHS) · Strength · Thermal conductivity · Waste electrical and electronic equipment (WEEE) · Weaving · Yarn

4.1 Introduction

Printed circuit boards (PCBs), sometimes referred to as printed wiring boards (PWBs), are the foundation upon which the architecture of today's high-speed electrical circuitry is built. First and foremost, PCBs provide a means for the fixed structural placement of the large variety of electrical components that make up modern electrical devices.

The breadth of the electronics industry, even when focused down to a PCB, represents a tremendous amount of technology. The disciplines involved are numerous and varied spanning many fields of science, engineering, and manufacturing. This chapter cannot hope to treat these interrelated subjects in any rigorous manner. Rather, we will simply attempt to highlight the vital role played by fiberglass in

A.V. Longobardo (✉)
AGY World Headquarters, Aiken, SC 29801, USA
e-mail: anthony.longobardo@agy.com

F.T. Wallenberger, P.A. Bingham (eds.), *Fiberglass and Glass Technology*,
DOI 10.1007/978-1-4419-0736-3_4, © Springer Science+Business Media, LLC 2010

contributing to the performance of PCBs. Where beneficial to the reader, specific resources within the technical literature will be suggested.

4.1.1 Printed Circuit Board Requirements and Their Implications for Fiberglass

Stripped of its electrical hardware and their associated connections, a rigid PCB is a reinforced polymer–composite substrate. Sandwiched within the board, the reinforcement material has tended to be multiple layers of finely woven fiberglass cloth. At the extremes of the cost and performance continuum other materials such as paper at the low end and high cost polymers such as aramid at the high end are used as the reinforcement layer. Fiberglass, however, has been the mainstay of the industry since the 1950s.

Figure 4.1 illustrates how fiberglass is integrated into a laminate for use as a PCB. Manufacture of a laminate is a series of relatively straightforward processes performed by different segments of a supply chain made up of the fiberglass yarn manufacturer (Fig. 4.1, panels 1–3), a weaver (panels 4–9), the laminator (panels 10 and 11), and finally the board manufacturer (panel 12). Excellent overviews specific to each individual segment can be found in [1, 2].

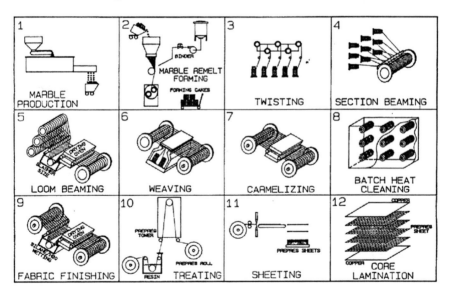

Fig. 4.1 Key process steps for making fiberglass yarn, fabric, and laminate boards (used with permission of AGY)

Glass is melted, in this example via a marble-making step, and formed into continuous filaments and wound into forming cakes as shown in panels 1 and 2. Twisting the filaments and winding them onto bobbins in panel 3 creates fiberglass yarn. Panels 4–9 illustrate the steps taken by the weaver to form and finish an appropriate fabric construction defined by the board's final application. The rolled fabric is then put through a treater (panel 10), which impregnates the fabric with an appropriate resin system. The impregnated fabric continues through an oven where the resin is cured and then either rolled or cut to size. At the stage shown in panel 11, the fiberglass-resin material is cut into sheets and called a prepreg.

At the board manufacturer, the prepreg is then interlaced sheet by sheet with copper foil to form a book. Books are then stacked together separated by steel press plates and placed in a heated hydraulic press. This is shown in panel 12 and more detailed rendering of this step is given in Fig. 4.2. After pressing, heating, and curing each book becomes a laminate board sometimes called a core. A well-done review of these process steps is presented by Kelley [3]. The board then undergoes a large number of subsequent steps to build up the electrical infrastructure that will support and connect all of the components called for in the PCB design. The reader is directed to a detailed presentation of each manufacturing process given in parts 2 and 6 in a handbook by Coombs [4].

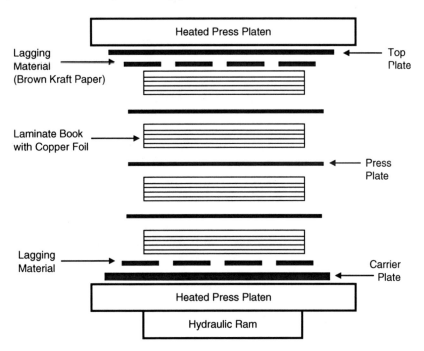

Fig. 4.2 Detailed illustration of a prepreg press, after Kelley [3], © The McGraw Hill Companies, Inc.

4.1.2 Fiberglass' Role in PCB Construction

The fiberglass in a PCB board acts to reinforce the polymer matrix to form a composite. A composite is an engineered material made from two or more constituent materials (a matrix and a reinforcement) having significantly different chemical and/or physical properties. While each component remains separate and distinct on a macroscopic level within the finished structure, the resulting product has improved end-use properties. As reinforcement, fiberglass contributes mainly electrical, thermal, and mechanical benefits to the surrounding polymer resin matrix. Figure 4.3 summarizes the components found in a laminate prepreg composite and their functions [5].

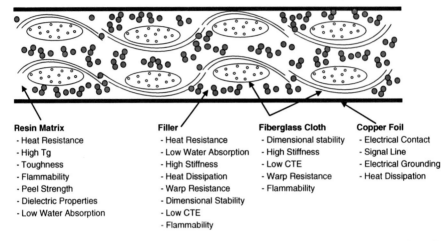

Resin Matrix	Filler	Fiberglass Cloth	Copper Foil
- Heat Resistance	- Heat Resistance	- Dimensional stability	- Electrical Contact
- High Tg	- Low Water Absorption	- High Stiffness	- Signal Line
- Toughness	- High Stiffness	- Low CTE	- Electrical Grounding
- Flammability	- Heat Dissipation	- Warp Resistance	- Heat Dissipation
- Peel Strength	- Warp Resistance	- Flammability	
- Dielectric Properties	- Dimensional Stability		
- Low Water Absorption	- Low CTE		
	- Flammability		

Fig. 4.3 Components and their function of a laminate prepreg. After Andresakis [5], © BR Publishing

It is clear from Table 4.1 that the fiberglass modifies the bulk mechanical and thermal properties of the laminate. Many of the processing steps to convert a laminate board into a PCB require excursions to temperatures near or above the maximum use temperature of the resin. In addition, thermal cycling leads to the development of transient thermal stresses due to thermal expansion differences between the PCB and some of the individual electronic components. Without reinforcement the structural integrity of a resin-only substrate would be unworkable.

Table 4.1 Key properties of fiberglass and resin used in a typical FR4 laminate board

	Elastic Modulus GPa	Tensile Strength MPa	CTE $\times 10^{-6}/\,°C$	Upper use Temp, °C	Dielectric Constant, Dk @ 10 GHz	Loss, Df @ 10 GHz
Epoxy-cyanate blended resin	3.3	61	20	135	2.80	0.0080
E-glass, D450 1080 weave	72	789	5.4	650	6.88	0.0062
Laminate board, 2 plys 1080 fabric (48.4% glass)	29	415	12	210	3.93	0.0078

PCB performance is critical to achieving the overall electrical requirements dictated by a device's design. New designs continue to push the boundaries of what is possible. Thus, over the past decade the general trends in PCB design have been [6, 7]

- Smaller, thinner overall dimensions
- Finer wire connections and higher circuit density, all of which enable continued miniaturization
- Faster operating frequencies, now approaching 77 GHz, to increase processing/device speed but without sacrificing signal efficacy
- A conscious move by the industry away from environmentally hazardous materials such as heavy metals (Pb, Hg, Sb) and halides (F, Cl, Br)
- The corresponding need for more thermally stable materials to cope with the required temperatures used in manufacturing
- Cost containment

All of these trends have implications for fiberglass as a PCB constituent.

4.1.3 Electrical Aspects

Fiberglass plays an important role in the electrical behavior of a PCB. Electronically, a PCB is an insulating element within a design. That is, it does not directly conduct current and acts to shield conductive parts of the circuitry from one another. In this way it forms the roadbed for the numerous electrical connections and vias that stretch across and between its surfaces. Insulating these connections from one another without interfering with functionality is therefore the primary technical challenge.

In many ways glass, specifically fiberglass, is uniquely suited to act as reinforcement in a PCB. It is an excellent insulator, it has good dielectric properties (relative permittivity and dissipation factor), it provides sufficient mechanical reinforcement, it has excellent thermal stability (in areas of both thermal expansion and flammability), as well as low sensitivity to moisture and chemical corrosion. The next few sections focus on key areas that need emphasis to better understand where further developments need to occur.

4.1.3.1 Dielectric Constant

Figure 4.4a illustrates what happens when a voltage, V, is placed across two electrodes of a parallel plate capacitor separated by a gap, d, in vacuum. The plates develop a static charge per unit area, σ, and a corresponding electric field, E, is induced. Permittivity, ε, relates to a material's ability to transmit (or "permit") an electric field. In the case of a vacuum, ε is known as the permittivity of free space, $\varepsilon_o = 8.8554 \times 10^{-12}$ F/m. In most cases the ε of air is taken to be approximately the same value as ε_o.

When a material is inserted between the electrode plates and into this static electric field, E, the atoms in the material become polarized. (Charges within the

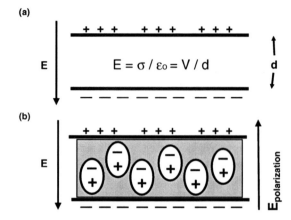

Fig. 4.4 (a) Parallel plate capacitor separated by air and (b) polarization of a dielectric when inserted between the plates

material physically shift as + attracts −. This is shown in Fig. 4.4b.) As a result a portion of the energy contained within the field, E, is consumed by physically displacing these charges. The effective field, as shown in 4.1, is thereby diminished (with the storing of charge) by a factor, k, which is called the relative permittivity (sometimes written as, ε') or dielectric constant. The dielectric constant of a material is also often referred to as Dk.

$$E_{\text{effective}} = E - E_{\text{polarization}} = \sigma / k\varepsilon_o \qquad (4.1)$$

At high frequencies, moving from the MHz to GHz range, the behavior of an electric field within the insulating layer becomes important because a PCB can become a waveguide. Skin effects and signal reflections can give rise to losses, which can degrade signal speed [8]. The velocity of an electrical signal is governed by the square root of the permittivity of the PCB laminate as shown in the following equation:

$$\text{Velocity} = \frac{\text{Speed of light}}{\sqrt{\text{Permittivity}}} \qquad (4.2)$$

The dielectric constant of fiberglass is important to the performance of a PCB because it influences how well electrical signals are shielded from one another. As the conductive lines within a circuit design are moved closer together they can experience interference known as cross talk. Since the Dk of fiberglass is higher than the resin portion of the composite (as shown in Table 4.1), minimizing this property is of critical importance to controlling cross talk interference.

4.1.3.2 Dielectric Loss

The above description relates to a static or DC field. Most electronics work with alternating current, AC. These relationships therefore become time- and, as a result, frequency dependent. How a material responds to the speed of an electric field's oscillation determines its frequency response Dk $f(\omega)$ and its loss characteristic, Df, sometimes called dissipation factor, or tan δ.

The amount of dielectric energy loss, W, can be quantified by the following expression:

$$W = K\omega v^2(\varepsilon' \tan \delta) \qquad (4.3)$$

where K is a constant, ω is the signal frequency, v^2 is the potential gradient, ε' is the relative permittivity, and $\tan \delta$ is the dielectric loss tangent. The quantity ($\varepsilon' \tan \delta$) is termed the loss factor [9].

The amount of signal energy used in most microelectronics is quite small and lower power inputs are the desired industry trend. Loss mechanisms within the materials that make up the circuitry consume some of this energy, thereby degrading signal integrity. Losses also generate heat within the device. Heat degrades performance and needs to be eliminated. Heat removal from electronic circuits also tends to increase cost. The dielectric dissipation factor of a PCB is a key criterion in designing the microwave circuitry in today's wireless devices. Excellent general treatments of the dielectric behavior of glass are given by Babcock [10] and Tomozawa [11].

4.1.3.3 Hollow Filaments

Hollow filaments are a type of defect found within fiberglass yarn. Molten glass naturally contains a small fraction of gaseous inclusions, known as seeds. Well-refined glass is virtually seed-free. However, if seeds exist in the glass they become attenuated within the fiber during forming. The resulting fiber is therefore hollow for a portion of its length depending upon the volume of the seed. Hollow sections can be millimeters in length.

Bundles of filaments make up a strand of yarn. Hollow filaments are problematic because they represent a reliability issue for the PCB. A fabric that contains significant numbers of hollow filaments has the potential to have one of these voids opened and exposed while drilling of the laminate board during manufacture. An exposed void in a fiber can act as a capillary. The capillary force can subsequently attract moisture (that can leach out conductive species from the surrounding material), or in some cases solder material, which can create a short circuit that allows an uncontrolled current path within the device. What makes this type of defect so serious is that it can lead to board failure at anytime during service life [12].

4.1.4 Structural Aspects

Apart from the critical electrical characteristics of fiberglass, certain mechanical attributes are also important for success as PCB reinforcement. Of particular relevance to fiberglass are the structural needs of the laminate. Properties such as mechanical strength, elastic modulus (stiffness), thermal expansion, and thermal conductivity are all key. In addition, woven fiberglass fabric brings along with it a few aspects that are also noteworthy.

4.1.4.1 Mechanical Strength

Composites are in many ways materials of compromise. Well-chosen components work in concert together to provide the best all-around performance across multiple end-use properties. One of the main reasons for choosing a composite material is improved mechanical strength. Since an applied load on a fiber-reinforced laminate board is shared between the lower strength polymer and the stronger fiberglass, the resulting strength is better than the resin alone. Table 4.1 gives some idea of the magnitude of the increase achievable with fiberglass in combination with an epoxy-type resin system.

The strength of glass, especially fiberglass, is a surface-controlled phenomenon. Flaws present on the glass surface act as stress concentrators and weaken the surface under an applied tensile load [13]. However, the ultimate strength of a PCB depends on factors beyond just the intrinsic strength of the glass fibers. For example, composite strength is a material property that obeys the rule of mixtures [14]. The rule states that a composite property can be estimated as the sum of each component's material property multiplied by its volume fraction. Properties such as elastic modulus, Poisson's ratio, and thermal expansion, as well as Dk and Df all follow this rule. Consequently, these properties are all found to scale directly with resin-to-fiber ratio. In a similar way, other means of varying glass content such as a fabric's weave pattern, weight (g/unit area), board construction (number of fabric layers) can all influence board strength [2, 6].

4.1.4.2 Elastic Modulus

The elastic modulus is the slope of the stress–strain curve measured during tensile testing a specimen. It is also called Young's modulus. The relationship follows that the steeper the slope, the higher the corresponding modulus. The elastic modulus can be thought of as the stiffness of the material being tested, i.e., its resistance to strain (deflection) per unit stress exerted. Stiffness retards board flex during manufacturing making the process more robust and less sensitive to the influence of processing temperatures.

Polymers have inherently low stiffness mainly due to their chain-like structures and relatively short range bonding (cross-linking). Silicate glasses, on the other hand, have excellent elastic modulus/stiffness characteristics. The large difference in behavior is easily recognized in Table 4.1. The resulting fiberglass-reinforced laminate benefits from significantly improved mechanical stiffness.

4.1.4.3 Thermal Expansion

Most materials respond to a positive temperature change by increasing their volume or expanding. The rate of expansion per degree change in temperature is known as the coefficient of thermal expansion or CTE. Thermal expansion is important to PCB laminate board performance because their manufacture relies on a variety of multi-step joining processes. These processes, such as soldering, frequently employ heat, which can create thermal stresses that can lead to a range of failure modes. This is critical since many of these joints are electrical connections and their integrity cannot be compromised.

Glass is the low CTE component in a PCB laminate. As can be seen in Table 4.1, adding fiberglass to resin in this case results in a 40% reduction in the laminate's CTE. And via the rule of mixtures this depends directly on the %-fraction of fiberglass cloth used. Any modification that decreases the amount of resin in the laminate will decrease thermal expansion; refer to Fig. 4.5. Another way this can be accomplished is by adding various silicate filler materials, such as talc, in order to decrease CTE as noted in Fig. 4.3.

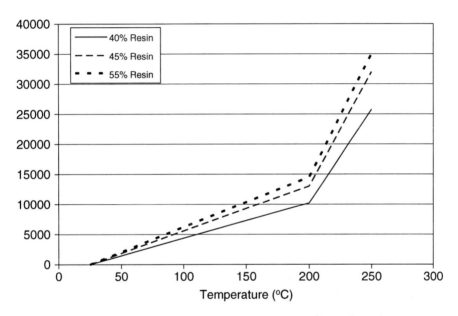

Fig. 4.5 Variation of thermal expansion of laminate as a function of increasing resin content

4.1.4.4 Upper Use Temperature

The upper use temperature of a prepreg or laminate board is governed by the resin system being employed rather than the fiberglass reinforcement. Processing temperatures used in the manufacture of the laminate prepreg as well as those employed in the final PCB will influence the selection of resin based on the transition temperature, T_g, of the polymer. The transition temperature can be thought of as the range in which an amorphous material (both glass and polymers fit this classification) changes from behaving as a stiff solid to a liquid by softening. E-glass has a T_g 400–500°C higher than most polymer resins used in PCB manufacturing.

Adding fiberglass to a composite construction increases the upper service temperature of the laminate. Another important aspect of laminate behavior is thermal degradation or breakdown of the resin system. Adding fillers to the resin can yield an improvement. These accommodations are important in high-end applications as well as during manufacture since process temperatures for steps, such as soldering, have increased significantly since the move away from lead-containing solders. References [15,16] by Kelley are excellent resources for more information on this topic.

4.1.4.5 Weave and Fabric Construction

Weaving fiberglass yarn into fabric is virtually indistinguishable, as a process, from weaving other common materials into cloth. The interlocking nature of the yarn strands make up a 3D network within the cloth that has a major impact on the properties of the resulting laminate [17]. References [1, 2] give good general overviews of the fabric styles used in PCB laminates, as well as their respective benefits. The varieties of weave patterns and how they are oriented within a prepreg lay-up represents an important way of modifying the end-use properties of the laminate board.

It has been recently noted in the literature that weave pattern also has an affect on the electrical characteristics of PCBs at frequencies >1 GHz. By virtue of being woven, the fabric has areas of fiber density that spans from zero (the voids between strands) to areas with two times the strand thickness (known as the knuckle, where the warp and weft cross). This distribution presents a non-homogeneous dielectric contour to propagating pairs of signals and their associated electromagnetic fields. This results in loss of signal integrity, a phenomena called skew. And when skew is caused by the profile of the fiberglass weave it is known as the fiber weave effect or FWE [18]. This problem can be addressed by treating the fiber bundle in a way that allows spreading of the strands into a more uniform weave pattern [19].

Hybrid fabric constructions are also being used in certain segments of the market. This includes all-glass hybrid constructions where different slivers are mixed to optimize fabric characteristics, as well as fabrics that are mixtures of glass and strong organic polymers such as cyclic olefin copolymers [20]. These are examples where the fiberglass fabric itself becomes a kind of composite to further improve end-use properties.

4.2 Glass Compositional Families

Glass property modification is ultimately a study of how chemical composition influences glass structure and thereby the resulting properties. Glass can be made from an amazingly wide variety of materials, both organic and inorganic as well as from certain metals, if cooled rapidly enough from the molten state. For a thorough background on this subject the reader is directed to references by Varshneya [21] and Shelby [22].

4.2.1 Improvements Initially Based on E-Glass

All commercial fiberglass compositions fall into a relatively narrow range of compositions known as silicate glasses. That is, they have silicon dioxide (SiO_2 or silica) as their majority component. E-glass, already mentioned here and throughout this volume, is the foundational silicate glass composition for fiberglass in PCB laminates.

4.2.1.1 E-Glass – The Industry Standard

E-glass is a borosilicate glass composed of three primary constituents, silica (SiO_2), calcium oxide (CaO), and aluminum oxide (Al_2O_3, alumina); refer to Table 4.2. These three components along with other minor ingredients have been optimized over the years into a range of formulations, many of them manufacturer specific, which represent an acceptable balance between forming characteristics, processing characteristics, performance properties, and cost [23].

Table 4.2 E-glass specification as per ASTM D-578-00, two typical E-glasses, and one at the limit

E-glass for printed circuit boards				E-glass for general applications		
Chemical	ASTM % by weight	Typical	Theo. limit	Chemical	ASTM % by weight	Typical
SiO_2	52–56	54.3	55.87	SiO_2	52–62	60
B_2O_3	5–10	6	10.00	B_2O_3	0–10	–
Al_2O_3	12–16	14	12.30	Al_2O_3	12–16	13
CaO	16–25	22.1	16	CaO	16–25	22.1
MgO	0–5	0.6	5	MgO	0–5	3.11
Na_2O and K_2O	0–2	1	0.82	$Li_2O + Na_2O + K_2O$	0–2	0.77
TiO_2	0–0.8	0.5	–	TiO_2	0–1.5	0.55
Fe_2O_3	0.05–0.4	0.3	–	Fe_2O_3	0.05–0.8	0.24
Fluoride	0–1.0	0.7	–	Fluoride	0–1.0	Trace
Properties						
Dk, 1 MHz	–	6.8–7.1	6.39		–	7.0
log 3 FT, °C	–	1200	1251		–	1260
References	[24]	[23]	[28, 29]		[24]	[23]

The textile industry has standardized E-glass into two general categories: (i) yarn for PCB and aerospace applications and (ii) common electrical and utilitarian applications, such as for electrical isolation or structural reinforcement [24]. Table 4.2 summarizes these compositional ranges. The primary difference between these two categories of E-glass involves the presence of boron trioxide (B_2O_3). E-glass for PCBs must contain between 5 and 10 wt% B_2O_3 to meet the standard, whereas glasses with less than 5 wt% can only be used for general purposes. A typical E-glass suitable for PCBs is included in the table for reference.

Boron in silicate glasses is unique in that it behaves as a flux by decreasing viscosity and thereby helping melting and fiber forming, but it does so without drastically breaking down the glass network structure like alkali or alkaline earth oxides. Boron, however, is the most expensive component in the glass and also one of the most volatile (tending to want to escape while melting). Fugitive emissions, like B_2O_3, are strictly regulated in the glass industry and often require costly emission controls.

Boron-free E-glasses made their appearance in the mid-1980s as an answer to these problems. Unfortunately, boron-free E-glass yarn was not acceptable to the PCB market because of a high level of hollow filaments. It has since been understood that this was due to poor refining and fluorine content. Some in the industry believe that there is no technical reason why a boron-free E-glass would not work for PCB applications.

4.2.1.2 Improving Dielectric Properties of E-Glass

Varying the amounts of specific oxides in E-glass will tend to influence Dk, Df, and nearly all thermal and structural properties like forming viscosity and elastic modulus. A summary of how a few relevant glass making oxides can influence some key E-glass properties is given by Aubourg and Wolf [25]. Their presentation highlights groups of oxides by function: (i) network (glass formers), (ii) modifiers (additions that only serve to modify the glass structure), and (iii) intermediates (additions that by themselves are not network formers but that can participate partially in glass structure).

Table 4.3 introduces a few representative low-dielectric constant glass compositions. Fused silica, the prototypical glass-forming material, has excellent dielectric

Table 4.3 Low dielectric constant fiberglasses

Fiber	Commercial		Experimental		Commercial		
	E-glass	D-glass	Low Dk-glasses		Low Dk-glasses		Silica
Weight %	Typical	Typical	Eastes	Li	NE-glass	L-glass	Quartzel®
SiO_2	54.3	72–76	62	61.03	50 to 60	52–60	99.999
B_2O_3	6	20–25	8	10.73	14 to <20	20–30	–
Al_2O_3	14	0–1	14	12.04	10 to 18	10–18	–
TiO_2	–	–	3	0.5	0.5 to 5	Trace	–
MgO	0.6	–	8	9.97	1 to <6	Trace	–
CaO	22.1	<1	–	2.98	2 to 5	4–8	–
ZnO	–	–	3	–	–	–	–
Li_2O	–	–	–	1.05	<0.3	–	–
Na_2O	0.8	<2	–	0.5	<0.3	Trace	–
K_2O	0.2	<2	–	0.4	<0.3	–	–
Fe_2O_3	0.3	–	–	0.35	Trace	Trace	–
F_2	0.7	–	2	0.45	<2	<2	–
Properties							
Dk, 1 MHz	6.8–7.1	4.1	5.47	5.61	4.4	<5	3.78
Dk, 10 GHz	~6.8	4.2	–	–	4.7	<5	3.78
Df, 1 MHz	0.0060	0.0005	–	0.0034	0.0006	<0.0005	0.0001
Df, 10 GHz	0.0070	0.0025	–	–	0.0035	<0.0050	0.0001
log 3 FT, °C	1200	1410	~1460	1244	<1350	<1340	>1800
Density, g/cc	2.54	2.14	–	2.42	2.3	~2.3	2.15
CTE, × 10^{-6}/°C	5.4	3.1	2.5	–	3.4	~3.9	0.54
Elastic modulus, GPa	72.3	51.7	–	–	57	62	69
Special process needed	No	Yes	Yes	No	Yes	Yes	Yes
References	[23]	[23, 30, 31]	[26]	[27]	[30, 32]	[31]	[23]

characteristics (Dk = 3.8, Df = 0.0001), but it is very challenging to melt and form into fiber that can be used in textiles. Moreover, the forming temperature involved (>1800°C for a viscosity of 10^3 P, or log 3 P) is at the limit of available technology regardless of cost. So while it is in some ways an ideal material, it is impractical.

E-glass is presented in the table as a benchmark composition. As noted above, the dielectric properties of glass are controlled by the polarizability of cations present in the glass. Glasses with higher amounts of polarizable species, alkali and/or alkaline earth oxides will tend to have stronger, and therefore poorer, dielectric behavior. However, it is the presence of these modifying ions that facilitates glass melting at acceptable temperatures, and therefore reasonable cost. For example, the temperature needed to reach a suitable forming viscosity (log 3 P) is in the range of 1200°C, more than 600° less than fused silica. Lower forming temperatures are viewed as desirable since they conserve energy, lower costs, and can be handled with traditional technology.

Nearly all important glass properties can be influenced to yield an improvement. Optimization, however, is a matter of balance and compromise. The two experimental low Dk-type glasses shown in Table 4.3 represent similar approaches to decreasing dielectric constant. The approach taken by Eastes [26] drastically lowers the CaO content from ~22 wt% to 0, essentially making the glass CaO-free. In contrast, MgO is taken to 8 wt%, which is 3 wt% above the maximum set by ASTM. In addition, both the network formers SiO_2 and B_2O_3 are increased. Based on what we know about these various influences, one would expect a significant decrease in Dk. On the other hand, we would also expect that the melting and viscosity behavior would also worsen (require more temperature), as well. Eastes compensates for this by adding TiO_2 and ZnO, which modifies viscosity and helps melting but has less impact on dielectric constant. The resulting Dk is reported as 5.47 vs. between 6.8 and 7.1 for a typical E-glass.

Li [27] takes a slightly different approach. Like the previous example, the network formers (silica and boron oxide) increase and the total amount of alkaline earths decrease in favor of MgO. However, in this case, the viscosity impact is offset by using small amounts of alkali, the majority of which is Li_2O due to its strong fluxing action and minimal impact on Dk and Df. The resulting Dk (5.61) is nearly as good as Eastes, but the forming temperature (1244°C) is much closer to standard E-glass (1200°C) compared to Eastes (1463°C).

These examples are two of many ways to approach the problem of improving upon the dielectric constant beyond E-glass. It should be noted as we close this section that the above glasses are not technically E-glass as per the ASTM standard since the alkaline earth contents were varied outside the allowable limits. To date these glasses remain experimental.

4.2.1.3 Challenges and Limitations

It should be obvious from the foregoing that attempting to improve the dielectric properties of E-glass can only be taken so far. Because it is a standardized glass family, the compositional limits imposed by the industry effectively restrict the available possibilities. Since the examples cited above are technically outside the commercial boundaries of E-glass, it is interesting to speculate how much improvement is possible within the limits of the standard.

Properties such as dielectric constant and viscosity can be estimated mathematically from glass composition [28, 29]. Looking back, Table 4.2 proposes a theoretical E-glass composition within the limits of the ASTM standard but adjusted for composition to simultaneously minimize Dk and forming viscosity. The results suggest that if network formers are taken to their maximums, CaO is set to its minimum and MgO to its maximum (to aid in melting), a Dk of ~6.4 would be expected at reasonable log 3 temperature of 1251°C. Based on this simple exercise it should be safe to say that one would not expect Dk values below 6.0 for any currently allowable variant of E-glass.

Given that fused silica has a Dk of 3.8, it is fair to consider it as a practical limit for silicate glasses. And, that the best that can be hoped for from an E-type glass is no better than 6.0. This begs the question what kind of glass would be needed to yield dielectric behavior between about 4.0 and 6.0 for Dk and a Df of <0.0050?

This question was addressed years ago by the fiberglass industry as a response to tightening requirements for dielectric behavior in high-performance PCBs. E-glass was recognized as unable to meet the requirements of certain emerging applications, driven mainly by computing, and a new compositional family was developed and commercialized. This family was termed D-glass for improved dielectric characteristics.

4.2.2 D-Glass and Its Compositional Improvements

4.2.2.1 D-Glass

Like E-glass, D-glass is a member of the borosilicate family. It is distinguished from most borosilicates in that it contains nearly the highest amounts of B_2O_3 and SiO_2 of any commercialized glass composition; refer to Table 4.3. It is nearly all network former (>95%). The reader should now recognize that a tightly bound network structure is a key attribute of low-dielectric constant glasses. Relatively small amounts of alkali and alumina are present to assist in melting and modifying the glass structure to control other properties.

D-glass offers a nearly ideal set of dielectric characteristics to the fiberglass manufacturer, as well as the downstream board maker. However, the composition also poses significant challenges in areas of glass melting, fiber forming, fabric weaving, and most importantly final end-use product performance. The following specific problems have been noted in the literature [30]:

- Glass melting – Highly networked glasses like D-glass are extremely difficult to melt with conventional furnaces due to its extreme temperature requirements.
- Homogenization – Highly viscous glasses, especially borosilicates that have a tendency to phase separate, tend to be difficult to make chemically uniform leading to viscosity-related defects in fiberization.
- Boron volatility – The high boron content magnifies the volatilization rate and complicates both achieving and maintaining homogeneity.

- Limited manufacturing options – Extreme temperatures coupled with difficult behavior severely limit the melting technologies available to continuously manufacture D-glass fibers.
- Hollow filaments – Viscous glasses are also difficult to refine and thus are more prone to hollow filaments and their associated performance issues.
- Poorer forming behavior compared to E-glass – Poor homogeneity and seediness tend to cause poor fiber drawing behavior, which decreases productivity.
- PCB-related difficulties – Laminated boards made with D-glass tend to have poorer chemical resistance to water. Researchers have also observed poor adhesion characteristics to the laminating resin, which causes board failure. Hole drilling has also been noted to be more prone to defects in boards made with D-glass.
- All of the above difficulties have negative cost implications.

Generally speaking, these shortcomings have been major obstacles to wider adoption of D-glass by fiberglass yarn makers and by the PCB manufacturing community.

4.2.2.2 Improvements Based on D-Glass

On the whole, D-glass only meets part of the requirements for a successful low-dielectric constant-type fiberglass. Just having excellent Dk and Df are insufficient. Being a commercial success requires more than just technical success. Primarily, the product must be manufacturable without undue complexity and expense, and from a customer's perspective it must process through their portion of the supply chain at least as well as E-glass.

Numerous attempts have been made by manufacturers to commercialize improved versions of D-glass. Only a few have been successful. Two of these glasses are presented in Table 4.3 for further consideration. Both glasses, like D-glass, are characterized by high SiO_2 and B_2O_3 content (>70 wt% total), as well as low alkaline earth oxide content (<11 wt%). Unlike D-glass, the amount of Al_2O_3 has been maintained at E-glass levels to minimize the likelihood of phase separation and improve chemical durability against water.

The glass composition family recently disclosed by Boessneck [31] and commercialized by the revitalized name L-glass^{TM} for "Low Loss," contains as much B_2O_3 (>20 wt%) as D-glass, which is a major benefit for Dk and Df. (Reported as Dk <5.0 and Df <0.005 at 10 GHz.) Moreover, it avoids the use of alkali additions that aid in melting but increase Dk by using a relatively limited amount of CaO (4–8 wt%). No MgO or other melting aids such as TiO_2 or ZnO are used. Most importantly, the resulting glass can be readily drawn into appropriate fiber forms suitable for PCB applications. L-glass for PCBs should not be confused with high lead-containing glasses developed years ago for their high Dk [32].

NE-glass, a fiber produced by Nitto Boseki [33], is a different variation on the theme. Specifically, boron content is limited to less than 20 wt% due to concerns over volatility during fiber forming. In addition, limited amounts of alkali are used, but only at levels that do not significantly degrade dielectric performance. Interestingly, NE-glass incorporates substantial amounts of TiO_2, which is claimed to improve melting, decrease forming viscosity, and improve dielectric loss behavior. For this embodiment they report Dk = 4.7 and Df = 0.0035 at 10 GHz. Numerous modifications to this approach can be found in the patent literature surrounding this family of glasses.

4.2.2.3 Challenges and Limitations

It should be recognized that D-glass and its variants share much of what chemically contributes to excellent dielectric behavior in glass. With that in mind, many of the technical hurdles identified in Section 4.2.2.1 are operative for these glasses as well. It is likely that producing commercial quantities of these types of glass fiber would not have been possible just 20 years ago. Commercialization of these improved D-type glasses indicates, however, that progress has been made in the area of manufacturing. These glasses, although difficult to produce, have benefited from incremental advances in furnace, as well as fiber-forming technology. A brief list of the challenges that remain is given below:

- Viscosity – As noted previously, low-Dk-type glasses necessitate very high temperatures to melt, refine, and form. This behavior is directly dictated by the required glass chemistry. Other than the rather modest modifications embodied by the improvements discussed above there is likely little more that can be done to improve further without lessening dielectric performance.
- Boron volatility – High processing temperatures and fugitive nature of boron present serious challenges to continuous melting of these types of glasses. Difficulties in maintaining the desired homogeneity and the specter of environmental regulation and abatement costs pose high levels of risk that must be overcome to be successful.
- Electric melting – Electric melting offers an alternative to traditional combustion melting approaches. The primary benefit is decreased reliance on flame heating which helps to lessen boron volatility. However, the electrical conductivity of low-Dk-type glasses is not well suited to resistive or Joule heating via electric boosting. Replacing overhead-fired gas burners with electrically powered resistive-heating elements such as SiC or $MoSi_2$ is a viable alternative. But, this method results in less-direct heating and the additional cost of electricity is not desirable.
- Glass strength – High amounts of boron in borosilicates (at levels much greater than E-glass, e.g., >10 wt%) also tend to impair glass strength [30]. Lower fiber strength can present problems during fabric weaving due to broken filaments. Broken filaments not only lead to potentially weaker fabric but also create fuzz on contact points during weaving which hurts productivity. Improving the strength of these fibers will likely come from improved processing techniques rather than an increase in the inherent strength of the glass.

- D-glass compositions are not covered by the existing ASTM E-glass specifications. And further, no governing body or association has established a specification specifically for D-type glass. While this may slow their adoption, many necessary innovations such as NE-glass have tended to be implemented well before standardization takes place.

Technological developments, like water or electrical current, tend to follow the path of least resistance. Looking back at how most technological goals are accomplished we see that obstacles are inevitably encountered and either overcome or worked around. Ultimately, achieving the goal involves a set of compromises; rarely is the target hit without multiple and varied attempts.

It is the opinion of the author that the D-type glasses now offered in the marketplace represent a realization of a significant goal in the manufacture of fiberglass for PCBs. Their dielectric performance will never match that of pure fused silica. This is a limitation imposed by chemistry and physics. But unlike fused silica these glasses are manufacturable, commercially available, and able to meet the PCB maker's current performance needs. The foregoing does not preclude the possibility for further incremental improvements. Rather it forms a firm foundation upon which to build.

4.3 Future Needs of the PCB Market

The electronic industry is highly dynamic. The fast pace with which new developments occur and the speed with which they can be adopted into commercial applications are astonishing. PCBs therefore play an increasingly multifunctional part of this rapid development cycle. To paraphrase the Interconnect Technology Research Institute (ITRI), the printed circuit board has been called the essential "glue" that holds together the critical components of today's advanced electronics [34]. Component-level development can no longer be advanced independently but must take an integrated and holistic approach that embraces goals beyond the subsystems involved. The PCB therefore will continue to play a critical part in meeting the technological needs of the industry.

4.3.1 The Electronics Manufacturer's Roadmap

Part of the success in managing the dynamic changes found in the electronics market must be attributed to the industry's foresight in creating and following technological roadmaps. A key roadmap covering the entire semiconductor industry is the International Technology Roadmap for Semiconductors (ITRS), which has been guided by five regional semiconductor industry associations (SIAs) [35].

Board designers and manufacturers have their own industry associations; one important group, the IPC (previously known as the Institute for Printed Circuits), has been leading the effort to compile a detailed biannual roadmap for the industry

[36]. The current roadmap for 2008–2009 covers numerous topics related but not limited to PCBs in a very in-depth manner. Some of these topics fall directly in line with the recent trends identified in Section 4.1.2 above. For this reason, only the primary themes of the roadmap that relate to PCBs will be presented.

4.3.2 What This Means for the Board and Yarn Makers

The search for more efficient use of space on a PCB is a growing problem because boards are being used to perform more functions. Generally speaking, the number of circuits per unit area is increasing so that boards are becoming crowded. The challenge to PCB manufacturers continues to be the need to achieve a high density of interconnects (termed HDI) and increasing the number of discrete conducting layers within a board (known as layer count). Since performance is directly linked to device speed (operating frequency) both HDI and layer count are critically important.

Figure 4.6 illustrates how higher board frequencies and their associated market segments depend on layer count. As expected, boards with the highest layer count (and therefore price) correspond to the highest levels of performance and represent the lowest volume market segment. This is the segment that predominantly benefits from new D-type fiberglass fabrics. Signal integrity and mechanical stability are paramount here and the increased cost of low-loss fiberglass compared to other more expensive alternatives can be more easily justified.

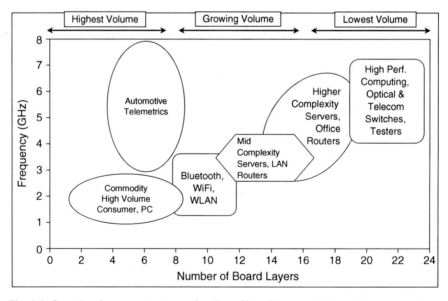

Fig. 4.6 Operating frequency range as a function of board layers and the market segments they cover (Provided courtesy of Park Electrochemical Inc.)

The overall thickness of the newest printed circuit boards is quite thin, typically 1.6 mm or less. Striving to fit more layers of reinforcing fabric into the same space entails the need for finer fibers. The current standard fiber has a nominal diameter of just over 5.3 μm, termed with a letter designation D [24]. The finest in current usage is a C-filament at 4.4 μm. BC-fibers at a nominal 4 μm are technologically possible and present room for improvement. B-filaments are possible at 3.7 μm, but there are concerns about respirability of fibers of this size. Diameters finer than B represent a technological barrier that has not been crossed.

Fine fabrics for higher layer counts are also becoming important for a technology called buried passives [37]. This type of PCB design replaces surface-mounted bypass style components such as capacitors and resistors and builds them within the board, between the dielectric layers. The main benefit here is to free up valuable board surface space as well as decreasing parasitic capacitive losses. The dimensional and surface quality tolerances of the participating layers are extremely tight and can prove challenging to control. Fine fiber development and diameter control are current areas of investigation.

As fine as these fabric interlayers are, there is another key process used to make HDI that is significantly impacted by just the presence of fiberglass in the board. Vias are electrical pathways that connect various interlayers through the thickness of the PCB (in the z-direction, rather than the planar x–y-directions). Over the years, creating vias in a PCB has steadily moved from mechanical drilling with fine bits to lasers. The benefits, accuracy, speed, hole quality, and repeatability are obvious and vitally important in achieving HDI.

Lasers, however, create holes by ablating away material as the beam passes along the z-axis. An epoxy-fiberglass laminate presents a problem for laser control because fiberglass is not uniformly distributed within the resin. The power required to burn through resin is much less than for fiberglass and therefore laser drilling can be a difficult process to control. Poor hole quality, a direct result of poor process control, can lead to vias that would not pass QC inspection. Fabric style (a more flat weave) is one way fiberglass can help to address this issue. Another approach is to use multiple-frequency lasers that can apply pulses of an appropriate energy to remove the required material at the required depth [38].

Also as board density increases, the dimensions of the metal interconnects must become smaller. Smaller, thinner metallic wires are fragile and as a result more prone to breakage. Building a PCB involves joining many dissimilar materials such as organic resin, metals such as copper, fiberglass, and silicon with the action of heat. Thermally induced stresses due to expansion rate mismatches are inevitable and must be minimized.

IC chips made from silicon have a very low CTE equal to about $2.5 \times 10^{-6}/°C$. The desired trend for fiberglass CTE is lower compared to standard E-glass. Driving CTE lower has the double impact of decreasing the total mismatch ($\Delta\alpha$) magnitude of the thermal strain, as well as moving the total material system closer to silicon. Progress has been made in this area since fibers such as L-glass and NE-glass, for example, have CTEs less than 4.0 compared to $5.4 \times 10^{-6}/°C$ for E-glass. CTE and dielectric behavior for these types of glasses usually move in the same direction, so further improvements in Dk will likely result in improvements in CTE as well.

The last major theme that we will cover is the impact environmental regulations have had, and will continue to have, on PCBs. Environmental regulations like all laws act locally within an area of jurisdiction. Two primary sets of regulations that pertain to PCB are (i) Waste Electrical and Electronic Equipment (WEEE) and (ii) Restriction of Hazardous Substances (RoHS) adopted by the European Union in 2005 and 2006, respectively [39]. Since electronics are produced for a global market, one region's imposition of restrictions can drive an entire industry in a new direction. An excellent overview of these regulations and how they affect PCB manufacturers was recently published by Freedman [40].

Of the two laws, RoHS has had the greatest impact on PCB fabrication. RoHS in effect banned the use of six chemicals, hexavalent chromium (Cr^{+6}), lead (Pb), cadmium (Cd), mercury (Hg), polybrominated biphenyls (PBB), polybrominated diphenyl ethers (PBDE), all of which have some application in the construction of typical PCBs. Two of the six restrictions had significant impact and were the especially disruptive. They were the move to lead-free soldering of electrical connections and the exclusion of brominated chemical additives, such as fire retardants, from laminate resin systems. Both WEEE and RoHS directives required major efforts in material development throughout the industry.

Since fiberglass contains neither lead nor bromine, it was not a direct subject of the legislation. However, because these two chemicals play significant roles in PCB construction the changes had important implications for fiberglass. Fabricators have begun to remove bromine from their product offerings as a general response, but there is some debate on whether this is necessary based on third-party testing. It is possible that the primary brominating agent, tetrabromobisphenol-A, may not be banned under RoHS.

The move to lead-free processing, on the other hand, has driven fabrication temperatures beyond the acceptable processing range for dicyandiamide-cured (dicy) epoxy resin systems. These resins were replaced by phenolic-cured epoxy, which have similar Dk but poorer loss characteristics. Phenolic-cured epoxy, however, is not the best choice for high-frequency applications and it is rarely paired with the newest low-Dk fiberglass. New resin systems that better match the dielectric capabilities of the glass and still meet the lead-free and bromine-free process requirements are currently available and continue under heavy development [41].

Replacing lead and bromine with alternate materials has tended to change or limit, usually for the worse, the flexibility that was available to the well-established fabricating processes used throughout the industry. All of these changes have also tended to increase complexity and drive up the cost of PCB manufacture. Fiberglass is a relatively inexpensive material on a per-weight basis, especially compared to alternate materials like aramid and PTFE. Fiberglass has been and will continue to be the backbone reinforcement of much of the PCB laminate market. Recent developments to improve dielectric properties will ensure that as the industry continues to push the envelope of performance, fiberglass will remain an important contributor to meeting their goals.

Acknowledgments We would like to extend our thanks to AGY for permission to participate in this work. We also wish to thank the following individuals for their kind and thoughtful help in the writing of this chapter: Doug Leys from Park Electrochemical Corp., Doug Sober from Kaneka Texas Corporation, Lee Ritchey from Speeding Edge, and Scott Northrup and WardAston with AGY.

References

1. P. D. Lyle and M. D. Kranjc, Continuous glass fiber reinforcements for printed wiring board applications, in *Printed circuit board materials handbook (electronic packaging and interconnection)*, M. W. Jawitz, ed., McGraw-Hill, New York (1997).
2. D. J. Vaughan, Fiberglass reinforcement, in *Printed circuit board materials handbook (electronic packaging and interconnection)*, M. W. Jawitz, ed., McGraw-Hill, New York (1997).
3. E. J. Kelley, Introduction to base materials, in *Printed circuits handbook*, Edition 6, Chapter 6, C. F. Coombs, ed., McGraw-Hill Professional, New York, p. 6.18 (2007).
4. C. F. Coombs, *Printed circuits handbook*, Edition 6, C. F. Coombs, ed., McGraw-Hill Professional, New York, Parts 2 and 6 (2007).
5. J. Andresakis, Materials for HDI, in *The HDI Handbook*, Chapter 5, H. Holden, ed., BR Publishing, Delhi, p. 187 (2009).
6. E. J. Kelley, Base materials performance issues, in *Printed circuits handbook*, Edition 6, Chapter 9, C. F. Coombs, ed., McGraw-Hill Professional, New York, p. 9.1 (2007).
7. H. Holden, Introduction to high-density interconnects, in *The HDI handbook*, Chapter 1, H. Holden, ed., BR Publishing, Delhi, p. 19 (2009).
8. K. Taylor, Controlled impedance, in *Printed circuits handbook*, Edition 5, Chapter 16, C. F. Coombs, ed., McGraw-Hill Professional, New York, p. 16.18 (2001).
9. N. Shibuya, High speed signal circuit in PWB, Circuit Technology, 4(3), 99–105 (1989).
10. C. L. Babcock, *Silicate glass technology methods*, John Wiley & Sons, New York, pp. 65–71 (1977).
11. M. Tomozawa, Dielectric characteristics of glass, in *Treatise on material science and technology, Vol. 12*, M. Tomozawa, et. al., eds., Academic Press, New York, pp. 283–345 (1977).
12. E. J. Kelley, Densification issues for base materials, in *Printed circuits handbook*, Edition 5, Chapter 9, C. F. Coombs, ed., McGraw-Hill Professional, New York, p. 9.7 (2001).
13. A. K. Varshneya, *Fundamentals of inorganic glasses*, Academic Press, New York, p. 412 (1994).
14. A. B. Strong, *Fundamentals of composites manufacturing: Materials, methods and applications*, Society of Manufacturing Engineers, Dearborn, p. 104 (1987).
15. E. J. Kelley, The impact of lead-free assembly on base materials, in *Printed circuits handbook*, Edition 6, Chapter 10, C. F. Coombs, ed., McGraw-Hill Professional, New York, p. 10.1 (2007).
16. E. J. Kelley, Selecting base materials for lead-free assembly applications, in *Printed circuits handbook*, Edition 6, Chapter 11, C. F. Coombs, ed., McGraw-Hill Professional, New York, p. 11.1 (2007).
17. A. B. Strong op cit, p. 66.
18. R. Gali, PCB dielectric materials for high-speed applications, Printed Circuit Design & Fab., December (2008).
19. R. Dudek, P. Goldman and J. Kuhn, Advanced glass reinforcement technology for improved signal integrity, Printed Circuit Design & Fab., February (2008).
20. B. Morin, Low dielectric fabrics for circuit board applications, Circuitree, 21, January (2008).
21. A. K. Varshneya, op. cit. Chapter 1 and 2.
22. J. E. Shelby, *Introduction to the glassy state*, Royal Society of Chemistry, Cambridge (1997).

23. F. T. Wallenberger, *Advanced inorganic fibers: Processes, structures, properties, applications*, Kluwer Academic Publishers, Boston, pp. 129–166, (2000).

24. ASTM Test Method D 578 Standard specification for glass fiber strands

25. P. F. Aubourg and W. W. Wolf, Glass fibers, in *Advances in ceramics, Vol. 18, commercial glasses*, D. C. Boyd and J. F. McDowell, eds., American Ceramic Society, Westerville, pp 51–63 (1986).

26. W. L. Eastes and D. S. Goldman, Glass compositions having low expansion and dielectric constants, US Patent No. 4,582,448, April 15 1986.

27. H. Li and C. A. Richards, Low dielectric glass and glass fiber for electronic applications, US Patent Application 2008/0146430 A1, June 19 2008.

28. V. A. Kharyuzov and A. P. Zorin, On the calculation of dielectric constant of glasses from composition, Elektronnaya Tekhnika, Ser. XIV(3), 65–69 (1967).

29. A. I. Priven, Calculation of the viscosity of glass-forming melts: IV. A unified method for calculating the viscosity of silicate and aluminate melts, Glass Phys. Chem., 24 (1), 31–40 (1998).

30. M. Matsumoto, et al., New low dielectric constant glass (NE-Glass) fiber woven materials for next generation board, Electronic Circuits World Convention 8, P2-3, Tokyo, (1999).

31. D. S. Boessneck, et al., Low dielectric glass fiber, US Patent Application 2008/0103036 A1 May 1 2008.

32. P. K. Gupta, Glass fibers for composite materials, in *Fibre reinforcements for composite materials*, Chapter 2, A. R. Bunsell, ed., Composite Materials series 2, Elsevier, Amsterdam, pp. 19–71 (1988).

33. S. Tamura, Low dielectric constant glass fiber and glass fabric made thereof, US Patent Application, 2003/0054936 A1, March 20 2003.

34. http://www.ipc.org/4.0_Knowledge/4.1_Standards/ITRI%2050000.pdf

35. For more information please visit http://www.itrs.net/home.html

36. For more information please visit http://www.ipc.org/default.aspx

37. E. J. Kelley, Chapter 9, ibid, pp. 9.13–9.14.

38. B. Forcier, "Laser Drillable E-glass Multilayer Materials. . ." The Board Authority, p. 67, July (2000).

39. For more information please visit http://www.rohs.eu

40. G. M Freedman, Legislation and impact on printed circuits, in *Printed circuits handbook*, Edition 6, Chapter 1, C. F. Coombs ed., McGraw-Hill Professional, New York (2007).

41. D. Leys, Private communication regarding new resins referencing www.parkelectro.com/parkelectro/images/n4000-13ep.pdf

Chapter 5
High-Strength Glass Fibers and Markets

Robert L. Hausrath and Anthony V. Longobardo

Abstract High-strength glass fibers play a crucial role in composite applications requiring combinations of strength, modulus, and high-temperature stability. Compositions in the high-strength glass group include S-glass and R-glass, which are used for applications requiring physical properties that cannot be satisfied by conventional E-glass. Additional compositions are also available for specialized applications requiring extreme performance in any one area. The main competition for high-strength glasses in the marketplace comes from carbon and polymer fibers. Ultimately, the product of choice is based on a compromise between cost and performance and will vary depending on the application.

Keywords Aerospace · Annealing point · Aramid · Armor · Ballistic · Basalt · Carbon fiber · Coefficient of thermal expansion (CTE) · Composite · Compressive strength · E-glass · Elastic modulus (E) · Elongation · Fatigue · Fiberglass · Fiber reinforced composite (FRC) · Flammability · Glass · High-strength glass (HSG) · HiPer-tex^{TM} · HS4 glass · Kevlar · K-glass · Nomex · Polybenzobisoxazole (PBO) · Pressure vessels · R-glass · Roving · S-glass · S-1 Glass^{TM} · S-2 Glass$^®$ · Softening point · Strength · Surface flaw · Tensile strength · T-glass · Thermal history · Thermal stability · U-glass · Ultra high molecular weight polyethylene (UHMWPE) · Yarn · Young's modulus (E) · Zylon

5.1 Attributes of High-Strength Glass

High-strength glass fibers are a small but important part of the overall glass fiber market, comprising only about 1% of the total market volume, but finding a necessary niche in applications ranging from aerospace and military defense to automobiles and structural reinforcements. These types of fibers were originally conceived to meet growing demand for reinforcement fibers in the aerospace and

A.V. Longobardo (✉)
AGY World Headquarters, Aiken, SC 29801, USA
e-mail: anthony.longobardo@agy.com

F.T. Wallenberger, P.A. Bingham (eds.), *Fiberglass and Glass Technology*,
DOI 10.1007/978-1-4419-0736-3_5, © Springer Science+Business Media, LLC 2010

defense markets, exhibiting properties that surpassed compositions in the E-glass family. The original patent for high-strength glass fibers, US Patent 3,402,055 [1], was put forth by Owens Corning in the late 1960s and expanded upon work they had done earlier in the decade with the United States Air Force developing fibers with increased physical properties, specifically higher tensile strength and higher temperature resistance.

The work done at Owens Corning was responsible for the emergence of a new classification of glass fibers known as "S" glass or high-strength glass (HSG). This family consisted of a ternary magnesium aluminosilicate glass composition ranging from 55.0 to 79.9% SiO_2, 12.6 to 32.0% Al_2O_3, and 4.0 to 20.0% MgO. The composition identified as Example 1 in Patent 3,402,055 indicates an optimum composition as 65.0% SiO_2, 25.0% Al_2O_3, and 10% MgO. This was a lower silica version of the composition originally created in the early 1960s and represented a compromise between properties and forming ability. Known as S-2 Glass®, this composition is still produced today by AGY and offered for a variety of high-performance applications. It should be noted that the term "S" glass is commonly used in reference to both high-strength glasses as a family and in particular when referring to the S-2 Glass composition of 65.0% SiO_2, 25.0% Al_2O_3, and 10% MgO [2–6].

High-strength glass fibers in today's markets are valued for their higher performance properties within their end use applications. These applications include composites requiring high tensile strength, high stiffness as shown through increased modulus of elasticity, and elevated thermal stability, i.e., the ability to retain properties at high temperatures. In general, these glasses fill the gap between lower priced and lower performance E-glass and higher priced and higher performance carbon and polymer fibers [7, 8]. Products are offered in a variety of filament diameters and forms including yarns, rovings, and chopped strands and include sizing chemistries that are specific to the application. For the scope of this chapter, however, HSG fibers will be defined as glass fibers exhibiting increased performance in one or more of the following areas as compared to other fiberglass compositions: tensile strength, modulus of elasticity, and high-temperature thermal and mechanical stability. Table 5.1 shows a variety of glass fiber compositions, of which R-glass and S-glass are considered HSG fibers.

5.1.1 Strength

Glass fiber strength is a property of particular interest for a wide range of applications in the composite market. The ultimate practical strength of the glass fiber is dependent upon several factors including chemical composition, forming conditions, applied sizings, process handling, and composite design parameters. The practical strength, however, will be much lower than theoretical calculations indicate since defects within the fiber and on the fiber surface act as stress concentrators that cause failure in the presence of an applied tensile stress. In the production environment, the optimization of glass melting and forming techniques helps to minimize

Table 5.1 Typical glass fiber compositional ranges (wt%) [9, 10] (used with permission of AGY)

	A-glass	C-glass	D-glass	E-glass	AR-glass	R-glass	S-2 Glass®
Oxide (%)							
SiO_2	63–72	64–68	72–75	52–62	55–75	55–60	64–66
Al_2O_3	0–6	3–5	0–1	12–16	0–5	23–28	24–25
B_2O_3	0–6	4–6	21–24	0–10	0–8	0–0.35	–
CaO	6–10	11–15	0–1	16–25	1–10	8–15	0–0.2
MgO	0–4	2–4	–	0–5	–	4–7	9.5–10
BaO	–	0–1	–	–	–	–	–
Li_2O	–	–	–	–	0–1.5	–	–
$Na_2O + K_2O$	14–16	7–10	0–4	0–2	11–21	0–1	0–0.2
TiO_2	0–0.6	–	–	0–1.5	0–12	–	–
ZrO_2	–	–	–	–	1–18	–	–
Fe_2O_3	0–0.5	0–0.8	0–0.3	0–0.8	0–5	0–0.5	0–0.1
F_2	0–0.4	–	–	0–1	0–5	0–0.3	–

the presence of flaws in the formed glass, but they cannot be eliminated. The following is a review of the flaw dynamics that determines glass strength for a particular composition.

Calculation of the theoretical strength of a glass surface can be narrowed down to the stress required to separate and pull apart two atoms. Orowan expressed the value of this stress as:

$$\sigma_m = \left[\frac{\gamma_f E}{a_0} \right]^{1/2},$$

where σ_m is the maximum stress at breakage, γ_f is the interfacial surface energy for a pristine glass surface, E is the Young's modulus, and a_0 is the inter-atomic separation distance [11]. Using values typical for silicate glasses of $E = 70$ GPa, $\gamma_f = 3.5$ J/m^2, and $a_0 = 0.2$ nm, the calculated strength is equal to 35 GPa. By comparison, typical values for steel are in the 400–450 MPa range, much lower in value than the calculated theoretical strength of glass. For any particular glass composition, the actual stress at failure will be lower than the calculated theoretical stress due to inherent flaws in the glass.

Applied loading of the fiber will result in stress concentration at specific points in the fiber where flaws exist either in the bulk or on the glass surface. Inglis first made this assumption, and later Griffith theorized that not all flaw geometries are equally likely to cause failure. For any given stress, a flaw must exceed a critical length before it will propagate. In addition, the failure would occur from the site of the most severe flaw. Figure 5.1 shows a two-dimensional schematic of a typical surface flaw on a glass article, where a is the flaw radius, L_0 is the crack length prior to loading, and S is the applied tensile stress. Using a surface energy approach, Griffith put forth an equation for calculation of the stress required for failure based upon some critical flaw length:

Fig. 5.1 Surface flaw cross section with applied tensile stress S

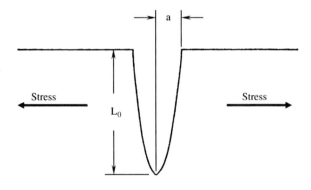

$$\sigma_f = [\sigma_a]_{\text{failure}} = \left[\frac{2\gamma_f E}{\pi L_c}\right]^{1/2},$$

where σ_f is the maximum stress at failure, γ_f is the interfacial surface energy, E is the Young's modulus, and L_c is the critical flaw length. A more detailed analysis of these calculations can be found in Varshneya [12] and Doremus [13].

The failure mode of glass fibers is important to understand the effect of fiber composition on strength. Table 5.2 shows properties for various glass fiber compositions including single-fiber tensile strength measured at various temperatures. Griffiths' equation for calculation of strength indicates that failure will occur in the presence of a flaw of a specific severity. It also indicates that the ultimate breaking strength will be dependent upon the interfacial surface energy (γ_f) and the Young's modulus (E), properties that are composition specific. It therefore relates the chemical composition of the glass directly to the achievable strength and shows that for equivalently prepared fibers with similar flaw distributions, the strength will depend directly on the glass composition. The room-temperature single-fiber tensile strengths in Table 5.2 range in value from 2415 MPa for D-glass to 4890 MPa for S-2 Glass with the various other glasses lying in between. The compositional differences in S-2 Glass and R-glass make these glass families superior in strength performance to the other glass families.

Glass fiber strength can be thought of in two ways: (i) the inherent strength associated with a specific glass composition as determined by single-fiber tensile testing and (ii) the finished product strength, which is a function of the glass chemistry, and also filament diameter, forming process, applied sizing, number of filaments in the strand, and product form. In the second instance, compiling comparable data can be difficult as one glass composition can be used to produce a vast number of finished products. As a result, the most effective way of comparing inherent glass strength is through single-fiber tensile testing. Figure 5.2 is a graphical representation of the tensile strength values shown earlier in Table 5.2. During a single-fiber tensile test, a filament is gripped at both ends and elongated at a constant rate until failure occurs. Depending on the glass composition, the fiber structure will strain differently as it is stretched, resulting in a change in the slope of the line as the fiber elongates until

Table 5.2 Properties of glass fibers [9] (used with permission of AGY)

	A-glass	C-glass	D-glass	E-glass[a]	AR-glass	R-glass	S-2 Glass®
Density (g/cm³)	2.44	2.52	2.11–2.14	2.58	2.7	2.54	2.46
Refractive index	1.538	1.533	1.465	1.558	1.562	1.546	1.521
Softening point (°C)	705	750	771	846	773	952	1056
(°F)	1300	1382	1420	1555	1424	1745	1932
Annealing point (°C)	–	588	521	657	–	–	816
(°F)	–	1090	970	1215	–	–	1500
Strain point (°C)	–	522	477	615	–	736	766
(°F)	–	1025	890	1140	–	1357	1410
Tensile strength							
−196°C, MPa	–	5380	–	5310	–	–	8275
ksi	–	780	–	770	–	–	1200
23°C, MPa	3310	3310	2415	3445	3241	4135	4890
ksi	480	480	350	500	470	600	709
371°C, MPa	–	–	–	2620	–	2930	4445
ksi	–	–	–	380	–	425	645
538°C, MPa	–	–	–	1725	–	2410	2415
ksi	–	–	–	250	–	310	350
Young's modulus							
23°C, GPa	68.9	68.9	51.7	72.3	73.1	85.5	86.9
Msi	10.0	10.0	7.5	10.5	10.6	12.4	12.6
538°C, GPa	–	–	–	81.3	–	–	88.9
Msi	–	–	–	11.8	–	–	12.9
Elongation percentage	4.8	4.8	4.6	4.8	4.4	4.8	5.7

[a]Listed properties are for borosilicate E-glass compositions.

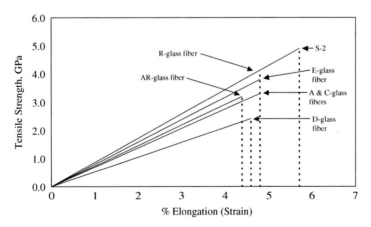

Fig. 5.2 Single-fiber tensile strength at 23°C for various glass fiber compositions

fracture. Although the rate at which the fiber is stressed remains constant, the value of the load increases as the test proceeds from time zero to the time of failure. The reported glass fiber strength is the maximum stress required to initiate failure.

Often, the strength of the fiber alone is not sufficient for understanding the critical fitness for use for potential applications. Weight is often a critical factor and thus the strength of a glass composition or product can be expressed as a specific tensile strength, which takes into account the density of the glass. The equation for specific strength (σ_{sp}) is shown below as:

$$\sigma_{sp} = \frac{\sigma_f}{\rho_a},$$

where σ_f is the tensile strength, ρ is the glass density, and a is the acceleration due to gravity. For S-2 Glass, the calculated specific strength is 203 km compared to E-glass with a value of 136 km. The resulting units work out to be in kilometers, but the intent of the calculation is to compare various fiber strengths to determine which fibers exhibit better strength performance with respect to their inherent mass.

Thermal history, drawing conditions, fiber diameter, mechanical history, and testing conditions are all key elements when comparing and understanding glass fiber strengths. The inherent structure of a drawn glass fiber will be different than the corresponding annealed bulk glass structure of identical composition. The quenching rate and the directional strain applied to a glass fiber during the drawing process will cause a stretching of the structural units within the fiber and alter its physical properties. In addition, the surface of the glass is quenched at a much higher rate than a comparable bulk specimen, resulting in some measure of compressive stress on the glass surface. As a result, the fiber will have higher tensile strength, lower density, lower Young's modulus, lower index of refraction, lower thermal conductivity, lower specific heat, and higher chemical reactivity [14, 15].

Just as fibers can differ in properties from equivalent bulk glasses, they can also differ from each other depending on melt history, drawing viscosity, drawing conditions, and subsequent handling. Continuous glass fibers are generally drawn at a viscosity between log 2.5 and log 3.5 P (Poise), and can be melted and conditioned at viscosities approaching log 1.7 [16]. The time the glass spends at each viscosity range in conjunction with the raw material form, whether it is pure raw materials or preformed glass marbles, will produce a glass fiber with varying degrees of homogenization and thermal history. Cameron demonstrated the relationship between melt treatment time and glass fiber strength using E-glass under various conditions and showed that the measured strength can vary between 5 and 15% depending on the glass thermal history [17].

The temperature at which a structure of an equilibrium liquid is frozen into the glass during forming is known as the fictive temperature (T_f). Changes in thermal history and forming conditions can therefore be characterized qualitatively in terms of fictive temperature. Strength measurements of silica fibers formed under different conditions with different fictive temperatures have shown that an increase in the fictive temperature resulted in an increase in glass strength [18]. The reported single-fiber tensile strengths, therefore, are dependent not only on glass compositions but also on the glass thermal history. This relationship is important when considering the difference between a single-fiber bushing and a production bushing with many hundreds of fiberizing tips. The thermal history and the forming conditions of the glass will likely differ to some degree and can produce products with either enhanced or reduced physical properties such as strength. Reported strength data for various glass compositions can therefore be difficult to interpret since the exact nature of the melting, forming, and testing parameters is rarely included.

5.1.2 Elastic Modulus

The elastic modulus or the Young's modulus (E) of glass compositions differs from fiber strength, as modulus is a bulk glass physical property and not affected by defects such as surface flaws and handling. As mentioned previously, the modulus will be affected by forming conditions and thus the as-formed fiber modulus will generally be lower than the modulus of an annealed bulk glass of equivalent composition. Elongation of the structure, lower glass density, and thermal history of the glass fiber all contribute to the decrease in the measured fiber modulus, which can be as much as 10% less than the corresponding annealed bulk glass value.

As shown in Table 5.2, room temperature measurements of the Young's modulus of E-glass fall around 72.3 MPa, while R-glass and S-2 Glass have values of 85.5 MPa and 86.9 MPa, respectively. Although both R-glass and S-2 Glass demonstrate approximately a 20% increase in the modulus compared to E-glass, applications exist for which the modulus is equally as important as the strength, and as a result, specialized compositions have been developed to improve the modulus of high-strength glasses. Research has indicated that there are essentially five pathways to achieving higher glass modulus and these are shown in Table 5.3.

Table 5.3 Methodologies for improving glass fiber modulus

Method	Description
1	The addition of BeO to silicate glasses; although toxic, BeO addition has the greatest impact on modulus of any modifier
2	An increase in the alumina content in silicate and borate glasses, with substitution for silica showing the most significant impact
3	The addition of rare earth elements such as Y_2O_3 and La_2O_3 into silicate glasses; in some cases with silica content lowered to 50 wt% or below
4	The addition of TiO_2, ZrO_2, or CuO to silicate glasses; often these elements are used in conjunction with other known modulus modifiers
5	The introduction of nitrogen into the silicate network forming oxynitride glass fibers; requires well-controlled forming conditions
[19, 20]	

Beryllia (BeO) additions to silicate glasses result in the greatest increase in modulus for a single modifier. Within the basic silica–alumina–magnesia ternary compositional range, Bacon [21] was able to increase the modulus of glass fibers to 112 GPa using BeO additions along with adjustment of the silica–alumina–magnesia ratio. One of the glasses had a composition of 50% SiO_2, 35.0% Al_2O_3, 7.5% MgO, and 7.5% BeO and a modulus that was 29% higher than S-2 Glass. The high field strength of BeO, its ability to tightly coordinate four oxygen atoms, is responsible for its ability to affect an increase in glass modulus [19]. Unfortunately, BeO is toxic and therefore an impractical additive to use commercially.

Increasing the Al_2O_3 content of the glass in place of SiO_2 can also have a significant impact on fiber modulus. In the previous example, Bacon not only added BeO to the ternary silica–alumina–magnesia composition but also increased the alumina content and decreased both the silica and magnesia contents with respect to S-2 Glass. For three-component compositions within the ternary silica–alumina–magnesia phase field, an increase in modulus can be achieved through shifting of the alumina/silica ratio toward higher alumina content. The change in coordination of the network as more alumina is introduced in place of silica can form a strong silicon–aluminum–oxygen network, with modifying oxides providing compaction to the structure and resulting in increased stiffness and modulus [20].

The addition of rare earth oxides such as Y_2O_3 and La_2O_3 are also known to increase glass modulus [22, 23]. Ternary silica–alumina–magnesia compositions incorporating La_2O_3 or Y_2O_3 can reach measured fiber moduli between 100 and 130 GPa, respectively. The major drawback of utilization of these oxides lies in their negative effect on glass density due to the high molecular weights of both Y_2O_3 and La_2O_3. When the specific modulus of a glass modified with these oxides is calculated, the result is typically a value equal to or less than the one for original unmodified glass. The specific modulus (E_{sp}) of a material is calculated using the equation below:

$$E_{sp} = \frac{E}{\rho_a},$$

where E is, in this case, the glass modulus, ρ is the glass density, and a is the acceleration due to gravity.

The inclusion of TiO_2 [18], ZrO_2 [18], ZnO [24], or CuO [25] has also been shown to improve glass modulus. Typically these modifying oxides are not included singularly, as BeO can be used, but in combination with other oxides known to increase modulus. Table 5.4 shows the chemical composition of M-glass, a commercially available high-modulus fiberglass. This glass includes TiO_2, ZrO_2, and BeO and has a modulus of 110 GPa [18].

Table 5.4 Composition of commercially available M-glass [15]

SiO_2	CaO	MgO	Li_2O	TiO_2	CeO_2	ZrO_2	BeO	Fe_2O_3
53.7	12.9	9.0	3.0	7.9–8.0	3.0	2.0	8.0	0.5

Glasses incorporating nitrogen into the structure, or oxynitride glasses, are also known to demonstrate high modulus values and can reach 243 GPa [26–28]. Nitrogen having three valence electrons forms an additional bond compared to oxygen within the structure, giving rise to a much tighter glass network and improved elastic properties. These glasses are melted under inert N_2 or reactive reducing atmospheres, making conventional melting of these materials difficult. Oxynitride glasses also require specialized boron nitride-coated carbon or molybdenum bushings to make fibers. A more detailed explanation can be found in Wallenberger [19] and in Part I in Chapter 1 in this volume.

5.1.3 Thermal Stability

Some applications of HSG fibers involve elevated temperatures and therefore the ability of the fiber to retain its physical properties at these temperatures becomes important. The softening point, the point above which a fiber will deform under its own weight, and the ability to retain tensile strength at temperature are two key properties of interest for most high-temperature applications. Table 5.2 includes data for various glass fiber types and includes softening point, tensile strength measured at −196°C, 23°C, 371°C, and 538°C as well as the Young's modulus measured at 23°C and 538°C. Both R-glass and S-2 Glass retain fiber shape at least 100°C above the other listed glass compositions as shown through the softening point data. For structural reinforcement applications, this translates to both R-glass and S-2 Glass being able to retain product form at approximately 100°C and 200°C greater than E-glass, respectively. The ability of the glass fiber to retain structural integrity and physical properties as the temperature increases can be thought of as thermal stability.

Table 5.2 also shows how glass fiber strength varies as a function of temperature. All of the glass fibers listed in Table 5.2 experience a drop in tensile strength as the temperature to which they have been exposed increases. At 371°C, the E-glass and R-glass values fall to 76 and 71% of their room temperature values, respectively, while S-2 Glass retains approximately 91% of its room temperature tensile strength. At 538°C, all three compositions exhibit tensile strengths that are roughly 50% of the room temperature measured value. Compared to E-glass strength at 538°C, R-glass shows a strength improvement of 24%, while S-2 Glass remains 40% higher in strength.

Figure 5.3 shows a comparison of E-glass and S-2 Glass tensile strengths versus both carbon and aramid fibers. It is important to note that the non-glass fibers completely degrade at temperatures approaching 600°C. At temperatures that exceed the glass annealing temperature, S-2 Glass will exhibit a slow phase separation process whereby the glass structure changes from a homogeneous state into that of two distinct phases: one phase composed of a continuous silica-rich network with a higher viscosity and a second phase composed of a silica-deficient network with a lower viscosity [29]. The presence of the higher viscosity phase allows S-2 Glass to remain intact at higher temperatures than other glass fibers.

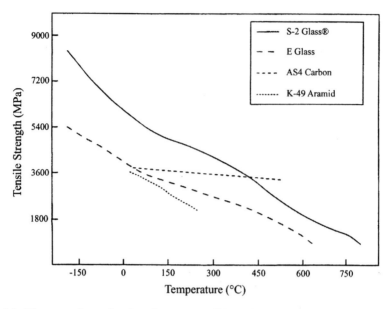

Fig. 5.3 Fiber strength as a function of temperature [9] (used with permission of AGY)

Elastic modulus is another property that will also vary with temperature. Table 5.2 lists modulus values for E-glass and S-2 Glass at both 23°C and 538°C. At elevated temperatures both fiber types exhibit an increase in modulus over their room temperature measured value. For S-2 Glass, the increase is only 2%, but for E-glass, the 538°C modulus is 13% better than the 23°C measured modulus. As mentioned previously, there is a difference in modulus between an annealed bulk glass and a quenched fiber of the same composition due to the decrease in density and the stretching of the glass structure that results during fiber formation. As the temperature of a glass fiber is increased to levels at or near its annealing temperature, the structure will begin to relax and shift toward a more compact structure, thus increasing the elastic modulus. The annealing point of E-glass is much closer to the test temperature of 538°C than the annealing temperature of S-2 Glass, therefore resulting in more thermal relaxation and a greater increase in modulus. All glass fiber compositions will show some degree of modulus increase at elevated temperatures. The glass softening point, however, will control the maximum use temperature of the fiber.

5.2 Glass Compositional Families

High-strength glass (HSG) fibers have traditionally been categorized into two different glass families: R-glass and S-glass. As research and development has continued and compositions have been developed to exploit markets requiring specific strength, modulus, and high-temperature properties, additional glasses, such as basalt, have become available that lie outside the ranges of either S-glass or

R-glass. The following is a review of the various commercially available high-strength fiberglasses in the marketplace.

5.2.1 S-Glass

The patent for S-glass, US Patent 3,402,055 assigned to Owens Corning [1], denotes the composition by weight as: 55.0–79.9% SiO_2, 12.6–32.0% Al_2O_3, and 4.0–20.0% MgO. S-2 Glass has a composition of 65% SiO_2, 25% Al_2O_3, and 10% MgO and is listed as Example 1 in the patent. As stated previously, S-glass is commonly referred to as having a composition equivalent to S-2 Glass. S-glass as it relates to a family of high-strength glass, however, really refers to glass compositions that fall within the limits of ASTM Specification D 578 [10]. As such, there are a variety of available glass fibers composed of silica, alumina, and magnesia that fall within this range. Table 5.5 shows the composition of various high-strength glasses including those in the S-glass family, R-glass family, and other high-strength glasses that fall outside of these classifications. R-glass and the additional high-strength glasses will be explored further in the following sections.

Table 5.5 High-strength glass fiber compositional ranges (wt%)

	Mineral	R-glass	Intermediate HSG			S-glass		
	Basalt glass[a]	R-glass (Vetrotex)	HiPer-tex™ (3B)	S-1 Glass™ (AGY)	K-glass (NGF)	HS4 glass (Sinoma)	T-glass (Nittobo)	S-2 Glass® (AGY)
Oxide (%)								
SiO_2	45.9–58.7	58–60	62.4	60–65	64	55–60	64–66	65
Al_2O_3	8.2–18.6	23.5–25.5	18.5	18–26	20.5	24–27	24–26	25
B_2O_3	–	–	–	–	2	–	–	–
CaO	5–16	8–10	8.6	7–10	–	–	–	–
MgO	1.3–14.2	5–7	9.6	9–11	13.5	11–16	9–11	10
MnO	0.1–.06	–	–	–	–	–	–	–
Li_2O	–	–	–	–	–	–	–	–
Na_2O + K_2O	3–7.5	–	0.7	< 1.0	–	–	–	–
TiO_2	0.9–3.2	–	0.01	–	–	–	–	–
ZrO_2	–	–	–	–	–	–	–	–
Fe_2O_3	2.8–15	–	0.05	–	–	–	–	–
P_2O_3	0.1–0.8	–	–	–	–	–	–	–
F_2	–	–	–	–	–	–	–	–
		[30]	[31]		[32]	[33]	[34]	[9]

[a]Basalt is a naturally formed mineral varying in composition depending on the deposit.

In addition to S-2 Glass, several other glasses fall within the definition of S-glass. T-glass from Nittobo, HS4 glass from Sinoma, and U-glass [32] from NGF are all glasses composed of silica, alumina, and magnesia that satisfy the scope of S-glasses. U-glass is not listed in Table 5.5 but has the same approximate composition as both S-2 Glass and T-glass. As such, there are similarities in the properties of these glasses as evidenced by their densities and softening points. HS4 glass is actually a lower silica version of S-2 Glass and therefore exhibits lower tensile strength and has a lower softening point as shown in Table 5.6. S-glasses at the higher end of the silica content are extremely high melting temperature glasses and require special melting technology. The lower silica content of HS4 glass indicates that it will have a slightly lower melting temperature compared to S-2 Glass, T-glass, and U-glass. The price of S-glass is generally substantially higher than that of E-glass as a result of the higher melting temperature and associated technology that is required to produce continuous fibers.

Table 5.6 Properties of high-strength glass fibers

	Basalt glass	R-glass (Vetrotex)	S-1 Glass[TM] (AGY)	HS4 glass (Sinoma)	T-glass (Nittobo)	S-2 Glass[®] (AGY)
Density (g/cm^3)	2.6	2.54	2.55	2.54	2.49	2.46
Softening point (°C)	up to 900	952	996	915	1050	1056
(°F)	1652	1745	1825	1679	1922	1932
Tensile strength						
23°C, MPa	3000–4840	4135	4454	4600	4650	4890
ksi	435–702	600	646	667	674	709
Young's modulus						
23°C, GPa	79.3–93.1	86.4	86.4	86.4	86.4	86.9
Msi	11.9–15.9	12.5	12.5	12.5	12.5	12.6
Elongation percentage	3.1	4.8	–	5.3	5.5	5.7
	[35]	[9]		[33, 36]	[34]	[9]

The modulus of glasses in the S family is approximately 20% greater than that of the more widely used family of E-glasses. These glasses are not considered to be ultra high-modulus glasses as they do not reach the 100 GPa plus range, but they represent an excellent compromise between high strength and high modulus. On the other hand, ultra high-modulus glasses have been developed using additions of other oxides to base S-glass compositions as detailed in Section 5.1.2. The combination of high strength, high elastic modulus, high softening point, and high elongation percentage (strain to failure) makes S-glass ideal for a variety of applications including ballistics, helicopter blades, pressure vessels, and aerospace composites [7].

5.2.2 R-Glass

R-glass is a fiber composition produced by Vetrotex derived from the quaternary SiO_2–Al_2O_3–CaO–MgO system [30]. The composition of R-glass is listed in Table

5.5. Compared to S-glass it has lower SiO_2 content, equivalent Al_2O_3, and the remainder is a combination of CaO and MgO. This combination allows the glass to retain a relatively high softening temperature, but the decrease in SiO_2 content and associated decrease in network connectivity also reduce the glass tensile strength. Table 5.6 shows that the room temperature strength of R-glass is only 85% of that for S-glass. In addition, although the modulus of R-glass is equivalent to that of S-glass, the elongation percentage or strain to failure is also approximately 85% of the S-glass value. At increased temperatures, however, the tensile strength of both R-glass and S-glass decreases to the point where they are equal as shown in Table 5.2.

5.2.3 Other High-Strength Glasses

Other glasses exist which exhibit high strength, high modulus, and high temperature stability that fall within the definition of high-strength glasses based on their properties but fall outside the chemical compositional definitions of either S-glass or R-glass. Both S-1 Glass[TM] produced by AGY and HiPer-tex[TM] produced by 3B the Fiberglass Company fall within a compositional range between R-glass and S-glass. Tables 5.5 and 5.6 show the quaternary SiO_2–Al_2O_3–MgO–CaO compositions and their respective properties. These glasses have higher SiO_2 content than R-glass and therefore demonstrate higher tensile strength but are nearly equivalent in both modulus and softening point. The inclusion of less than 1% alkali in these glasses distinguishes them between both R-glass and S-glass, and is likely added to aid in melting behavior.

K-glass is a composition produced by NGF and represents a modified S-glass composition with the inclusion of 2% B_2O_3 as shown in Table 5.5. The addition of boron in the glass structure not only lowers the melting temperature but also modifies the silicate glass network sufficiently to alter the glass' properties. The tensile strength of K-glass is lower than that of S-glass and closer to the high-strength end of E-glass, but the modulus is only slightly lower than that of S-glass [32]. Other glasses not shown in Table 5.5 also exist with modified S-glass compositions including VMP, VMP-1, and a high-strength glass made by Stekloplastik Co. in Russia [37]. The SiO_2, Al_2O_3, and MgO levels in these glasses all fall within the S-glass range, but these glasses are also modified by the inclusion of non-trace levels of TiO_2 and/or ZrO_2.

Based on their published claims, basalt glasses can be considered a member of the high-strength fiberglass family. Basalt is a naturally occurring mineral that is mined from various deposits around the world and in some cases melted directly into glass fibers. As a mined mineral, the chemical composition can vary greatly depending on the deposit, and as a result, the glass properties of the fiber formed from basalt are also highly variable. Table 5.5 shows the wide range of chemical compositions that fall within the basalt mineral definition. Basalt fibers formed from deposits in the European North of Russia can have high thermal stability up to 900°C

[38]. Basalt Fiber & Composite Materials Technology Development (BFCMTD) is a company located in China and produces basalt fibers from Chinese mined basalt deposits. Table 5.6 lists some properties associated with these fibers and highlights their variability.

5.3 High-Strength Glass Fibers in Perspective

High-strength glass fibers are but one type of high-performance material that a system designer or an engineer has available to them. Selecting one over another is a critical task, which embodies taking all aspects of the constituent fiber's contribution to performance, process suitability, and financial impact into account. Ultimately, the final selection is to some extent a matter of compromise. High-strength glass fibers are typically intermediate materials with regard to their properties. That is, at times they may not be the best in class for a particular property, but they are often selected because of a combination of desirable attributes that make them the best overall choice.

A high-strength material that can be offered as a continuous fiber has a significant advantage over other types of high-strength materials. Glasses and polymers are good examples of materials that are easily made into continuous fibers. Other solids, like metals and ceramics, are limited in their processability by their physical chemistry and are difficult to fiberize into fine fibers suitable for yarns.

Monolithic solids such as steels or ceramics are also less flexible with regard to the range of possible end-product forms. Continuous fibers can be woven into fabrics, which are pliable and can be shaped. Taking this one step further, they are ideal for use as reinforcements in many types of polymer resin composites. Fiber-reinforced composites (FRCs) are highly regarded engineering materials because they offer, among other properties, excellent strength and stiffness per unit weight. One disadvantage to composites is that they are typically expensive compared to non-composite materials. However, designers frequently resort to composites because of a lack of alternative materials to meet their design requirements.

Table 5.7 summarizes the key property and cost profiles of a number of competitive reinforcement fibers used in high-performance FRCs. The fibers selected for the table represent commercially available fibers that have found use in high-strength and/or high-modulus applications across a wide range of markets.

5.3.1 The Competitive Material Landscape

There are three main categories of fibers being used in high-performance FRCs today. They are (i) carbon fibers, (ii) organic or, so-called, polymer fibers, and (iii) fiberglass. Each of these material types has strengths and weaknesses that make them excellent candidates for certain applications and a poor choice for others. A brief summary of each category follows.

Table 5.7 Comparison of properties for a variety of high-performance fibers

Category	Property	Units	Glass		Aramid	Carbon	UHMWPE	PBO
			E-glass	S-2 Glass®	K49	T700SC	Spectra® 1000	Zylon® HM
Physical	Density	g/cm³	2.58	2.46	1.45	1.8	0.97	1.56
Mechanical	Tensile strength	MPa	3445	4890	3000	4900	2830	5800
Mechanical	Tensile modulus	GPa	72.3	86.9	112.4	230	103	280
Mechanical	Spec. tensile strength	km	136	203	211	278	298	379
Mechanical	Spec. tensile modulus	x10³ km	2.9	3.6	7.9	13.0	10.8	18.3
Mechanical	Compress. strength	MPa	1080	1600	200	1570	170	561
Mechanical	Strain to failure	%	4.8	5.7	2.40	1.5	2.8	2.5
Thermal	CTE	10⁻⁷/°C	54	29	-48.6	-38	1300	-60
Thermal	Softening point temperature	°C	846	1056	Oxidizes > 150	Oxidizes > 350	Melts at 147	Decomposes at 650
Price	Approximate price	$/kg	2	20	30–50	40–60	80	120
Reference					[41]	[49]	[44]	[48]

Data presented are typical values and thus will vary depending upon fiber denier
Dick Holland of Composite One, private communication

5.3.1.1 Carbon Fibers

Carbon fiber is one of the most important high-strength, high-stiffness synthetic fibers in use today. It has found application in a variety of military, aerospace, automotive, and consumer applications. Carbon fibers are produced from fibers made from one of three primary precursor materials: rayon, carbon pitch, or PAN (polyacrylonitrile). The fibers are then heat treated using a process call pyrolysis, which turns the carbon-rich precursor material into pure carbon. An excellent general overview of carbon fibers has been given by Fitzer et al. [39].

PAN-based carbon fibers are recognized as superior to pitch- and rayon-based alternatives when strength is required. Pitch and rayon fibers are more expensive but give better modulus performance. PAN fibers are partially amorphous in structure. Pitch-based fibers have a much higher level of crystallinity, but are weaker due to a tendency for structural defects. More than 90% of all carbon fibers are PAN type [40].

Compared to fiberglass, carbon fibers have a much lower density (1.8 vs. > 2.45 g/cm^3). The best carbon fibers have strength and elastic moduli that exceed that of even S-glass. And as a result, the specific tensile strength and moduli are superior to S-2 Glass by a wide margin (+ 37% for specific tensile strength and 3.6 times for specific modulus). See reference [19] and Part I in Chapter 1. In compression, the strength of carbon fiber is comparable to that of S-2 Glass. Carbon's high stiffness also translates into a more brittle fiber that has lower strain to failure (1.5% elongation compared to 5.7% for S-2 Glass). Consequently carbon fiber has poor impact resistance. In addition, carbon fibers being what they are will burn on exposure to flame, making them unacceptable for applications that are at risk of fire. Finally, they cost about twice as much per unit weight as S-2 Glass "Dick Holland of Composite One, private communication."

5.3.1.2 Polymer Fibers

Organic fibers are a very broad category of materials. They span a vast array of chemistries all having their own key characteristics, but only a few have the required combination that would qualify them as high-performance fibers. This subsection focuses on three of the most widely used organic fibers.

Aramids are a class of polymers that take their name as an abbreviation for **ar**omatic poly**amide**. There are two main families of aramids from which commercial fibers are derived. The first is known as *meta*-aramid (*m*-aramid), which is the basis for the Nomex® family of fibers manufactured by Dupont. These fibers are not used for their strength but rather are well known for their thermal stability and find broad use in applications where temperature insulation is required [40].

The second is *para*-aramid, of which DuPont's Kevlar® is the most widely recognized commercial fiber. Other competing materials are sold under the names Twaron® and Technora® [41]. The chemical name for this class of aramid is polypara-phenylene terephthalamide, which is better known as a *para*-aramid (or *p*-aramid). *para*-Aramid forms extremely strong linear chain-like molecules, which are hydrogen bonded together. Both *m*- and *p*-aramids are produced using a solvent spinning technique. para-*Aramid* molecules develop orientation during the drawing process that gives them enhanced tensile strength [40, 41].

p-Aramid has a lower density compared to carbon fibers (1.45 vs. 1.8 g/cm^3). Kevlar-type aramid fibers have tensile strengths that are less than that of either E- or S-glass. Elastic modulus, on the other hand, is significantly better than that of S-glass (112.4 vs. 86.9 GPa). Owing to its low density, the specific tensile strength and moduli properties of Kevlar are excellent compared to that of S-2 Glass but inferior to that of carbon. Kevlar also has a negative thermal expansion, which tends to give good dimensional stability with temperature. Impact resistance is comparable with fiberglass.

One of aramid's major shortcomings is compressive strength, which is one-eighth the value of S-glass. Its high stiffness also translates into a lower strain to failure (2.4% elongation compared to 5.7% for S-glass). Kevlar, like many polymeric materials, is also known to be susceptible to strength degradation after UV exposure [42]. It is also hygroscopic (i.e., it absorbs moisture up to 4.5 wt%) [43]. While water by itself does not tend to degrade strength, in combination with temperature and UV, Kevlar has been shown to weaken dramatically. Like carbon fibers, aramid costs about twice as much per unit weight as S-2 Glass (Dick Holland of Composite One, private communication).

Another important type of polymer that has found significant use in fiber-reinforced composites goes by the term ultra high molecular weight polyethylene (UHMWPE). As the acronym implies, many thousands of ethylene monomer units[$(CH_2-CH_2)_n$ where *n* can equal up to 250,000 monomer units in a single molecular chain] are specially processed, resulting in molecular weights on the order of several million[44]. Two commercial brands of UHMWPE fibers being used in composites today are trade named Spectra® [45] and Dyneema® [46].

UHMWPE is not as strong in tension as either E-glass or S-2 Glass fibers (σ_f.~2830 MPa compared to 3445 MPa for E-glass). Its tensile modulus, however, is impressive at 103 GPa compared to 86.9 GPa for S-2 Glass. What is most remarkable is this material's density. Spectra 1000 has a density 2.5 times less than S-2 Glass. This results in extraordinarily high specific tensile properties. Refer to Table 5.7. While the practical tensile strength of UHMWPE is only 58% of that of S-glass, on a density-normalized basis, it has 47% better (specific) tensile strength. The advantage low density brings with it is even more apparent when specific tensile modulus is considered (298 $\times$ 10^{3}km). This type of material also has generally very good chemical and mechanical resistance due to low surface energy [41, 47].

Unfortunately, UHMWPE fibers suffer from the same shortcomings as aramid, i.e., poor compressive strength (only ~10% of S-2 Glass' value) and a very low upper use temperature (below 150°C) [41, 47]. This temperature also limits the kinds of resin systems that could be used with UHMWPE, since many cure temperatures are in this same range. In addition, the inert nature of the fiber surface makes bonding fibers to a resin matrix a challenging task [40, 41]. Lastly, all of this performance comes at a significant cost. The current price of the best UHMWPE fibers, like Spectra and Dyneema, are five times greater than that of S-glass (Dick Holland of Composite One, private communication).

The final polymeric high-performance fiber that will be considered is polybenzobisoxazole, or PBO in short. Like *p*-aramid and UHMWPE, PBO is also composed of extremely rigid rod-like structural units. These molecular units are capable of forming exceptionally ordered structures that result in extremely rigid networks. This rigidity results in PBO having the highest tensile modulus and superior thermal stability of any organic fiber [48]. The best-known commercially available PBO fiber is Zylon® produced by Toyobo [49].

Comparative data of Zylon are presented in Table 5.7. As expected, the tensile strength is remarkably high (5800 MPa) compared to the next two strongest fibers (carbon 4900 MPa and S-2 Glass 4890 MPa). At 280 GPa the tensile modulus is also extraordinarily high. Zylon is 22% stiffer than carbon and 3.2 times stiffer than S-2 Glass on an absolute basis. When taken in combination with Zylon's low density (1.56 g/cm^3 compared to 2.46 g/cm^3), both specific tensile strength and tensile modulus are beyond all other competing fibers. Unlike other rigid-rod polymeric fibers such as aramid and UHMWPE, its compressive strength is also very good. As already noted, Zylon also has the highest thermal stability of any organic fiber and is second only to glass.

However, Zylon does have an Achilles heel. PBO is very sensitive to sunlight (especially in the UV and visible wavelengths) [40, 41, 48, 49]. Zylon loses more than 60% of its strength within months when exposed to sunlight [49]. Both temperature and moisture have been found to accelerate the degradation mechanisms [41, 49]. As a result, without significant improvement, PBO's long-term commercial viability is questionable at this point in time. In addition, the inert nature of the fiber's surface also limits compatibility with polymer resin matrices common in hard armor applications [41]. This is generally true for high-performance polymeric fibers. Like aramid and UHMWPE, Zylon has found use in soft armor applications. The cost of Zylon is the highest of the commercially available fiber materials. At about $120 /kg it is several times the cost of other high-performance fibers, such as HSG, aramid, or even carbon.

Just as a final note, there is ongoing development of a new fiber material similar to PBO known as PIPD, or commercially as M5. This material is similar in chemistry and has many of the same properties as PBO and has shown promise as an improvement upon Zylon [41].

5.3.1.3 The Importance of Specific Properties

Normalizing important mechanical properties like tensile strength and elastic modulus based on density is a valuable way of comparing candidate materials. Since both weight and mechanical properties are such important design criteria for composites, being able to look at them in a combined fashion allows an engineer to quickly evaluate alternatives. Selecting a material with a better specific property means that (all other considerations being equal) the part will be stronger or stiffer for the same design weight. On the other hand, it is also possible to consider making a part smaller, lighter, or with less reinforcement (and therefore potentially less expensive) with a better performing material with no loss in that property.

5.3.2 Inherent Advantages of Continuous Glass Fibers

It should be clear from reviewing Table 5.7 that the landscape for high-performance fibers is highly diverse and competitive. Each material type has its strengths and weaknesses. Fiberglass brings with it a unique set of advantages that set it apart from the rest of the field. These can be summarized as follows:

- *Thermal stability* – fiberglass is very stable and resists degradation at temperatures below its softening point for years. S-glass has a softening point of 1056°C and the highest use temperature.
- *Flammability or oxidation resistance* – since glass is inorganic, it does not burn making it a strong candidate where non-inflammability is a key requirement. Carbon and the polymer-based fibers, even Zylon, will eventually burn when exposed to flame.
- *Compressive strength* – both E- and S-glass exhibit excellent compressive strength. This is critically important for structural composites, which must bear load as part of their design criteria. This is one of glass' key advantages.
- *Cost* – compared to carbon and polymer fibers, fiberglass is relatively inexpensive. This is due to economy of scale in the case of E-glass where large furnaces can produce a few hundred tons per day. S-2 Glass, which is also made in a continuous but more specialized process, requires higher temperatures and thus costs more to operate. But it is still about half the price of aramid on a per kilogram basis.
- *Strength and modulus per unit price per kilogram* – on a cost per unit weight basis, HSG fibers represent the best value per dollar spent.

5.4 Markets and Applications

The markets that use high-strength fiberglass are very diverse. All of them have settled on fiberglass for at least one if not many reasons. The following is a brief overview of a few of the primary markets and some selected applications where various fiberglass products can be found. Figure 5.4 is a generalized illustration of just a few examples within those markets.

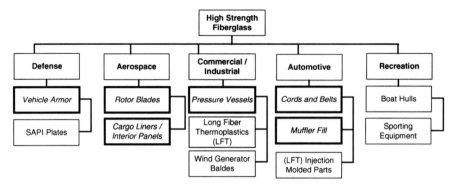

Fig. 5.4 High-strength fiber market overview with selected applications in *bold boxes*

5.4.1 Defense – Hard Composite Armor

Today's soldiers are equipped with the best systems that technology can provide and that money can buy. On the battlefield, mission success is often just as dependent on defensive systems as they are on the strengths of the offensive weapons being employed. Armor systems are a crucially important part of the defensive safeguards that our military uses to preserve assets and keep our soldiers safe.

5.4.1.1 Application Overview

Armor can be broadly classified into two types: (i) soft armor employed in personal protection equipment such as ballistic clothing (vests, helmets, etc.), which typically face threats such as small arms fire and shrapnel and (ii) hard armor that is used to protect against more dangerous threats such as large caliber projectiles and explosives. Hard armor further branches into light armor for vehicular protection and heavy armor found on tanks. Fiberglass-reinforced resin composites, especially those using S-type glass, are particularly important in the light armor arena. Heavy armor is dominated by steel [50].

Armor system developers face the challenge of simultaneously satisfying the three Ps that make up what is known in the military establishment as the "iron triangle" [51]. The three vertices of the triangle are traditionally represented as payload, performance, and protection. Figure 5.5 is a modified rendition of the iron triangle that takes cost into consideration and combines performance and protection as a broader category. The tension between weight (payload) and performance (survivability) and cost is readily evident. For a design to be optimized, all three must be balanced, or driven as close to the center of the triangle as possible. An article by Rush [52] gives a good general overview of this topic.

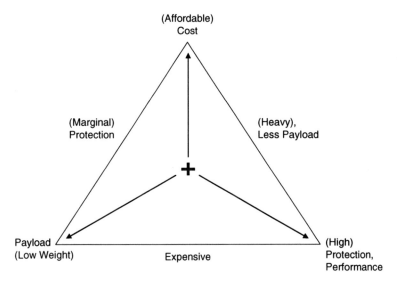

Fig. 5.5 A revised iron triangle for armor systems

5.4.1.2 Critical Fitness for Use Properties

In its simplest form, armor is intended to prevent a puncture from a projectile. Today's systems are very sophisticated. Current designs employ multiple layers of differing materials all combined to stop a projectile threat. Since the direct impact of a projectile can be defeated only by dissipating its kinetic energy, the most effective means is to break up the projectile into numerous, smaller fragments. In light armor designs, this is done with a striker plate typically made out of a dense ceramic. The fragmented projectile and the crushed ceramic material then need to be slowed down. This key part of the design uses a multilayered HSG fiber and phenolic resin composite as a catcher mitt of sorts. The excellent strength and modulus combined with a well-designed composite matrix allow for significant deformation without rupture.

Ballistic panels are evaluated on a number of criteria. One of the critical measures is known as the V_{50} or the velocity at which a projectile will penetrate an armor component 50% of the time [41]. V_{50} is known to be dependent on fiber strength and modulus, fabric design, resin properties like shear modulus and durability, and composite properties such as areal density, panel thickness as well as shape.

S-glass offers a very favorable weight–cost–performance relationship. Compared to aramid, a composite backing plate with S-2 is 33% thinner for the same V_{50} performance. In addition, fiberglass' excellent compressive strength aids in load support, thus allowing armor that can be structural and not just dead weight. Low flammability and smoke-generating behavior is also important in combat situations. Looking at the ease of manufacture, glass fabric is also much easier to cut, mold, and shape than many competing fiber materials.

5.4.1.3 Market Trends and Future Needs

The threats facing our military seemingly continue to escalate. With each improvement in armor protection, our adversaries make their weapons more deadly. The tactics being used are also highly situational favoring, at times, certain types of armor over others. Threats such as IEDs (improvised explosive devices) and EFPs (explosively formed projectiles) are just two examples.

Armor designers are always looking for better mechanical properties in the materials they have available. Greater strength and elastic modulus are highly desirable. In addition, there is always an emphasis on weight reduction. Sometimes the availability of metallic armor comes under pressure, which creates opportunities for composite solutions for designs. Hybrid composites, those that use glass–polymer fabrics, are also of keen interest today because of synergistic benefits that glass can add to organic fiber constructions. Thermal stability and compressive strength are good examples of properties that can be improved by blending with fiberglass [52, 53].

5.4.2 Aerospace – Rotors and Interiors

Again, it is interesting to note that both the US Air Force and NASA supported a significant portion of the early research and development in high-performance FRC technology [40]. Aircraft rely on lift to fly; so very strong lightweight materials have been a consistent R&D focus for both the USAF and aerospace companies.

5.4.2.1 Application Overview

Like the defense market there are numerous specific applications where fiberglass is employed. Two of the more significant applications will be highlighted here. The first is rotor blades for rotary winged (helicopters) and propellers for fixed wing aircraft (airplanes). Aircraft blades are an extremely high-stress application because of the rotational speeds at work and the applied loads. While a propeller mainly drives the aircraft through the air, a helicopter's rotor functions as both driving surfaces and lift surfaces (since it acts as a wing). The span of a helicopter rotor is also much larger than that of a prop on a conventional airplane, hence further magnifying any applied loads.

The second primary application area is interior aircraft surfaces, such as flooring, furnishings, and cargo liners. In uses such as these, strength vs. weight and cost are the primary drivers for meeting the design specification. All three of these performance drivers are foremost in the aerospace industry, especially for commercial airliner manufacturers such as Boeing and Airbus.

5.4.2.2 Critical Fitness for Use Properties

The key glass properties for the rotor/propeller application are strength and stiffness (elastic modulus), fatigue resistance, and resistance to impact damage. Strength and stiffness are obvious property requirements based on the stress-loading expectations involved. Better strength and stiffness allow designers to make lighter and more efficient designs.

Fatigue is a material's response to cyclical loading where cracks can propagate in a non-uniform manner weakening a part over the course of its lifetime. Fatigue damage is cumulative and is the reason for the very thorough and costly inspection procedures that are used for maintaining aircraft. Fiberglass composites have excellent fatigue resistance compared to metallic alternatives.

High-strength fiberglass is frequently used on leading edge applications for most if not all composite rotor designs manufactured in the United States. The leading edge of a prop undergoes tremendous wear from abrasion due to particulates in the air as well as being at constant risk of impact damage. HSG edge treatments are very well suited to withstand the punishment linked with this application.

The interior and functional surfaces inside an aircraft also undergo hard use during their service life. Maintenance is less routine and the materials that make up the interior surfaces, flooring, and cargo liners are all expected to handle heavy wear and tear of normal operation. In these instances, especially for cargo liners, providing tough, lightweight construction is a key requirement. Non-flammability and low smoke generation behavior are also mandated by safety regulators and are non-negotiable. Impact and abrasion resistance is important to any area of the interior where repeated use is a concern. And cost is also, of course, an important point of consideration. Looking at these combined requirements in light of what has been discussed about FRC makes a strong case for HSG.

5.4.2.3 Market Trends and Future Needs

Commercial aircraft will continue to come under pressure to be more fuel efficient while bearing increasing levels of payload. These facts, in turn, will continue to force designers to focus on higher specific strength and modulus materials. The weight and performance linkage will remain at the top of designers' wish lists. Currently getting weight out of a design takes higher precedence over cost containment, if the improvement can be justified. There is little doubt that composite material consumption by the aerospace market can be expected to steadily increase as a result.

An example of where fiberglass-reinforced composites are making inroads into new areas of aircraft design is metal hybrid composite structures, specifically a material called GLARE®. GLARE is a glass fiber-reinforced aluminum composite. The material is constructed of high-strength fiberglass sandwiched between two or more layers of aircraft quality aluminum. When compared to standard aluminum alloys, these structural panels are very strong, extremely stiff, lightweight, fatigue, and impact resistant while providing excellent fire and lightning strike resistance. The product is manufactured by Stork Fokker AESP and is currently being used on the Airbus A380 in a few key fuselage applications [54]. Blast-resistant air cargo/luggage containers using materials like GLARE are also under development by Galaxy Scientific [57].

5.4.3 Automotive – Belts, Hoses, and Mufflers

Fiberglass-reinforced composites have been used in automotive applications for more than 50 years. Probably the most famous is the Chevrolet Corvette body, which has been made of fiberglass since 1953. In the past five decades a multitude of uses have been exploited by carmakers; just a few will be highlighted here.

5.4.3.1 Application Overview

The places within an automobile that contain fiberglass might be surprising to the uninitiated. Many injection-molded plastic parts used on the body, such as mirrors, bumper components, moldings, as well as within the engine compartment (intact manifolds) are made from fiberglass-reinforced, injection-molded thermoplastics. Many rubber components such as hoses, belts, and tires are also reinforced with fiberglass.

Fiberglass-reinforced rubber uses a special combination of multiple glass yarn strands plied together and coated with a resorcinol formaldehyde latex (RFL) compound that both binds and protects the fiberglass and also acts as the interface with the rubber matrix. All cord applications require the reinforcement to have long life under adverse conditions [32].

Another area where fiberglass might go unnoticed is as muffler fill. The muffler in a car or a motorcycle tames what would otherwise be loud exhaust noise from the engine. Most modern mufflers are highly tuned sound-canceling instruments. The muffler chamber is filled with fiberglass or sometimes steel wool to help absorb stray sound waves that are not readily canceled. The fiber filling also produces back-pressure (flow restriction) within the muffler, which is a stress and must be withstood during operation.

5.4.3.2 Critical Fitness for Use Properties

The purpose of adding cord reinforcement to rubber is to help strengthen and preserve functionality over the designed lifecycle of the part. Cyclical loadings in multiple directions at high speeds and temperatures and in difficult environments are all par for the course when it comes to timing belts and tire cord. Cords used in automotive hoses are typically there to improve burst resistance. The advantages that fiberglass contributes are high strength and stiffness, dimensional stability, excellent creep resistance, fatigue resistance, good chemical and moisture resistance, and low heat generation. E-glass makes up the majority of the market volume for reinforcement cord because of cost. The remaining portion is high-strength S-type glass because of the application's performance requirements.

Mufflers require a different set of characteristics. They operate at temperatures approaching fiberglass' upper use limit, and subject the fill material not only to a steady stream of hot exhaust gases but also to a continuous series of sonic pulses. Hot exhaust gases are reactive and therefore the fill material must be very inert and resistant to the imposed temperatures and not tend to coalesce or sinter. The constant pounding of sound waves at high temperatures is a unique environment that can easily breakdown a material and blow it away. HSG glass is suitable on all three counts.

5.4.3.3 Market Trends and Future Needs

Trucks and automobiles will continue to be pushed to achieve better fuel economy. Decreasing engine displacement, augmentation with hybrid technologies such as alternate fuels, electricity, or fuel cells as well as decreasing vehicle weight are all key forces currently in play. Smaller vehicles must be stronger in order to maintain safety standards, so improved materials are needed. New materials are typically expensive.

It is clear that carmakers are caught up trying to satisfy numerous sets of divergent requirements. On one front, car customers want high quality at a low price. Regulators want state-of-the-art safety, high fuel economy, and low emissions from new vehicles. And the business community wants manufacturing efficiency and profitability. Over the past 30 years, they have done an admirable job of balancing these goals but the goalpost is continuously moving. Fiberglass-reinforced composites have played an important part in getting them where they are today. And composite materials will continue to play a vital role in getting the most out of the new engineering designs now and in the future.

5.4.4 Industrial Reinforcements – Pressure Vessels

A significant portion of the cost of energy is transporting it to the place where it will be used. Natural gas (NG) has the capability to offset some of the world's dependence on petroleum [56]. Because the majority of oil is used to make gasoline for automobiles, NG has been slow to be adopted as a fuel for transportation, especially in the United States. Years of cheap oil, low levels of interest, and a lack of infrastructure have all conspired to keep natural gas as a home heating product rather than a more widely used fuel. More efficient means of transporting natural gas is a key to diversifying the sources of energy we have available.

5.4.4.1 Selected Application Overview

The United States has a respectable natural gas pipeline network throughout the country. The network has the best coverage in the mid-west with the majority concentrated in and near Texas and the Ohio valley area. This means vast amounts of the country are without an effective means of natural gas supply. Without adequate local pipelines, overland transportation becomes the primary means for moving NG from a supply point to a distributor or an end user, so called off-pipeline transport.

Off-pipeline natural gas distribution is primarily accomplished by compressing and moving the gas in specially designed and certified high-pressure cylinders. Internal pressures of at least 3000 psig are needed to optimize tank utilization for methane, making safe transport a critical concern. Pressure vessel design must take into account both the strength requirements of the tank and its weight since heavy tanks cannot transport as much gas as lightweight tanks.

CNG tanks are divided into four types [56]. Type I or all-metal tanks are typically made from steel. Tanks made from steel are heavy, even when thinned and wrapped with a unidirectional fiberglass reinforcement to improve strength (Type II). All-composite tanks (Type IV) have polymer liners and are wrapped primarily with carbon fiber in the interest of weight and strength. They are very costly but they are finding application in vehicular-use retrofits where space (and thus size) is a concern. Weight is also important for bus applications where tanks are placed on the roof. Low weight is vital to minimize shifting the bus' center of gravity upward.

CNG tanks made from aluminum and wrapped with a high-strength glass (Type III) are an acceptable compromise for applications for which both weight and cost are sensitive such as bulk gas transport (delivery trucks, etc.). Their use is becoming more widespread [57]. High-pressure hydrogen tanks are also an area where glass-wrapped pressure vessels will find some utility. These will be all-composite Type IV tanks since H_2 contributes to metal embrittlement.

5.4.4.2 Critical Fitness for Use Properties

High-strength fibers like S-type glass or carbon are ideal for meeting the strength and weight requirements of a Type IV fiber-reinforced CNG tank. Their cost, however, has opened opportunities for the lower cost, high-strength fiberglass such as R-glass or S-1 Glass to break into the market.

5.4.4.3 Market Trends and Future Needs

The industrial market clearly needs a higher strength alternative to E-glass that has a price point lower than that of S-glass. Most industrial and consumer applications are very price sensitive. In some respects this has limited the wider use of S-type glass beyond the specialized and niche. R-glass and some of the near S-glass compositions like S-1 Glass and HiPer-tex fit that bill. Other applications that are benefiting from these types of glasses are long fiber-reinforced thermoplastics (LFRT) for pultrusion, high-strength injection molding, and wind-powered electrical generators.

5.5 Concluding Remarks

Fiberglass is often used in applications that require good strength and stiffness properties. Standard E-glass offers the best value proposition for good properties at a relatively low cost. When situations require significantly higher strength and stiffness capabilities, there are alternative high-strength fiberglass products available that can meet not only purely mechanical challenges but also thermal. Compared to other high-performance fibers, high-strength fiberglass can often compete with less complexity and at a lower cost.

Acknowledgments We would like to extend our thanks to AGY for granting permission to participate in this work. We also wish to thank the following individuals for their kind and thoughtful help in the writing of this chapter: Dick Holland of Composite One, and Scott Northrup, David Fecko, R. J. Fisher, Larry Huey, and Sudhir Hublikar all with AGY.

References

1. R. S. Harris, D. Tiffin and G. R Machlan, Glass composition, US Patent 3,402,055 (1968).
2. K. L. Loewenstein, *The manufacturing technology of continuous glass fibers*, American Elsevier Publishing Company, Inc., New York, p. 29 (1973).
3. J. G. Mohr and W. P. Rowe, *Fiber glass*, Van Nostrand Reinhold Company, New York, p. 207 (1978).
4. D. M. Miller, *Glass fibers, composites, Vol 1, engineered materials handbook*, ASM International, Materials Park pp. 45–48 (1987).
5. A. K. Varshneya, *Fundamentals of inorganic glasses*. Academic Press, Boston, p. 3 (1994).
6. J. E. Shelby, *Introduction to glass science and technology*, Edition 2, The Royal Society of Chemistry, Cambridge, UK, p. 268 (2005).
7. S. N. Loud, Jr., "Advanced composites applications of S-2 glass fiber," reinforcing the future; Proceeding of the Thirty-Fourth Annual Conference of the Reinforced Plastics/Composites Institute, (1979).
8. S. J. Walling, S-2 Glass fiber: Its role in military applications, Fifth International Conference on Composite Materials: ICCM-V, San Diego (1985).
9. D. Hartman, M. E. Greenwood and D. M. Miller, "High Strength Glass Fibers," AGY Technical Paper (1996).
10. ASTM Test Method D 578 Standard specification for glass fiber strands.
11. J. E. Shelby, op. cit. p. 191.
12. A. K. Varshneya, op. cit. Chapter 18.
13. R. H. Doremus, *Glass science*, John Wiley and Sons, New York, p. 291 (1973).
14. W. H. Otto, Compaction effects in glass fibers, J. Am. Ceram. Soc., 44 (2) 68–72 (1961).
15. Y. Z. Yue, J. deC. Christiansen and S. L. Jensen, Determination of the fictive temperature for a hyperquenched glass, Chem. Phys. Let., 357, 20–24 (2002).
16. F. T. Wallenberger, *Advanced inorganic fibers: processes, structures, properties, applications*, Kluwer Academic Publishers, Norwell, p. 93 (2000).
17. N. M. Cameron, Relation between melt treatment and glass fiber strength, J. Am. Ceram. Soc., 49 (3) 144–148 (1966).
18. C. A. Richards and H. Li, An industry perspective of factors that affect fiber glass strength, *Glass Res.*, 11 (2) 18–20, 29, (2002).
19. F. T. Wallenberger, op. cit. pp. 133–145.
20. V. I. Kostikov, *Fibre science and technology*, Springer Verlag, Berlin, pp. 188–190 (1995).
21. J. F. Bacon, High modulus, high temperature glass fibers, Appl. Polym. Symp., 21, 179–200 (1973).
22. J. F. Bacon, The Kinetics of Crystallization of Molten Binary and Ternary Oxide Systems and their Application to the Origination of High Modulus Glass Fibers" (NASA CR-1856), United Aircraft Corporation, East Hartford, Connecticut, (1971).
23. S. Inaba, S. Fujino and K. Morinaga, Young's modulus and compositional parameters of oxide glasses, J. Am. Ceram. Soc. ,82 (12) 3501–3507 (1999).
24. F. T. Wallenberger, S. D. Brown and G. Y. Onoda, ZnO-modified high modulus glass fibers, J. Non-Cryst. Solids, 152, 279–283 (1993).
25. A. Lewis and D. L. Robbins, High-strength, high modulus glass fibers, J. Polym. Sci. C, 19, 117–150 (1967).
26. D. R. Messier, E. J. DeGuire and N. Katz, Oxynitride glass fibers, US Pat. 4609631 (1986).

27. J. Kobayashi, M. Oota and H. Minakuchi, Oxynitride glass and the fiber thereof, US Pat. 4957883 (1990).
28. H. Osafune, S. Kitamura and T. Kawasaki, Oxynitride glass, method of preparing the same and glass fiber, US Pat. 5576253, (1996).
29. P. B. McGinnis, High temperature glass fibers, US Pat. 6809050 (2004).
30. R Continuous Filament Glass Fibres; MSDS; Saint-Gobain Vetrotex International; Chambery Cedex, France (June, 2007).
31. D. A. Hofmann and P. B. McGinnis, Owens Corning, "Composition for high performance glass, high performance glass fibers, and articles therefrom, US Pat. App 0009403 (2008).
32. NGF EUROPE. (2004). *Glass cord for rubber drive belt reinforcement. [Brochure]*. NGF EUROPE Limited, Lea Green, St. Helens, England.
33. HS2 Glass Fiber, HS4 Glass Fiber, Chopped Strands; MSDS; Specialty Fiberglass Division Sinoma Science-Technology Co. Ltd.: Nanjing, China. http://corporateportal. ppg.com/NR/rdonlyres/75A02A67-ED04-497E-9607-7D0D5FCB423F/0/HighStrengthHS2-HS4MSDS.pdf (accessed 4-13-09).
34. http://www.nittobo.co.jp/business/glassfiber/sp_material/t-glass.htm
35. Basalt Fiber & Composite Materials Technology Development. Retrieved 4-14-09, from http://www.basaltfm.com/eng/fiber/info.html
36. G. Gardiner, The making of glass fiber, Compos. Technol. [Online] April (2009).
37. Y. I. Kolesov, M. Y. Kudryavtsev, and N. Y. Mikhailenko, Science for glass production, Glass Ceram., 58 (5–6), 197–202 (2001).
38. N. N. Morozov, V. S. Bakunov, E. N. Morozov, L. G. Aslanova, P. A. Granovskii, V. V. Prokshin, and A. A. Zemlyanitsyn, Materials bases on basalts from the European north of Russia, Glass Ceram., 58 (3–4), 100–104 (2001).
39. E. Fitzer, A. Gkogkidis, and M. Heine, Carbon fibers and their composites (a review), High Temp. High Pressures, 16, 363–392, (1984).
40. Committee on High-Performance Structural Fibers for Advanced Polymer Matrix Composites, available at the National Academies Press, http://www.nap.edu/catalog.php? record_id=11268 .
41. R. A. Lane, High performance fibers for personnel and vehicle armor systems, AMPTIAC Q., 9 (2), 3–9 (2005).
42. Kevlar® K29 is a trademark of I. E. DuPont de Nemours, http://www2.dupont.com/Kevlar/ en_US/assets/downloads/KEVLAR_Technical_Guide.pdf
43. R. E. Allred, Aramid fiber composites, in *Handbook of composite reinforcements*, S. M. Lee, ed., John Wiley and Sons, New York, pp. 5–24 (1992).
44. L. Pilato and M. J. Michno, *Advanced composite materials*, Chapter 3, Springer, Heidelberg, pp. 75–96 (1994).
45. SPECTRA® is a trademark of Honeywell International, Inc. http://www51.honeywell. com/sm/afc/common/documents/3.1_SpectraFiber1000.pdf
46. DYNEEMA® is a registered trademark of Toyobo Co. Ltd., Japan http://www.toyobo. co.jp/e/seihin/dn/dyneema/index.htm
47. T. F. Cooke, Fiber reinforcement, high performance, in *Handbook of composite reinforcements*, S. M. Lee, ed., John Wiley and Sons, New York, pp. 217–232 (1992).
48. S. Kumar, Ordered polymer fibers, in *Handbook of composite reinforcements*, S. M. Lee, ed., John Wiley and Sons, New York, pp. 470–493 (1992).
49. ZYLON® is a registered trademark of Toyobo Co. Ltd. Japan http://www.toyobo. co.jp/e/seihin/kc/pbo/Technical_Information_2005.pdf
50. M. A. Meyers, *Dynamic behavior of materials*, John Wiley and Sons, New York, pp. 597–606 (1994).
51. S. Magnuson, Military services ponder future of their war-worn trucks, Nat. Def. Mag., April (2009), http://www.nationaldefensemagazine.org/ARCHIVE/2009/APRIL/Pages/Military ServicesPonderFutureofTheirWar-WornTrucks.aspx

52. S. Rush, The art of armor development, High Perform. Compos. Mag., January, (2007), http://www.compositesworld.com/articles/the-art-of-armor-development.aspx

53. D. Fecko, High strength glass reinforcements still being discovered, Reinfor. Plast., 50 (4), 40–44 (2006).

54. GLARE® is a registered trademark of Stork Fokker AESP, refer to http://www. storkaerospace.com/eCache/DEF/17/751.html

55. Containers designed to withstand explosions to experience school of hard knocks, Air Safety Week, January 4, (1999). See http://findarticles.com/p/articles/mi_m0UBT/ is_1_13/ai_53518407/

56. J. G. Ingersoll, *Natural gas vehicles*, The Fairmont Press Inc., Lilburn (1996).

57. http://www.neogas.us/technology/cpv-technology.html

Part II
Soda–Lime–Silica Glasses

Chapter 6
Compositions of Industrial Glasses

Antonín Smrček

Abstract The principles behind commercial glass manufacture are discussed in terms of production-related considerations: meltability, workability, refining, and economics. Examples of the implementation of these principles are given to explain their importance and their technical impact. The historical development of the key commercial glasses are charted over the centuries up to the present day, providing insight into how and why we have arrived at today's commercial glass compositions and detailing their strengths, weaknesses, and variations.

Keywords Industrial · Compositions · Commercial · Economics · Meltability · Workability · Raw materials · Refinning · Historical · Flat · Container · Utility · Technical · Lead · Crystal

6.1 Guidelines for Industrial Glass Composition Selection

Mass-produced glasses (specifically flat, container, fiber, and utility glasses) are known as industrial glasses and constitute about 95% of glass production volume. Production costs are the primary factor determining their composition. Specific glass compositions do not generally arise from market demands or standards, although they can occur in certain glass groups such as lead crystal, technical glasses, and glass fibers. Other glasses such as those for medical and electronic applications are not industrial glasses because they are melted in very small volumes and their task is to provide specific properties, therefore cost does not play a key role in defining their composition.

A. Smrček (✉)
Teplice, Czech Republic
e-mail: sklarakeramik@seznam.cz

F.T. Wallenberger, P.A. Bingham (eds.), *Fiberglass and Glass Technology*,
DOI 10.1007/978-1-4419-0736-3_6, © Springer Science+Business Media, LLC 2010

6.1.1 Economics

Glass composition mainly contributes to production costs in two ways:

- Raw material costs
- Melting costs

Raw material prices are given ex works, i.e., exclusive of transport to the glass factory and any treatment of raw materials (cullet). The price of raw materials varies with time and location, which is why economically optimal compositions will also vary with time and location.

Melting costs are affected by specific fuel consumption as well as by the cost of energy per tonne of glass melted. In cases such as when a new composition requires changes in flue gas cleaning or when environmental legislation changes, such costs are also included in the cost per tonne of melted glass. Changes in the furnace lifetime can also be included in melting costs; however, furnace lifetimes are controlled by several factors in addition to the effects of glass composition. Increasing the forming machine speed as a result of viscosity changes may also be considered in clear-cut cases. Changes to the rate of rejected ware may be included if they demonstrably result from changes in glass composition. In this case the costs for a tonne of net production are compared. It is the best practice to carry out detailed economical analysis comparing situations before and after the change.

6.1.2 Demands on the Glass Melt

All glass works demand that their glass is cheap, i.e., raw materials and melting costs are low, the rejection rate is low, and the glass should be able to be formed quickly. Customer demands are concentrated on form and dimensions of products, i.e., their aesthetics, which is not primarily related to the glass composition. The absence of stones and minimal bubble contents are demanded of industrial glasses. For container glass, a maximum of one stone per tonne of glass melted is average and a maximum of 200 bubbles per kilogram of glass melted is acceptable. Demands on flat glass as well as for other kinds of glass are stricter. Such demands can be achieved by appropriate measures during melting and the suitable addition of refining agents.

Color or the lack of it is demanded to some degree in all glass products. The level of color in industrial glasses is chiefly affected by their iron content. Individual glass types typically contain the following levels of iron:

Top quality crystal glass	0.010 wt% Fe_2O_3
Common pressed crystal glass	0.025 wt% Fe_2O_3
White container glass	0.040 wt% Fe_2O_3
Decolorization limit	0.100 wt% Fe_2O_3

The higher the iron content the more difficult it is to decolorize the glass. Corresponding doses of decolorizing agents become more difficult to establish, and a visible grayish color occurs. This demand must be taken into account in the course of the choice of composition and raw materials.

Color demands are specific as far as colored container glass, flat glass, and filter glass are concerned. In many cases, the demand is specified numerically by light transmittance at a given wavelength. For example, some colored container glasses require UV light transmittance to be as low as possible. Taking into consideration that the required color is achieved using elements or complexes (amber glass) sensitive to redox conditions, only qualitative guidelines can be used for their choice. The necessary amount of colorants must be obtained by means of laboratory analyses or during the operation. Color utility glass is colored often by a combination of different coloring agents; the color is evaluated visually and old, established recipes are often used.

Chemical resistivity, which is required by all types of glass products, is achieved by suitable compositional formulation. This is especially so in technical glasses and also in container glass, utility glass, and fiber glass. Chemical resistivity can be increased by suitable surface treatments; however, this can be more difficult than to melt glass with a suitable composition. Chemical resistivity can be estimated using the factors given in Table 6.1. Chemical resistivity is decreased markedly by high alkali contents, while it is increased by CaO, MgO, and other bivalent oxides, also by higher contents of SiO_2. Additions of Al_2O_3 increase chemical resistivity strongly, which is one reason that practically all industrial glasses contain at least 1–2% Al_2O_3.

Table 6.1 Factors for calculation of glass properties from [1]

Property	Liquidus temperature/°C	Chemical resistivity 0.01 N HCl	Density/g cm^{-3}
$b_o = b_l$ for	975.1	−0.507	2.2221
Al_2O_3	17.11	−0.156	0.0034
$Fe_2O_3 + Cr_2O_3$	8.75	−0.061	0.0183
CaO	21.93	−0.072	0.0139
MgO	9.55	−0.030	0.0098
Na_2O	−15.85	0.167	0.0081
K_2O	−6.67	0.083	0.0058
SO_3	−3.24	0.402	0.0056
BaO	14.59	−0.053	0.0209

Other demands on glass composition and properties are different in various glass branches. Container glass and flat glass customers have no other demands. Utility glass is focused on product aesthetics, nevertheless easy grindability, gold adhesion, and other subjectively defined properties occur. Specific property requirements are imposed on technical glasses (thermal expansion, electric resistivity, etc.).

Good workability is a demand which occurs in glass works, especially in container glass production and in fiber production. It means a glass with a suitable viscosity curve declination: this can be pre-calculated when designing the glass. Such glasses must also have as few as possible defects, which cannot be pre-calculated.

Legislative requirements include environmental and hygienic regulations often compiled without experience in a given situation. For example, an environmental campaign against lead glasses toward the end of the twentieth century resulted in regulations requiring the absence of Cr^{VI}, Pb, and Hg in container glasses. When proposing a glass composition the use of As, Sb, Cd must be avoided; and use of Pb and other elements (Cl, F, etc.) can be problematic. The issue is complicated: nonpermissible amounts of As have been found in water arising from cutting of lead glass and in extracts from grinding waste.

6.1.3 Meltability

The batch for a glass with a certain composition can be melted more rapidly or more slowly at higher or lower temperatures, respectively, and may require more or less heat for glass-forming reactions. These parameters are functions of

– Glass composition
– Raw material selection
– Particle size of raw materials
– Furnace design and furnace charging system

The effect of glass composition is expressed through the viscosity. The lower the viscosity, the more rapidly the melting and refining progress; when the viscosity is the same the melting occurs at the same velocity (although the temperature differs). Log (viscosity / dPa s) = 2.0 is usually taken as glass-melting viscosity. This viscosity fixed point, and others, can be calculated according to factors derived statistically from measurements of a range of industrially produced glasses, for example, using [1]. The calculation of characteristic temperature (property) is carried out according to (6.1).

$$V = b_o + b_i.x_i \qquad (6.1)$$

where V = property, b_o = constant, b_i = factors for individual oxides in glass, and x_i = content of oxide in glass in mass %. Factors for individual oxides contained in industrial glasses are given in Table 6.2. Similar factors have been published by Lakatos et al. [2, 3].

Theoretical calculation shows [4] that input increasing by 1.08% is necessary for increasing of melting temperature by 10°C. However, practical statistics indicate that the necessary increase is 2.20%, and the specific value depends on design and adjustment of the individual furnace.

Bingham and Marshall [5] referred to possible decreases in melting temperature of container glasses by changes in composition. Decreasing of melting composition has positive environmental effects because less NO_x is formed at lower combustion temperatures [5]. Decreases in melting temperature can have three main uses: (i) fuel input and thereby temperature is decreased and furnace output and glass quality do not change; (ii) fuel input is not changed, the temperature increases, and

Table 6.2 Factors for calculation of viscosity fixed points from [1]

Property: Log (η/dPa s)	Transformation point 13.3	Softening point 11	Littleton point 7.65	Immersion point 4.0	Melting point 2.0
$b_o = b_l$ for	605.7	653.1	855.6	1339.6	2155.3
Al_2O_3	2.928	3.578	3.248	5.486	9.03
$Fe_2O_3+Cr_2O_3$		3.179	2.116	−6.775	−25.71
CaO	3.015	3.067	0.198	−6.792	−22.17
MgO		−0.066	0.686	−3.811	−14.28
Na_2O	−7.284	−6.876	−9.469	−15.780	−28.96
K_2O	−7.718	−7.433	−7.096	−7.984	−16.81
SO_3	10.220	9.031	2.283	3.426	−36.11
BaO		−0.189	−3.056	−14.194	−29.55

furnace output rises. Quality does not change and savings are in increased output; (iii) temperature, input, and output are maintained and glass quality improves by melting at lower viscosities.

Devitrification can be a serious problem during forming, and as a result the liquidus temperature must be below the lowest temperature in the furnace (feeder). Today this demand is defined as the lowest possible tendency to devitrify and includes low liquidus temperature as well as low crystallization tendency and velocity. Such requirements may be met using glass with an appropriate composition. The demand can be defined numerically as the difference between liquidus temperature and gob (forming) temperature for container, utility, and technical glasses. Theoretical considerations and also practical knowledge show that glass melts containing a greater number of different compounds are often preferred because they can have lower melting temperatures and lower crystallization tendencies, thus requiring a lower liquidus temperature and lower to zero velocity of crystal growth.

The dissolution of sand grains is a critical process in practically all glass melting. Grain size of sand influences melting rates; sands with smaller grain size are melted more quickly. Batch charge should never contain a less meltable raw material than sand. For this reason materials such as corundum (Al_2O_3) or ZrO_2 are not suitable as raw materials. From a melting point of view (if higher Fe_2O_3 is not a concern) it can be significantly better to use phonolite, nepheline syenite, or basalt than feldspar. In cases where difficult-to-melt raw materials must be used (e.g., colorants) they must be added as finely divided particles. The meltability may be improved by using pre-reacted raw materials, e.g., CaO instead of $CaCO_3$, and pre-sintered raw material mixtures.

6.1.4 Workability

Workability means the ability of glass to be formed quickly, thus allowing higher machine speeds, with minimal product defects, which both have significant economic effects. Forming speed depends on the amount of heat which must be removed during forming. This is given partially by the specific heat of the glass melt, but above all by the temperature range over which forming may take place. Temperature ranges are given for each forming process by viscosities corresponding to the beginning and end of the forming. Because specific heat varies only a little with composition, the forming velocity depends on the character of glass viscosity curve. The workability interval depends on the kind of processing; according to some authors it is limited by viscosities of log $(\eta/dPa\ s) = 3$ to 5 on the one hand and log $(\eta/dPa\ s) = 4$ to 7,6 on the other hand.

To compare glass melts, the equation derived by Vaughan and published by Lyle [6] in 1954 and updated in 1984 [7] is often used. The equation was compiled from statistical data and is described in (6.2).

$$v_{rel} = \frac{(t_L - 450) \times 100}{t_L - t_A + 80} \tag{6.2}$$

where v_{rel} = relative machine speed, t_L = Littleton point temperature, t_A = annealing temperature, both in °C. The equation applies in the range from 95 to 130% of relative velocity. Values for common container glasses tend to be 105 ± 10. A significant disadvantage of the equation is that all glasses with transformation temperatures of 520°C have $v_{rel} = 100$ regardless of viscosity characteristics. When comparing glass melts (6.2) is generally practicable. Alternatively the beginning and end of forming may be defined by the viscosity of the nearest viscosity point from Table 6.1 and to compare the length of forming interval of both glass melts. The immersion point is usually taken as the upper immersion limit and the Littleton point or softening point as the lower limit. The relative change (%) of the forming interval length will be approximately equal to the relative change of forming characteristics. However, a new glass composition does not automatically allow faster machine speeds: it must be implemented by qualified personnel.

Every glass engineer can recall cases when production rates have changed suddenly without any obvious reason or when the reject ratio changed suddenly. In most cases changes to composition, temperature, machine conditions, or personnel account for such behavior. Nevertheless, a few cases remain for which the cause cannot be determined. Such changes have been ascribed to the term "workability" or "formability," which is always a mysterious property. Formability is discussed only in connection with automatic machine production. Poole [8] finds excessive forming of wrinkles and especially hair fissures, form prints on a product, problems with cutting off and form filling, and emphasized cutting off as symptoms of poor workability of the glass melt. Poor workability is also associated with increases in the number of rejects due to unequal distribution of wall thickness, warping of bottles, and decreased strength. In some cases "workability" does not depend on glass composition. Poole [9] considers glass inhomogeneity one cause of poor workability. He found round or elongated formations of 0.05–0.5 μm on electron micrographs of glasses exhibiting poor workability, whereas similar formations were not observed in glasses with good workability. Ultra-microinhomogeneities affect the glass melt viscosity and increase the tendency to form fissures. Their formation can be explained by the thermal history of the glass melt and may also arise from the use of coarser raw materials such as feldspar. Other causes of poor workability can include the rheological properties of surface layers, into which sulfur and carbon can diffuse from mold lubricants. To improve formability, melting at higher temperatures followed by rapid cooling is recommended, as is using a composition with the lowest acceptable liquidus temperature.

6.1.5 Choice of Raw Materials

Selection of glass composition and raw materials must be carried out concurrently because both factors interact with one another. Raw materials' data must be used to select the most suitable raw materials in terms of composition, guarantee of compliance, grain size, price (including transport), and delivery condition. Also the prospects of delivery from individual suppliers must be taken into account. At present, free-of-care delivery from specialized suppliers is preferred.

It can also be advantageous to source raw materials locally. Some time ago a glass works melting glass for amber and green containers replaced glass sand by building sand coming from a nearby deposit. Although the company bought sand for the same price (with the same grain size), the higher feldspathic component of the sand allowed 1–2% of alkali to be supplied to the final glass, reducing batch Na_2CO_3 requirements. Also waste raw materials can be used – waste saltcake ($Na_2SO_4 \cdot 10 H_2O$) was delivered for the price of transport only to glass the works and was used instead of sodium sulfate.

The simplest criterion for raw materials choice is the type of glass to be melted. Raw materials with the lowest Fe_2O_3 contents must be used for melting crystal glass, whereas using fine materials for melting container glass is unnecessary. In the case of special raw materials (such as lithium raw materials) it is important to quantify their effect (for example, decreasing melting temperature) and to consider whether the same or similar effect can be achieved more cheaply in another way (for example, by increasing Na_2O content). When rationalizing many "traditional" glass compositions it is important to ask "why this oxide is used at all?"

A basic criterion is raw material grain size [10]. Mono-dispersive grain size with an average grain size of 0.20–0.25 mm is recommended, with no grains larger than 1.0 mm and with a maximum 2% of grains under 0.1 mm. Raw materials which melt or are dissolved easily may be coarser (for example, limestone or dolomite). Raw materials that melt or dissolve slowly (for example, Al_2O_3) must be finely divided. Regarding sodium carbonate, heavy soda ash (average grain size 0.3–0.4 mm) is preferable to light soda (0.10 mm) because it is less dusty during handling. Colorants that dissolve slowly in the melt and added in small amounts must be as finely divided as possible to ensure homogenized distribution in the glass well.

Fluxes as batch additives are often discussed. These raw materials, when added to the batch at levels of up to 1 wt%, accelerate reactions between glass-forming raw materials. Their effect is in forming an aggressive and fluid liquid phase at relatively low temperatures. This molten phase etches the surface of sand grains and joins batch compounds together so that glass-forming reactions are accelerated.

Fluorides (CaF_2, NaF), fluorosilicates (Na_2SiF_6), or chlorides were used as fluxes most frequently in the past. Results in operation were not unambiguous because fining and flow in a furnace, not melting itself, determine furnace output. It is possible that fluxes will be used in the future within highly loaded furnaces. However, such fluxes will not contain fluorine or chlorine because environmental regulations permit them only in furnaces with neutralized emissions. However, more ecologically acceptable sulfates also accelerate melting. Na_2SO_4 becomes molten during the early stages of batch melting and because it is a surface active material it is concentrated at the surface of sand grains, forming a surface film and increasing wettability; sand flotation stops and melting is accelerated. Sulfates will probably continue to remain part of many batches, but the amount added must be chosen carefully.

Batch moisture also accelerates melting: it joins soda grains with sand, avoids batch decomposition, and decreases dusting. Optimal batch moisture is considered to be 2–4 wt%. Also sulfides are included in accelerators such as blast furnace slag. Sulfides do not accelerate primary reactions but they do accelerate refining so their overall effect is similar.

6.1.6 Cullet Effect – Glass Melt Production Heat

In-house cullet has always been re-used. Only in recent decades has the use of recycled foreign cullet, i.e., cullet arising from external sources, become widespread. Collection and processing techniques [10] were developed enabling average recycling rates of container glass in the EU of about 60%, while some countries exceed 80%. Many green glass furnaces operate with 80–90% cullet in the batch charge. Recycling of flat glass is becoming more widespread; however, it does not exceed about 25%, mainly from in-house cullet. Recycling of technical glass is problematic, especially of TV screens, because they contain both valuable and toxic raw materials and form ecologically harmful waste.

Cullet recycling has ecological advantages: lower environmental loading due to reduced sand mining and soda production and reducing common waste. Glass works can make savings in raw material and fuel costs by using cullet. Two potential cases should be considered when using cullet:

– using cullet with the same composition as the melted glass
– using cullet with a different composition to the melted glass

In the first case, the cost of cullet (for a tonne, on input into the furnace, including processing and manipulation) must be compared with the raw materials' price necessary for melting of 1 tonne of glass, also on input into the furnace. For the glass works, use of cullet is particularly advantageous if it is cheaper or the same price as the corresponding amount of raw materials.

The profit from reducing the glass melt production heat compensates for any risk of production losses due to impure cullet. Glass melt production heat is the heat necessary for melting of glass from raw materials without heat losses. This heat consists of three parts:

- Heat of reaction, which differs for various glass melts. When melting from cullet is zero; its share of production heat is about 20%.
- Heat required to raise the batch to the selected temperature, which is about 70% of production heat.
- The balance is heat losses due to water evaporation from the batch charge and exhaust gas heat losses.

According to measured values [4] a decrease in production heat requirement is about 1.0 GJ/t for soda lime glass melts with total replacement of batch by cullet. This corresponds to heat savings of about 1.7 GJ/t in common container glass furnaces. Specific heat consumption decreases by approximately 0.3% for every 1% increase in batch cullet for soda lime glass manufacture.

In the case that cullet with a different composition is used, its composition must be known and such cullet must be considered as a further raw material when calculating and preparing the batch. Its effect on heat savings is the same as cullet with an identical composition.

The basic source of the success of cullet is its quality; the better it is, the larger the acceptable levels of batch cullet. Ceramic impurities and metals interfere because they cause stones and glass melt coloring. Small additions of glass with a different composition do not necessarily cause problems (for example, one or two pieces of utility glass in a container batch charge) because they homogenize in the furnace. On the contrary, a truck full of lead crystal or other very different glass in container glass cullet will not be homogenized and will cause significant problems in the furnace.

6.1.7 Glass Refining

Refining is a critical phase of the glass-melting process, because an undesirable content of bubbles is the most frequent defect occurring during melting. The velocity of refining in a tank furnace depends on temperature (viscosity) of glass and its distribution, on furnace design affecting the flow, on glass composition, and on the addition of refining agents.

Designing refining agents and especially their level of addition is the most difficult part of glass melt design, because there are few meaningful calculation rules. It is recommended that suitable amounts of refining agents are developed through laboratory trials, or to optimize additions during operation. There is no direct proportionality between the amount of refining agent added and the quality of refining; however, decreasing the glass viscosity always assists refining.

In the past, a range of refining agents was used: As_2O_3, Sb_2O_3, $NaCl$, Na_2SO_4, and others. Environmental regulations limit the use of arsenic and antimony as well as of chlorides to special glasses, predominantly utility glasses. Sodium sulfate (or other sulfates such as $CaSO_4$ or $BaSO_4$) is the typical refining agent, which is complemented by CeO_2 (cerium concentrate) in some cases. Refining by sulfate occurs by the following reaction:

$$nSiO_2 + 2Na_2SO_4 + C = 2Na_2O \cdot nSiO_2 + 2SO_2 + CO_2$$

When using larger amounts of sulfate, carbon must also be added in the form of coal or coke dust (6–15% of the sodium sulfate content). If small additions of Na_2SO_4 are used, carbon impurities from the batch or a reducing furnace atmosphere are sufficient to reduce carbon supply in the batch. Batch sodium sulfate levels occur within broad limits, usually from 0.5 to 2.0 kg for every 100 kg sand in the batch.

Additions of 40–120 g As_2O_3 for every 100 kg of batch sand were typically used for arsenic refining, although larger doses have been used. Arsenic was used in a mixture with saltpeter ($NaNO_3$) which oxidizes As^{3+} to As^{5+} at lower temperatures and which then decomposes at higher temperatures. Doses for antimony refining were typically 100–500 g for every 100 kg of batch sand, and again the doses fluctuate. CeO_2 doses are recommended to be from 50 to 2000 g for 100 kg of batch sand.

The task performed by the refining agent is not only to refine the glass but also to oxidize it. This function depends on balancing the glass redox state into a neutral position. The glass redox state may be considered in terms of the Fe^{2+}/Fe^{3+} ratio, and a semi-empirical method for its calculation was proposed by Manring and Diken [11]. This calculation method is known as chemical oxygen demand (COD) and compares the oxidizing and reducing effects of constituent raw materials, both of stoichiometrically expressed and analytically determined effects. Analysis is carried out by oxidation with potassium bichromate and results are given in ppm C as shown in Table 6.3. The reducing capability of raw materials varies according to their source and locality, which is why analytical determination of COD is advisable.

The calculation is expressed in so-called sulfate units. The oxidizing ability of Na_2SO_4 is equal to 1, and the reducing ability of compounds is calculated according to reaction equations such that 1 kg of C corresponds to 23.7 kg Na_2SO_4 to achieve neutrality. The calculation is based around a batch recipe for 2000 kg sand, as given in Table 6.4.

Table 6.3 Reducing ability of raw materials (COD) as ppm C

Raw material	Manring–Diken [11]	Bláhová–Lederer [12]
Sand	150	500–530
Limestone	150	40–60
Soda ash	75	–
Feldspar/Neph. Syenite	260	68
Dolomite	60–150	60–150
Phonolite	390–590	390–590
Blast furnace slag	10 000	–
Carbon/graphite	650 000	580 000–680 000

Table 6.4 Example of calculated redox values after Manring and Diken [11]

Raw material	Charge/kg	Carbon content/ppm	Carbon total/kg
Sand	2000	500	1.00
Soda ash	670	75	0.05
Feldspar	120	70	0.01
Limestone	200	100	0.02
Dolomite	450	100	0.05
Graphite	5	650 000	3.25
Sodium sulfate	34	0	0
Glass yield	3860	Graphite total	4.38
Batch moisture	3.0%		

Carbon sulfate equivalent $= 4.38 \times 23.7 = -103.8$
Sulfate-oxidizing potential $= 34 \times 1 \quad = +34.0$
Moisture-oxidizing potential $= 4 \times 3 \quad = +12.0$
Batch redox potential $\qquad\qquad\qquad -57.8$
Glass redox number relative to 1 kg glass is
$-57.8 \times \frac{2000}{3860} = -29$.

On the basis of calculated redox number the maximum solubility of SO_3 in glass and the color stability are shown in Fig. 6.1. Manring and Diken [11] divide glasses according to the value of reduction number into four groups:

(1) Oxidized flint glass. This is white glass with the highest SO_3 contents, such that small additional reduction is necessary and SO_3 is discharged. Such glass is not advantageous from a refining point of view because a substantial part of the added SO_3 dissolves in glass and is not involved in refining.

(2) Reduced flint glass. This is always white glass, well refined, with a low tendency to form secondary fine seed.

(3) Stable amber. This forms in the field of minimal SO_3 solubility in the glass. In this field, small change in redox conditions does not matter, the amber is stable, and bubbles are not formed.

(4) Unstable amber. This glass is too strongly reduced. It is darkly colored but bubbles occur if small additional oxidation occurs.

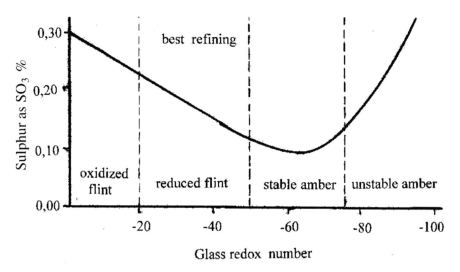

Fig. 6.1 Solubility of SO_3 in glass as a function of redox number after Manring and Diken [11]

The batch shown in Table 6.4 has a redox number of −29.9 and lies in the reduced flint field which provides optimal refining. The COD calculation according to Manring and Diken [11] is a good tool for solving problems with amber and sulfate refining in container and flat glasses; however, it is only semi-empirical. One weakness is in considering the contribution from moisture, the effect of which the authors [11] do not explain. The calculation does not include furnace atmosphere effects but it confirms the crucial effects of reducing material volume. As a result, graphite must be weighed precisely, more so than sulfate. It is likely that the values of Manring and Diken [11] cannot be directly transferred to other operational conditions but calculations for different alternatives on a given furnace offer useful comparisons.

6.2 Industrial Glass Compositions

6.2.1 Historical Development

Compositions of glasses melted at "forest works" were broadly identical for hundreds of years, differing only as a result of the available raw materials based on location and according to practice at individual glass works. More or less clear white glass was produced, and only the best glass works melted smaller amounts of colored glass by adding appropriate raw materials to the basic composition. The products from leading glass works were superior owing to the higher purity of the raw materials which they employed.

The main difference in output quality and composition arose from the alkali raw material and thus in Na_2O and K_2O contents of the glass. Coastal countries (Italy, France, England) used the ash generated from burning seaside plants – so-called barilla – which contained mainly Na_2O. Continental countries (Germany, Austria, Bohemia) used the ash from burning locally sourced wood, which was rich in potash. Such a distinction is imprecise because barilla was exported to Austria from Italy and Spain, but raw materials were not exported in the opposite direction. Wood ash, in addition to containing potash, contained many impurities (including P_2O_5), further complicating the historical development of glass compositions. The composition of continental (potash) glasses melted from fourteenth to seventeenth centuries was roughly as follows [13]:

SiO_2	50–60 wt%
Al_2O_3	2–3 wt%
Fe_2O_3	0.4 – 0.7 wt%
P_2O_5	1–4 wt%
CaO	15–25 wt%
MgO	2–4 wt%
Na_2O	0.5–2.5 wt%
K_2O	7–16 wt%
MnO	0.5–2 wt%
SO_3	0.2–0.4 wt%
Cl	0.3–0.5 wt%

Dietzel's sum $\Sigma(RO + R_2O)$ was 30–40 wt%, an amount which improved glass meltability to the extent that glasses could be readily melted even in primitive furnaces. The relatively high contents of Al_2O_3 in these glasses arose from impurities in the sand and from the products of pot corrosion. Manganese originated from wood ash, and it was also deliberately added as a decolorizer. Chlorine and P_2O_5 also originated from wood ash. Some of the alkali may have been replaced by CaO, and for this reason its content is relatively high.

In the seventeenth century a period of experimentation began which has continued to the present day. Increasingly educated glass makers experimented by adding many kinds of imported raw materials or interesting pieces of stone, regardless of their cost. As a result glasses containing BaO, B_2O_3, As_2O_3, and other elements (Au, Cu) appeared. Later glasses were colored using newly discovered elements (Se, U, Nd, Ce), and by the end of the nineteenth century most of the elements in the Periodic Table had been tried. Use of these sometimes very expensive raw materials was possible thanks to relatively high price of glass products but their use was restricted to top quality glass production.

Glass development continued toward ever more pure raw materials. For example, double refining of ash potash removed P, Cl, and higher contents of Mn from the glass; selection of high-purity silica decreased Fe and Al contents. By the end of the eighteenth century almost all glass works had converted to using limestone (lime) as a source of CaO, which had previously been supplied as an impurity in other raw materials. The shortage of natural alkali raw materials led to the use of Leblanc soda at first and later, cheaper sodium sulfate. Sodium sulfate became the dominant alkali raw material up to the end of the nineteenth century; this is the chief reason why sodium glasses prevailed and the potassium content of glass was reduced.

In parallel, a number of new types of glass also arose. In the seventeenth century, lead crystal from England became very popular. In Bohemia new black and opal glasses were developed. The development of technical glass began in the nineteenth century – indeed the origins of optical glasses have been linked with the works of Faraday and Fraunhofer [14]. It has been reported that the first wide-ranging trials concerning the effects of different compounds on glass properties were carried out by Harcourt [14]. However, typical glass works continued to use traditional compositions described in the glass masters' book and passed down from one glass master to his successor. The Schott company in Jena was the first to produce technical glasses on the basis of experimentally derived compositions at the end of the nineteenth century. From that time, production of glass began to divide into individual production branches with different glass compositions.

6.2.2 Flat Glass

The oldest handmade flat glasses are classic potassium glasses, yet sodium glass prepared using sodium sulfate has prevailed since 1880. Sodium sulfate became, for a time, the main alkali raw material for flat glass production. In the early twentieth century sodium sulfate was replaced by soda ash. Flat glass is made by various methods, from hand making to Fourcault drawing and other systems (for example, Pittsburgh) and in modern times the production of almost all flat glass has been carried out by Pilkington's Float process. Only a small fraction of flat glass is now produced by casting. Although there were opinions that each production system required its own glass composition, the survey shown in Table 6.5 demonstrates that compositional differences arise rather from practices at individual works than from the demands of the forming technology. Values given in Table 6.5 have been derived as averages from a wide body of published glass analyses from Germany, France, Great Britain, Belgium, Italy, USA, Japan, Russia, Holland, Sweden, Czech Republic, Hungary, Austria, and Finland [15].

It is evident from Table 6.5 that cast glass is strongly related to Float glass in its composition; drawn Fourcault glass, on the other hand, contains approximately 1 wt% more alkali than the others. The SiO_2 content is broadly constant across the groups, on average $72.02 \pm 0.42\%$. The content of Al_2O_3 is 1.00 ± 0.13 wt%. Iron contents are relatively high in all groups: extra "white" glasses with lower iron contents have been produced only in recent years. Small contents of MnO occurred in some older samples that were decolorized using manganese-bearing raw materials. In addition SO_3 contents are lower in the most recent samples: this is the result of switching to reducing refining conditions. Small amounts of K_2O originate from feldspar and from impurities contained in the sand.

Table 6.5 Average composition of flat glass 1951–2000 [15]

Glass: No. of analyses	Cast 41	Fourcault 53	Pittsburgh 35	Float 27	Average 156
SiO_2/wt%	71.46	72.01	72.48	72.12	72.02
Al_2O_3/wt%	0.98	1.06	1.13	0.83	1.00
Fe_2O_3/wt%	0.10	0.07	0.08	0.09	0.08
CaO/wt%	9.91	7.47	7.98	8.64	8.51
MgO/wt%	2.97	3.72	3.72	4.03	3.61
Na_2O/wt%	13.92	14.98	13.76	13.56	14.03
K_2O/wt%	0.25	0.42	0.18	0.31	0.29
SO_3/wt%	0.42	0.34	0.31	0.27	0.33
MnO/wt%	0.17	–	0.03	–	–
ΣRO/wt%	12.88	11.19	11.75	12.67	12.12
ΣR_2O/wt%	14.17	15.30	13.94	13.87	14.32
Dietzel	27.05	26.49	25.69	26.64	26.44

All four groups of flat glasses are formulated to give approximately the same value of Dietzel's sum ($\Sigma R_2O + \Sigma RO$) of 26.44 ± 0.61 wt%. In this only the Pittsburgh glasses are different. However, there exist greater differences in the amounts of individual oxide groups: $\Sigma RO = 12.12 \pm 0.79$ wt% and $\Sigma R_2O = 14.32 \pm 0.63$ wt%.

Statistical analysis has identified fluctuations in composition within each glass group (between individual producers and over a shorter time period). Variability in average composition was found [15] to be

SiO_2	± 0.64 wt%
Al_2O_3	± 0.21 wt%
Fe_2O_3	± 0.02 wt%
CaO	± 0.49 wt%
MgO	± 0.34 wt%
Na_2O	± 0.45 wt%
K_2O	± 0.08 wt%
RO	± 0.43 wt%
R_2O	± 0.41 wt%
Dietzel's	± 0.49 wt%

These fluctuations demonstrate that flat glasses have one common composition and differences arise from conditions at individual manufacturers, for which the composition of flat glass may vary within certain undefined limits.

Glass makers began to give their attention to the composition of flat glass after the development of machine production by Fourcault drawing, when they tried to find optimal compositions with low tendencies to crystallize, with good workability, and of course with a low price. The development of Fourcault glass compositions is given in Fig. 6.2. Each point is the average from a number of analyses from one country (11 countries) over a 5-year period [15]. Flat glass began to be

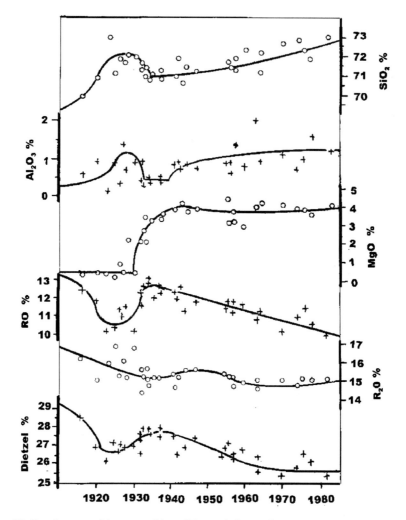

Fig. 6.2 Development of the composition of Fourcault drawn glass

produced by Fourcault drawing in 1913 and this system became the main production process after 1919. The composition of handmade flat glass was used
during early production, but such glass crystallized easily in the *débiteuse*. As a
result the RO content was decreased to 9–11 wt% and R_2O content was increased
to 16–18 wt%.

Such glass was readily meltable and formable but its chemical resistivity was
poor. Methods were therefore identified to prevent crystallization with lower alkali
contents. Additions of small amounts of BaO, B_2O_3, and ZnO were tested but were
not cost-effective. A solution was found around 1924 through the addition of MgO
which was probably the only cost-effective oxide commonly used at that time. As

a result a raw materials mixture of limestone and dolomite began to be used. The MgO content was decreased quickly to approximately 4 wt% with small fluctuations, and the alkali content was decreased as a result. The RO content reached its maximum in the mid-1930s and decreased from that time. The Al_2O_3 content decreased at first to its minimum (0.3 wt%) and then increased stepwise to provide increased chemical resistance up to about 1 wt%, with further increases in Float glasses.

Prior to 1940 flat glass was decolorized similarly to other glasses by using brown stone or manganese ore. During the Second World War a lack of manganese ore occurred and this method of decolorizing was interrupted. After the end of the war manganese ore was available again but had been replaced by other materials and was not widely re-introduced.

6.2.3 Container Glass

Container glass is traditionally the cheapest type of glass. Demands on its quality are more moderate and it is melted using less pure raw materials. Principally, the composition does not depend on the forming process (except in the case of the difference between glass for suction and feeder machines), but it may differ in color.

White container glass constitutes approximately one-third of global glass production and one-half of container glass production. These glasses have always simple soda lime compositions. The average composition from the period between 1980 and 2000 is shown in Table 6.6, which includes glasses from Germany, France, Great Britain, Italy, Russia, USA, Canada, Austria, Hungary, Bulgaria, Poland, Czech Republic, Slovakia, and others (32 countries in total) [16–19].

The content of SiO_2 in white container glass is roughly the same as that in flat glass in order to provide acceptable chemical resistivity. Aluminum was added as a partial substitution for SiO_2, however, Dietzel's sum did not change in the course

Table 6.6 Average composition of container glass 1980–2000 [16]

Glass: No of analyses	White 67	Green 57	Amber 55	Average 179
SiO_2/wt%	72.28	71.21	71.75	71.75
Al_2O_3/wt%	1.83	2.28	2.03	2.05
Fe_2O_3/wt%	0.09	0.36	0.26	n/a
CaO/wt%	9.69	9.30	9.30	9.45
MgO/wt%	1.60	2.27	1.98	1.95
Na_2O/wt%	13.47	13.37	13.83	13.55
K_2O/wt%	0.54	0.74	0.57	0.62
SO_3/wt%	0.21	0.13	0.02	n/a
ΣRO/wt%	11.36	11.56	11.30	11.40
ΣR_2O/wt%	14.00	14.00	14.39	14.17
Dietzel	25.36	25.66	25.69	25.57

or such changes. The iron content is rather high; however, it has been decreased to 0.050 wt% Fe_2O_3 at leading producers. The RO content is relatively constant. Limestone-only and dolomite-only batches are not used practically. However, the MgO content fluctuates. Attempts have been made to decrease the alkali content, and the K_2O present in these glasses originates from feldspar. Sulfate is used for refining, and thanks to the high melting temperatures and addition of reducing agents the SO_3 content of the glass is relatively low. Some container glasses are refined/decolorized by CeO_2 additions although decolorizing using selenium is more common. Due to the growth in container glass recycling, hundredths to tenths of 1% of impurities may occur (PbO, BaO, etc.).

A specialty of white container glasses was small deliberate additions of BaO and B_2O_3. Additions of BaO occurred during the twentieth century at levels of 0.1–0.8 wt% and on average 0.34 wt%. Barium probably arose from $BaSO_4$ used as a refining agent or melting accelerant. Boron additions were found in 5% of analyzed glasses at levels of 0.2–1.0 wt%, and on average 0.69 wt%. Boron oxide was added mainly into pharmaceutical glasses, possibly to improve its chemical resistivity or its appearance.

The development of white container glass through the twentieth century is shown in Fig. 6.3. The SiO_2 content remained at about 72.5 wt%; Al_2O_3 increased at first due to impure raw materials but from 1920 to 1930 it decreased thanks to purer raw materials. It increased again after 1950 because of deliberate addition to improve chemical resistance. The Fe_2O_3 content has slowly decreased. Since about 1930, the RO content has risen. MgO has been added to white container glass since about 1925. Initially its level of addition increased quickly, but it has varied in recent years and levels of 0–4 wt% MgO now occur. The content of other oxides is consistent in all analyses to date. The alkali content has gradually declined and glasses from the most industrialised countries typically contain 1 wt% less alkali than glasses from less industrialized countries.

Brown amber glass constitutes about one-quarter of all container glass. The amber color arises from iron and sulfur atoms and is generated in the glass by melting with the addition of reducing batch materials (carbon, graphite, etc.). Individual glasses differ in color intensity or tone based on commercial factors. The average composition is given in Table 6.6. Characteristic of amber glass is its lower Al_2O_3 content, reputedly because higher Al_2O_3 and Fe_2O_3 contents cause a dirty brown color. This is why the composition of amber glass is very close to the composition of white container glass, from which it differs only in terms of the necessarily higher iron content (0.26 wt%) and lower SO_3 content (0.02–0.05 wt%).

Brown amber glasses have been melted as utility glasses for many years. The first amber bottles originate from the beginning of the twentieth century; however, it only became widely produced in the second half of the twentieth century. Over the years the sum $SiO_2 + Al_2O_3 = 73.7$ wt% has been maintained. The composition has copied that of white glasses: introducing MgO, trials to decrease alkali content, and the corresponding increase in RO. Contents of 0.01–0.03 wt% Cr_2O_3 have occurred in recent years due to cullet recycling.

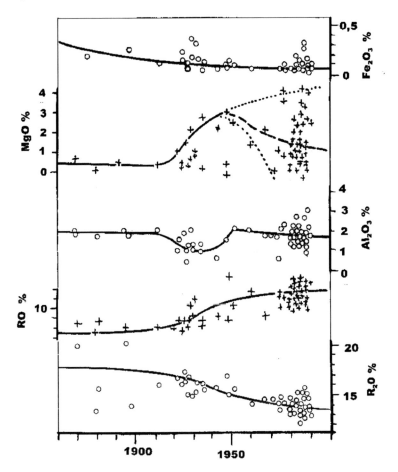

Fig. 6.3 Development of white container glass

Green container glass constitutes about one-quarter of container glass. In terms of the glass plant, the production of green glass is the simplest because demands on its purity (iron content) are not great. This is why it is melted with high contents of mixed cullet of various colors. Its basic composition is very close to the other container glass as shown in Table 6.6. The iron content in green glasses is higher (0.36 wt% Fe_2O_3) due to addition of alumina-bearing raw materials (phonolite, feldspar). Also the Al_2O_3 content of green glass is greater. The MgO content varies, similarly to all container glasses, but other components of green container glasses correspond with white container glass and amber glass. Due to the use of lower grade cullet, traces of impurity can be expected (PbO, BaO, etc.). Green glass is refined using sulfate and is melted under oxidizing conditions.

Most green container glasses are colored by chromium (and iron). Their contents vary although 0.15 ± 0.04 wt% Cr_2O_3 is used for melting common emerald green glass. Light green bottles are colored using iron only. Prior to 1960, so-called olive green was colored using a combination of iron and manganese [20, 21]. Additions of 0.50–1.1 wt% MnO and about 0.40 wt% Fe_2O_3 were used according to the required color.

In terms of historical development significant variation of the composition occurred with balancing of compositions on a global scale after 1970. This is particularly evident for the aluminum content. Before about 1960, glasses almost free from aluminum and also high-aluminum glasses occurred (see Fig. 6.4), although more recently the aluminum content decreased to about 2 wt% Al_2O_3. This corresponds with a decrease in iron content. The RO content varied very significantly and

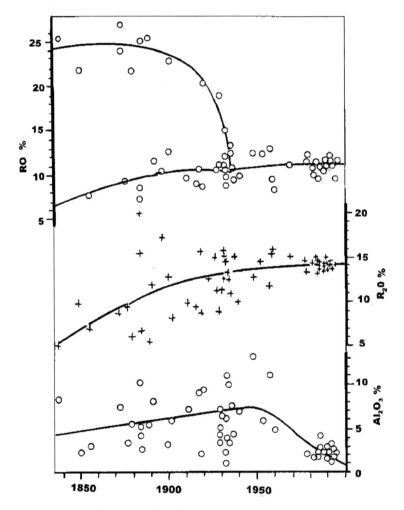

Fig. 6.4 Development of green container glass

irregularly and was divided into two groups before 1930. Glasses from Germany, Bohemia, Hungary, and the USA had low RO content (8–11 wt%). Glasses from France, Italy, and England contained about 25 wt% RO and the content decreased stepwise up to 1930, when compositions became uniform and have not changed greatly since. It is interesting to note that MgO was added into green glass (probably unwittingly) in the nineteenth century. At the beginning of the twentieth century, very high iron contents (5 wt% Fe_2O_3) were decreased to 1 wt%, and after 1960 it decreased further to about 0.40 wt% Fe_2O_3. The alkali content varied significantly and before 1930 it increased from 6 to 13 wt% R_2O, later being reduced somewhat. Also the Dietzel's sum varied, initially being increased to 30 wt%, although it has been approximately constant since 1930.

In the nineteenth century, so-called French bottle glasses with 7–7.5 wt% Al_2O_3 with high RO contents (about 26 wt%) and with only 5–6 wt% alkali were used across Europe [21]. These glasses are cheap and meltable; however, they had to be ladled at high temperatures and to be processed rapidly so that they would not begin to crystallize (Dietzel's sum was about 32 wt%). For this reason the composition was modified by increasing the alkali content to 8 wt% R_2O and 9 wt% Al_2O_3 in Germany at the end of the nineteenth century. Alkalis were added cheaply to the glass by using mineral raw materials and the glass melted well in the newly introduced Siemens tank furnaces [21]. The alkali content was increased upon the appearance of new machines, firstly from Owens, later IS, and aluminum contents were reduced accordingly. Rising demand for white glasses was another reason for the decrease in Al_2O_3 content, because such glasses cannot be melted from cheap mineral raw materials with high iron contents. The Czech company Mühlig did not follow this trend for lower aluminum: the company used a suitable German high-aluminum composition on Owens machines. In 1927 König [22, 23] carried out laboratory trials and began to increase the aluminum content in glass. Aluminum and alkalis were added as cheap, finely crushed phonolite. This made it possible to significantly decrease the addition of alkali raw materials (only 10 kg sulfate for 100 kg glass). A melting temperature of 1460°C was applied in the furnace; melting occurred without failures, and energy demands did not change. Hence the aluminum content was increased to 17 wt% Al_2O_3 and bottles were produced from this composition all year. It was observed that glasses with extremely high aluminum content are not resistant to acids. Therefore, the aluminum content was decreased to 13–14 wt% Al_2O_3 and such glass was melted from 1929 to 1947. Later the aluminum glasses were examined by Kříž [24] who decreased the Al_2O_3 content to 10.3 wt%, introduced a combination of CaO + MgO, increased alkali content to 14.2 wt%, and increased Dietzel's sum up to an optimal value. Cheap aluminum glasses melted with high proportions of mineral raw materials are one prospective possibility for container glass development.

Brown glasses also belong to the past. They were colored by a combination of iron and manganese; these glasses were replaced by amber glasses after 1950. High-aluminum glasses melted for several years with little or no soda were advantageously colored by manganese [25]. Additions of 3–8 wt% MnO and about 1.5 wt% Fe_2O_3 were used.

Today, container glass is produced in more than 100 glass works around the world, and a strong tendency toward a unified composition is evident. This tendency is supported by the high levels of glass recycling leading to melting glasses with >60 wt% of cullet in the batch charge. Compositional unification is made easier by recycling since changes in composition do not occur, but compositional changes within one glass works are more difficult. High recycling levels also input low levels of impurities into the glass (PbO, BaO, etc.).

Even though average compositions of separate glass groups are relatively constant, individual variations occur due to the large number of producers and smaller production units in comparison with flat glass:

SiO_2 $\pm$ 0.68 wt%
Al_2O_3 $\pm$ 0.41 wt%
Fe_2O_3 $\pm$ 0.07 wt%
CaO $\pm$ 0.69 wt%
MgO $\pm$ 0.66 wt%
Na_2O $\pm$ 0.59 wt%
K_2O $\pm$ 0.20 wt%
RO $\pm$ 0.67 wt%
R_2O $\pm$ 0.63 wt%
Dietzel's $\pm$ 0.58 wt%

It may be noted (Table 6.6) that the average compositions of all three color groups of container glass are virtually the same (excluding Fe_2O_3, Cr_2O_3, and SO_3). Glasses have Dietzel's sums of 25.57 wt% and are similar to flat glasses.

6.2.4 Lead-Free Utility Glass

Colorless glasses for common uses rank among this group, which includes common pressed glass (sometimes molded for technical use), machine-processed utility glass, drinking glass, lighting glass, and glass bricks. Borosilicate glasses are sometimes used for the same applications. The glass is made by smaller producers, which is why composition can vary and there is little published compositional information. Demands are subjective: high-quality glass. Glass must be colorless (low Fe_2O_3 and good decolorization) and free from stones, bubbles, and cord. Demands for handmade glass include easy formability on the glass pipe and easy cutting.

The simplest and most widely used of these glasses is colorless pressed glass. It is melted in relatively large furnaces and is processed by machines. The average composition from 1980 to 2000, along with average fluctuation at individual producers, is given in Table 6.7. Analyzed samples originated from Germany, France, Italy, UK, USA, Czech Republic, and 15 other countries. Oxide additions of BaO and B_2O_3 are often used. BaO was present in 44% of analyzed samples at levels of 0.1–5.6 wt%, and on average 1.39 wt%. Boron oxide was added to 31% of samples from traces to 2.2 wt%, and on average 0.70 wt% B_2O_3. Additions of 0.20 wt% Li_2O and 0.84 wt% ZnO occurred sporadically.

Table 6.7 Average composition of colorless pressed glass 1980–2000 [17, 19]

Compound	Average composition/wt%	Fluctuation/wt%
No. of analyses 68		
SiO_2/wt%	72.07	± 1.22
Al_2O_3/wt%	1.27	± 0.31
Fe_2O_3/wt%	0.031	± 0.008
CaO/wt%	7.55	± 1.25
MgO/wt%	1.56	± 0.58
Na_2O/wt%	13.67	± 0.95
K_2O/wt%	2.07	± 1.42
SO_3/wt%	0.25	
ΣRO/wt%	10.19	± 0.92
ΣR_2O/wt%	15.73	± 0.94
Dietzel	25.93	± 1.08

The historical development of colorless pressed glasses is shown in Fig. 6.5 and average compositions/fluctuations in Table 6.7. The SiO_2 content decreased in recent decades to achieve the constant sum of $SiO_2 + Al_2O_3 = 73.4$ wt%. Aluminum oxide was first added from the impurities, then it was decreased (by using pure raw materials) to 0.40 wt% Al_2O_3, but this negatively impacted on chemical resistance. This is why the Al_2O_3 content increased after 1970. Aluminum is often added using pure hydrated alumina. Low iron content is a basic demand for pressed sodium lime-stone crystal, its content was decreased from 0.20 to 0.030 wt% Fe_2O_3 on average within the last 100 years. Nevertheless, leading countries are able to produce pressed crystal with less than 0.020% Fe_2O_3.

The RO has fluctuated, although in recent years its variability has lessened. MgO addition fluctuates markedly according to the availability of pure dolomite and according to the subjective opinions of furnace managers and operatives. The R_2O content has not changed markedly and is significantly higher than in other glasses (glasses were melted having 13.5–20 wt% R_2O). Machine-processed glasses have, similar to white container glasses, lower alkali contents. Subjective opinion also plays a role. Perhaps this is crucial when adding K_2O. Machine-processed glasses contain only about 0.40 wt% K_2O, hand-pressed glasses contain 1–3 wt% K_2O and occasionally more. Reputedly this is because of better workability and also perhaps because these glasses do not crystallize. At the present time, these glasses are refined by Na_2SO_4 and its level of addition must be chosen carefully. Refining by antimony or by cerium oxide is used for glasses of higher quality. Decolorizing by erbium instead of selenium has recently been used.

Modified colorless utility glasses are "improved" from a commercial or subjective production point of view by the addition up to 0 5 wt% BaO (sonorous crystal); 5 wt% PbO, or ZnO and other oxides. Such glasses are produced individually, at low levels.

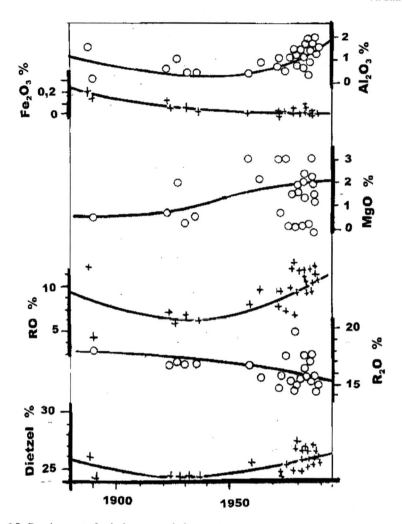

Fig. 6.5 Development of colorless pressed glass

High-aluminum utility glasses were proposed by Hatakka [26], who melted glass with following composition:

SiO_2	56.58 wt%
Al_2O_3	13.27 wt%
Fe_2O_3	0.028 wt%
B_2O_3	0.75 wt%
CaO	4.92 wt%
MgO	0.07 wt%
Na_2O	15.02 wt%
K_2O	9.36 wt%

The characteristics of this glass were unusual: RO = 4.99 wt%, R_2O = 24.38 wt%, Dietzel = 29.37 wt%. Its chemical resistivity was acceptable due to its high aluminum content. The high potassium content guaranteed low liquidus temperatures. Aluminum was added to the glass by large additions of feldspar so that foaming occurred during melting, which was removed by the addition of CaF_2. The glass was melted and processed; however, this represented the end of development because melting with high feldspar additions is always difficult and demanded low iron contents that cannot be reached.

6.2.5 Technical Glass

Products with technical uses are known as technical glasses because there are technical property requirements, for example, glasses in which controlled thermal expansion behavior is demanded but the appearance is less important. For these reasons iron content is not significant. However, some products formed from technical glass melts are used as glass ovenware, which is why their appearance must be good (color, bubbles). Volumes of produced technical glass melts are no more than 1–2% of global glass production, but the products are important. A specialty of technical glasses is the use of more expensive and unusual raw materials (B_2O_3, ZrO_2, SrO, BaO, ZnO, etc.). There are perhaps tens of technical glass melts and their compositions are not fully recorded so it is not possible to refer to all kinds. The most complete sources of information are Volf's books [27, 28], one [28] describing mainly classical laboratory glass melts in detail.

Hard alkali limestone technical glasses were used in laboratories before the introduction of borosilicate glasses. Later they were used for the production of less loaded devices (funnels, Petri dishes, etc.) or for blood bottles after dealkalization. They are also used for production of fluorescent tubes. The average composition calculated on the basis of analyses coming over several years and published by Volf [28] is given in Table 6.8.

Table 6.8 Average composition of alkali limestone technical glasses according to Volf [28]

Glass No. of analyses	Hard glass 16	Soft apparatus glass 9
SiO_2/wt%	72.93 ± 2.19	69.21 ± 1.72
Al_2O_3/wt%	1.84 ± 0.89	4.07 ± 0.60
CaO/wt%	8.03 ± 2.06	6.74 ± 1.17
MgO/wt%	2.03 ± 1.64	1.11 ± 1.28
Na_2O/wt%	13.09 ± 2.64	14.16 ± 2.66
K_2O/wt%	1.25 ± 1.76	3.74 ± 2.86

Hard glasses have contents of RO $= 10.06$ wt%, $R_2O = 14.34$ wt%, and Dietzel's sum 24.40 wt%. From 1.0 to 4.5 wt% B_2O_3 are present in five glasses (31% of total), whereas 1.9 wt% BaO is present in one glass (6% of total).

A specialty of mainly older glasses in this group was a high SiO_2 content (75–79 wt%), which gave the glass good chemical and heat resistance. The Al_2O_3 content was increased later to inhibit matting in the flame and to increase chemical resistivity. The RO content lies between 8 and 13 wt%, older glasses do not contain MgO, and more recent glasses contain up to 3.5 wt% MgO. The R_2O content lies between 13 and 15 wt%. About half of the published compositions are without K_2O and 0.5–6 wt% K_2O are contained in the remaining compositions. High K_2O additions belong to the past and advanced glasses had less than 2 wt% K_2O.

The average composition of hard alkali limestone technical glasses is very close to white container glass. The only significant difference is the RO content being 1.3 wt% lower and the Dietzel's sum being 1 wt% lower. Due to the K_2O content it is rather close to pressed glass but has more alkali. All three groups are closely related.

Soft apparatus glasses are intended mainly for glass blowing purposes such as for apparatus construction. To this group also belong glasses for fluorescent tubes and tubes in general, lamp chimneys, and insulating flasks as well as decorative glasses. The main property of such glass melts is nonmatting during processing above the glassblowing burner. Matting occurs as a result of alkali volatilization and may be prevented through the addition of aluminum oxide to the glass [28]. The average composition of hard apparatus glasses according to Volf [28] is given in Table 6.8.

Soft apparatus glasses have an average RO $= 8.10 \pm 0.86$ wt% and $R_2O = 17.90 \pm 1.95$ wt%. The Dietzel's sum is on average 26.00 ± 1.41 wt%. The large spread in values shows that the composition of this group is nonuniform. Between 1.2 and 2.5 wt% B_2O_3 was found in two glasses (22% of total), whereas 2.5 wt% BaO occurred in one glass (11% of total).

The speciality of this glass group is a high Al_2O_3 content as well as high alkali content. The main alkali is always Na_2O, but K_2O contents can be up to 7 wt%. Additions of BaO and B_2O_3 are rather rare. These glasses are peculiar as they do not correspond with glass from previous groups.

Glasses for lamp chimneys and insulation flasks are harder due to the higher content of SiO_2 and lower alkali content [28]. Decorative glasses have lower RO and higher R_2O contents (up to 24.5 wt%, always including K_2O). The Dietzel's sum is about 29.5 wt% [28].

Neutral glasses are intended mainly for pharmaceutical applications such as ampules for syringes. Tubes are the primary products in general. The main demand is chemical resistance because syringe solutions are sterilized at 121°C and the glass must not interact with the contents at this temperature and glass flakes must not release from the surface. Such properties are achieved through high additions of B_2O_3 and Al_2O_3 with BaO and ZnO also being added. On the other hand, CaO and alkali contents are lowered and MgO is added only rarely. From industrially produced glasses ("5.0" type) there belong here laboratory glass G20 and Jenatherm produced by the Schott company in the past, a Czech glass called Sial, and also other brand names. Also laboratory apparatuses were produced: for example, "Jenaer Glas" is known for household use in the kitchen as well as glass pipelines and water glasses. However, their properties were surpassed by 3.3 glasses, so that classical neutral glasses are no longer produced.

The composition of 5.0 glass is given in Table 6.9 according to Volf [28]. The composition differs significantly from soda lime glasses because RO = 4.60 wt%, $R_2O = 7.0$ wt%, and Dietzel's sum is on average only 11.6 wt%.

Table 6.9   Composition of borosilicate glasses 5.0 and 3.3

Type of glass	Neutral	Pharmaceutical	Hard 3.3	Opal
Literature	Volf [28]	Biavatti [29]	Volf [28]	Volf [28]
No. of analyses	4	6	8	
SiO_2/wt%	74.5–75.7	71.12 ± 2.58	80.34 ± 0.36	75.13 ± 2.28
B_2O_3/wt%	7.0–8.7	11.58 ± 0.84	12.79 ± 0.19	12.58 ± 1.56
Al_2O_3/wt%	4.3–6.2	4.90 ± 1.40	2.29 ± 0.29	1.97 ± 0.87
CaO/wt%	0.5–1.0	1.28 ± 0.15	–	0.88 ± 0.57
BaO/wt%	3.5–4.2	1.68 + 1.12	–	2.64 ± 0.99
Na_2O/wt%	6.5–7.5	7.32 ± 1.04	3.71 ± 0.28	5.38 ± 0.85
K_2O/wt%	–	1.48 ± 0.62	0.72 ± 0.42	0.20 ± 0.20
ZnO/wt%	–	0.67 ± 0.46	–	1.94 ± 0.74

The "3.3" glasses have very low thermal expansion and thus high thermal shock resistance (expansion coefficient $\alpha = 3.3 \times 10^{-6}°C^{-1}$). Their other key property is excellent chemical resistance. They are used for production of laboratory apparatus, of technical as well as utility glasses, tubes, and devices. They have been marketed since 1915 as Pyrex, Duran, Razotherm, Simax, and other brands. Average compositions are given in Table 6.9. The specialty of these glasses is that they contain only trace levels of alkaline earths. The R_2O content is also low, on average 4.41 ± 0.23 wt%, the same value as the Dietzel's sum. Small spreads show that producers melted identical compositions. The reason for this is that the excellent properties of 3.3 glasses are limited to a narrow compositional range [28]. The 3.3 glasses have the highest SiO_2 content of all commercial glasses. Additions of Al_2O_3 depress the tendency for crystallization and phase separation, with 2.8 wt% the upper limit [28]. Small changes in Al_2O_3 content are made so that the sum of $SiO_2 + Al_2O_3$ remains consistent. The iron content can be up to 0.10 wt% Fe_2O_3. RO are present only as impurities in amounts not exceeding 0.2 wt% as they cause phase separation in these glasses. High B_2O_3 contents enable such glasses to be melted. The alkali content is low with K_2O being preferred over Na_2O [28]. However, cheaper, purely Na_2O glasses for household use have been melted in recent years.

Borosilicate glass for pharmaceutical containers is a group of advanced glasses used for the production of ampoules and medical bottles. These are produced in the classical way on IS machines or mechanically from tubes, where slightly different compositions are used. The compositions were published by Biavati et al. [29].

Chemical resistance is emphasized in these glasses, as well as good workability. Average compositions are given in Table 6.9. The compositions of these glasses lay between neutral glasses and classical hard 3.3 glasses. A part of SiO_2 is substituted by a relatively high content of Al_2O_3. Because the highest temperature resistance is not required, RO and alkalis can also be added. The RO content is increased to 2.95 $\pm$ 1.17 wt% and R_2O content to 8.80 $\pm$ 0.66 wt%, so that the Dietzel's sum is 11.75 $\pm$ 1.68 wt%.

Glasses intended for tube production have lower alkali contents and higher SiO_2 contents because alkali diffusion to the surface during tube processing must be prevented. The alkali content was lowered by degrees and substituted by CaO [29]. This example shows that even very sophisticated glasses may be improved.

Thermally resistant opal borosilicate glasses have been used instead of porcelain vessels in recent decades, thus making them utility glasses. The products are formed mainly by pressing, thus the glass melt must have the correct viscosity curve. The compositions of such glasses were published by Volf [28] and include preliminary phosphor opals, sometimes with fluorine additions, as given in Table 6.9. The glass melts for such products are slightly softened 3.3 glasses with lower SiO_2 contents, with RO additions, and with slightly higher alkali contents. BaO additions were found in three analyses (37% of total) and ZnO in two cases (25% of total). Phosphorous pentoxide as an opacifier was present in all samples; this was supported by additions of 0.59–1.85 wt%, on average 1.15 wt% of F. The RO content was 1.87 $\pm$ 0.99 wt%, the R_2O content was 5.54 $\pm$ 0.76 wt%, and the Dietzel's sum 7.39 $\pm$ 0.76 wt%. For products made from opal borosilicate glass, resistance against sudden temperature change to 110°C is required as the products are not usually thick-walled. Chemical resistance can be only moderate, but the opacity must be deep and uniform. For this, additions of RO are necessary, and B_2O_3 and Al_2O_3 additions are also favorable [28].

Production of glasses for vacuum cathode-ray tube (CRT) screens is now a thing of the past and their compositions are problematic when disposing of old screens. Nevertheless this is an example of complex constructed glass. Generally, the composition may be divided into black-and-white CRT, color screens (panels), and color cones (funnels). The composition of black–white CRT was about the same at all producers, and for color CRT each producer used slightly different compositions. Average compositions according to Volf [28] and Bernardo et al. [30] are given in Table 6.10. Recent data from Dahmani [31] give average funnel composition and different producers according to the waste glass. From the high spreads it is notable that compositions of individual producers were quite different and published analyses are always not complete.

One-half of analyzed compositions contain SrO and some CRT glasses are lead free. For funnels, the aluminum content fluctuates significantly and all such compositions exhibit high lead contents of up to 22.5 wt% PbO. Strontium is present only as an addition, perhaps coming from in-house cullet. All three glasses are different

Table 6.10 Composition of CRT television glass

Type	Black-and-white CRT	Color CRT cones		Cones (average)
Author/s	Volf [28]	Volf [28]	Bernardo [30]	Dahmani [31]
SiO_2/wt%	64.5–69.1	61.48 ± 1.26	55.94 ± 4.79	53.0 ± 2.3
Al_2O_3/wt%	3.2–5.6	3.13 ± 0.98	3.76 ± 1.32	3.2 ± 2.1
CaO/wt%	0–2.5	2.32 ± 1.73	3.55 ± 1.46	2.5 ± 1.5
MgO/wt%	0–0.3	0.89 ± 0.62	1.63 ± 1.12	1.7 ± 1.2
BaO/wt%	11.4–12.8	8.89 ± 5.94	1.35 ± 0.49	1.3 ± 1.3
SrO/wt%	–	9.52 ± 1.31	(0.70)	1.0 ± 1.0
PbO/wt%	–	–	16.06 ± 6.00	22.5 ± 1.1
Na_2O/wt%	7–8.8	8.42 ± 1.23	6.67 ± 1.00	6.3 ± 0.5
K_2O/wt%	6.4–7.1	8.30 ± 1.65	9.30 ± 2.48	8.1 ± 0.7
Li_2O/wt%	0.6–1.5	–	–	–
CeO_2/wt%	–	0.22 ± 0.06	–	–
ZrO_2/wt%	–	–	–	0.25 ± 0.25
Sb_2O_3/wt%	0.4	–	–	0.3 ± 0.2
F/wt%	0.9	–	–	–
ΣRO/wt%	13.5	16.83 ± 1.87	6.63 ± 2.45	9.5
ΣR_2O/wt%	15.6	16.72 ± 2.23	15.97 ± 2.56	14.4
Dietzel's/wt%	29.1	33.55 ± 0.63	22.58 ± 3.76	23.9

principally – containing BaO, SrO and BaO, and PbO. The RO content as well as Dietzel's sum are different while R_2O contents are about the same.

The glass for black–white CRT tubes was refined the classical way – by adding antimony oxide with fluorine as the accelerator. Glass for screens was refined by CeO_2 and cones were refined by antimony oxide. A mixture of colorants was added to black–white screens to give a gray tint.

TV bulbs were machine formed which is why a suitable viscosity curve was required. Li_2O was added reportedly for its viscosity modification. Also a precise coefficient of thermal expansion was required for sealing of metal anodes and for welding of cone and screen together. This glass also required specific electrical insulating ability. High additions of heavy metal (Ba, Pb, and especially Sr) were used for attenuation of emerging radiation. CeO_2 was used not only as a refining agent, but its addition stabilizes the glass to prevent browning caused by electron irradiation.

In the past, light bulbs were produced from lead glass melts. Thanks to suitable glass bulb design the bulb glass does not require a high electrical insulating ability. Pure glass melts without defects or tint and with the usual properties are sufficient. Easy and quick forming on machines is a requirement. It is not now known why a dolomitic glass composition which was suggested in the USA was used for light bulb production around the world (in modifications with and without

BaO). The glass composition was slightly modified after the introduction of ribbon machines. Compositions of three such options according to Volf [28] are given in Table 6.11.

Table 6.11 Glass composition for light bulbs according to Volf [28]

Glass type	Barium glasses	Dolomitic glasses	Ribbon machine glasses
SiO_2/wt%	70.5	72.8	71.5
Al_2O_3/wt%	1.5	1.3	2
CaO/wt%	5.5	5.1	6.6
MgO/wt%	3.5	3.5	2.8
BaO/wt%	2	–	–
Na_2O/wt%	16.5	16.9	15.5
K_2O/wt%	1	0.9	1
ΣRO/wt%	11.0	8.6	9.4
$ΣR_2O$/wt%	17.5	17.8	16.5
Dietzel's	28.5	26.4	25.9

All glasses for light bulbs are very close in composition to white container glasses and to pressed glasses. The glass processed on ribbon machines generally corresponds to white container glass.

Thermometer glasses are a very small part of glass production, but a great deal of research effort had to be expended to identify a glass that would provide reliable thermometers. When thermometers were made from common glass they were inaccurate and nonreproducible. The so-called depression of zero point occurred. This is caused by nonuniform glass shrinkage upon cooling and "ageing" with time. Schott found that the main cause was the presence of two alkali metals and developed special thermometer glass that has been melted with only small changes to the present day. In addition to requirements for volume stability, resistivity against water vapor is also demanded. The composition of the most widely used thermometer glass, Schott 16 III, is [28]

SiO_2	67.5 wt%
Al_2O_3	2.5 wt%
B_2O_3	2 wt%
CaO	7 wt%
ZnO	7 wt%
Na_2O	14 wt%

The composition is rather similar to white container glass, except that some RO is substituted by ZnO. Zinc is added to avoid the influence of water vapor during processing on the glass blow burner.

Other technical glasses, for example, quartz glass, water glass, foam glass, sealing glass, optical glass, microporous glasses, and glass ceramics, are only a small part of glass production. Their composition is regulated by special rules, for this reason they are not discussed in depth here.

6.2.6 Lead Crystal

The best quality utility glass is called "crystal," which corresponds in its purity with mountain crystal (pure crystalline quartz). Glass crystal is usually decorated by cutting and engraving and the product is highly aesthetic.

The first "crystal" products were made from white, mainly potassium glass and they were called "Czech crystal." After the discovery of lead glass in England in the seventeenth century it was found that it outperformed common soda potassium glass in terms of its attractive form and hence the term "English crystal" was originated. Melting of lead glasses spread around the world incrementally and, at same time, the term "crystal" became the general name for the best quality (and the most expensive) utility glass. Efforts to protect the customer led the EU to accept a (terminological) standard that specified various kinds of lead crystal as given in Table 6.12. This method of classification is only general because almost every producer has its own composition. According to the standard, crucial characteristics are composition (content of chosen oxides), density, and refractive index. The last two factors depend on composition and are used for quick verification.

Table 6.12 Categorization of lead glass according to EU regulations [10]

| Quality class | Metal oxides/wt% | Demands | | |
		Density, g/cm^3	Refractive index, n_D	Vickers hardness
High lead crystal	$PbO \geq 30$	≥ 3.00	≥ 1.545	–
Lead crystal	$PbO \geq 24$	≥ 2.90	≥ 1.545	–
Crystalline	$\sum (ZnO + K_2O + BaO + PbO) \geq 10$	≥ 2.45	≥ 1.520	–
Crystal glass	$\sum (K_2O + BaO + PbO) \geq 10$	≥ 2.40	$\geq$	550 ± 20

Lead crystal is popular because higher PbO contents (and a certain degree of BaO) increase the refractive index and the product, decorated by suitable cutting, reflects and diffracts the light giving "spark and fire." Perfect quality glass is of course demanded, with no color tone, no bubbles and stones, and with good visual homogeneity. It is melted mostly in small tank furnaces and occasionally pot furnaces and is processed by machines.

Lead crystal (24 wt% PbO) is the standard product. The glass is mainly of the K_2O type. It is decolorized by nickel oxide and refined by As_2O_3 and saltpeter. Top-quality raw materials are used so that Fe_2O_3 content is 0.012–0.015 wt%, sometimes less. CaO and MgO are essentially eliminated. The composition according to Smrček and Voldřich [10] is given in Table 6.13. B_2O_3 and ZnO occur in only one sample and were perhaps used for light compositional improvement. High lead crystal is rarely produced. As shown in Table 6.13, it is a ternary SiO_2–PbO–K_2O glass without other additions.

Table 6.13 Composition of lead crystal according to Voldřich [10]

Type: No. of analyses	24% PbO 3	30% PbO 1
SiO$_2$/wt%	60.20 ± 0.26	55.4
CaO/wt%	0.67 ± 0.89	0.1
Na$_2$O/wt%	2.80 ± 1.13	0.1
K$_2$O/wt%	10.80 ± 1.20	11.6
PbO/wt%	24.70 ± 0.53	32.3
B$_2$O$_3$/wt%	(0.8)	0.3
ZnO/wt%	(1.5)	–
As$_2$O$_3$/wt%	0.13 ± 0.06	0.2

Crystalline (special crystal glass) is one of the transition categories between lead and nonlead crystal from the point of view of properties and price. The glass melt is used for automatic production of thin-walled products; nevertheless, it is also used for cutting. For achieving the required refractive index, the content of stabilizers must be increased to about 15 wt%. Glasses intended for refining by buffing and chemical polishing may contain up to 12 wt% of PbO complemented by BaO and/or ZnO. It is necessary to use a potassium composition and to decolorize be means of NiO. Automatically produced glasses decorated by painting exhibit PbO additions of up to 3 wt% and the content of CaO + BaO can be up to 10 wt%. In glasses in which soda prevails over potassium, they are decolorized by selenium [10].

Crystal glass is the cheapest "lead" glass in which lead is not present at all. It is used for machine production of drinking glasses. The simplest solution to achieving a content of 10% (K$_2$O + BaO + PbO) is to use predominantly K$_2$O combined with up to 5 wt% BaO, whereas increasing the stabilizer content enables the alkali content to be decreased. Crystal glass is refined by antimony oxide and decolorized by selenium.

Other lead glasses are produced in small amounts. This is due in part to the toxicity of lead raw materials and for the necessary safety precautions. In the past, lead tubules for electro-technical purposes were produced (for example, lamp bulbs and electron tubes), windows for radioactive areas with high PbO contents, and colored lead glasses used by artists.

In the late twentieth century, when unrepresentative results of medical trials of lead leaching from glass into beverages (wine) were published, public opinion began to turn against lead glasses. A number of research projects to develop "lead-free crystal" were launched. During these trials, lead was substituted by barium, zirconium, and other elements. But it was not possible to find an entirely lead-free glass which would have the same appearance (refractive index, dispersion) as well as the same good workability (meltability, viscosity, grindability) as lead glass at a comparable cost. More detailed study of lead leaching from the glass showed that it is at its highest from new, unused glasses: leaching is evidently decreased by chemical polishing and by washing in warm water. In addition the typical residence time of a beverage in glass before its consumption is not long enough for leaching out of a significant amount of PbO. Lead leaching can be eliminated by glass surface treatment. This is done by surface dealkalization or by adding protective coatings. There is little sense in being afraid of lead leaching from lead beverage glasses.

6.2.7 Colored Glasses

Except for colored container glass, colored and opacified glasses constitute only a tiny fraction of glass production; nevertheless, there are maybe hundreds of their recipes. As required, small amounts are melted in pots and this is one of the reasons for a shortage of colored glasses. Filter glasses, signal glasses, welding glasses, and sunglass filters are produced to provide color that is defined exactly in trichromatic coordinates. Utility glass and glass jewelry make use of a variety of traditional colors.

The starting glass composition does not correspond to nonlead crystal (see Section 6.2.6) with addition of coloring agents. Traditionally, higher K_2O contents have been used in almost all recipes. Saltpeter is used for to provide stronger oxidation and the glass is "improved" by additions of B_2O_3 or PbO and sometimes ZnO. Recipes for melting ruby or opacified glasses are quite specific. Glass makers tend to be dependent on very old traditional recipes (see [32]) or on recipes derived from laboratory-based experimental studies.

6.3 Example Glass Compositions

6.3.1 Perspectives

The compositions of the various glass melts given in this work represent only average values. Data spread and compositions in the various branches show that glass may be melted within very broad compositional limits. Average values may not be optimal. The optimum always depends on local costs (raw materials, energy) as well as on production conditions (melting temperature) and on customer demands. All such conditions vary in locality and also in time. Also the conservatism of glass makers cannot be ignored when evaluating compositions: it is especially evident in older compositions as well as influencing the advertising and promotion of raw material producers.

In every case it is advantageous to rationalize the composition used, especially the calculations, using laboratory melting. In this way the composition can be optimized taking present conditions into consideration. It is always necessary to verify the effects of each added oxide.

It is not possible to accurately predict the development of glasses. It is likely that energy prices will continue to rise, together with costs of energy-consuming man-made raw materials (soda). Customer demands on product quality (color, purity) will be rather higher so demands on raw material quality cannot often be decreased. Nevertheless, using local raw materials may be of advantage. It would be necessary to face unreasonable consumer demands (container glass color). Glass recycling will be expanded to about 80% re-use in container glass. Cullet use will also be higher in flat glass. The glass composition must be constructed with as low as possible melting temperature (log (η/dPa s) = 2). Low liquidus temperatures will also be advantageous. The viscosity curve must provide the fastest possible forming. If fuel prices grow faster than the soda price, it will be necessary to soften the glass so that fuel may be saved by decreasing the melting temperature.

Slightly increasing the Al_2O_3 content may provide better chemical resistivity and optimization of the CaO/MgO ratio will be necessary. Potassium oxide will be eliminated from pressed and utility glasses, as will B_2O_3 and BaO. Efforts to decrease B_2O_3 contents will dominate for borosilicate fibre glasses.

Environmental laws prohibit using compounds of fluorine, chlorine, arsenic, antimony, cadmium, and other metals, and make it more difficult to use lead, barium, and other compounds. From this reason, refining will be provided largely by sulfates, which the batch contents of which must be balanced precisely. Decolorizing of good quality glass will be performed by erbium and CeO_2 and selenium + cobalt will be used in mass production.

There are no clear indications at present of the imminent development of fundamentally new glass compositions that could be produced industrially on a cost-effective basis. However, one group of glasses which arguably merits a rebirth is high-aluminum glasses.

6.3.2 Practical Examples of Container Glass Batch Charge

In this example batch calculation we study the modification of an original white container glass. Glassworks use a very simple batch, as shown in Table 6.16. Sand–feldspar–dolomite–soda and fining agents with high doses of sulfate require a minimum of handling. However, it offers poor chemical resistance, a green color, and bubbles. The objective is to save on raw materials or fuel, and machine speed should not be reduced.

The compositions of the available raw materials are listed in Table 6.14. Relative prices are considered as a ratio and one must use the current local prices. Batch compositions are shown in Table 6.16. The theoretical glass compositions produced are shown in Table 6.15. Comparison with a typical white container glass (Table 6.6) shows that the glass has a high Dietzel's sum and a low content of Al_2O_3. Purely dolomitic glasses are not common. To improve the color, sand "40" with lower iron content than sand "70" has been used. Improvement of chemical resistance is obtained by the addition of Al_2O_3 via feldspars, which also lead to an increase in the content of K_2O. Furthermore, the ratio of $CaO : MgO$ has been varied. For each of the proposed alternatives the predicted properties have been calculated (shown in Table 6.15) according to Table 6.1 factors and Lyle's formula (6.2).

Table 6.14 Composition of raw materials used/wt %

Material	Sand-70	Sand-40	Dolomite	Limestone	Feldspar	Soda
SiO_2	99.93	99.35	0.50	0.23	78.08	–
Al_2O_3	0.26	0.25	0.48	0.36	12.36	–
Fe_2O_3	0.050	0.028	0.031	0.044	0.15	0.002
CaO	0.06	0.05	30.52	55.41	0.37	–
MgO	0.05	0.04	21.18	0.46	0.14	–
Na_2O	0.01	0.05	–		1.64	57.67
K_2O	0.01	0.07	–	–	6.96	–
SO_3	–	–	–	–	0.01	–
Cost	3.0	4.0	4.0	5.0	20.0	100.0

Table 6.15 An example of proposed container glass/wt%

Glass	Origin	1	2	3	4	5	6	7
SiO_2	71.17	71.46	71.90	71.90	72.26	71.90	69.91	69.91
Al_2O_3	0.94	2.00	2.20	2.20	2.20	2.20	2.20	2.20
Fe_2O_3	0.05	0.04	0.04	0.04	0.04	0.04	0.04	0.04
CaO	7.39	7.50	9.11	10.03	9.75	9.53	10.02	12.02
MgO	5.13	3.50	2.00	0.50	1.00	1.00	2.49	0.50
Na_2O	14.95	13.80	14.00	14.58	14.00	14.58	14.58	14.58
K_2O	0.38	0.70	0.75	0.75	0.75	0.75	0.75	0.75
ΣRO	12.52	11.00	11.11	10.53	10.75	10.53	12.52	12.52
ΣR_2O	15.33	14.50	14.75	15.33	14.75	15.33	15.33	15.33
Dietzel's	27.85	25.50	25.86	25.86	25.50	25.86	27.85	27.85
$T_t/°C$	519	528	532	530	534	529	530	536
$T_d/°C$	574	583	587	586	589	584	586	592
$T_{soft}/°C$	712	730	728	722	727	722	723	722
$T_{sink}/°C$	1036	1064	1054	1045	1054	1046	1038	1032
$T_{melt}/°C$	1486	1547	1526	1510	1526	1514	1481	1466
$T_{liq}/°C$	975	995	1005	1002	1010	996	1021	1045
Chem. Res/ml	1.18	0.90	0.73	0.91	0.82	0.93	0.85	0.77
RMS	99.6	103.0	104.4	103.9	105.2	103.3	103.9	106.4
$T_{sink} -$ $T_{liq}/°C$	60.6	68.6	49.3	42.9	44.9	50.6	16.9	–13.6

Table 6.16 Batches and compositions of selected glasses (batch/100 kg glass)

Batch	Origin	3	5	6
Sand/kg	66.95	59.75	59.74	57.75
Feldspar/kg	5.26	16.02	16.02	15.94
Limestone/kg	–	18.59	14.74	11.68
Dolomite/kg	24.00	–	4.15	11.33
Soda/kg	24.21	24.09	24.09	24.09
Sulfate/kg	2.10	?	?	?
Price	30.23	30.61	30.58	30.62
SiO_2/wt%	71.17	71.90	71.90	71.06
Al_2O_3/wt%	0.94	2.20	2.20	2.20
Fe_2O_3/wt%	0.050	0.049	0.048	0.048
CaO/wt%	7.39	10.39	9.53	10.02
MgO/wt%	5.13	0.14	1.00	2.49
Na_2O/wt%	14.95	14.17	14.17	14.18
K_2O/wt%	0.38	1.16	1.16	1.15
T_{melt}/°C	1486	1512	1518	1487
T_{liq}/°C	975	1010	999	1024
Chem. Res./ml	1.18	0.86	0.89	0.81
RMS	99.6	103.4	103.3	103.8
$T_{sink} - T_{liq}$/°C	60.6	37.4	50.6	16.7

Terms used in Table 6.15 are as follows: T_t = transformation point (log (η/dPa s)) = 13.6/°C, T_d = deformation point (dilatometric)/°C, T_{soft} = Littleton softening point (log (η/dPa s)) = 7.6/°C, T_{sink} = sink point (log (η/dPa s)) = 4/°C, T_{melt} = melting point (log (η/dPa s)) = 2/°C, T_{liq} = Liquidus temperature/°C, Chem. Res. = chemical resistivity (consumption in ml of 0.01 N HCl in aqueous leachate), and RMS = relative forming speed after Lyle (6.2).

For glass 1, Dietzel's sum, the contents of RO and R_2O, and the ratio of CaO : MgO are all reduced, and Al_2O_3 increases. This change has led to an increase in melting temperature of 60.5 K and chemical resistance has improved by 25%. The glass is "shortened," so that the forming rate rises by $\sim$3%.

Glass 2 has a slightly increased Dietzel's sum, R_2O and CaO contents, and reduced MgO content. The melting temperature is still high – about 40 K higher than the origin glass, but chemical resistance is improved by about 40%, the glass melt is further shortened, and forming speed RMS increased by about 5%. However, the difference between sink temperature and liquidus temperature is only 49.3 K.

Dietzel's sum has been modified in glasses 3 and 5, which maintain the same R_2O content as the original glass. They present no raw material cost savings but the CaO : MgO ratio varies. Glass 3, using mostly limestone, exhibits an increase in melting temperature of 23.8 K, and chemical resistance increases by about 25%. The forming rate rises 4.4% and the difference between sink and liquidus temperature is an acceptable 42.9 K.

Glass 4 is another variant on this theme. Its melting temperature is $\sim$40 K greater than the initial glass, it is "shorter," the relative speed of forming rises by 5.6%, and the temperature difference between sink and liquidus is 45 K. Reducing Dietzel's sum to below that of the original glass cannot provide the same or lower melting temperatures.

In the last two glasses (6 ,7) the original glass is modified only by changes to the ratio of CaO : MgO and increasing the content of Al_2O_3. Glass 6 lowers the melting temperature by 5 K, increases the speed of forming by 4.4%, and chemical resistance by about 30%; however, the temperature difference from the sink point to liquidus is only 17 K, which may result in devitrification. Similar performance is provided by glass 7, which reduces the melting temperature by 20 K, increases the speed of forming almost 7%, but the liquidus temperature is 14 K above sink point, hence the risk of devitrification occurring is significant.

From the alternatives examined, glasses 3, 5, and 6 may be promising. Their batch charges, final compositions, prices, and properties are calculated in Table 6.16. The relative price of raw batch was calculated according to the prices in Table 6.14. Since the price of soda exceeds the cost of other raw materials, changes to other raw materials have little effect on the batch cost. The new batches vary from the original by 1.2% at most. Table 6.16 shows the calculated properties of glasses 3, 5, and 6. All three batches improve the chemical resistance (by 25–31%) and increase production rate by about 4%. The original glass is already unusually high in alkali and it is not feasible to change the composition to provide a lower melting point and thus save fuel. The most advantageous glass is No.6, which does not change the melting temperature and provides an increased forming speed. A result of its use is a good regulation in the feeder and a good standard of forming.

Fining is provided by sulfate and the optimum amount has been calculated according to Manring and Diken [11], using COD values from Table 6.3. The original glass has a Redox number of 32.1, which means that it is a heavily oxidized flint glass. The glass melt is oversaturated by SO_3 and hence it is susceptible to the formation of secondary bubbles. Therefore, the sulfate addition is reduced to 1 kg sulfate per 100 kg sand (in the case of glass No. 6, 0.58 kg sulfate per 100 kg glass). To decompose excess sulfate will require graphite added in quantities of 0.087/100 kg glass. This provides a Redox number of –24.1, which is a reduced Flint, i.e., the best fining regime. The amount of sulfate used will allow the amount of soda to be reduced as 1 kg sulfate is equal to 0.737 kg soda. Thus the batch for glass No. 6 (per 100 kg of glass) is as follows:

Sand	57.75 kg
Feldspar	15.94 kg
Limestone	11.68 kg
Dolomite	11.33 kg
Soda	23.66 kg
Sulfate	0.58 kg
Graphite	0.087 kg

This calculation is made for dry materials and corresponds to the values in Table 6.16. To verify these calculations it is appropriate to carry out laboratory experiments of meltability and measurement of glass properties with various additions of sulfate.

References

1. A. Smrček and J. Ryšavý, *Proceedings of the International Congress on Glass*, Leningrad, USSR, Volume 3b, 45 (1989).
2. T. Lakatos, L. G. Johansson and B. Simmingsköld, Glass Technol., 13 (3), 88 (1972).
3. T. Lakatos, L. G. Johansson and B. Simmingsköld, Glastekn. Tidskr. 27 (2), 25 (1972).
4. A. Smrček, Tavení skla (Glass melting) Ed. Czech Glass Society, Jablonec, Czech Republic, (2008).
5. P. A. Bingham and M. Marshall, Glass Technol., 46 (1), 11 (2005).
6. A. K. Lyle, Ind. Eng. Chem. Res., 46 (1), 166 (1954).
7. A. K. Lyle and F. V. Tooley, Section 1. Glass composition design and development, in *Handbook of glass manufacture*, Third Edition., F. V. Tooley, ed. Ashlee Publ. Co. Inc., New York, pp. 1–17 (1984).
8. J. P. Poole, Glass Ind., 48, 129 (1967).
9. J. P. Poole, Forming process. Sborník XI. sklář kongredu, Prague, Czechoslovakia, 2, 309 (1977).
10. A. Smrček and F. Voldřich, Sklářské suroviny. Informatorium Prague, (1994).
11. W. H. Manring and G. M. Diken, J. Non-Cryst. Solids, 38 and 39, 813 (1980).
12. M. Bláhová and J. Lederer, Provozní aplikace změn oxidačně redukčních poměrů. Průmyslové sklo, bulletin VÚSU Teplice 2 (1986).
13. A. Smrček, Složení historických skel. In. Drahotová,O.: Historie sklářské výroby v českých zemích I. díl, Praha, s. 412 (2005).
14. C. R. Kurkjian and W. R. Prindle, J. Am. Ceram. Soc., 81 (4), 795 (1998).
15. A. Smrček, Glass Sci. Technol., 78 (4), 173 (2005).
16. A. Smrček, Glass Sci. Technol., 78 (5), 230 (2005).
17. A. Smrček, Glass Sci. Technol., 78 (6), 287 (2005).
18. A. Smrček, Glastech. Ber. 63 (10), 309 (1990).
19. A. Smrček, Glastech. Ber. 65 (7), 192 (1992).
20. M. H. Chopinet, Verre, 8 (6), 38 (2002).
21. G. Keppeler, J. Soc. Glass Technol., 21, 415 (1930).
22. W. König, Glast. Ber. 21 (12), 255 (1943).
23. W. König, Glast. Ber. 21 (12), 260 (1943).
24. M. Kříž, Personal communication, In.:Historie sklářské výroby v českých zemích, II. díl/1, ACADEMIA Prague, (2003).
25. M. H. Chopinet, Verre, 8 (5), 63 (2002).
26. L. Hatakka, Collected papers XIV International Congress on Glass, s.29 (1986).
27. M. B. Volf, Chemical approach to glass, Elsevier, Amsterdam (1984).
28. M. B. Volf, Technická skla a jejich vlastnosti. SNTL, Prague (1987).
29. A. Biavati, G. Branchi and R. Dall'Igna, Verre, 7 (2), 34 (2001).
30. E. Bernardo, G. Scarinci and S. Hreglich, Glass Sci. Technol., 78 (1), 7 (2005).
31. B. Dahmani, Verre, 7 (2), 43 (2001).
32. J. Kocík and J. Nebřenský, Fanderlik, I., Barvení skla, 2. vyd. SNTL, Prague (1978).

Chapter 7
Design of New Energy-Friendly Compositions

Paul A. Bingham

Abstract In order to be energy efficient, environmentally friendly and sustainable, commercial glass production in the 21st century must evolve and some of the technologies and methodologies that will make this possible are discussed. Development and implementation of energy-efficient and environmentally friendly soda–lime–silica glass compositions are discussed in terms of environmental and legislative requirements; the reduction of melting energies and atmospheric emissions; glass properties and the effects of individual glass components and raw materials; and technologies that can help glassmakers to meet new requirements. This in-depth treatment provides detailed step-by-step analysis, with appropriate examples, of the opportunities for compositional reformulation, new raw materials, new melting and abatement technologies, and some of the practical and economic effects that such changes will provide.

Keywords Composition · Conductivity · Corrosion · Design · Devitrification · Durability · Efficiency · Energy · Environment · Glass · Liquidus · Melting · Models · Pollution · Properties · Raw materials · Refining · Reformulation · Soda-lime-silica · Viscosity

7.1 Introduction

Commercial soda–lime–silica (SLS) glass has evolved continuously and has hitherto been the cheapest acceptable formulation given the available raw materials. However, global uncertainties regarding fuel supply, energy costs, and ever more stringent environmental legislation are making glassmakers think again about the way in which glass is made. Environmental legislation, taxation, and changing customer attitudes are driving glassmakers to reconsider all measures that could reduce their energy consumption and environmental impact. These measures can take a number of forms: some may be beyond their direct control (for example, the

P.A. Bingham (✉)
Department of Engineering Materials, The University of Sheffield, Sheffield S1 3JD, UK
e-mail: p.a.bingham@sheffield.ac.uk

F.T. Wallenberger, P.A. Bingham (eds.), *Fiberglass and Glass Technology*,
DOI 10.1007/978-1-4419-0736-3_7, © Springer Science+Business Media, LLC 2010

return and reuse of glass containers – a system that has worked well in many countries for decades but is still not implemented globally). Glassmakers can, however, control the way that they make glass and hence control energy use and environmental impact. Measures include increased cullet use, chemical reformulation, new or alternative raw materials, batch consolidation, furnace and burner design changes, furnace insulation, and waste heat recovery.

This work considers methods of reducing the energy demand and environmental impact of glassmaking. The available technologies and opportunities for achieving this are discussed, with particular attention to raw materials and compositional modification. Some options have been available for many years while others have only been developed recently. For various reasons (chiefly economic) many technologies have yet to be widely implemented. However, all methods, new and old, should be evaluated as part of any strategy for the manufacure of glass in the 21st century.

7.2 Design Requirements

While commercial SLS glass compositions have remained broadly similar for many years, a continual process of development and refinement has in fact been taking place. To date, this process has been driven by new furnace designs and by the availability and cost of raw materials. However, raw material costs are no longer the only factor driving glass formulation and a re-examination of the pertinent factors is therefore timely.

The first step in the design of a new glass composition is to prepare a detailed specification of property requirements. Specifications may lack technical detail and be commercial in nature, for example "same glass but at lower cost." In order to meet all of the design requirements for SLS glass, a careful balancing act must be performed between competing factors. Lyle and Tooley [1] justified the standardization of glass compositions on the basis that good glasses meet the same criteria:

- Make use of available raw materials
- Decrease costs
- Improve melting and fining characteristics
- Increase machine productivity
- Improve chemical durability and resistance to weathering
- Provide color and light-protective characteristics
- Improve resistance to thermal shock
- Achieve combinations of these general specifications

Environmental legislation has become far more stringent and emission limits will become still tougher in the future. Penalties for non-compliance can be severe and glassmakers are now obliged to give more consideration to their environmental impact. We should therefore add a further consideration to the design criteria for the manufacture of SLS glass in the 21st century:

- Comply with environmental legislation

7.2.1 Commercial Glass Compositions

The composition of SLS glass varies with application. SLS glasses that are produced in high volumes are container, flat (float), domestic, and A-glass. Their compositions generally fall within the ranges described in Table 7.1. Minor components (not shown) presently include transition metals, Se, Cl, Zr, P, Ba, Zn, and/or certain lanthanides.

Table 7.1 Typical commercial SLS glass compositions

Weight%	Container	Float	Domestic	A-glass
SiO_2	70–73	70–74	70–74	71–73
Al_2O_3	1–2	0–1	1–2	0.5–2.5
B_2O_3	0	0	0	0–0.5
Na_2O	12–15	12–14	12–14	12–15
K_2O	0–1	0–0.5	0–2.5	0–1.5
MgO	0.5–2	2–4	1–2	1–4
CaO	10–12	7–10	7–9	9–11
Fe_2O_3	0–1	0–1	0–0.05	0–0.5
TiO_2	0–0.1	0–0.1	0–0.05	0–0.1
SO_3	0–0.5	0–0.5	0–0.5	0–0.5
Cr_2O_3	0–0.3	0–0.3	0	0

7.3 Environmental Issues

The efficiency of SLS glass melting in terms of specific energy consumption (SEC) and specific emissions generation has improved greatly over the past century. However, further improvements are still possible and relevant considerations are summarized in this section.

7.3.1 Specific Energy Consumption

Specific Energy Consumption (SEC) is fundamental to the determination of the energy savings available to glassmakers. The theoretical minimum SEC for producing one tonne of SLS glass from raw batch is 2.3–2.6 GJ [2–5]. As shown in Fig. 7.1, furnace efficiencies have improved markedly over the past century. The average SEC in practice is now ~5 GJ/tonne; however, this is still twice the theoretical minimum. Substantial scope therefore exists for further improvements to be made in furnace, burner, insulation and, fuel technology.

7.3.1.1 Energy Efficiency

The total energy supplied to a glass furnace, E_T, is consumed in three processes: E_I, E_{II}, and E_{III}. Changes in furnace design, insulation, fuel, burner, and heat exchange technologies largely determine E_{II} and E_{III}. However, E_I is principally a function of raw materials, glass composition, and temperature.

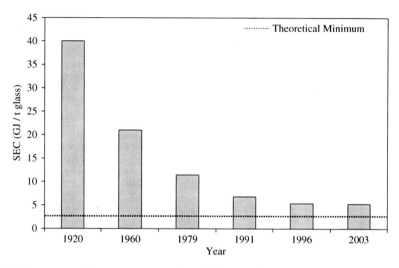

Fig. 7.1 Average specific energy consumption (SEC, gross basis) for SLS glass furnaces. Data sources: [2, 3, 6, 7]

(E_{I})	Energy to melt the batch and maintain the glass melting temperature.
(E_{II})	Energy lost from the furnace structure.
(E_{III})	Energy removed by the exhaust gases.

Most large furnaces now use regenerative heat recovery systems [2]. Waste gas temperatures after regeneration can be ~550°C, which still constitutes a 40% loss of total heat input. Steam heat recovery cannot be widely used since steam has few applications in glass factories. However, when waste heat from several furnaces can be combined, power generation is feasible [2]. Batch or cullet preheating by waste gas typically raises batch temperatures to 275–375°C, giving energy savings of 8–15% [8–10] and using consolidated or treated batch further increases efficiency [9, 11]. SEC of SLS glass furnaces was recently studied by applying various energy-saving technologies to a 250 tpd furnace, as shown in Fig. 7.2.

The technology types listed in Fig. 7.2 are as follows:

(A) Recuperator with air preheat to 770°C.
(B) Ceramic recuperator with air preheat to 1100°C.
(C) Regenerator/twin-bed burners with air preheat to 1280°C.
(D) Air preheat to theoretical maximum of 1480°C.
(E) Regenerator with complete batch preheat to 350°C.
(F) Oxy-fuel, no preheat.
(G) Oxy-fuel with batch preheat to 250°C.
(H) Thermochemical recuperator.
 (I) Regenerative with 100% cullet, melt at 1450°C.

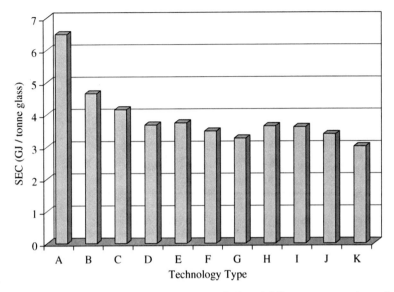

Fig. 7.2 Specific energy consumption following installation of different energy-saving technologies. Data adapted from [8]

(J) Regenerative, 100% cullet, melt at 1350°C.

(K) Regenerative, 100% cullet, melt at 1450°C and cullet preheat to 350°C.

The authors of the above study, Beerkens and Muysenberg [8], concluded that reducing flue gas heat losses and increasing batch cullet levels were among the best methods of reducing SEC. The above technologies have been evaluated and discussed in depth [2, 6, 9, 12, 13].

7.3.2 Atmospheric Emission Limits

The operation of a glass melting furnace results in several gaseous and particulate wastes, and environmental release of these wastes is controlled by legislation. The abatement of atmospheric emissions can be both problematic and expensive. Table 7.2 shows the relevant European Union (EU) emission limits as of October 2006. Note that SO_2 limits vary depending upon fuel type.

7.3.3 Pollution Prevention and Control

EU guidance details the Best available technologies (BAT) for pollution prevention and control [2, 12] and Drake [13] provides discussion.

7.3.3.1 Furnace Design

Furnaces run continuously for ~10 years with practical lifetimes of 5–14 years [6]. Furnace design is critical to efficiency, as summarized in Table 7.3. Regenerators

Table 7.2 EU limits on atmospheric emissions from glass melting operations producing >20 tonnes of glass per day [12]

Pollutant	Maximum allowable concentration/mg m^{-3}
NO_x annual average	500
NO_x daily average	700
Particulate matter	30
SO_2 – No recycling of EP/filter dust	Gas 500 Oil 1200
SO_2 – Recycled EP/filter dust	Gas 800 Oil 1500
Chlorides (as HCl)	30
Fluorides (as HF)	5
As, Co, Ni, Se, Cr(III,VI), Sb, Pb, V	5

Table 7.3 Best available technology furnace designs

Furnace	Principle of operation	Advantages	Disadvantages
Cross-fired regenerative	Waste gas preheats regenerative chambers via which combustion air is preheated	Combustion air preheated to 1400°C. High efficiency. Economy of scale	Tendency to produce higher NO_x emissions than end-fired regenerative
End-fired regenerative	Similar to above	Lower cost than above	Less T control than cross-fired
Recuperative (unit melter)	Waste gases preheat combustion air via a heat exchanger	–	Combustion air preheated to ~800°C. Small furnaces
LoNO$_x$ melter	Preheated combustion gas and raw materials	Lower temperatures. Low NO_x emissions	>70% cullet required in raw materials
Electric melter	Resistive heating by submerged electrodes	Low waste gas volume. Reduced abatement. High efficiency	High cost of electricity. Net CO_2 balance less favorable
Pot furnaces, day tanks	Not continuous, specific batches	–	Small (<20 tpd)
Flex melter	As above. Combination of electricity and gas	–	Small (<20 tpd)
Oxy-fuel fired	Higher percentage of O_2 used than 21% from air	Reduces waste gas volume by up to 85%. Potential for reduction of on-site fossil fuels	Waste gas cooling. Needs infrastructure, electricity/CO_2 penalty

Adapted from [12]

and recuperators recover heat from exhaust gases and use it to preheat cold incoming air prior to combustion [10, 14, 15]. These technologies have contributed to efficiency, and regenerative furnaces in particular are now widely used for large-scale glass manufacture [2, 16, 17].

7.3.3.2 Carbon Dioxide

Industry guidance [2, 3, 12] stresses the need to reduce energy consumption and CO_2 emissions. While CO_2 is not yet a proscribed emission its generation is directly linked to global warming. Carbon trading schemes and carbon taxes are now in place or being considered in many countries. The CO_2 emitted by the glass industry arises principally from three sources:

(a) Fuel CO_2 results from combustion of fossil fuels either directly in the furnace or in fossil fuel-fired power stations to provide electrical boost energy. Assuming a SEC of 5 GJ/t (=1389 kWh/t) glass, and a conversion rate of 0.19 kg CO_2/kWh [18], a 250 tpd gas-fired furnace with no electric boost will generate (250 × 1389) × 0.19 = 66 tonnes CO_2/day.

(b) Batch CO_2 is generated by decomposition of carbonate raw materials and carbonaceous batch components, e.g., cullet contamination and reducing agents such as coke. Based on recent factory input data [19], average batch CO_2 contents are ~17% for container glass and ~28% for flat glass. Therefore 250 tpd furnaces will typically generate (250 × 0.17) = 42.5 tpd batch CO_2 for container glass and (250 × 0.28) = 70 tpd batch CO_2 for flat glass.

(c) The glass industry emits CO_2 indirectly through its use of electricity, the majority of which is generated at fossil fuel-fired power stations. Electric boost is typically rated at 5% of furnace energy input for SLS container glass [2], additional to this are site electricity demands.

The solubility of CO_2 in SLS glass is ~5 g m^{-3} at 1500°C and 1 atm [20]. Decreases in furnace temperature increase the CO_2 solubility of glass but this will have a negligible effect on CO_2 retention as a fraction of CO_2 generated. However, changes to raw materials or glass composition can reduce batch CO_2 levels and furnace temperatures, and hence CO_2 emissions.

7.3.3.3 Oxides of Nitrogen

Oxides of nitrogen (NO_x) are greenhouse gases which contribute to global warming. NO_x is generated during glass melting via three mechanisms:

(1) Thermal NO_x is formed by high-temperature oxidation of nitrogen. Molecular nitrogen and oxygen in combustion air dissociate into their atomic states and participate in several reactions (extended Zeldovich mechanism). Thermal NO_x increases exponentially with increasing temperature (Fig. 7.3). At temperatures above ~1,100°C it is the dominant NO_x generation mechanism, becoming yet more dominant when combustion air is preheated.

(2) Fuel NO_x originates from N-bearing fuels (coal, oil). Nitrogen is released as free radicals, forming free N_2 or NO by oxidation during the initial stages of combustion and the combustion of nitrogen present in the resulting char.

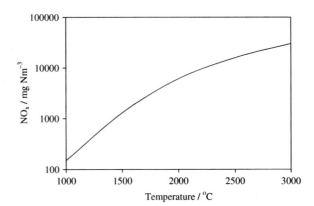

Fig. 7.3 Generation of thermal NO_x as a function of furnace temperature. Data adapted from [2]. Note that Nm^{-3} denotes Normal m^{-3} (see List of Abbreviations)

(3) Prompt NO_x is formed by the reactions between N, O, and hydrocarbon radicals. This process consists of hundreds of reactions. Prompt NO_x is generally less important than thermal NO_x at glass furnace temperatures.

Solubility of N_2 and NO_x in SLS glass at 1500°C is of the order of 10^{-5} wt% [20]. As with CO_2, the dissolved N_2 content is so low that increasing the N solubility in glass through reformulation has a negligible impact. However, large decreases in NO_x emissions can be achieved using a range of technologies. As with reformulation for energy efficiency, little consideration has yet been given in the literature to reformulation of SLS glass for NO_x reduction, yet a reduction in upper furnace temperature by only ∼30°C would decrease thermal NO_x generation by 20% [21].

The best available technology (BAT) guidance recommends both primary and secondary techniques for NO_x control [2, 12]. Primary measures reduce NO_x formation by removing the simultaneous presence of nitrogen and oxygen at high temperatures. Secondary measures allow NO_x to form during combustion and aim to chemically neutralize it. These technologies are discussed in depth elsewhere [2, 3, 12, 22, 23] and summarized here.

(A) Primary techniques

- Prevention of air ingress into furnace. 10% NO_x reduction.
- Sealed burners. Prevent inspiration of cold secondary air and remove oxygen from flame root. Reduces NO_x by up to 15%.
- Reduced air/fuel ratio. 40% NO_x reduction is possible.
- LoNOx® melter. Features shallow bath refining and raw material preheating to achieve emissions of < 1 kg NO_x per tonne of glass melted.
- FENIX (Faibles Emissions de NO_x) system. Involves optimization of combustion conditions.
- Oxy-fuel melting. Fuel is burned in pure O_2, avoiding heating of N_2 in combustion zone. Reduces NO_x by up to 90%.

- Low NO_x burners. 30% NO_x reduction is possible.
- Staged combustion. Can be applied by burner block sealing; waste gas recirculation; or staged provision of combustion air. Systems are inexpensive to install and run but do not reduce particulate emissions and temperature control is more difficult. Reductions of 35% NO_x feasible.

(B) Secondary techniques

- 3R process. The 3R (reaction and reduction in regenerators) process involves injection of fuel into the waste gas. This fuel pyrolyses forming radicals which react with NO_x to form $N_2 + H_2O$. Reductions in NO_x of up to 85% may be possible; however, this method typically carries a fuel penalty of 4–8%. Reburn technologies provide NO_x reductions of up to 60%.
- Selective non-catalytic reduction (SNCR). Ammonia is reacted with waste gas. Most effective at temperatures of 800–1100°C. Reductions in NO_x of ≤70% may be possible. Practical difficulties restrict SNCR to recuperative furnaces.
- Selective catalytic reduction (SCR). Utilizes a catalyst, commonly titanium or vanadium oxides impregnated into a metallic or ceramic host material, to promote the reduction of NO_x by injected ammonia, as with SNCR. NO_x reductions of up to 70% may be possible. Drawbacks include high capital and operating costs, and poisoning and blockage of catalysts. Electrostatic precipitators (EPs) are required to protect the catalyst from dust in the gas stream.

7.3.3.4 Oxides of Sulfur

Atmospheric sulfur dioxide directly affects vegetation, contributes to acid rain, and can exacerbate asthma or lung disease. SO_x emissions by the glass industry are generated from two sources: sulfur-containing fuels and sulfur-containing batch materials such as sulfates. Natural gas contains little sulfur (typically 5 mg m^{-3}) whereas fuel oils contain considerably more (heavy fuel oil < 1 wt% S and gas oil <0.1 wt% S). Switching from oil to gas fuel therefore greatly reduces SO_x generation. Current emission limits from oil- and gas-fired furnaces reflect this difference in fuel sulfur content (see Table 7.2).

The BAT for reducing SO_x emissions is again a combination of techniques including the use of low-sulfur fuels, minimizing the levels of batch sulfate, recycling of filter dusts, and installation of waste gas treatment plants. These plants generally consist of electrostatic precipitators (EPs) or fabric (bag) filters to remove particulates and dry, semi-dry, or wet scrubbing to remove the gaseous species SO_3, SO_2, HF, and HCl and heavy metals such as Se and Pb compounds [2, 12, 13]. The scrubbing medium used is usually $Ca(OH)_2$, Na_2CO_3, NaOH, or $NaHCO_3$ [24]. For acid gas removal, dry scrubbing typically removes 50–60%, semi-dry scrubbing removes 60–85%, and wet scrubbing can remove 85–99% [2]. One primary method for SO_x reduction that has received surprisingly little interest is the use of one or more different refining agents including sulfates. Through the carefully controlled release of O_2 and SO_2 throughout melting and refining, it may be possible to minimize batch sulfates and SO_x emissions, i.e., achieve the same quality of refining with less sulfate (see Sections 7.4.4.1 and 7.5.11).

7.3.3.5 Volatilization and Particulates

Volatilization during glass melting leads to the formation of particulates, increased corrosion of refractories, plugging of regenerators and flue gas systems, and depletion of volatile components from the melt [25]. The sub-micron-sized (0.03–0.5 μm) dust particles resulting from volatilization and their subsequent condensation from different glass furnaces producing different glass types are shown in Table 7.4. Deposits from SLS furnaces consist mainly of alkali sulfates although Se compounds (decolorizing agents) are also present in samples from container glass furnaces. Species evolved from borosilicate and E-glass furnaces include $NaBO_2$ and HBO_2 [26].

Table 7.4 Comparison of dust emissions from glass furnaces

Emission	Container SLS	Float SLS	Borosil	E-glass
Emission/mg m^{-3}	125–200	120–180	1500–2000	800–1400
Specific emission/kg t^{-1}	0.2–0.4	0.3–0.45	0.5–0.9	3.5–5.0
SiO_2/%	0.5–2.0	0.5–1.5	< 2	1.0–3.0
Al_2O_3/%	0.2–1	< 1	< 2	< 1.5
B_2O_3/%	< 1	< 0.1	35–40	55–70
MgO/%	1.0–3.0	1.0–3.0	1.0–3.0	0–1
CaO/%	2.0–7.0	2.5–4.0	2.0–5.0	2
Na_2O/%	25–40	35	25–30	2.0–5.0
K_2O/%	1.5–2.5	0–1	5.0–10	10.0–20.0
SO_3/%	45–55	45–55	8.0–15	2.0–5.0
Cl, F/%	< 1	< 1	< 1	2

Data adapted from [25]

Cable and Chaudhry [27] studied the effects of glass composition, temperature, and pressure on volatilization from three SLS glasses. The relationship between M, the total loss of volatile matter, and time t can be described by a model involving melt diffusion and a first-order reaction for surface losses. A linear relationship exists between $\log (M)$ and $(1/T)$, when M is the total loss of volatile matter and T is absolute temperature. Owing to the duality between volatilization and particulate emissions this relationship therefore also applies to particulates [28]. Some of the results of Cable and Chaudhury [27], shown in Table 7.5, demonstrate that the effects of furnace temperature on volatilization and particulate generation are substantially greater than the effects of changing composition when the two are considered separately. The predicted melting temperature, the temperature at which $\log (\text{viscosity/dPa s}) = 2$ (see Section 7.4.1) changes by more than 100°C over the range of glasses studied, yet the increase in volatility can be mitigated by a decrease

Table 7.5 Glass compositions and total volatilized mass loss, M, as a function of temperature in a static air atmosphere

Glass	Glass G1	Glass G2	Glass G3
SiO_2/wt%	74.0	69.95	68.8
Na_2O/wt%	16.0	20.67	20.3
CaO/wt%	10.0	9.46	9.28
Al_2O_3/wt%	0	0	1.69
T (log $(\eta/dPa\ s) = 2)/°C$	1415	1302	1328
M (mg cm^{-2}) at (time/h)$^{\frac{1}{2}}$ = 10, $T = 1400°C$	8.5	18.5	17
M (mg cm^{-2}) at (time/h)$^{\frac{1}{2}}$ = 10, $T = 1450°C$	13	35	33

Data adapted from [27]. Viscosity-at-melting data modeled using [29]; see Section 7.4.1.1

in temperature of roughly half that amount. Therefore, from an environmental perspective, any increase in volatilization arising from changes in glass composition can be more than offset by the reduction in furnace temperatures that such compositional changes would make possible. This means that the net effect of changing the glass and/or batch compositions to be more energy friendly can be to reduce volatilization and particulate generation.

Particulates arise from three key sources:

(1) Volatilization of glass components
(2) Carryover of small batch particles
(3) Fuel ash

Typical particulates are alkali-rich but may also contain compounds of B, S, halides, transition metals, Pb, and Se [24, 25, 30–33]. Specific emission is strongly dependent on temperature, hence the higher emissions from borosilicate and E-glass furnaces as shown in Table 7.4. A decrease of 100°C in furnace temperature can reduce volatilization and particulate emissions by ~50% (Fig. 7.4). The BAT for minimizing particulate formation and emission includes raw material modification (chemically and physically), reduction of melting temperature, modification of burner position, gas firing, and installation of treatment plant such as a fabric (bag) filter or EP.

Alkali volatility increases with alkali ionic size. At equimolar alkali contents, volatilization losses at 1400°C were 3.0 Li_2O, 5.1 Na_2O, and 12.6 K_2O [26]. Fluorides and chlorides arise from batch impurities, particularly from Na_2CO_3, Na_2SO_4, and fluoride or chloride melting accelerants/fining agents, if added. Heavy metals such as Pb and As may arise from external cullet contamination. The transition metals Cr, Ni, and Co are widely used as colorants and decolorizing agents in SLS glass manufacture. Selenium is deliberately added to clear container glasses

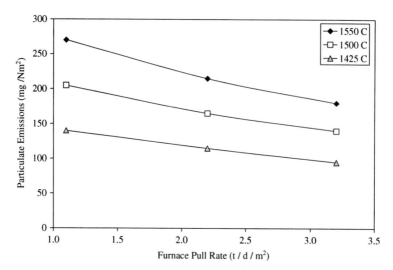

Fig. 7.4 Effects of furnace pull rate and temperature on particulate emissions. Data adapted from [2]. Note Nm^{-3} denotes Normal m^{-3} (see List of Abbreviations)

as a decolorizing agent and to some flat glasses (particularly those for solar control) as a colorant. The BAT for controlling emissions of fluorides, chlorides, and heavy metals is a combination of acid gas scrubbing, and EP or fabric (bag) filtration. Reduction in melting temperatures and careful selection of raw materials can also contribute to reduced emissions of these components.

Significant steps have been taken toward the development of accurate mathematical models for evaporation processes in industrial glassmaking (for example, see [25, 34]). Such models can help to steer reformulation of glass batches/compositions as part of integrated glass optimization strategies. Ideally it will be possible to model the effects of changes to glass batches and the final glass composition in order to reduce emissions, refractory corrosion, and formation of glass defects through process optimization.

7.4 Fundamental Glass Properties

In order for any new glass to meet the design requirements described in Section 7.2 and provide the production efficiency discussed in Section 7.3, it is necessary to understand the relevant melting requirements and the pertinent composition – property relationships. Often these can act in opposition, requiring the design of compositions providing the most acceptable balance between competing factors. Optimization is therefore a key aspect of compositional design. One vital aspect of this "design toolbox" is the availability of models that allow prediction of key glass properties based on their chemical composition. For the key properties discussed in this section a range of models have been developed; some are more accurate, and some less so. A web-based repository of a number of these models in computer spreadsheet form has recently been made available by Fluegel [35].

7.4.1 Viscosity–Temperature Relationship

The viscosity–temperature (η–T) relationship is the principal factor determining the melting temperature, and therefore the melting energy requirement, of glass. At typical melting temperatures, SLS glass has a viscosity of $\sim 10^2$ dPa s or P; this is roughly the same viscosity as molasses at room temperature. The melting point, T_M, is the average temperature required to produce a good quality of glass in a reasonable time (hours). The fiber-forming temperature T_F, corresponding to a viscosity of 10^3 dPa s, is the temperature at which fibers are typically formed in the fiberglass industry. The working point, T_W, corresponding to a viscosity of 10^4 dPa s, is the upper temperature at which the glass can be formed into shapes, for example, bottles and jars. The Littleton softening point, T_S, corresponds to a viscosity of $10^{7.65}$ dPa s and is the temperature at which a fiber can deform at a specific rate under its own weight. T_S is the low temperature boundary for shape forming. The annealing point, T_A, 10^{13} dPa s, and strain point, T_{Str}, $10^{14.5}$ dPa s, are the upper and lower boundaries of the glass transformation range, which determines annealing temperatures. Varshneya [36] defined the annealing point as the temperature at which a glass releases 95% of its internal stress within 15 min. Critical viscosities for SLS glassmaking and typical corresponding temperatures are described in Table 7.6 and illustrated in Fig. 7.5.

Two simple but important methods of measuring the η–T profile of glass and its interactions with other properties are used in the glass industry [1, 37, 38]. These methods are used only for indicative purposes.

(1) Relative machine speed (RMS), the relative average speed at which articles can be produced using a particular composition. Compositional changes have increased RMS (see Table 7.7):

$$\text{RMS} = \frac{S - 450}{(S - A) + 80} \qquad (7.1)$$

Table 7.6   Viscosity set points for a typical SLS glass

Log (η/dPa s)	Term	Temperature/°C
2	Melting temperature	1440
3	Fiber-forming temperature	1180
4	Working point	1020
7.65	Softening point	740
13.0	Annealing point	540
14.5	Strain point	500

Fig. 7.5
Viscosity–temperature (η–T)
curve for typical SLS glass

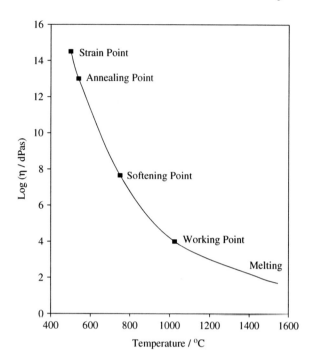

Table 7.7 Relative machine speed (RMS) of Russian SLS container glass since 1950

Year	SiO_2/wt%	R_2O/wt%	$RO+R_2O_3$/wt%	RMS/%
1950	72.1	15.5	12.5	103.0
1964	72.0	15.0	13.1	104.0
1972	72.4	14.7	13.2	106.0
1977	72.2	14.4	13.3	107.0
1998	72.0	14.0	14.0	108.5

Data adapted from [38]

where S = softening point/°C, defined as log (η/dPa s) = 7.65 and A = Annealing Point/°C, defined as log (η/dPa s) = 13.0.

(2) Working range index (WRI), the temperature difference between the softening point S and the annealing point A. WRI is used as an indicator of the working range and should not be confused with the actual working range. For most SLS container glasses, WRI > 160°C [38]:

$$WRI = (S - A) \tag{7.2}$$

7.4.1.1 Viscosity Models

Several models exist for viscosity–temperature behavior of SLS systems [39]. A common approach uses the Vogel–Fulcher–Tammann (VFT) equation which describes the viscosity of Newtonian fluids (7.3):

$$\text{Log } \eta = A + \frac{B}{T - T_0} \tag{7.3}$$

where η = viscosity, A and B are temperature-independent constants, T_0 is a temperature-dependent constant, and T is temperature. Several empirical models use the VFT equation [29, 35, 40–50]. Through data regression, values have been obtained for A, B, and T_0. The most applicable models are those developed by Lyon [42], Cuartas [47], Lakatos et al. [29, 43–46], and Fluegel [50]. The merits of each model are a subject for debate elsewhere; however, the models of Lakatos et al. have been widely used and comparisons by others [35, 50, 51] confirm their accuracy. Lakatos' VFT factors [29] are summarized in Table 7.8. Predicted and measured values may differ due to volatilization, measurement errors [50], and model imperfections. Figure 7.6 shows the effects of replacing 1 wt% SiO_2.

Table 7.8 VFT factors from Lakatos [29] as weight-for-weight replacement of SiO_2 (note that B_2O_3 has first- and second-order terms)

	Lower wt%	Upper wt%	A	B	T_0
SiO_2	59.52	77.02	−1.713	6237.013	149.4
MgO	0	6.0	−5.890	5621	−212
CaO	4.48	13.0	−0.64	−6063	771
BaO	0	16.54	−0.26	−2103	109
PbO	0	12.22	0.50	−2544	82
ZnO	0	9.38	−1.60	−376	96
Li_2O	0	3.0	3.18	−11518	−1329
Na_2O	10.41	17.0	1.62	−6601	50
K_2O	0	8.7	−0.66	−541	−236
Al_2O_3	0	8.26	0.87	1521	140
B_2O_3	0	14.37	4.65	−15511	1203
$[B_2O_3]^2$	N/A	N/A	−16.27	40999	−2765

7.4.2 Devitrification and Crystal Growth

If, during melting and forming, glass is subject for a sufficient length of time to temperatures that produce crystallization, then manufacturing problems may arise. Liquidus temperature (T_{Liq}), the temperature below which crystal growth is positive, and crystallization rates within the crystallization range are both functions of glass composition. The standard method of measuring T_{Liq}

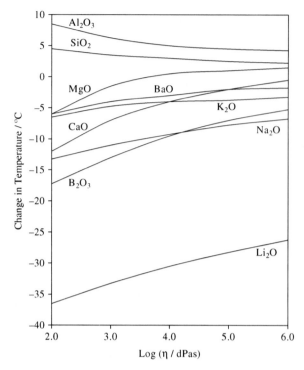

Fig. 7.6 Effect of increasing (by weight) 1 oxide/100 SiO_2 on temperature of different viscosities. Data adapted from [29]

requires a temperature-gradient furnace. Knowledge of the phase diagram for each glass-forming system and the effects upon phase relationships of compositional changes are vital to the accurate measurement and prediction of crystallization behavior.

Phase relationships in multi-component commercial SLS glass systems are more complex than in simple SiO_2–Na_2O–CaO ternaries since minor components such as MgO, Al_2O_3, K_2O, Fe_2O_3, and SO_3 all modify and shift phase field boundaries and temperatures. Figure 7.7 illustrates the phase diagram for the ternary SiO_2–CaO–Na_2O system. The shaded region broadly describes typical SLS glass compositions. Container glass compositions usually occur within the devitrite ($Na_2O \cdot 3CaO \cdot 6SiO_2$) primary phase field and float glasses occur close to the silica–devitrite–wollastonite (SiO_2–$Na_2O \cdot 3CaO$ $6SiO_2$–$CaO \cdot SiO_2$) boundary [52].

7.4.2.1 Methods of Avoiding Devitrification

The liquidus temperature, T_{Liq}, of container and float glass is typically 990–1050°C [21, 52, 54, 55]. T_{Liq} should be at least 10–20°C below the working point (T_W, for

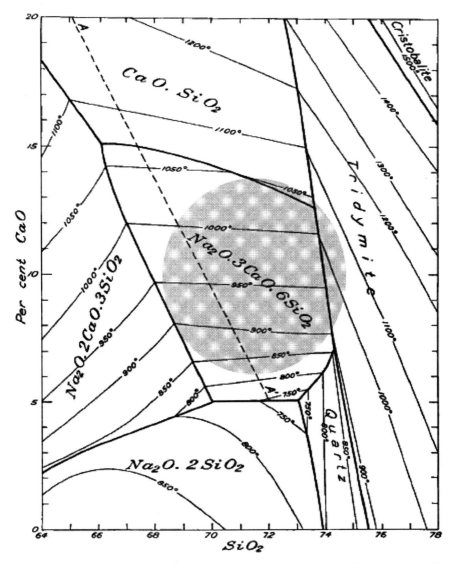

Fig. 7.7 Phase diagram (wt%) for part of the SiO₂–CaO–Na₂O system [53]. Reproduced with permission of Wiley-Blackwell Publishing Ltd. *Shaded region* broadly describes SLS glass compositions (Na₂O is by difference)

container glass this is the temperature at which log $(\eta/dPa\ s) = 4$ to minimize the possibility of crystal nucleation and growth during forming [21]. The devitrification index (D) has long been used to estimate the likelihood of devitrification problems occurring [1, 38]. A positive value of D indicates relative freedom from devitrification while a negative value of D indicates increasing likelihood of devitrification, particularly if the glass is fed to the forming machine at relatively low temperatures or high viscosities, for example, during the manufacture of large articles:

$$D = \text{WRI} - 160°C \tag{7.4}$$

where WRI = working range index, as defined in (7.2). Values of D vary; however, +15 is now common. Zhernovaya et al. [38] reported that in 1977 values of D for container glasses were +29 (USA) and +39 (USSR).

ΔT_{FL}, the difference between forming temperature (T_F) and liquidus temperature (T_{Liq}) is an important criterion in the development of SLS glass formulations. This criterion has been used successfully in the reformulation of container glass [21] and fiberglass [56–60] compositions. According to phase relations in, for example, the devitrite ($Na_2O \cdot 3CaO \cdot 6SiO_2$) primary phase field, the replacement of 1 wt% SiO_2 by Na_2O could be expected to have only a small effect on T_{Liq}. However, T_F decreases, therefore ΔT_{FL} decreases. Compositional changes may be sufficient to produce a negative value of ΔT_{FL}, increasing the risk of devitrification problems during forming. Whether this would actually cause problems is dependent on the particular conditions for a given furnace, forming, and handling operation. This author has received anecdotal evidence that some manufacturers have operated with values of ΔT_{FL} close to zero. There exists, therefore, an element of uncertainty surrounding devitrification and forming that can only be fully assessed on an individual basis. Jones and Ohlberg [61] stated that the preferred ΔT_{FL} for flat glass production is +50°F. Typical devitrification problems in production can be very expensive and manufacturers require surety of the avoidance of devitrification before any new composition can be implemented. The criteria ΔT_{FL} and D provide useful methods for the estimation of the possibility of devitrification before any glass is melted.

As SiO_2 content increases above ~74 wt% or decreases below ~68 wt%, T_{Liq} increases rapidly as composition shifts away from the devitrite ($Na_2O \cdot 3CaO \cdot 6SiO_2$) primary phase field (see Fig. 7.7). Such increases in T_{Liq} will dramatically increase the likelihood of devitrification. This suggests that $74 > SiO_2 > 68$ wt% provides sensible approximate limits for many of the more practicable reformulations of SLS glass. Should the RO content remain constant, R_2O ($Na_2O + K_2O + Li_2O$) contents of ~12–20 wt% accompany these SiO_2 limits, although economic considerations currently determine that R_2O contents of ~11–14 wt% are most common. Application of the same principle to the RO (divalent oxides) content indicates that upper RO limits of ~13 to ~18 wt% may be practicable, although the exact mixture of monovalent and divalent oxides determines ΔT_{FL}. This initial assessment of the SiO_2–CaO–Na_2O ternary phase diagram therefore presents an approximate compositional range within which reformulation efforts should focus.

7.4.2.2 Liquidus Models

Within a defined compositional range, properties such as viscosity and chemical durability vary smoothly and can be modeled using a single mathematical description. However, this is not the case for devitrification, for which smooth changes only occur within each primary phase field. It is therefore more difficult to accurately model liquidus temperature using a single function, as separate functions may apply within each primary phase field. The relevant part of the phase diagram, shown in Fig. 7.7, illustrates sharp discontinuities at the boundaries between devitrite ($Na_2O \cdot 3CaO \cdot 6SiO_2$) and (i) SiO_2 (quartz, tridymite, and cristobalite fields) and (ii) $Na_2O \cdot 2CaO \cdot 3SiO_2$. The boundary between devitrite and wollastonite ($CaO \cdot SiO_2$) provides a more linear change.

Hrma et al. [52] suggested that the best strategy for modeling phase relations is to begin by using a global liquidus model, followed by the development of local models as appropriate. Global models cover more than one primary phase field, whereas local models each apply to only one primary phase field. Global models are certainly the more practical models to the glass technologist, although the accuracy of local models may be greater within a given phase field. Dreyfus and Dreyfus [62] developed a non-linear model using neural network processing. Liquidus models have been reviewed and discussed by Hrma et al. [52] and by Bonetti [63, 64]. These reviews both indicate that Cuartas' [47] model is among the more accurate of the global models when applied to commercial SLS compositions while retaining compositional flexibility. Cuartas' model [47], described in wt% in (7.5), covers a wider range of both compositions and components than the majority of the other models, including terms for SiO_2, B_2O_3, Al_2O_3, Fe_2O_3, MgO, CaO, BaO, Na_2O, K_2O, MnO, Cr_2O_3, and SO_3. However, it does not provide terms for Li_2O and some other oxides, and its accuracy is unproven:

$$T_{Liq} = [(1170.87 \times SiO_2) + (1848.99 \times Al_2O_3) + (897.33 \times Fe_2O_3) + $$
$$(1503.91 \times CaO) + (652.96 \times MgO) - (2011.99 \times Na_2O) - $$
$$(1606.03 \times K_2O) - (500.61 \, BaO) - (756.77 \times B_2O_3) - $$
$$(1148.21 \times MnO) - (3473.44 \times Cr_2O_3) + (1832.68 \times SO_3)]/SiO_2$$

(7.5)

Lakatos and Johansson [46] developed a global model for T_{Liq}, based on 30 compositions, which incorporate many commercial SLS compositions. However, components are limited to SiO_2, Al_2O_3, Na_2O, K_2O, MgO, and CaO. The model is described in (7.6); components are considered in wt%. Standard deviation is 9.9°C and the model's validity is limited to 3.5 wt% MgO:

$$T_{Liq} = 1002.0 + (14.54 \times Al_2O_3) - (17.11 \times Na_2O) - (8.73 \times K_2O) + $$
$$(22.51 \times CaO) + (7.58 \times MgO)$$

(7.6)

Other models include those by Sasek et al. [65], Babcock [66], Backman et al. [67], Karlsson et al. [68], and Hrma et al. [52]. Babcock's model [66] provides good accuracy within its range of compositional applicability according to an assessment by Hrma et al. [52]; however, Babcock's model considers only SiO_2, Al_2O_3, CaO, and Na_2O and therefore has limited applicability to industrial glass compositions. Whichever model is used, the compositional range of validity must be carefully considered. Lithium oxide is neglected in most T_{Liq} models, including Cuartas' [47] and Lakatos and Johansson's [46], and most existing models fail to present a combination of predictive accuracy and compositional flexibility. Indeed, liquidus temperature models that consider a large number of components including those that could be beneficial for reformulation, such as Li_2O and B_2O_3, do not presently exist, indicating that further research and development for the construction of new liquidus models, or modification of existing ones, would be beneficial.

7.4.3 Conductivity and Heat Transfer

7.4.3.1 Specific Heat Capacity

The specific heat capacity of glass is more closely an additive function of composition than any other property [54]. Sharp and Ginther [69] developed a series of additive factors which, when applied to the model described in (7.7), yield the mean heat capacity, C_m, of silicate glasses. The authors claimed that the model was accurate to within $\sim 1\%$ at temperatures of 0–1300°C. The model was expanded and improved by Moore and Sharp [70]:

$$C_m = \frac{(aT + C_0)}{(0.00146T + 1)} \tag{7.7}$$

where a and C_0 are compositional constants and T is temperature in °C. Values of a and C_0 are shown in Table 7.9. Within the range of existing and potential practicable future SLS glass compositions, calculations using (7.7) and measured values confirmed that changes in specific heat resulting from compositional changes would be not greater than perhaps 2 or 3%. The temperature dependence of heat capacity must also be considered since the effects of compositional changes may allow furnace temperatures to be reduced. The monotonic increase of heat capacity with temperature means that a decrease in furnace temperature will reduce heat capacity, thus contributing to reductions in melting energy. The net change in heat capacity resulting from any compositional changes must therefore be calculated on an individual basis as part of any strategy to reduce melting energy.

Table 7.9 Factors for calculation of mean specific heat capacity of SLS glass using (7.7). Data adapted from [69, 70]

Oxide	a	C_0
SiO_2	0.000468	0.1657
Al_2O_3	0.000453	0.1765
MgO	0.000514	0.2142
CaO	0.000410	0.1709
Na_2O	0.000829	0.2229
K_2O	0.000445	0.1756
B_2O_3	0.000598	0.1935
Fe_2O_3	0.000380	0.1449
Mn_3O_4	0.000294	0.1498
SO_3	0.00083	0.1890

7.4.3.2 Thermal Conductivity and Optical Properties

The rate of heat transfer through molten glass is crucial to productivity, and there is considerable interest in modeling heat transfer with the aim of optimizing melt energies [71]. Thermal conductivity results from three mechanisms:

A. Conduction determined by phonon or "true" conductivity, k_c.
B. Convection determined by viscosity, density, furnace design, and heat capacity.
C. Radiation determined by photon or "optical" conductivity, k_r.

The value of k_c is relatively insensitive to temperature and composition: the replacement of 10 mol% SiO_2 by CaO in a 77 SiO_2–5 CaO–18 Na_2O glass resulted in a change in k_c from 0.933 to 0.905 W m^{-1} K^{-1} at 0°C [72]. Choudhary and Potter [73] developed a model for predicting k_c as described by the additive model (7.8):

$$k_c = \sum A_i C_i \tag{7.8}$$

where A_i = coefficient for component i and C_i = concentration of component i. Their comprehensive set of coefficients provide values of k_c within 10% of measured values. Convection (B) is a function of viscosity, density, and specific heat capacity, which themselves are functions of composition. Convection is also dependent on furnace design: thermal gradients within the furnace set up convection currents, which increase mass transfer and improve homogenization.

Conduction (A) dominates < ~600°C, and radiation (C) dominates > ~1000°C [71]. During heating and melting of glass, heat is continuously radiated and re-absorbed, with a steady state being achieved when steady heat flow occurs. Seward and Vascott [74] and Endrys et al. [75] noted that this process can be described using an effective radiative thermal conductivity, k_{eff}, shown in (7.9):

$$k_{eff} = \frac{16n^2 \sigma T^3}{3\alpha_r} \tag{7.9}$$

where n is the refractive index of the melt, σ is the Stefan–Boltzmann constant, T is absolute temperature, and α_r is the average absorption coefficient of the melt over the wavelength range of the radiation source. Radiative heat transfer is therefore strongly dependent on the optical absorption characteristics of the melt at visible and near-IR wavelengths. Absorption inherent to the silicate network renders SLS glass opaque at wavelengths outside the range ~0.3–5 μm. Absorption bands occurring at 3–5 μm are due to the vibration of ppm quantities of hydroxyl –OH groups in the glass (see Section 7.5.12). However, at furnace temperatures the majority of radiation occurs at 0.3–3 μm. The presence of colorants which absorb energy within this range of 0.3–3 μm, therefore, has a strong effect on heat transfer. The most widely used colorants in container and flat glass manufacture are Fe, (Fe+S), and Cr, although transition metals, lanthanides, and certain other elements can also produce coloration. As illustrated in Fig. 7.8, radiative heat conductivity of SLS glass varies.

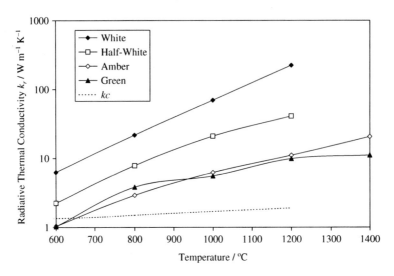

Fig. 7.8 Thermal conductivity of SLS glasses. Data adapted from [75]

Float and container glass contain Fe from raw material contamination; green glass contains added Cr and Fe; and amber glass obtains its characteristic color from the coordination of Fe^{3+} with sulfur in its reduced S^{2-} oxidation state [76]. The visible and near-IR optical absorption characteristics of many colorants in SLS glass have been well documented at room temperature [77–79], but only a small amount of data has been obtained at high temperatures [73, 75].

While chromium usually occurs as Cr^{3+} in SLS glasses, the oxidation states of both iron and sulfur are strongly dependent on furnace conditions. Indeed reducing conditions are necessary to form the Fe^{3+}–S^{2-} amber chromophore. In oxide melts containing iron the Fe^{2+}/Fe^{3+} ratio increases under reducing conditions. Fe^{2+} ions exhibit two broad, strong optical absorption bands at ~ 1 μm and ~ 2.2 μm. Radiation conductivity therefore decreases as the Fe^{2+}/Fe^{3+} ratio increases. Due to their low thermal conductivity, heat transfer through deeply colored glasses is poor and electrical boost heating may be required in critical parts of the furnace; this can be costly. The effect of any change in composition of SLS glass on k_c is negligible by comparison with its effect on k_r, which depends on the transition metal content of the raw material.

It is important for the production of clear glasses in particular to ensure that their Fe content is minimal and that the Fe which is present occurs mainly in its oxidized Fe^{3+} form ($Fe^{2+}/Fe^{3+} \approx 0.25$ is typical). The situation is somewhat more complex for colored glasses, in which the color is necessary and therefore the lower k_r is unavoidable. This may be compounded by the lightweighting of containers (as suggested by, for example, in [3, 80]), a process in which lower vessel thicknesses lead to substantial energy-per-article savings. However, this could feasibly lead to increased colorant levels in some glasses to achieve the same optical density. There is no simple solution to this problem, although polymeric coatings and labels may be one answer. The near-IR region in which the visual impact is nil may also offer opportunities. Decreasing the Fe^{2+}/Fe^{3+} ratio and/or the –OH content could allow manufacturers to partially offset the effects of increased colorant contents on k_r. Endrys et al. [75] calculated that an increase in k_r of up to ~5% would occur in green glass with the addition of 150 ppm of water.

While being vitally important for optical applications, a change in refractive index is not neccessarily considered to be important for many commercial SLS glasses. However, A-glass is widely used in fiber-reinforced composite materials [81, 82]. The refractive indices of fiber and matrix are often matched in order to provide a visually homogeneous product. It may therefore be advantageous that the refractive index of any new glass should match that of its predecessor in order to avoid changing the resin. Additive models for calculation of the refractive index have been reviewed in depth by Volf [39] and by Bonetti [83].

7.4.3.3 Electrical Properties

Electrical conductivity is important for any producer that utilizes either electrical boost or all-electric melting. A number of furnaces fired by fossil fuel combustion, particularly smaller ones, utilize an electrical boost capability which can provide increased melting rates and higher pull rates [84–86]. As noted in Section 7.4.3.2, electrical boost may be needed to melt strongly colored glasses. Increasing electric boost is also associated with a reduction in specific energy consumption (SEC). For a typical 200 tpd furnace, every 100 kW of continuous electrical boost can produce ~3.5 tonnes of extra glass per day with a SEC of ~2.5 GJ/tonne [2]. The energy required to melt this extra glass is only half of the ~5 GJ/t required for fossil fuel-fired furnaces although, since the majority of electrical energy originates from fossil fuel-fired power stations, the net environmental benefit is debatable. Furthermore, the high cost of electrical energy means that electrical boost is used only sparingly.

All-electric melting offers a number of advantages over conventional fossil fuel-fired furnaces. These include lower furnace emissions (but no net saving – the emissions take place at the power station); increased energy efficiency; greater control of furnace conditions; and higher quality of glass produced [2]. All-electric glass melting has been widely carried out [84–87] and all-electric SLS container glass has been documented as early as 1952 [2]. However, a recent trial at a UK float glass plant demonstrated that, while high-quality glass can be made, all-electric melting of float glass could not yet be considered to be economically viable [6]. Cold-top all-electric melting systems provide a number of environmental benefits. A blanket of unmelted batch floats on the surface of the molten glass. The blanket provides batch preheating and high thermal insulation, prevents batch carryover and batch volatilization, and eliminates the formation of fuel-derived SO_x, NO_x, and CO_2. Structural heat losses, which account for up to 40% in a fossil fuel-fired furnace, are minimal: electric furnaces are ~85% efficient on a delivered basis [6]. However, despite all the advantages of electric melting, the high cost of electricity has hitherto restricted its use to smaller operations. Reduced furnace lifetimes and the inability to melt glasses requiring reducing conditions (such as olive green and amber colors) also limit its appeal.

Volf [39] discussed modeling of low-temperature electrical resistivity based on composition. The high-temperature resistivity is equally important, particularly if electric boost is required. Varshneya et al. [88] measured the high-temperature resistivity (ρ) of a range of container and float glasses, obtaining values of $\rho = 3\text{–}8 \ \Omega$ cm at 1450°C, increasing to $\rho = 11\text{–}45 \ \Omega$ cm at 1000°C. Fluegel [49] discussed the effects of individual glass components on electrical resistivity of commercial glasses and developed a simple additive model for predicting ρ. At present, the model only provides factors for Al_2O_3, CaO, Li_2O, Na_2O, and K_2O. Fluegel [35] also discussed a far more comprehensive model that predicts resistivity at 1000°C, 1200°C and 1400°C for a wide range of silicate glasses and component oxides.

Mazurin and Prokhorenko [89] developed a second-order model for calculation of the resistivity of SLS-type glasses at 600°C, 900°C, and 1300°C using the VFT equation (7.10):

$$\text{Log } \rho = A + \frac{B}{T - T_0} \qquad (7.10)$$

Their model applies to glasses containing SiO_2, Na_2O, K_2O, MgO, CaO, and Al_2O_3 within the following compositional range (mol%): $69 - 81$ SiO_2; $0 - 4.5$ Al_2O_3; $12 - 20$ R_2O; $4 - 15$ $R'O$, where R = ($Na_2O + K_2O$) and R' = (CaO + MgO). This model gave remarkably good agreement with experimental values for two commercial float glasses [88]. No factors were published for other components such as Li_2O, B_2O_3, or BaO, but it is hoped that this model will be expanded in the future. Calculation of ρ is performed in three steps:

(i) Log (ρ/Ω cm) is calculated at 600°C, 900°C, and 1300°C.
At 600°C

$$R_{600} = 5.09 - 0.313 \times C_{NK} + 0.154 \times C_{CM} + 0.00776 \times (C_{NK})^2 - 0.00478$$
$$\times C_{NK} \times C_{CM} - 0.00194 \times (C_{CM})^2 + 0.0334 \times C_{Al} - 0.00234 \times (C_{Al})^2$$
$$-0.0211 \times C_{Mg} - 0.00179 \times (C_{Mg})^2 + 0.212 \times C_K - 0.0104 \times (C_K)^2$$

At 900°C

$$R_{900} = 3.984 - 0.285 \times C_{NK} + 0.0578 \times C_{CM} + 0.00646 \times (C_{NK})^2$$
$$-0.00144 \times C_{NK} \times C_{CM} - 0.00189 \times (C_{CM})^2 + 0.0312 \times C_{Al} - 0.00135$$
$$\times (C_{Al})^2 - 0.000275 \times C_{Mg} - 0.0038 \times (C_{Mg})^2 + 0.1686 \times C_K - 0.0126$$
$$\times (C_K)^2$$

At 1300°C

$$R_{1300} = 2.984 - 0.226 \times C_{NK} + 0.0255 \times C_{CM} + 0.00429 \times (C_{NK})^2$$
$$+0.000535 \times C_{NK} \times C_{CM} - 0.003 \times (C_{CM})^2 + 0.0289 \times C_{Al} - 0.000307$$
$$\times (C_{Al})^2 + 0.00439 \times C_{Mg} - 0.00226 \times (C_{Mg})^2 + 0.0685 \times C_K - 0.00177$$
$$\times (C_K)^2$$

where C_{NK} is the sum of the mol% concentrations of Na_2O and K_2O; C_{CM} is the sum of the mol% concentrations of CaO and MgO; C_{Mg}, C_{Al}, and C_K are the mol% concentrations of MgO, Al_2O_3, and K_2O.

(ii) Constants of the VFT equation (7.10) are calculated as follows:

$$M = (R_{600} - R_{1300})$$

$$N = (R_{900} - R_{1300}$$

$$T_0 = \frac{(M \times 873 - N \times 1173)}{(M - N)}$$

$$B = \frac{(R_{900} - R_{1300}) \times (1573 - T_0) \times (1173 - T_0)}{400}$$

$$A = \frac{R_{600} - B}{(873 - T_0)}$$

(iii) Log (ρ/Ω cm) at any temperature in the range 600–1400°C can be calculated using the VFT equation (7.10).

7.4.4 Interfaces, Surfaces, and Gases

7.4.4.1 Refining

Refining is the removal of gaseous inclusions from molten glass. These can occur as seed < ~0.4 mm diameter, often present in clusters; as larger bubbles which do not exhibit clustering; and as non-spherical blisters [16]. The quality of the final glass depends on good refining; however, quality requirements vary by application. The acceptable bubble and seed content of float glass is orders of magnitude lower than container glass and domestic glass. Bubbles are removed by a combination of two mechanisms: (a) buoyancy effects causing the bubbles to rise to the surface of the melt and (b) dissolution of the gas within the melt. Buoyancy effects can be described using Stokes' law (7.11):

$$V_S = \frac{2 g \Delta \rho r^2}{9 \eta} \tag{7.11}$$

where V_S = bubble velocity, g = gravitational acceleration, $\Delta \rho$ = difference in density between bubble and molten glass, r = bubble radius, and η = melt viscosity. However, as noted by Shelby [90], a more accurate representation of bubble rise within a molten glass is provided by (7.12):

$$V_B = \frac{3}{2} V_S = \frac{g \Delta \rho r^2}{3 \eta} \tag{7.12}$$

Sulfate has now become the standard refining agent for SLS glass (see Section 7.5.11). The solubility of SO_3 in SLS glass is dependent on several factors including furnace atmosphere, melting temperature, melting time, and glass composition. New melting and refining technologies offer the opportunity to reduce emissions and increase refining rates, thereby reducing furnace energy requirements. Müller-Simon and Gitzhofer [91] have recently calculated sulfur mass balances for modern SLS glass furnaces. Excess sulfur, i.e., that which is not bound within the glass, was 2–8 times the amount of bound sulfur. They concluded that the most promising method for minimizing sulfur emissions is, in most cases, to reduce the amount of added (batch) sulfur to a point at which it does not affect glass quality. A recent report confirms that several alternative methods of refining are now being considered [92], described here under (i–iii). To that list we should also add (iv) alternative refining agents, as described below.

(i) Pressure-related refining. Methods include pulsed pressure [93], overpressure [94, 95], underpressure [96], and vacuum [97]. Cable and Chaudhry [27] demonstrated that under reduced pressures, alkali volatilization substantially increases and stronger foaming may occur.

(ii) Alternative refining gases. These include He [98], H_2O [99], N_2 [100], and O_2 [101]. The economics of using He or O_2 may be prohibitive, although if the furnace is oxy-fuel fired then the required on-site O_2 production may make O_2 refining more attractive. It is possible that water-enhanced refining may increase volatilization and refractory corrosion.

(iii) Physical refining methods. Mechanical [92] and microwave [102] refining have been studied. Ultrasonic refining received interest in the 1990s [103] and while small-scale laboratory tests and modeling proved successful, severe difficulties were encountered on a pilot (1.4–3.6 tpd) scale. Several probe materials failed under the extreme conditions encountered in the forehearth. If these problems can be overcome then ultrasonic refining will once again be a highly promising technology. A recent patent application [104] describes the use of ultrasonic methods to reducing foaming in a glass furnace. Centrifugal force has also been successfully demonstrated as a means of rapid refining [105].

(iv) Alternative refining agents. Sodium sulfate is now widely used as the chief refining agent for SLS glasses. However, there is substantial evidence that different sulfates or mixtures thereof may provide superior refining to using Na_2SO_4 alone (see Section 7.5.11). Older methods of refining have included the use of nitrates (see Section 7.5.20) and some authors have suggested using a combination of sulfate and nitrate batch additions, which may provide a net reduction in environmental impact over sulfate-only refining. The refining behavior provided by the oxides of some multivalent elements including CeO_2 and MnO_2 has also been studied (see Section 7.5.17.1). Substantial scope therefore exists for considering alternative refining agents or mixtures thereof to maximize bubble removal while minimizing the net generation and emission of SO_x and NO_x.

7.4.4.2 Refractory Corrosion

Conditions within glass tanks are highly corrosive, therefore the basic requirements for glassmaking refractories are that they do not affect the glass product and provide sufficient service lifetimes. Loss of molten glass and refractory contamination of glass occur when refractories become corroded. Corrosion affects production efficiency and is responsible for yearly decreases of ~2% in furnace efficiency. In addition to glass-contact refractory corrosion, corrosion also occurs in the furnace crown and within regenerative or recuperative systems due to volatilization of alkalis, water, and sulfates. Volatilization from glass melts can also increase particulate generation and emission, leading to greater demand for abatement technology. Glassmaking refractories have been treated by Begley [106] and updated by Davis [107].

A continual flow of glass through the furnace ensures that corrosion is continuous and system equilibrium is not attained. Corrosion of a refractory by molten glass is governed by three key factors:

(a) Diffusion of refractory components into the glass melt
(b) Their solubility within the melt
(c) Convection, by which refractory-rich glass is transported away from the interface

Corrosion is not uniform throughout the glass melting tank but is most vigorous at the atmosphere/refractory/molten glass interface. If foreign metallic particles enter the furnace they sink to the bottom due to their greater density and cause refractory corrosion known as downward drilling. Melt/refractory interaction mechanisms have been reviewed by Cable [108].

Corrosion testing is used to assess the resistance of refractories and the aggressiveness of glasses under melting conditions. A standard test is the static finger corrosion test in which a cylindrical refractory finger is suspended in molten glass for the required time and temperature [106, 109]. Tests for vapor resistance, important above the meltline and within regenerators and recuperators, are also commonly used [106].

The introduction of measures to reduce furnace temperatures and melting energies could substantially decrease refractory corrosion rates, thereby reducing the occurrence of discarded ware and potentially increasing furnace lifetimes. The corrosion rate of refractories can be described [110] as an Arrhenius function as shown in (7.13):

$$K = K_0 \exp\left(-\frac{E}{RT}\right) \tag{7.13}$$

where K = corrosion rate, K_0 = constant, E = corrosion activation energy, R = gas constant, and T = temperature. A small decrease in furnace temperature provides a substantial decrease in refractory corrosion rate [111]. One study concluded that the corrosion rate, C, of fusion-cast and sintered refractories decreases with temperature so that $C_{1550°C}/C_{1500°C} \sim 2$ and $C_{1500°C}/C_{1400°C} \sim 3$ [112]. Begley [106] stated the rule of thumb that in normal operation the refractory corrosion rate doubles for every 50°C increase in temperature.

Kato and Araki [113] studied the effects of temperature and glass composition upon the corrosion by molten glass of zircon (34% SiO_2 – 65% ZrO_2) and ZrO_2 (95% ZrO_2 – 2.5% SiO_2 – 1.5% Al_2O_3) refractories. These two materials typify glass-contact refractories [106, 107]. Test specimens were immersed in glass in Pt crucibles and held at temperature for 24 h. Several glasses were studied including float, borosilicate, and E-glass. Glass compositions, viscosities, and corrosion at 1500°C are shown in Table 7.10.

Table 7.10 Glass compositions (wt%), viscosities, and corrosion losses from ZrO_2 and zircon refractories at 1500°C according to [113]

	Glass A	Glass B	Glass C	Glass D	Glass E1	Glass E2
SiO_2	71.4	64.1	69.3	60.0	54.0	54.2
B_2O_3	0	9.0	17.8	4.5	8.9	8.2
Al_2O_3	1.6	4.2	2.6	14.3	14.2	14.1
MgO	3.8	6.8	0	2.3	4.2	0.4
CaO	8.9	0	0	8.1	18.1	22.1
BaO	0	0	0	0	0	0.5
Li_2O	0	0	1.2	0	0	0
Na_2O	13.6	11.5	2.5	4.3	0.3	0.3
K_2O	0.8	4.5	6.7	6.5	0.3	0.2
Tests carried out at 1500°C						
Log (η/dPa s)	1.80	1.88	2.05	1.98	1.57	1.55
Loss/% (ZrO_2)	12	11	6	6	10	12
Loss/% (Zircon)	61	37	24	32	16	23

The rate of corrosion of the ZrO_2 refractory is controlled by Zr transport into the glass rather than ZrO_2 dissolution [113]. The result of the different driving forces for corrosion is the differences in dependency upon temperature and upon glass composition, as shown in Fig. 7.9. Kato and Araki [113] applied the Noyes–Nernst equation to the corrosion rate of a ZrO_2 refractory. This is controlled by transport of dissolving material (7.14):

$$X = \frac{(C_s - C_0)\,D}{d} \tag{7.14}$$

where X = corrosion rate, C_s = concentration of the saturated glass, C_0 = concentration in the bulk glass, D = diffusion constant, and d = effective film thickness for diffusion. The value of D/d broadly correlated with melt viscosity and it was assumed that (7.15) can be used to model ZrO_2 corrosion:

$$X = k\,(C_S - C_0)\,\eta^m \tag{7.15}$$

where η = viscosity of the molten glass and k and m are constants that were obtained by performing regression analysis using the values of X and C_S determined from (7.14). The resulting relationship is expressed in (7.16):

$$X = 9.8\, C_S\eta^{-0.58} \tag{7.16}$$

Corrosion of the zircon refractory was studied with the same glasses and over a wide range of temperatures. A plot of corrosion loss, X, from the zircon refractory as a function of test temperature is shown in Fig. 7.9.

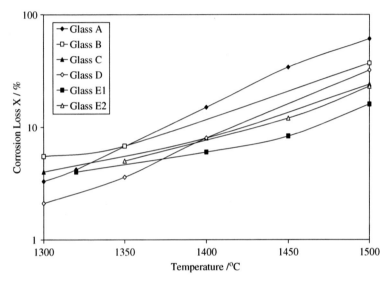

Fig. 7.9 Corrosion loss from a zircon refractory as a function of temperature and glass composition (values given in Table 7.10). Data adapted from [113]

The range of glass compositions and viscosities studied by Kato and Araki [113] was wide by comparison with considerations for reformulating SLS glasses. Even so, the corrosion rates shown in Fig. 7.9 are at most a factor of ~3.8 different for glasses A and E1, i.e., float glass and E-glass. Given that reformulation of SLS glass will involve small changes by comparison, it can be expected that the effects of small compositional changes on refractory corrosion rates will be similarly small. Experimental evidence supports this view: using a static finger corrosion test similar to that described by Dunkl [109], Bingham and Marshall [21] demonstrated that fluxline corrosion of a typical AZS refractory, tested for 24 h at 1370°C, was the same, within measurement errors of ± 3%, for a number of reformulated SLS container glasses as for the representative baseline glass.

The available published scientific and industrial evidence confirms that temperature dominates refractory corrosion behavior as expected on the basis of (7.13). It appears likely that SLS glass reformulation to reduce melting temperatures will give a net reduction in refractory corrosion, potentially increasing furnace lifetimes and thereby resulting in considerable economic benefits. Further research in this area would be helpful.

7.4.4.3 Surface Energy

Surface energy is involved in several processes of importance to SLS glass manufacture, in which context "surface tension" usually refers to the melt/atmosphere interface or meltline. Other interfaces also occur at the refractory/melt/atmosphere triple point and at the glass/refractory boundary. Surface energy is generally considered in glassmaking when $\eta < 10^8$ dPa s. Measurement may be performed by a number of techniques including drop weight, ring method, bubble pressure, sessile drop, pendant drop, and fiber elongation [54, 114–116].

It is important to maintain low surface tension during glass melting and forming in order to promote fusion and refining, and also to reduce the potential for cord formation. Constituents with low partial surface tension (for example, SO_3, B_2O_3, K_2O, Rb_2O, Cs_2O, PbO) tend to migrate to the surface, whence they can volatilize. Their presence in the melt reduces cord formation, which is increased by the presence of more refractory oxides such as SiO_2, Al_2O_3, MgO, CaO, or ZrO_2. Historically the use of halides and sulfates in the batch has reduced surface tension, which promotes fusion and refining. The removal of halides from SLS batches on the basis of environmental considerations (see Sections 7.3.2 and 7.3.3.5) means that this function is now provided by sulfate. The addition of 1 wt% SO_3 as Na_2SO_4 reduces the surface tension of SLS glass at 850°C from 309 to 266 mNm^{-1}, enhancing wetting of unreacted sand particles by the molten salt [36]. The presence of –OH groups in the furnace atmosphere can also substantially reduce surface tension.

Many mathematical descriptions of the dependence of surface tension in oxide glasses on composition and temperature exist; these have been analyzed in detail [39, 54, 72, 114–116]. Descriptions for surface tension occur in the form of additive calculations as in (7.17):

$$\gamma = \sum_i a_i X_i \tag{7.17}$$

where for the ith component, a_i is the surface tension factor and X_i is the molar percentage. Recently Kucuk et al. [116] obtained the regression factors shown in (7.18) for estimating the surface tension of a silicate melt at 1400°C with concentrations given in mol%.

$$
\begin{aligned}
\gamma \left(\tfrac{mN}{m}\right) = {} & 271.2 + 1.48[Li_2O] - 2.22[K_2O] - 3.43[Rb_2O] + 1.96 \\
& [MgO] + 3.34[CaO] + 1.28[BaO] + 3.32[SrO] + 2.68 \\
& [FeO] + 2.92[MnO] - 1.38[PbO] - 2.86[B_2O_3] + 3.47 \\
& [Al_2O_3] - 24.5[MoO_3]
\end{aligned}
\tag{7.18}
$$

7.4.5 Chemical Durability

The chemical durability of a glass is its resistance to dissolution by other media. This is usually water, although container glass may come into contact with mildly acidic and alkaline media. Aqueous dissolution of silicate glasses, as discussed by Newton [117], Doremus [118], and Paul [119], is presently believed to proceed in the following stages:

Stage (1): Ion exchange. Mobile alkali (R^+) ions in the glass surface are replaced by hydronium (H^+) ions, thus maintaining charge balance; this reaction is described in (7.19):

$$\equiv Si - O - R + H_3O^+ \rightarrow \equiv Si - OH + R^+ + H_2O \qquad (7.19)$$

Stage (2). Diffusion. Protons diffuse inward and alkali ions outward into solution. The diffusion zone becomes rich in SiO_2. This diffusion zone continues to move into the bulk of the glass.

Stage (3). Network hydrolysis and dissolution. As more alkali ions enter aqueous solution, its pH increases and the silica-rich zone becomes hydrolyzed according to (7.20):

$$\equiv Si - O - Si \equiv +OH^- \rightarrow Si - OH+ \equiv Si - O^- \qquad (7.20)$$

Glass corrosion mechanisms are not yet fully understood. A recent study of the corrosion of alkali borosilicate nuclear waste glasses and ancient soda–lime–silica glasses [120] indicated that a continual process of dissolution and re-precipitation occurs. Standards, principally ISO (International Standards Organization), USP (United States Pharmacopoeia), ASTM (American Society for Testing and Materials), and EP (European Pharmacopoeia) allow classification of SLS glass into types suitable for specific uses. The USP classifications will be considered here; however, EP and ASTM are equally applicable. Chemical durability is based on the extraction of an equivalent of alkali expressed as Na_2O per gram of glass grains tested, usually in $\mu g/g$. Testing may be carried out at 98°C or at 121°C.

Type I.   Applies only to borosilicate glasses.
Type II.   Soda–lime glass de-alkalized by surface treatment.
Type III.   Describes container glass. Neutralization of the test leachate must use less than 8.5 ml of 0.02 N sulfuric acid.
Type NP.   General-purpose SLS glass used for non-parenteral applications. Neutralization of the test leachate must use less than 15.0 ml of 0.02 N acid.

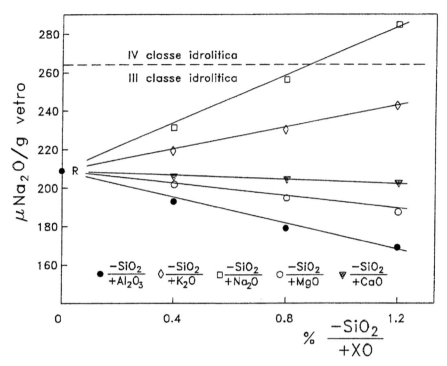

Fig. 7.10 Effects of wt% SiO₂ substitution on hydrolytic durability of SLS glass [83]. Reproduced with permission of Stazione Sperimentale del Vetro

Bonetti [83] illustrated the effects of substituting SiO₂ by various oxides upon hydrolytic durability according to Durability Class, as shown in Fig. 7.10. Small changes in durability can be acceptable providing that the glass remains within its existing classification, normally USP Type III. However, changes in Durability Class may be unacceptable.

Chemical durability of flat glass decreases rapidly when SiO₂ <~69 wt% [121]. This can be mitigated by careful reformulation [21]. A simple example is to increase Al₂O₃ content [61, 121, 122]; however, the resulting increase in viscosity may be unacceptable. Modeling of SLS glass durability guides compositional development although the range of accurate models is presently limited. Durability can be modeled (i) thermodynamically using standard free energies of hydration or (ii) empirically using regression analysis of measured data. This discussion focuses upon method (ii) because the thermodynamic approach can be less accurate over the relatively narrow compositional ranges with which this study is concerned. The reader is referred to the work of Paul [119], Newton [117], and Jantzen [123] for further information on the thermodynamic approach.

Bonetti [83, 124] reviewed several regression-based models for aqueous durability of experimental SLS glasses. Recently Sinton and LaCourse [125] devised a model from a set of 28 float and container glass compositions. The model of Lakatos and Simmingsköld [126], which is applicable to a wide range of SLS compositions, has been adapted [83] and is described in (7.21) and (7.22), where Q is the quantity of Na_2O extracted in $\mu g/g$ glass:

$$Q = 155y \qquad (7.21)$$

$$\begin{aligned} \text{Log } y = 0.4343[&- 1.35 + 16.85 \times Na_2O + 10.15 \times K_2O - 3.89 \times CaO - 6.16 \\ &\times MgO - 4.33 \times BaO - 65.44 \times Al_2O_3 + 1351.29 \times (Al_2O_3)^2 - \\ &10224.11 \times (Al_2O_3)^3 + 4.89 \times Li_2O - 4.33 \times BaO - 9.25 \times ZnO - \\ &4.77 \times PbO] \end{aligned}$$

$$(7.22)$$

7.4.6 Density and Thermo-mechanical Properties

SLS glasses typically have $\rho \sim 2.5$ g cm^{-3}. Density affects the volume of glass produced for a given raw material mass (cost). For example, glass (A) $\rho = 2.500$ g cm^{-3} and glass (B) $\rho = 2.510$ g cm^{-3}, a difference of $\sim 0.4\%$. Glass (A) fills 0.4% more volume than composition (B) and 0.4% more articles can be produced from the same mass of glass. Additive models have been reviewed for the estimation of SLS glass density, thermal expansion coefficient, and mechanical properties [39, 127]. Bonetti [127] concluded that Huggins and Sun [128] provided one of the most accurate density models. The thermal expansion coefficient α is usually quoted for 20–300°C or 50–300°C. The α_{20-300} of SLS glass is typically $\sim 90 \times 10^{-7}$°C^{-1}. By contrast, pure SiO_2 has $\alpha_{20-300} = 6 \times 10^{-7}$°C^{-1} and borosilicate ovenware has $\alpha_{20-300} = 33 \times 10^{-7}$°C^{-1}. Generally, compositional changes that increase viscosity also decrease α and increase thermal shock resistance. Reformulation of commercial SLS glass within the range 74 > SiO_2 > 68 wt% (see Section 7.4.2.1) can be expected to result in variations in α_{20-300} of the order of $\pm 10 \times 10^{-7}$°C^{-1}.

Thermal shock resistance, ϑ, is the ability of glass to sustain without damage a rapid change in temperature [39]. The most important factor in determining ϑ is α. However, tensile strength (P), Young's modulus (E), Poisson's number (μ), and thermal conductivity (k, see Section 7.4.3.2) all contribute to ϑ. The compositional ranges with which this study is concerned mean that changes in P, E, and μ can be neglected in many cases. Only changes to α and to a lesser extent k are likely to measurably affect ϑ. Models for predicting tensile strength (P), Young's modulus (E), Poisson's number (μ), and hardness have been discussed by Volf [39]. Mechanical properties are all affected by compositional changes, but assessment of property data [72] indicates that these effects are mostly small over the range of compositions that may be practically considered for reformulation of commercial SLS glass, namely 74 > SiO_2 > 68 wt% (see Section 7.4.2.1).

7.5 Design of New SLS Glasses

Typical SLS glass batches consist of recycled glass (cullet), sand, limestone, dolomite, soda ash, saltcake, feldspar, nepheline syenite, and/or blast furnace slag. One must consider the economics of batch selection: analysis by weight and cost reveals substantial differences between minerals and processed chemicals, as shown in Fig. 7.11.

Fig. 7.11 Typical SLS glass batch breakdown by (**a**) weight and (**b**) cost

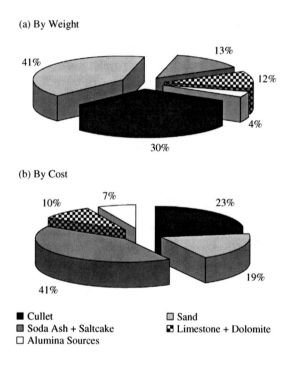

(a) By Weight

41%   13%   12%   4%   30%

(b) By Cost

10%   7%   23%   41%   19%

■ Cullet   ▨ Sand
▩ Soda Ash + Saltcake   ▧ Limestone + Dolomite
□ Alumina Sources

7.5.1 Batch Processing, Preheating, and Melting

Melting of SLS batches occurs in four stages [129]:

1. Initial melting involving conversion of batch to liquid with undissolved sand and gaseous inclusions.
2. Dissolution of sand grains.
3. Refining or bubble removal.
4. Homogenization.

Melting energies are reduced by more rapid progression of these stages. Stages 1 and 2 can be accelerated by a number of methods, chiefly raw materials choice.

In complex SLS batches the presence of soda ash and limestone leads to formation of the double carbonate $Na_2Ca(CO_3)_2$ at 480–510°C, which is a key species during melting [130]. This compound forms a molten eutectic with Na_2CO_3 at 785°C; the double carbonate $Na_2Ca(CO_3)_2$ melts at 813°C; and the liquids begin to react with SiO_2. Sodium carbonate melts at 851°C; however, in the presence of SiO_2 it decomposes at lower temperatures and reacts with SiO_2 at ~780–900°C to form Na_2SiO_3 + CO_2 [26]. A fusion reaction occurs between SiO_2, CaO, and the molten phase at 812°C [129]. Limestone decomposes at ~990°C; but decomposition may begin at temperatures as low as 500°C [26]. Reaction with SiO_2 occurs at 1010°C when $CaSiO_3$ is formed. In batches containing Na_2CO_3, $CaCO_3$ is involved in the additional decomposition routes described above. Dolomitic batches melt more readily than limestone-only batches, although the melting rate is unchanged [26]. Reactions involving MgO tend to occur at lower temperatures than the corresponding reactions involving CaO.

The treatment of raw materials prior to melting is vitally important to the efficient production of homogeneous glass, and modern batch plant ensures that batches reach the furnace in optimum condition [11, 16]. Batches comprise well-mixed raw materials with similar grain sizes, although it is important that cullet is coarser than the raw batch so that it does not interfere with normal batch reactions [11]. Batch processes have been finely tuned, however, a number of treatments exist that will increase melting efficiency.

Batch consolidation, which has been discussed in depth by Bauer and Tooley [11], is a broad term encompassing batch treatments ranging from simple batch wetting to sophisticated thermal and physical pre-treatments. In its simplest forms, consolidation reduces demixing and batch segregation but it can also reduce dusting and volatilization [11, 131, 132] and increase rates of melting and refining. Some consolidated batches are used in conjunction with batch preheating technologies, which increase furnace efficiencies and reduce emissions. Consolidated batch may occur in a number of forms:

(i) Pellets. Batch is wetted and formed into spheres.
(ii) Briquettes. Dry batch is formed into shapes by passing through rollers.
(iii) Granules. Similar to (ii) but heating drives partial reactions, evolving some batch CO_2 and forming liquid phase binders.
(iv) Extrusion. Similar to (ii) but with water added as a binder.

Liquids other than water may be added to SLS batches either as surfactants or binding agents [133]. Caustic soda-rich pellets can provide some advantages over water-formed pellets and normal batch, although these may be limited [134]. Water-formed pellets can provide lower batch-free times than standard batches.

Preheating of air prior to combustion is now standard practice. However, preheating of batch materials has received only limited implementation despite the development of several suitable technologies [6]. There is sufficient heat in exhaust gases, following regeneration, to heat batch to 300–400°C [14]. This increases furnace pull rates and significantly decreases SEC. Industrial trials showed a decrease of ∼23% in SEC by preheating to 300°C [14]. Further industrial trials [10] support these results, confirming that batch preheating does not have an adverse effect on dust generation. Furnace crown temperatures decreased from 1590–1600°C to 1530–1560°C, reducing NO_x generation and emission, in addition to reducing SEC.

7.5.2 Cullet

Cullet is waste glass suitable for remelting and arises from rejected ware and from external sources such as kerbside collection and bottle banks. Cullet-free SLS batches require ∼1.2 t of raw materials to make 1 t of glass. Of the additional 200 kg approximately 185 kg is CO_2. Cullet use leads to savings in fuel CO_2 and batch CO_2 and increases batch reaction rates, with the greatest acceleration observed with the smallest cullet particle size, < 0.15 mm [133]. However, the effects of particle size on refining and conditioning times must also be considered. When cullet is mixed with batch, the slowest reaction time ensues when particles are of equal size; however, acceptable results are obtained when large-sized cullet is mixed with batch [135]. Cullet does not require energy input to satisfy enthalpies of decomposition or entropic energy associated with destruction of crystallinity. Theoretical calculations [4] confirmed by measurements [2, 9], indicate a decrease in melting energy of ∼0.3% for every 1 wt% cullet addition, shown in Fig. 7.12. In practice other factors including furnace age, type and size, pull rate, glass composition, and air and batch preheating systems all influence SEC, creating substantial data spread as shown in Fig. 7.13.

The chemical and physical composition of cullet that is acceptable varies for different SLS glass types and colors. Cullet specifications are generally stricter for

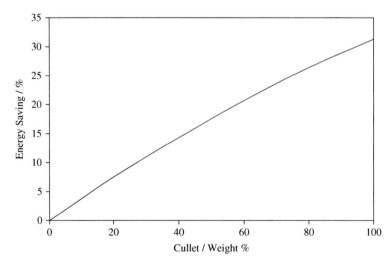

Fig. 7.12 Energy saving with batch cullet content. Adapted from [2, 4]

Fig. 7.13 Specific energy consumption (SEC) based on cullet content for 126 container glass furnaces, from [9]. Reproduced by permission of Deutsche Glastechnische Gesellschaft

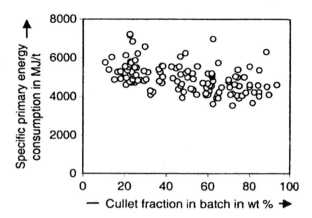

flat glass than for container glass. Some flat glass producers only use in-house cullet while others accept foreign cullet, providing that it is of an acceptable quality. A recent UK government report [136] defined the acceptable variabilities in UK flat glass cullet composition for cullet that is to be remelted to make flat glass, as shown in Table 7.11. Éidukyavichus et al. [137] detailed the compositional variability of a number of SLS cullet compositions and its effect on measured viscosities and surface energies.

Table 7.11 Acceptable limits of UK SLS flat glass cullet composition for remelting in SLS flat glass manufacture. Data adapted from [136]

Oxide	Minimum/wt%	Maximum/wt%
SiO_2	70.0	73.0
Al_2O_3	0.0	1.5
MgO	3.5	4.5
CaO	8.0	9.7
$Na_2O + K_2O$	13.4	14.6
Fe_2O_3	0.0	0.2

Compositional variability of cullet results in small variations in viscosity, liquidus temperature, and the other properties discussed in Section 7.4. This variability is carried through to the final glass when cullet is remelted. As a result it is natural that some manufacturers slightly "over-melt" glasses [138]. While remelting cullet provides energy savings, practicable limits are placed upon batch cullet levels: little or no high-quality SLS glass is produced using 100% cullet. This is for a number of reasons:

Ideally, given a plentiful and consistent cullet supply, manufacturers would melt glass using 100% cullet and thereby maximize savings in melting energy and batch CO_2. However, a number of factors determine that this is not practically achievable.

- Refining. Removal of bubbles requires the addition of refining agents such as sulfates or slag. This in turn means that any additions must not alter the final glass composition, so sand, limestone, etc. must also be added at an appropriate level.
- Redox. Redox conditions are essential to produce (a) acceptable refining behavior and (b) the required glass color.
- Color, which depends on redox conditions to control redox reactions, particularly the Fe^{2+}/Fe^{3+} equilibrium.

Glassy sand arising from rejected furnace-ready cullet [139] is one recent development that could offer increased cullet reuse efficiency. Crushed cullet with a grain size of 0.1–0.8 mm can be reclaimed from rejected cullet; processed to remove unwanted organics, ceramic, and metal inclusions; and re-introduced to the furnace as a raw material. Glassy sand has been successfully introduced at ∼10–20% into a green container glass furnace [139]. Its introduction substantially reduced the number of inclusions per tonne of glass from ∼12 to ∼3.

7.5.3 Silica, SiO$_2$

7.5.3.1 SiO$_2$ Raw Materials

Silica sand is the chief ingredient of SLS glass and the lowest cost. Glass grade sand is usually > 99 wt% SiO_2, with minor impurities including Al_2O_3, CaO, MgO, Na_2O, and K_2O. Allowable impurity levels vary depending on the type of glass being manufactured [140]. Coarse sand grains increase batch-free times and can lead to batch "stones." Fine grains lead to refining problems by introducing fine bubbles during melting, which are difficult to remove fully. The sand grain size distribution largely determines batch-free times, since other components melt and form eutectic phases well below the melting point of quartz sand, ∼1710°C. Glassmaking sand is usually supplied within a tightly controlled particle size range which ensures that especially coarse or fine particles are omitted.

Mined and synthetic silicates such as wollastonite ($CaSiO_3$), anorthite ($CaAl_2Si_2O_8$), diopside ($MgCaSi_2O_6$), or enstatite ($MgSiO_3$) can partially replace sand but also limestone and dolomite in glass batches, decreasing batch CO_2, bubble content, melting energy, and furnace residence times. Wollastonite melts at ∼1540°C, anorthite at ∼1550°C, diopside at ∼1390°C, and enstatite at ∼1560°C, all of which are substantially lower than the melting temperature of silica. Diopside was investigated by Mirkovich [141] as a potential replacement for nepheline syenite in SiO_2–Al_2O_3–Na_2O–K_2O fiberglass. It provided increased melting rate and a reduced energy requirement.

Unfortunately the compositional variability, abundance (= cost), and undesirable levels of iron in many naturally occurring silicates presently limits their widespread use. However, synthetic alkaline earth silicates have been available for some time [142–145]. These synthetic materials are now produced on an industrial scale [146].

The net energy requirement and consequent CO_2 emissions associated with their use can be greater than using virgin raw materials since synthetic materials must be heated twice (once to produce and once to melt). Nevertheless, synthetic silicates can provide glassmakers with extra flexibility, and may see increased use in future glassmaking.

7.5.3.2 SiO₂ Effects on Glass Properties

Of the network forming (NWF) oxides, i.e., oxides that form glasses alone without the addition of other components, silica is by far the most technologically important. In pure silica glass, Si^{4+} is tetrahedrally coordinated with four O^{2-} ions. Each O^{2-} links two Si^{4+} providing a strong three-dimensional network. Variations in Si–O bond angle and rotation are responsible for the loss of long range order that arises when crystalline silica is melted. The strong (59%) covalency of the Si–O bond leads to very high melting temperatures. These properties also cause molten silica to exhibit high crystallization resistance upon cooling. Silica glass also has low density (2.20 g cm^{-3}), low thermal expansion ($\sim$5.5 $\times$ 10^{-7}°C^{-1}), high thermal shock resistance, and exceptional chemical durability.

If it were not for its very high melting temperature, pure SiO_2 would provide an excellent material for glass manufacture. Other components must therefore be added to decrease the melting temperature. These components break Si–O–Si bonds through the introduction of non-bridging oxygens (NBO) that are associated with network modifier (NWM) cations such as alkali metals or alkaline earth metals. Silica contents of roughly 70 wt% provide an acceptable balance between melting temperature, viscosity, crystallization behavior, and chemical durability. Considering that silica glass has low density and low raw material cost, any significant change in silica content that may result from changing SLS glass formulations will not only affect properties, but also the raw material cost per unit volume of glass produced. As discussed in Section 7.4.2.1, SiO_2 contents of $\sim$68–74 wt% present a reasonable range for the formulation of commercial SLS glasses.

7.5.4 Soda, Na₂O

7.5.4.1 Na₂O Raw Materials

Sodium carbonate, Na_2CO_3, also known as soda ash, is the chief Na_2O source for SLS glass. Sodium is also provided to SLS glass through the use of sodium sulfate and alumina sources. Soda ash is a man-made batch component and therefore one of the most expensive. Natural deposits of trona, or sodium sesquicarbonate, ($Na_3HCO_3CO_3\cdot2H_2O$), are mined although much of the world obtains its soda ash from the ammonia–soda process. The chief stages of this process are described in (7.23) and (7.24). Sodium carbonate is very pure as a result of its production method: a typical analysis is 99.8 wt% Na_2CO_3, 0.05 wt% Na_2SO_4, 0.06 wt% NaCl, and 5 ppm Fe_2O_3 with 0.1 wt% LOI:

$$NaCl + NH_3 + CO_2 + H_2O \rightarrow NH_4Cl + NaHCO_3 \downarrow \qquad (7.23)$$

$$NaHCO_3 \rightarrow Na_2CO_3 + CO_2 \uparrow + H_2O \uparrow \qquad (7.24)$$

Alternatives to the use of carbonates as alkali and alkaline earth sources include hydroxides and halides; however, the widespread use of halides is now restricted by environmental legislation (see Sections 7.3.2 and 7.3.3.5). In theory the use of hydroxides can lead to substantially lower melting energies due to their lower enthalpies of decomposition, as illustrated in Fig. 7.14. However, a number of concerns arise, including material handling, increased OH content of the furnace atmosphere (and hence its effects on refractories in the furnace and regenerators), and the energy expended in manufacturing hydroxides, making any feasibility assessment highly complex.

Fig. 7.14 Molar enthalpy of decomposition at 296 K of raw materials from [4]. *Filled markers* = alkalis, *open markers* = alkaline earths. Reproduced with the permission of Deutsche Glastechnische Gesellschaft

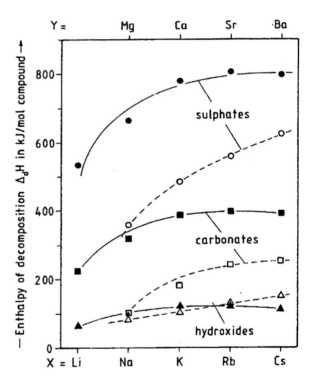

The economic viability of using NaOH instead of Na_2CO_3 was considered by Place [147]. He concluded that, at the time, the plant necessary to store and handle NaOH would be repaid by batch cost savings within 1–2 years. Cable and Siddiqui [134] considered the effects on melting behavior of using caustic soda, NaOH, as a batch component in SLS glasses. Caustic soda in aqueous solution produces slurry when added to the rest of the batch, raising concerns about its practicality. Preprocessing of the batch by heating/pressing to form pellets led to further difficulties:

$CaCO_3$ is converted to $Ca(OH)_2$ and NaOH to $\frac{1}{2}$ Na_2CO_3. Bhuiyan and Cable [148] observed that sodium peroxide, Na_2O_2, performed well as a refining agent in two SLS glass-forming melts at 1475°C.

7.5.4.2 Na₂O Effects on Glass Properties

Adding Na_2O to silica glass replaces Si–O–Si bridging oxygens with Si–O⁻ and O⁻Na⁺ non-bridging oxygens. This depolymerization of the network results in lower melt viscosities. Soda is the chief melting flux in SLS glass. Increasing the Na_2O content of SLS glass can also decrease T_{Liq} although this is dependent on the primary phase. However, increased alkali content reduces chemical durability, as shown by Bonetti's data [63], reproduced here in Table 7.12. Bonetti detailed a stepwise replacement of SiO_2 by Na_2O. Increasing the Na_2O content from 13 to 15 wt% decreased T_{Liq}, T_{CMax} (the temperature of maximum crystallization rate), crystallization rate (R_C), and viscosity. The primary phase field was not stated although the observed decrease in T_{Liq} suggests that it may be tridymite or wollastonite. Bingham and Marshall [21] showed that increasing the Na_2O content of a container glass from 13.4 to 15 wt% by replacing SiO_2 resulted in a decrease of 43°C in log (η/dPa s) = 2 and a decrease in T_{Liq} of 5°C. However, chemical durability was substantially impaired. The addition of Na_2O to SLS glass rapidly decreases its aqueous durability and increases thermal expansion. Na_2O additions to reduce melting temperature must consider not only these factors but the raw material cost.

Table 7.12 Effects of replacing SiO_2 by Na_2O (wt/wt). Adapted from [63]

	Glass 1	Glass 2	Glass 3	Glass 4	Glass 5
SiO_2/wt%	72.5	72	71.5	71	70.5
Al_2O_3/wt%	2.5	2.5	2.5	2.5	2.5
Na_2O/wt%	13	13.5	14	14.5	15
CaO/wt%	10.5	10.5	10.5	10.5	10.5
MgO/wt%	1.5	1.5	1.5	1.5	1.5
R_C μm/min	7.6	7.4	7.2	7.0	6.85
T_{CMax}/°C	970	960	950	940	931
T_{Liq}/°C	1105	1090	1076	1061	1046
Log η @ T_{Liq}	3.597	3.638	3.680	3.723	3.763
T [log (η/dPa s) = 7.6]/°C	742	738	733	728	723

7.5.5 Calcia, CaO

7.5.5.1 CaO Raw Materials

Limestone and limespar are calcium carbonate, $CaCO_3$. Dolomite, $MgCO_3 \cdot CaCO_3$, is the main source of MgO but also contributes CaO. Most SLS glass batches contain limestone and dolomite. Calcium can also be provided by $CaSO_4 \cdot xH_2O$, alkaline earth silicates, and some alumina sources. The use of $Ca(OH)_2$ [131, 149] and CaO [6, 149] has been reported.

Kazazoglu et al. [150] studied batch-free times for float glass batches. Varying the dolomite/limestone ratio resulted in non-linear effects as shown in Fig. 7.15. Minimum batch-free times occurred at MgO/CaO ∼0.25 and ∼0.5. Consequently it may be possible to decrease batch-free times and increase melting efficiency by modifying the dolomite/limestone ratio.

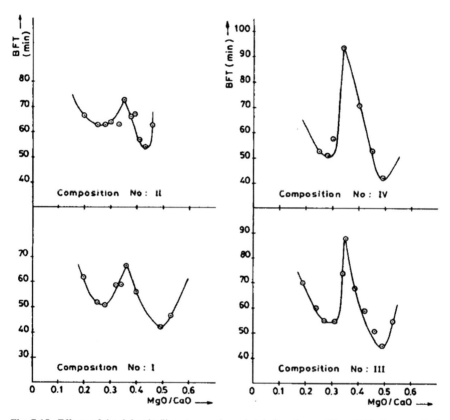

Fig. 7.15 Effects of the dolomite/limestone ratio on batch-free times of four SLS glasses melted at 1450°C, from [150]. Reproduced by permission of Deutsche Glastechnische Gesellschaft

7.5.5.2 CaO Effects on Glass Properties

Commercial SLS glass contains 7–12 wt% CaO (see Table 7.1). Some CaO may be replaced by MgO to tailor the viscosity and crystallization behavior. The effect of CaO as a melting flux is well known, although it also increases surface tension. Structural strengthening resulting from the introduction of Ca^{2+} improves chemical durability and increases thermal expansion coefficient and hardness. However, CaO additions must be limited as they increase T_{Liq} (Fig. 7.7) and crystallization rate.

Bingham [151] found that increasing the CaO content of container glass from 10.9 to 12.9 wt% by replacing SiO_2 decreased $\log (\eta/\mathrm{dPa\ s}) = 2$ by $\sim50°C$ but increased T_{Liq} by $\sim75°C$, and would therefore increase the chance of devitrification occurring. The CaO content must be limited – while it reduces viscosity, too much CaO will result in a glass which readily crystallizes.

7.5.6 Magnesia, MgO

7.5.6.1 MgO Raw Materials

Magnesia is provided by dolomite ($MgCO_3 \cdot CaCO_3$). It can also be provided by $MgCO_3$, $MgSO_4 \cdot xH_2O$, alumina sources and alkaline earth silicates, magnesia (MgO), and burnt dolomite ($MgO \cdot CaO$) [149].

7.5.6.2 MgO Effects on Glass Properties

The addition of MgO to the SiO_2–Na_2O–CaO system considerably reduces T_{Liq} and the crystallization rate [55, 152–154]. As a result, many SLS glasses contain MgO (see Table 7.1). Figure 7.16 illustrates the strong effect of MgO on T_{Liq} of SLS glasses. Within the tridymite primary phase field, increasing replacement of CaO by MgO on a molar [153] and a weight [55] basis reduces T_{Liq}. A minimum T_{Liq} occurs at ~2.5 wt% MgO. When the primary phase field changes from tridymite through

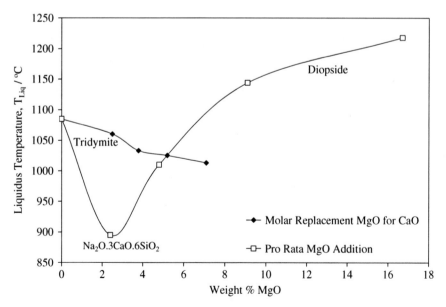

Fig. 7.16 The effect of MgO on the liquidus temperature of SLS glass. Data adapted from [153]. Primary phase fields are marked

$Na_2O \cdot 3CaO \cdot 6SiO_2$ to diopside with increasing MgO addition, then T_{Liq} increases again [55, 153].

MgO mildly decreases high temperature viscosity when replacing SiO_2 (see Fig. 7.6), but its effect is relatively small. When replacing CaO by weight, MgO increases viscosity [55]. However, when MgO replaces CaO on a molar basis a non-linear change in viscosity occurs, with a minimum occurring when both are present in equimolar quantities. This is indicative of a mixed alkaline-earth effect, similar to the well-known mixed alkali effect. Volf [26] reached the same conclusion, noting that the viscosity minimum is associated with a minimum batch-free time. Surface tension is increased when MgO replaces SiO_2, while chemical durability is only slightly impaired [55]. Mechanical properties are improved by adding MgO [26].

7.5.7 Alumina, Al_2O_3

7.5.7.1 Al_2O_3 Raw Materials

Pure Al_2O_3 is expensive and has a very high melting point, 2050°C, therefore its use in SLS glassmaking decreases melting rates and increases melting energies. By comparison, naturally occurring Al_2O_3-rich silicate raw materials such as feldspar and nepheline syenite melt at ~1120–1250°C [16] and contain no chemically bound CO_2 or water, providing a combination of fluxing behavior, melting acceleration, and reduced melting energy. Nepheline syenite and feldspar are alkali aluminosilicate minerals and provide the key sources of alumina for SLS glass. They also provide some of the alkali requirement of the final glass, thus reducing the amounts of expensive alkali carbonates that are needed and the atmospheric emission of CO_2 from their decomposition. Other aluminosilicate minerals that have been considered for SLS glass manufacture include phonolite, trachite, obsidian, and pumice; however, all have high iron contents which preclude their use in all but green glass compositions [133]. Sheridanite, a hydrated magnesium aluminosilicate mineral, has recently been studied as a partial replacement for feldspar and dolomite in SLS glasses [155]. Initial laboratory trials indicated reduced melting temperatures and times using sheridanite.

Blast furnace slag (BFS) has also been successfully used as a batch component in SLS glass manufacture since the 1920s [133] and its use in container and flat glass manufacture is now widespread. Glass-grade BFS is a calcium aluminosilicate material with low iron content. It also contains sulfur, a powerful refining agent, which is present as a mixture of sulfide (S^{2-}) and sulfate (SO_4^{2-}) species. In addition to its refining action, the sulfide can aid the production of amber glass. BFS also provides CaO and MgO to the glass, thereby reducing the requirement for carbonate raw materials and decreasing the batch CO_2 content. As with nepheline syenite and feldspar, BFS has a substantially lower melting point than silica and when used correctly it can substantially improve melting rates and decrease batch-free times and bubble contents [156, 157]. These improvements have contributed to the removal of arsenic oxide, sodium nitrate, and fluorides from glass batches on environmental and/or economical grounds, and decreases in batch sodium sulfate levels. Typical chemical analyses of alumina-bearing minerals are shown in Table 7.13; data has been adapted from West-Oram [133], Doyle [16] and Marriott et al. [157]

Table 7.13 Typical compositions of aluminous raw materials

	Soda feldspar/wt%	Potash feldspar/wt%	Nepheline syenite/wt%	Blast furnace slag (BFS)/wt%
SiO_2	69.2	65.4	60.1	36.0
Al_2O_3	18.7	18.7	23.5	13.0
MgO	–	–	0.1	8.5
CaO	1.8	0.5	0.3	40.0
Na_2O	7.2	3.4	10.6	0.3
K_2O	2.8	11.1	5.1	0.3
TiO_2	–	–	–	0.6
MnO	–	–	–	0.5
Fe_2O_3	0.11	0.06	0.075	0.23
SO_4^{2-}	–	–	–	0.02
S^{2-}	–	–	–	0.7
C	–	–	–	0.2

Commercial BFS as a raw material provides decreased viscosity during the early stages of melting, more rapid homogenization, and decreased surface tension when compared with nepheline syenite, hydrated alumina, and calcined alumina [158]. McBerthy [159] observed that hydrated alumina, $Al(OH)_3$ provides a melting rate approximately equal to that of feldspar, but pure Al_2O_3 is more slowly dissolved. Recent full-scale trials of BFS as a partial and full replacement for feldspar in a green container glass furnace have demonstrated the benefits of BFS over feldspar [157]. Total fuel consumption decreased by 1% (Fig. 7.17) and an increase in furnace pull of 3% above the previous maximum was achieved with improved glass quality and reduced seed count (Fig. 7.18). This was accompanied by a reduction of 25% in electrical boost energy, resulting in a 5% reduction in total energy cost.

7.5.7.2 Al_2O_3 Effects on Glass Properties

When substituted by weight for any commercial SLS glass component, Al_2O_3 substantially increases viscosity (see Fig. 7.6). This can lead to melting and refining difficulties if the Al_2O_3 content is too high. However, the addition of Al_2O_3 substantially improves chemical durability, expands the glass-forming region, and suppresses the tendency for crystallization to occur. As shown in Fig. 7.19, Al_2O_3 is particularly effective in decreasing T_{Liq} and the rate of crystallization when added in small amounts up to $\sim$2 wt% [154, 160–162]. Figure 7.20 illustrates that an optimum aqueous durability is attained at 1.5–2 wt% Al_2O_3 in commercial SLS glasses [163], although higher Al_2O_3 contents may be desirable in some cases to offset the

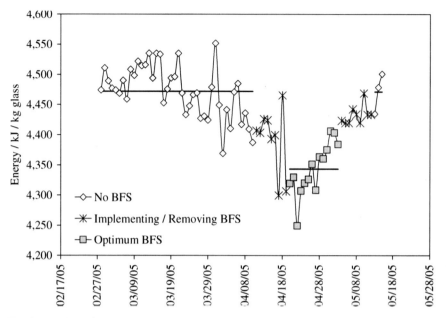

Fig. 7.17 Effect of BFS on energy and fuel consumption of a 250 tpd SLS glass furnace. Data adapted from [157]

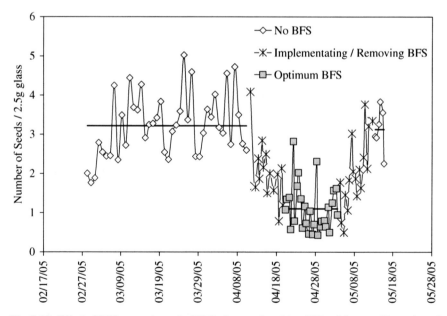

Fig. 7.18 Effect of BFS on seed count of SLS glass produced in a 250 tpd furnace. Data adapted from [157]

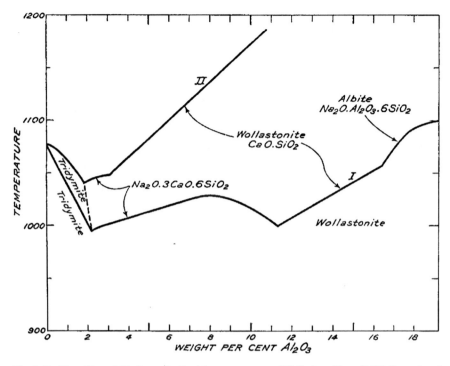

Fig. 7.19 The effect of Al_2O_3 on the liquidus temperature of SLS glass. From [160]. Reproduced with the permission of Wiley-Blackwell Publishing Ltd

effects upon chemical durability of a low SiO_2 content [61]. Substitution of Al_2O_3 for SiO_2 on a weight basis has only minor effects on density and thermal expansion, whereas mechanical properties are improved and surface tension increases sharply.

7.5.8 Potassia, K_2O

7.5.8.1 K_2O Raw Materials

Potassia originates in commercial SLS glasses mainly from the alumina sources. Potassium carbonate, K_2CO_3, is a direct source although it can be costly. Potassium carbonate is available in anhydrous or hydrated (15% H_2O) forms. The hydrated form is easier to handle [133]. Potassium nitrate, KNO_3, and potassium bicarbonate, $KHCO_3$, have both been used in lead crystal glass manufacture [133]. The use of nitrates in SLS glass manufacture has become limited due to the emission of NO_x. Potassium chloride, KCl, has been used occasionally in glassmaking; however, removal of Cl from batches has made the use of KCl redundant.

Fig. 7.20 Effects of Al$_2$O$_3$
on chemical durability of SLS
glass, from [163].
Reproduced with permission
of Wiley-Blackwell
Publishing Ltd

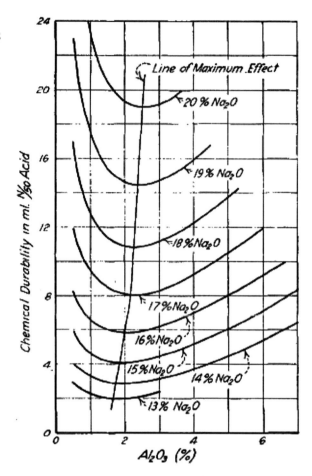

7.5.8.2 K$_2$O Effects on Glass Properties

The addition of K$_2$O to SLS glass on a weight basis has a similar, but less pronounced, effect to Na$_2$O. It decreases viscosity at high and lower temperatures (see Fig. 7.6), providing a "longer" glass. Potassia strongly decreases surface tension and is more volatile than Na$_2$O. Partial replacement of 2 wt% Na$_2$O by K$_2$O in dolomitic SLS glasses results in a shift of primary phase from tridymite (SiO$_2$) to diopside (MgCaSi$_2$O$_6$) when RO content is 12 wt%. The primary phase remains devitrite (Na$_2$O·3CaO·6SiO$_2$) when RO content is 10 wt% and diopside when RO content is 16 wt% [164]. Partial replacement of Na$_2$O by K$_2$O in these glasses resulted in decreases in T_{Liq} by 2–8°C and a substantial increase in log (η/dPa s) = 2 of 20–40°C [164]. Thermal expansion decreased slightly. On a molar basis these effects are expected because total molar alkali content decreases. However, chemical durability was substantially improved [164]. This behavior is attributable, at least in part, to the mixed alkali effect which occurs when two or more alkali types are present [26]. Sen and Tooley [165] concluded that minimum alkali extraction occurs at a ratio of 2.6 K_2O:1 Na_2O by weight.

The K_2O content of a container glass was increased from 0.4 to 3.0 wt% by replacing SiO_2 in a recent study [21]. Refractory corrosion at constant temperature was unaffected by this increase in alkali content, while log (η/dPa s) = 2 decreased by 40°C and T_{Liq} decreased by 25°C. The addition of K_2O also demonstrated the advantages of utilizing the mixed alkali effect. Chemical durability decreased, with the amount of 0.02 N acid required to neutralize the leachate solution increasing from 6.69 to 7.99 ml. However, this change was small by comparison with the effect of increasing Na_2O content of the same base glass by 1.6 wt%. This leachate solution required 10.10 ml of 0.02 N acid to effect neutralization.

7.5.9 Lithia, Li_2O

7.5.9.1 Li_2O Raw Materials

Lithia can be provided by minerals such as spodumene ($Li_2O\cdot Al_2O_3\cdot 2SiO_2$) or petalite ($Li_2O\cdot Al_2O_3\cdot 8SiO_2$), or by man-made chemicals (Li_2CO_3). Typical compositions of selected lithium-bearing raw materials are shown in Table 7.14. As a result of its high cost, lithium has not yet been widely implemented as a raw material for SLS glass manufacture, despite its many documented advantages. This has been an economic decision which may be subject to change given a changing global economic climate.

Table 7.14 Typical compositions of lithium-bearing raw materials

Oxide	Spodumene/wt%	Petalite/wt%	Li_2CO_3/wt%
SiO_2	64–75	74–77	–
Al_2O_3	18–27	15.5–16.5	–
Li_2O	5.0–7.5	3.0–4.0	40.4
Na_2O	0.20–0.25	0.7–1.5	0.15
K_2O	0.20–0.35	0.4–0.6	0.05
MgO + CaO	–	0.7–1.0	0.04
Fe_2O_3	0.1	0.06	0.002
P_2O_5	0–0.25	–	–
CO_2	–	–	59.6

7.5.9.2 Li_2O Effects on Glass Properties

Lithium is a strong melting flux and accelerant for SLS glass and small additions produce large decreases in melt viscosity (see Fig. 7.6). Tang and Frischat [166] studied SLS container glasses with molar substitution of Li_2O for Na_2O producing

Li_2O contents of 0.09–0.43 wt%. Chemical durability was substantially improved upon addition of only 0.18 wt% Li_2O: the volume of 0.01 M HCl required to neutralize leachate solutions decreased by 10%. This improvement is attributable to the mixed alkali effect and to the high field strength of Li^+ by comparison with Na^+, which strengthens the glass network. The decreases in furnace temperature made possible by the addition of Li_2O also provide a decrease in refractory corrosion rates, as shown by Tang and Frischat [166]. The addition of 0.26 wt% Li_2O led to a decrease in refractory corrosion rate from 2.35×10^{-2} to $\sim 1.4 \times 10^{-2}$ mm h^{-1} at a constant viscosity of log $(\eta/dPa\ s) = 2.24$. The effects of small (0.010 and 0.019 wt%) additions of Li_2O (as Li_2CO_3) to SLS container glass was studied by Franklin and Klein [167], who observed that even at such low levels, batch reactions shifted to lower temperatures and were also completed at lower temperatures. Lithium increases surface tension (see also Section 7.4.3.3) [114–116].

Addition of larger amounts of Li_2O to dolomitic flat glass compositions produced substantial decreases in viscosity and T_{Liq} [164]. Replacement of 2.1 wt% Na_2O by Li_2O in a 12 wt% RO glass reduced log $(\eta/dPa\ s) = 2$ by 87°C and T_{Liq} by 20°C. In their study, Bingham and Marshall [21] observed a decrease of 16°C in log $(\eta/dPa\ s) = 2$, but an increase of 17°C in T_{Liq} upon replacement of 0.25 wt% SiO_2 by Li_2O in an SLS container glass. While its effects on viscosity are predictable with reasonable accuracy (see Section 7.4.1.1), the effect of Li_2O on T_{Liq} is strongly affected by the primary phase of the system and the oxide for which it is substituted. Lithium decreases density and thermal expansion coefficient.

7.5.10 Boric Oxide, B_2O_3

7.5.10.1 B_2O_3 Raw Materials

The most significant boron source is Turkey; minerals include rasorite, ulexite, colemanite, tincal, kernite, szaibelyite, and hydroboracite. Some of these minerals are widely used in the manufacture of E-glass [19]. In addition the processed materials boric acid (HBO_3) and borax may be used. Borax is available with three different levels of chemically bound water: $Na_2B_4O_7 \cdot 10H_2O$, $Na_2B_4O_7 \cdot 5H_2O$, and $Na_2B_4O_7$. Borax is hygroscopic, making normal batch handling more difficult. The stable form, $Na_2B_4O_7 \cdot 5H_2O$, is often used because of its compatibility with other raw materials that might pick up moisture and so create caking problems. In the 1920s and the 1930s, borax was trialled as a container glass batch constituent by manufacturers in the United States and United Kingdom. As a result of these trials and associated academic studies, low levels (0.3–1.5%) of B_2O_3 were incorporated [168]. During the 1960s and the 1970s engineering improvements were made and the resulting batch simplification led to the removal of B_2O_3.

7.5.10.2 B₂O₃ Effects on Glass Properties

B_2O_3 is one of the most effective melting accelerants and fluxes (see Fig. 7.6) when added to SLS glasses at levels of up to a few weight percent. Important studies of the effects of B_2O_3 in SLS glass [169–172] have demonstrated that B_2O_3 reduces surface tension, viscosity, and liquidus temperature; initiates glass formation at lower temperatures; and increases the rates of melting and refining. Its addition also reduces thermal expansion and improves mechanical properties [26].

Boron strongly reduces T_{Liq} of SLS glasses [169, 170, 172] and modifies phase field boundaries by shifting the devitrite phase field to lower SiO_2 and higher CaO contents [169], as shown in Fig. 7.21. In addition to the beneficial effects of B_2O_3 itself, it may indirectly allow further reductions in furnace temperature to be made by allowing the CaO content to increase slightly without negatively impacting upon ΔT_{FL}. Crystallization rates are dramatically decreased by B_2O_3: the crystal growth rate of $SiO_2–Na_2O–K_2O–CaO$ glass was halved by the addition of 0.66 wt% B_2O_3 [170].

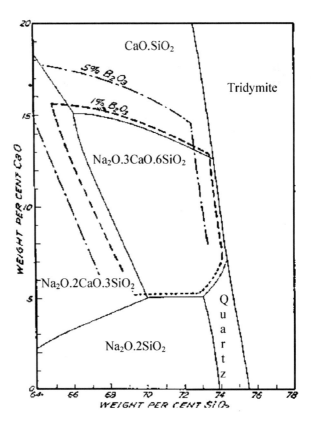

Fig. 7.21 The effects of B_2O_3 on the $SiO_2–CaO–Na_2O$ phase diagram of SLS glass, from [169]. Reproduced with the permission of Wiley-Blackwell Publishing Ltd. Primary phases marked

The Owens-Illinois Glass Company [172] studied the weight-for-weight addition of B_2O_3 to flat glass compositions as a replacement for SiO_2. Addition of 1.1 wt%

B_2O_3 presents the dual benefits of reducing $\log(\eta/\text{dPa s}) = 2$ by 28°C and reducing T_{Liq} by 31°C. Therefore the critically important value of ΔT_{FL} can be maintained.

Allison and Turner [171] carried out a thorough examination of the melting and refining behavior of SLS glasses with various additions of B_2O_3. The isothermal melting and refining rates were significantly increased at all temperatures by progressive addition of B_2O_3. Their results for several different SLS glass types at 1500, 1450, and 1400°C have been converted to provide isoviscous data by modeling the viscosity using Lakatos [29] (see Section 7.4.1.1). The generated data indicate that in most cases, the addition of low levels (< 2 wt%) of B_2O_3 increases melting and refining rates in excess of that which would be expected based on the decrease in viscosity alone. Stevenson [168] calculated that the addition of 1 wt% B_2O_3 to SLS glass would provide a reduction in melting temperature of 20°C that would allow a reduction in thermal NO$_x$ emission of $\sim 20\%$ given a flame temperature of 2100°C.

Aqueous chemical durability is little affected by small additions of B_2O_3. Results for container glass and 10 wt% dolomitic flat glass suggest a small increase [168, 172] while results for 12 wt% dolomitic flat glass suggest a small decrease [172]. Acid resistance of the flat glass is also increased.

Boron is one of the more volatile glass constituents, evidenced by its loss from borosilicate and E-glass, which are melted at higher temperatures than SLS glasses (Table 7.4). As discussed in Section 7.3.3.5, volatilization losses can be more than offset by the reductions in furnace temperature that accompany the addition of B_2O_3.

7.5.11 Sulfate, SO₃

Sodium sulfate has for a long time been the chief refining agent/melting accelerant used in SLS glass manufacture. Other sulfur sources include blast furnace slag (see Section 7.5.7.1). The melting point of Na_2SO_4 is 884°C, therefore it melts during early melting stages. It can enhance formation of wollastonite, $CaO \cdot SiO_2$, at 850–950°C [130]. Molten sulfates have limited solubility in silicate melts, in which they (i) act as surfactants that dissolve sand grains and (ii) form SO_2 and SO_3 which aid refining. Batch sulfate therefore increases melting rates [130, 173, 174]. Its refining action described in (7.25) is a result of thermal decomposition which begins at ~ 1200°C becoming vigorous and efficient at 1450°C [26]:

$$2\,Na_2SO_4 \rightarrow 2\,Na_2O + 2\,SO_2 + O_2 \qquad (7.25)$$

Sulfate has almost completely replaced other refining agents such as arsenic and antimony due to its lower toxicity, lower cost, and its beneficial effects on melt rate. Kloužek et al. [175] used evolved gas analysis (EGA) to study the role of sulfur compounds during SLS batch melting and their results underline the complexities of the reactions involved. Both C and CO act as reducing agents during melting; the ratio of C to Na_2SO_4 and kinetic factors determine the extent of sulfate reduction; and subsequent reactions between Na_2SO_4 and Na_2S and thermal decomposition of the remaining Na_2SO_4 accelerate refining and sand dissolution. Hence the addition of blast furnace slag or carbon (coal, coke) to sulfate-refined batches has become widespread as a means of further accelerating refining.

The predominance of Na_2SO_4 as a refining agent has arisen primarily on economic grounds. However, ammonium sulfate $((NH_4)_2SO_4)$, anhydrite $(CaSO_4)$, gypsum $(CaSO_4 \cdot 2H_2O)$, and barytes $(BaSO_4)$ have all been used as historical sources of sulfate [133, 176, 177]. The practice of using $BaSO_4$ was carried out during the 1950s [176, 177], although the addition of BaO, typically ~0.5 wt%, to container glass had by then been widespread for many years [178]. The use of $BaSO_4$ was phased out during the 1960s and 1970s, probably due to improved furnace efficiencies [179, 180]. Landa et al. [181] have recently added Epsom salt $(MgSO_4 \cdot 7H_2O)$ to grey automotive flat glass as a refining and oxidizing agent giving enhanced Se retention. Anecdotal evidence exists for the widespread use of $CaSO_4$ and $CaSO_4 \cdot 2H_2O$ until relatively recently; however, only a few studies exist on the comparative refining efficacies of different sulfates. Gottardi et al. [182] studied the addition of $MgSO_4$, $CaSO_4$, and $BaSO_4$ to simple SLS glass batches melted at 1300°C and at 1500°C. These sulfates decompose at substantially different temperatures and $BaSO_4$ provided a lower bubble count in the resulting glass than either $MgSO_4$ or $CaSO_4$ (shown in Fig. 7.22). No detailed study of the relative refining efficacies of different alkali sulfates has, to this authors' knowledge, ever been published. Guy [183] compared the effects of Na_2SO_4, $CaSO_4$, $BaSO_4$, and $FeSO_4$ in laboratory SLS glass melts and concluded that at 1400°C refining is more rapid using Na_2SO_4 but at 1440°C both $CaSO_4$ and $BaSO_4$ provide superior refining.

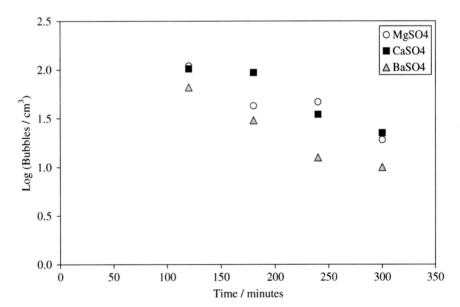

Fig. 7.22 Bubble content of SLS glass melts as a function of batch sulfate type. Data adapted from [182]

Conroy et al. [173] studied the relative batch-free times of simple SiO_2–Na_2O–CaO glasses refined with 0.6 wt% of Na_2SO_4, $CaSO_4$, $BaSO_4$, $PbSO_4$, $SrSO_4$, and $(NH_4)_2SO_4$, showing that batch-free time increased with decreasing SO_4^{2-} content. This trial did not determine the individual refining efficacies of each refining agent based on equal SO_4^{2-} contents, and therefore it is not possible to extract such information; however, the authors noted that the melting rate change was not linear and was also affected by the nature of the compounds used, lending qualitative support to the results of the other trials discussed above.

Refining of SLS glass can be tailored, to an extent, by using different sulfates or mixtures of sulfates. This can be partly attributed to the temperatures at which the different sulfates melt, as shown in Table 7.15. Römer et al. [184] noted similar benefits and also advocated the use of "preferred sulfates" including $MgSO_4$, $CaSO_4$, $SrSO_4$, $BaSO_4$, and K_2SO_4 for the refining of aluminosilicate glasses.

Table 7.15 Melting temperatures of some inorganic sulfates

Sulfate	Melting temperature/°C
$Al_2(SO_4)_3$	770
Li_2SO_4	859
Na_2SO_4	884
K_2SO_4	1069
$MgSO_4$	1124
$CaSO_4$	1450
$BaSO_4$	1580

A strong case exists for using alternative sulfates, or mixtures of different sulfates, in order to provide optimum silica dissolution and refining while minimizing SO_x generation. It is conceivable that in doing so, this would allow furnace temperatures to be decreased and/or pull rates to be increased while retaining the same quality of refining. A comparable conclusion is provided by Doyle [16], who notes that a mixture of different nitrates is preferable to a single nitrate for the refining of lead crystal glass.

The solubility of SO_3 in commercial SLS glass chiefly depends on furnace atmosphere, temperature, and residence time. Glass composition also influences SO_3 solubility [185, 186]. Ooura and Hanada [186] investigated the solubility of SO_3 in a range of ternary SiO_2–Na_2O–RO (R = Mg, Ca, Ba) glasses. Solubility increased dramatically with increasing Na_2O content and with increasing alkaline earth cation size. These effects can be broadly described by a linear relationship linking log (SO_3) capacity and the sum of the normalized partial cation field strengths of the constituent cations [187]. However, the relatively small compositional changes available for practical reformulation of SLS glasses suggest that sulfur capacities will vary only slightly following any reasonable degree of chemical reformulation.

7.5.12 Water, H_2O

Glassmaking involves charging a mixture of chemically and physically different solids with a range of particle sizes. As a result, batch segregation may occur unless handled properly. The presence of water during batch mixing and charging is essential to provide a homogeneous, well-mixed batch. Batch wetting has evolved into a practical way of maintaining the chemical uniformity of the batch, reducing furnace carryover, and dusting problems. Batch wetting fulfills three requirements [26]: (i) prevention of dusting; (ii) promotion of adhesion of flux to sand grains; (iii) suppression of batch segregation. Liquids added to batches have in the past included caustic soda, sodium silicate, alcohols, and surfactants [26, 133]. However, water has become by far the most widely used due to its low cost.

Typical SLS glass batches contain water which is usually added to the sand. Container glass batches may have 4.5 ± 0.5 wt% water and flat glass 5.0 ± 0.5 wt% water added. The improvement in resistance to batch segregation that occurs upon adding water to a typical SLS glass batch is evidenced by increasing batch angle, θ. This ensures that batch does not become segregated before it enters the furnace. Lehman and Manring [188] demonstrated that θ increased from 16.5 to 40.7° when 4 wt% water was added to SLS batch. It may also be advantageous to maintain batch temperatures within the range 35.4–109°C, since $Na_2CO_3 \cdot H_2O$ is stable at these temperatures. At lower temperatures $Na_2CO_3 \cdot 7H_2O$ and $Na_2CO_3 \cdot 10H_2O$ may form, which tie up water intended for moistening the batch and requiring extra energy to drive off during melting [188]. However, this does not yet appear to be a particularly widespread practice. Examination of the microstructure of wet batch revealed that substantial dissolution of soda ash occurred [188]. Upon recrystallization this produced the bonding medium that leads to agglomeration. This intimate mixing ensures decreased melting times, improved refining, and increased homogeneity.

A measurable decrease in viscosity occurs at temperatures near T_g with only ppm water contents [189] although smaller decreases occur at melting temperatures. Differences in crystallization behavior; density; and mechanical, optical, and electrical properties are also caused by the addition of ppm levels of water [90, 190]. Commercial SLS glasses melted in gas-fired or oil-fired furnaces typically contain 280–400 ppm of water. However, glasses melted electrically contain lower levels (100–200 ppm) and those melted in oxy-fired furnaces contain higher levels (350–650 ppm) [190–193]. The water content of sulfate-refined SLS glass can vary as a function of redox conditions, reaching a maximum value that corresponds to the minimum sulfur solubility which occurs under mildly reducing conditions [191]. The solubility limit of water in SLS glass is 1000–1200 ppm by weight at temperatures of 1000–1400°C [90]. Water is dissolved in SLS glass as –OH groups which can be detected by infrared (IR) spectroscopy. Characteristic absorption peaks occur at 2.7 μm (X–OH stretch where X = Si, H), 3.6 μm (H bonding), 4.2 μm (H bonding), and 6.2 μm (H–O–H bending). Overtones occur at 1.41, 1.9, and 2.22 μm [190, 194].

The presence of water in glass batches, melts, and atmospheres is, however, not entirely beneficial. Batch water contributes to the corrosion of furnace refractories, particularly those in the crown, by reacting with alkali species in the batch to form alkali hydroxides, which readily volatilize at furnace temperatures [25, 195, 196]. The addition of 4 wt% water to batch sand equates to 28 kg water per tonne of glass produced. The energy to raise 28 kg water from an entry temperature of 20°C to steam at a typical exhaust gas temperature of 550°C is 98 MJ. This is ~2% of the typical SEC of ~5 GJ/tonne for SLS glass, therefore careful control and minimization of batch water content are required in order to balance the need for batch uniformity with melting efficiency and fuel economy.

7.5.13 Chlorides and Fluorides

Fluoride and chloride have long been used in SLS glassmaking as melting accelerants and refining agents. As discussed by Volf [26], halides in glass batches (i) decrease surface energy and (ii) form eutectics and increase solid state reactions. Hong et al. [129] studied the effects of melting accelerants Na_2SO_4, NaCl, and $NaNO_3$ on batch fusion, concluding that NaCl was most effective due to formation of a $NaCl–Na_2CO_3$ eutectic phase at 636°C, which readily attacks and dissolves silica grains. Fluoride strongly decreases high temperature viscosity: replacement of 1 wt% SiO_2 by F_2 in float-type SLS glasses resulted in a reduction in melting temperature, $\log (\eta/dPa\ s) = 2$, of 33°C and in softening point ($\log (\eta/dPa\ s) = 7.65$, of 20°C [197]. This fluoride addition also improved chemical durability. Fluoride enjoyed widespread use as a component of SLS glasses until the 1960s, before which it was typically present at 0.2 wt% [178]. Chloride has a solubility of ~1 wt% in SLS glass and the addition of NaCl increases melting rates through the formation of eutectics. According to Fluegel [50], Cl decreases high temperature viscosity but less strongly than F.

Atmospheric emissions of F and Cl compounds are now limited by legislation (see Sections 7.3.2 and 7.3.3.5) and as a result they were largely removed from SLS batches during the 1960s and the 1970s [37, 179, 180]. Given the high volatility of halides, their use could lead to increased corrosion of furnace and heat exchanger refractories, and this, coupled with the other restrictions, suggests that in most cases the deliberate addition of halides to commercial SLS glasses will not be considered in the short term.

However, the use of abatement technologies can dramatically reduce emissions and the filter dust can be reintroduced to the furnace as a raw material (see Section 7.5.19). Kircher [31] noted that even when acid gas scrubbing measures are used, emissions of HF and HCl remained problematic, suggesting that reduced emissions at source (for example, due to lower furnace temperatures) and improvements to abatement technology would be beneficial. All-electric melters do not suffer from the same emissions problems as fossil fuel – fired furnaces and the F_2 and Cl_2 are retained in the melt [133]. It is therefore conceivable that the use of low levels of

F_2 and/or Cl_2 in SLS glass manufacture may one day become more attractive again, should the development of new melting and abatement technologies permit it.

7.5.14 Baria, BaO

Historically, BaO has been used in SLS glass manufacture through the use of $BaSO_4$ as a refining agent [133]. This began in the 1920s and was still widely practised into the 1990s [37, 177, 178, 198]. The BaO contents of these glasses are typically 0.5 wt%. Barium oxide is widely used in glasses for display applications and also in lead-free crystal glass. In small amounts, BaO causes little refractory corrosion, however, at high levels (>30 wt%) it can be highly corrosive. It mildly decreases viscosity (see Fig. 7.6) and increases the meltability of SLS glasses. Barium has the effect of making "long" glasses; that is, the working range is extended, allowing more time to work the glass in the critical temperature range. Surface tension is decreased by BaO: this effect is related to the low field strength of the Ba^{2+} cation. It also has the effect of increasing refractive index, density, and thermal expansion when replacing CaO. Chemical durability, on the other hand, is impaired slightly by such replacements.

The addition of BaO to dolomitic SLS glasses was studied by the Owens-Illinois Glass Company [199]. Stepwise additions of 1, 3, and 5 wt% BaO were made for SiO_2, Na_2O, and (MgO·CaO) in glasses with analyzed starting compositions of 73.6 SiO_2, 0.2 Al_2O_3, 7.1 CaO, 4.9 MgO, 13.9 Na_2O, and 0.2 SO_3. Liquidus temperature decreased markedly upon substitution of SiO_2 by BaO, and decreases of ~20°C, 45°C, and 60°C in T_{Liq} occurred at 1, 3, and 5% BaO additions, respectively. However, partial replacement of (MgO·CaO) or Na_2O resulted in large increases in T_{Liq} of up to ~70°C and 265°C at 5% BaO addition. Melting temperatures (defined as log (η/dPa s) = 2) exhibited a minimum at 3 wt% replacement of SiO_2 by BaO; however, the replacement of (MgO·CaO) and Na_2O both led to large increases in melting temperatures. The addition of BaO did, however, produce a substantially "longer" glass, decreasing log (η/dPa s) = 7.65 by ~10°C for only 1 wt% replacement of SiO_2 or MgO CaO. Only when replacing Na_2O did the glass become "shorter." The effects on aqueous chemical durability were small at 1 and 3% additions; however, density was substantially increased.

7.5.15 Zinc Oxide, ZnO

In terms of its effects on physical properties, zinc oxide behaves similarly to the alkaline earth oxides in SLS glasses, generally falling between MgO and CaO in terms of its effects on viscosity (Fig. 7.6), refractive index, and thermal expansion when considered on a weight percent basis. However, in other aspects ZnO behaves differently. For example, it dramatically improves aqueous chemical durability. Its effects in dolomitic SLS glasses have been studied by the Owens-Illinois

Glass Company [199], who investigated wt% replacement of SiO_2, (MgO·CaO), and Na_2O by 1.2, 3.3, and 5.4 wt% ZnO. Replacement of 1 wt% SiO_2 decreased log $(\eta/dPa\ s) = 2$ by $\sim 16°C$ and T_{Liq} by $\sim 28°C$; however, upon further ZnO additions T_{Liq} increased. Replacement of (MgO·CaO) and Na_2O led to increases in viscosity and T_{Liq}, but substantially increased resistance to both water and dilute acids. Density is increased by the addition of ZnO, regardless of the oxide that it replaces. Given the substantial cost of zinc raw materials, large additions of ZnO may not be feasible. However, the beneficial effects of small ZnO additions on melting and liquidus temperatures are noted. Zinc additions can improve clarity and brightness, according to Volf [26]. It can also aid color striking in ruby glasses. However, depending on thermal history and sulfur content, ZnO additions can also lead to opacity in glasses which are melted under reducing conditions due to the formation of ZnS [26].

7.5.16 Strontia, SrO

Strontium is not widely used in glass manufacture owing to the high cost of $SrCO_3$, the main source of SrO. However, SrO has received some use in glasses for display applications [200]. Few studies of the behavior of SrO in SLS glasses have been made; however, substitution of SrO for CaO on a weight basis in SLS glasses with wt% compositions of 74.8–77.8 SiO_2, 0.2 Al_2O_3, 14–17 Na_2O, 8 RO where $R =$ Ca, Sr has been examined by the Owens-Illinois Glass Company [201]. The substitution of 8 wt% SrO for CaO increases T_{Liq} by 45–95°C and shifts the cristobalite field toward lower RO and SiO_2 contents. Volf [26] also noted that T_{Liq} of SrO-containing glass is higher than T_{Liq} of corresponding CaO-containing glasses. The effect on viscosity of replacing CaO by SrO was to "lengthen" the glass; increasing the temperature corresponding to log $(\eta/dPa\ s) = 2$ by $\sim 25°C$ and decreasing the temperature corresponding to log $(\eta/dPa\ s) = 7.65$ by $\sim 15°C$ [201]. Replacement of CaO by SrO also resulted in marked decreases in chemical durability toward both water and dilute acid, and notable increases in the coefficient of thermal expansion and density. Surface tension is increased by the replacement of CaO by SrO. The technological benefits of using SrO in SLS glasses appear to be limited; a case that is strengthened by economic considerations.

7.5.17 Multivalent Constituents

7.5.17.1 Colorants and Refining Agents

Transition metals, particularly Fe and Ti, occur in all commercial SLS glasses at some level due to their presence as a contaminant in the raw materials, chiefly those obtained from mineral sources. Iron and chromium oxides may be deliberately added to SLS batches as iron oxide (FeO, Fe_2O_3, or Fe_3O_4), chromium oxide (Cr_2O_3), or chromite ((Fe, Mg)O·(Cr,Al)$_2O_3$) at < 1 wt%, in order to produce the green color widely found in container glass and in some tinted flat glasses. The iron redox ratio, Fe^{2+}/Fe^{3+}, determines the final color of the glass and affects heat transfer during melting (see Section 7.4.3.2). Chromium is relatively stable in SLS glass as Cr^{3+}, although the highly toxic Cr^{6+} should be avoided by controlling melting conditions. Other colorant transition metal oxides including CoO, NiO, or CuO may also be used at low levels in tinted or colored glass batches. In addition to restrictions on the atmospheric emission of Cr, Co, and Ni [12], the use of colorants present additional environmental interest through their effects on heat transfer during melting and consequently upon specific melting energies.

Oxides of multivalent elements have been widely utilized as refining agents in glassmaking. Historically, oxides of As and Sb have been used [26, 133, 202, 203]. The refining action of As and Sb [90] is summarized below, which shows the oxidation of As_2O_3 to As_2O_5 by reaction with batch nitrates (7.26) and the subsequent release of O_2 into the melt at high temperature (7.27):

$$4\,RNO_3 + 2\,As_2O_3 \rightarrow 2\,R_2O + 2\,As_2O_5 + 4NO \uparrow + O_2 \uparrow \qquad (7.26)$$

$$As_2O_5 \rightarrow As_2O_3 + O_2 \uparrow \qquad (7.27)$$

The use of As and Sb in any glass is being consigned to the past owing to their prohibition by environmental legislation [12]. However, several other, less toxic components exist which can release oxygen at high temperatures and some of these have enjoyed prior use as refining agents.

CeO_2 and MnO_2 have been used for many years as colorants/oxidizing agents/refining agents in the manufacture of certain glasses [26, 78, 90, 133, 176, 177, 198], although widespread use of CeO_2 may be prevented by its relatively high cost. Cerium can also be particularly useful as a decolorizing agent and as a colorant for increased UV absorption. In commercial SLS glasses iron is the principal color contaminant, and it is desirable to oxidize the strong blue–green Fe^{2+} to the weaker yellow–green Fe^{3+} by the reaction shown in (7.28):

$$Ce^{4+} + Fe^{2+} \rightarrow Ce^{3+} + Fe^{3+} \qquad (7.28)$$

Historically, manganese dioxide, MnO_2, fulfilled the role of oxidizing agent, undergoing a similar mutual redox interaction with iron, as described in (7.29):

$$Mn^{3+} + Fe^{2+} \rightarrow Mn^{2+} + Fe^{3+} \qquad (7.29)$$

Decolorized glasses are prone to change color over time, following exposure to high-energy UV photons in sunlight [78]. This phenomenon is known as solarization and is due to the partial reversal of the redox reactions shown in (7.28) and (7.29). This is demonstrated by the results of King et al. [204], which show that 8 years' simulated sunlight upon a commercial float glass containing 0.2 wt% CeO_2 resulted in spectral changes in the ranges 300–425 nm and > 800 nm. These spectral changes are associated with shifts in the Ce^{3+}/Ce^{4+} and Fe^{2+}/Fe^{3+} ratios and signify solarization. Cerium-doped sodium silicate glasses exhibit improved resistance to solarization by comparison with undoped glasses; however, the presence of more than one multivalent species may compromise this behavior [205]. Solarization can be particularly problematic for Mn-decolorized glasses, for which purple coloring is due to the presence of Mn^{3+} caused by the partial reversal of (7.29) [78]. This problem led to the disuse of Mn as the main decolorizer in commercial SLS glasses [176, 177, 198]. Alternative decolorizers such as Se have been used for some time; however, it may be feasible to use MnO_2 in colored glass batches.

In addition to their role as decolorizers, CeO_2 and MnO_2 can be useful refining agents due to decomposition and liberation of oxygen at high temperatures. CeO_2 decomposes to form $Ce_2O_3 + O_2$ at 1400°C [26]. However, its strongest refining occurs when used in conjunction with other refining agents, particularly sulfate [26]. Johansson [202] compared the nitrate-assisted refining actions of CeO_2, As_2O_3, and Sb_2O_3 in lead alkali silicate glasses. At 1400°C, CeO_2 was found to be less effective than Sb_2O_3 but nearly as effective as As_2O_3. Decomposition of MnO_2 begins at 530°C [26], and minor refining actions have also been noted due to the thermal decomposition of other oxides, including Fe_2O_3 and Pb_3O_4 [90] and SnO_2 [184, 202].

Sulfur, in addition to the refining action provided by its sulfate (SO_4^{2-}) form, can occur in its S^{2-} sulfide form when glass is melted under sufficiently reducing conditions. Sulfide, in conjunction with Fe^{3+}, provides the well-known amber chromophore, centered at $\sim$415 nm, which is necessary for the production of amber glass [206, 207].

Selenium is an expensive, volatile, yet widely used batch component, the atmospheric release of which is restricted by legislation [12]. It is therefore important to reduce the volatilization of Se as far as possible. Selenium, together with ppm levels of CoO, has become a widespread batch addition for decolorizing SLS glass [208–210], although Se is also used to produce colored automotive glasses [181, 211, 212]. Selenium occupies several oxidation states in SLS glass; these include Se^{6+}, Se^{4+}, Se^0, and Se^{2-}. Elemental Se^0 provides a pink color and Se^{2-} a brown color due to iron polyselenide (FeSe) formation. As with sulfur, the retention and speciation of selenium in oxide melts are dependent on furnace conditions [209, 210]. Selenium is added to SLS glass batches through a combination of cullet, recycled filter dust, and batch selenium (usually metallic Se or sodium selenite, Na_2SeO_3). Typical Se retention in the final glass is less than 20%, although this can be increased toward $\sim$50% through the use of oxidizing agents and oxidizing melting conditions [181, 210–212]. The chemical and physical form in which Se is introduced can also influence its retention. Knapp [213] indicated that alternative Se sources such as Na_2SeO_3, $ZnSeO_3$, ZnSe, or $BaSeO_3$ can provide greater Se retention in SLS glass than using metallic Se.

7.5.17.2 Effects on Physical Properties

The presence of Fe_2O_3 in dolomitic SLS glasses led to only small changes in high temperature viscosity and T_{Liq} in one study [214]. However, Fluegel [50] has indicated that at levels of < 1 wt% Fe_2O_3 can have a moderate fluxing effect similar in magnitude to that provided by CaO. This apparent discrepancy may be due to accuracies of measurement and compositional differences. Regardless of this disagreement the strong coloration caused by iron usually limits its addition to < 1 wt%, and therefore its effects on viscosity and T_{Liq} are small.

Cerium has been added to glasses in the past to impart a straw-yellow coloration. It also increases UV absorption, similarly to titanium. Indeed, cerium–titanium couples have been used for just this purpose. Technologically, cerium raw materials can be moderately expensive. Cerium increases refractive index, decreases thermal expansion coefficient, and substantially decreases viscosity when added at low levels [35, 50].

Titanium dioxide is a colorant when used in very high concentrations over 5 wt% [26]; however, such concentrations are impractical owing to its relatively high cost. Indeed, the addition of relatively low levels of titanium into iron-containing glasses can improve UV-absorbing properties. This effect only occurs when iron is present in addition to titanium. Container glass contains typically 0.1% Fe_2O_3 which colors the glass a pale green. Increased levels of TiO_2 (normally present at 0.03–0.05% TiO_2 in container glass) would greatly improve UV absorption, which is very desirable for certain products such as beers and other drinks which can be degraded by UV radiation. Even small amounts of titanium decrease markedly the high-temperature viscosity [35, 50], thus shortening refining times and allowing glasses to melt at lower temperatures. However, it can shorten the working range at lower temperatures. TiO_2 decreases thermal expansion and improves chemical durability, particularly with respect to attack by acids. Titanium is well known as a nucleating agent in glass ceramics, so it is expected to increase the devitrification tendency if present in large amounts.

7.5.18 Other Compounds

For the majority of the remaining elements in the Periodic Table and their oxides, there exist several reasons for their omittance from further consideration. Many of them can be incorporated into glass, however, they tend to be prohibitively expensive, particularly toxic and/or prohibited by legislation. Some cause unacceptable coloration, while others have negative effects on properties.

Calcium phosphate is a major raw material for bone china, and phosphates are produced in huge quantities, both intentionally and as by-products from other manufacturing processes. Other potential phosphate-rich raw materials include meat and bone meal ash arising from incineration operations. It is a low-cost source of P_2O_5; however, its variability has not been fully assessed and its use may present additional difficulties.

The solubility of P_2O_5 in SLS glasses is poor and the addition of > ~1.5 wt% leads to phase separation and opacity. The replacement of 1 wt% of SiO_2 by P_2O_5 in SLS container glass [215] and in dolomitic flat glass [197] causes no change in high-temperature viscosity. The container glass exhibited a decrease of 15°C in T_{Liq} [215] and a similar effect has been observed in sheet glass. No change in T_{Liq} was observed for dolomitic flat glass with the addition of 1 wt% P_2O_5 [197] but a shift in primary phase field from diopside to tridymite did occur. Chemical durability and density of SLS glasses exhibit only small changes with such addition of P_2O_5.

Lead oxide has the effect of significantly reducing viscosity and increasing the working range when added to SLS-type glasses. Its behavior is similar to that of K_2O when compared on a wt% basis (see Fig. 7.6). Lead silicate glasses show low devitrification tendency, and PbO reduces liquidus temperature when added to SLS glasses. Unfortunately PbO-containing glasses aggressively attack furnace refractories at high temperatures and PbO is prone to reduction to its metallic form, further attacking refractories [26]. PbO greatly increases density, thus fewer articles could be made from the same weight of glass. Legislation prohibits the use of PbO in SLS container glass and strongly restricts Pb-bearing atmospheric emissions (see Section 7.3.2). This combination of undesirable effects, combined with the relatively high cost of PbO raw materials and their high toxicity, means that the use of PbO in commercial SLS glass manufacture will probably receive little support for the foreseeable future. Bismuth behaves similarly to lead in oxide glasses and its use is not currently prohibited by legislation; however, the high cost of bismuth raw materials and many of the same problems that can occur using PbO also serve to preclude bismuth from SLS glass manufacture.

Rubidium and cesium oxides provide behavior similar to that of the other alkali oxides; however, on a weight basis their effects become smaller with increasing molecular mass. The high cost of Rb and Cs raw materials and the high volatility of these alkalis at furnace temperatures [26] prohibit their widespread use in glassmaking.

Beryllium oxide is highly toxic and causes berylliosis, while increasing high-temperature viscosity if replacing CaO, MgO, or even Al_2O_3. Its raw materials are also expensive, precluding its use in commercial SLS glassmaking.

Zirconia is not deliberately added to SLS glass but is usually present at levels of ~0.01 wt% due to contamination from furnace refractories and silaceous raw materials. At these levels ZrO_2 has little effect on properties. Zirconia gives rise to large increases in chemical durability but also to high-temperature viscosity and liquidus temperature, which precludes its use in SLS glasses.

Tin is present at low levels on float glass as a result of contamination from the float bath. As a deliberate addition, its low solubility can lead to phase separation and opacity on cooling, hence its widespread use in opal glazes. Tin dioxide decreases viscosity in the low temperature region [26], increasing the working range. Its addition increases density. Römer et al. [184] described the use of low levels of SnO_2 as a high-temperature (> 1500°C) refining agent for aluminosilicate glasses. The refining action is generated by the reduction of Sn^{4+} to Sn^{2+} and the associated release of oxygen at 1600–1900°C. The use of SnO_2 as a refining agent in SLS glass manufacture may be debatable owing to the lower furnace temperatures that are available.

7.5.19 Recycled Filter Dust

Strict legislational limits on the atmospheric release of SO_x and other volatiles such as Cl and heavy metals have necessitated the installation of waste gas treatment plant at many glass furnace operations. These plants typically consist of electrostatic precipitators (EPs) or fabric (bag) filters to remove particulates and dry or semi-dry scrubbing to remove the gaseous species SO_3, SO_2, HF, HCl, and Se compounds [2, 12]. The scrubbing medium used is usually $Ca(OH)_2$, Na_2CO_3, or $NaHCO_3$ and the reaction product is therefore largely a mixture of sulfates, halides, and selenides of calcium or sodium. Approximately 3 kg of filter dust is captured for every tonne of SLS glass melted [32], and the simple yet elegant solution is to re-introduce it to the batch. The re-introduction of filter dust into glass batches has been particularly well documented in Europe [24, 30, 32, 216]. However, implementation is not entirely straightforward: the oxidation state of the filter dust can be reduced [30, 32] depending on the type and particle size of the absorbent/s, the partial pressures of O_2 and H_2O in the waste gas stream, and on temperature and time of reaction. Moreover, handling of filter dust can be difficult owing to its hygroscopic nature and small particle size. Therefore, while reuse of filter dust is both welcome (from environmental and economic perspectives) and widely practised, it must be carried out carefully and with appropriate consideration.

7.5.20 Nitrates

Thermal decomposition of nitrate salts during glass melting releases O_2 and NO_x. Nitrates, particularly $NaNO_3$ and KNO_3, have therefore been used in the past, in conjunction with other multivalent species, as refining agents [26]. Indeed, nitrates are viewed as a necessary constituent to enable the refining action of As or Sb [90]. The addition of $NaNO_3$ has a minor effect as a melting accelerant [130]. Nitrates have become little used in SLS batches because sulfates decompose at higher temperatures than nitrates. As a result, the refining action of sulfates occurs later in the melting process, providing greater refining efficiency than nitrates. However, a combination of sulfate and nitrate refining agents might, in some circumstances, provide greater benefits than using one or the other alone. Amrheim et al. [217] described the introduction of ~0.1 wt% of dry float glass batch as nitrates of sodium, magnesium, calcium, or potassium. The nitrate suppresses early volatilization of sulfur species, making a greater percentage of batch sulfur available when it is required during melting and refining stages. This in turn allows for a reduction in batch sulfur levels and consequently in SO_x emissions. The introduction of these batch nitrates, it was thought, may allow changes in air/fuel ratio that would offset the increase in NO_x generation.

7.6 Glass Reformulation Methodologies

The first step in any glass reformulation process is, as discussed in Section 7.2, to fully define the goal/s, to establish design criteria and glass property requirements, and to define the constraints that are placed upon the reformulated batch. Constraints vary widely and may, for example, include practical considerations such as the number of spare batch hoppers for a particular furnace. It is therefore likely that each manufacturer will have different requirements, priorities, and constraints, and therefore reformulation methodologies can be expected to vary.

Once requirements and constraints have been established, the development of a new or modified SLS glass composition may proceed. Table 7.16 shows a compositional checklist which enables a rapid initial assessment of the potential effects of adding a prospective component to a typical SLS glass formulation upon several of the key properties discussed in Section 7.4. The important property requirements occupy columns and components occupy rows. The effect of increasing the abundance of each component upon each property is represented by the degree of shading of the box. White signifies acceptability, and the darker the shading, the greater the likelihood of problems occurring if the abundance of that component is increased. Black shading signifies that this component is completely unacceptable, for example, on the basis that the deliberate addition of a component is prohibited by legislation. In this way, unacceptable changes can be rapidly rejected and development work can focus upon options that may be (a) technologically sound and (b) worthy of further consideration on logistical and economic grounds.

Once the initial starting compositions have been established, the next steps are to model the key properties of these hypothetical glass compositions using appropriate models (many of which have been described in this work) and to make an early assessment of their potential economic viability. For example, a model of the viscosity–temperature profile for a glass to which 5 wt% Li_2O has been added would, by itself, suggest that this should be carried out. However, economic constraints must be considered at such an early stage to ensure that "common sense" is applied and the Li_2O content is greatly reduced to a level that is more economically favorable.

7.6.1 Worked Examples and Implementation

Reformulation of industrial glass should use a considered, scientific approach involving desk-based development and modeling, then laboratory-scale trials, prior to any larger scale trials and, ultimately, implementation. It is insufficient to establish the target glass composition and confirm that the physical properties of the final

Table 7.16 Component/parameter checklist for individual SLS glass components

Overall Rating	Raw Material Costs	Environmental Concerns/ Legislation	Viscosity	Devitrification	Thermal Expansion/ Shock	Optical Properties	Refractory Corrosion	Chemical Durability	Hardness/ Abrasion Resistance	Density	Surface Tension
SiO_2			▓								▒
B_2O_3	▒	▒					▒				
P_2O_5	▓		▓	▒							
Li_2O	▓				▒		▒		▒		
Na_2O					▒			▒	▒		
K_2O	▓				▒			▒	▒		
MgO					▒						
CaO					▒						
SrO	▓				▒						
BaO	▒	▒			▒						
ZnO	▒										
PbO		██					▓		▒	▓	
Al_2O_3			▓								▒
TiO_2						▒					
ZrO_2	▓		▓								
CeO_2	▒					▒					
Bi_2O_3	▓						▓	▒		▓	
F, Cl		██					▓				
Others	▒	▒	▒	▒	▒	▒	▒	▒	▒	▒	▒

Key: ☐ Acceptable ▒ ▓ ██ Unacceptable

glass are satisfactory: several other factors must also be considered. For example, the introduction of a new raw material can affect melting and refining behavior. Batch redox may change and levels of other refining/redox-active components such as saltcake and coke may require further adjustment to provide the correct level of refining while minimizing SO_x emissions and avoiding foaming or reboil. The iron content of the reformulated glass may be different from the original glass, therefore adjustment of the levels of decolorizing agents such as Se (see Section 7.5.17) may be required. These examples illustrate the complexities of reformulation and the requirement to consider every factor.

Thus far the primary aspects of SLS glass manufacture have been discussed from the standpoint of energy efficiency and environmental considerations. In this section, three worked examples are given. These provide a practical viewpoint to complement the scientific aspects previously discussed. Models and computer programs now aid the process of calculating optimal glass batches (see, for example, [218–221]), and computer programs are now widely used within the glass industry for this task. Here we consider three different reformulation strategies which present the flexibility required by the glass industry, given differences in company strategy, priorities, and size, together with practical constraints such as the availability of batch plant. These three approaches here are termed compositionally constrained, batch-constrained, and unconstrained reformulation. Here they are applied to SLS glass compositions; however, the principles can be applied to any commercial glass formulation.

(1) Compositionally constrained reformulation. Implementation of a new raw material while maintaining the original glass composition.
(2) Batch-constrained reformulation. Reformulation using only those raw materials that were used originally while providing a new glass composition.
(3) Unconstrained reformulation. Reformulation utilizing one or more new raw materials and providing a new glass composition.

Properties of the resulting glass that may vary as a result of reformulation according to these approaches, 1, 2, or 3 are shown in Table 7.17

For any given glass composition, a number of possible raw material permutations exist. As expected, the lowest cost raw material permutation is generally utilized. Cullet optimization is a key tool for minimizing specific energy consumption (SEC), as discussed in Section 7.5.2. However, the substantial majority of glass manufacturers have already optimized batch cullet levels so far as their existing melting technologies, infrastructure, and cullet supply will allow. In addition, the presence of cullet in any SLS glass batch presents a challenge for the unconstrained reformulation strategy described above, namely that the cullet composition will differ from the reformulated target glass composition. As a result batches will need to compensate for this difference as discussed, for example, by Khaimovich [222]. On the other hand, the chemical composition of cullet itself varies, therefore a degree of manufacturing tolerance is already "built-in." For the purposes of simplicity in the illustrative worked examples discussed here cullet has been omitted.

Table 7.17 Physical properties and parameters that may (✓) and may not (×) vary as a function of the reformulation approach employed

Methodology	1. Composition	2. Batch	3. Unconstrained
Raw materials selection	✓	×	✓
Viscosity–temperature relationship	×	✓	✓
Liquidus temperature and crystallization rate	×	✓	✓
Specific heat capacity	×	✓	✓
Thermal conductivity	✓	✓	✓
Optical properties	✓	✓	✓
Electrical properties	×	✓	✓
Refining and melting rate	✓	✓	✓
Refractory corrosion	✓	✓	✓
Surface energy	✓	✓	✓
Chemical durability	×	✓	✓
Density	×	✓	✓
Thermal expansion	×	✓	✓
Mechanical properties	×	✓	✓

7.6.1.1 Reformulation Constrained by Composition

Compositionally constrained reformulation is the introduction of one or more new batch raw material which produces a glass with the same chemical composition as the original one. The chemical composition of the new raw material/s will almost certainly differ from those previously in use, necessitating modification of the proportions of each raw material in the new, modified glass batch. A practical example of compositionally constrained reformulation is the introduction of blast furnace slag (BFS) into SLS glass batches as discussed in Section 7.5.7.1. This reformulation route will be discussed here.

It is rarely possible to produce *exactly* the same glass composition when measured to the second and third decimal place in wt% and usually a compromise must be found. For example, compositional differences between the Al_2O_3 sources feldspar, nepheline syenite, and BFS make it necessary to first ensure that the proportions of the major oxides in the new glass composition match as closely as possible those in the old glass composition. As shown in the reformulated batch in Table 7.18, slight differences in the CaO and K_2O contents of the new glass arise from fundamental differences in the contents of these oxides in the feldspar and BFS material replacing it, which are carried through to the final glass. These differences will have a negligible effect upon the physical properties of the final glass, which can be assumed, for production purposes, to have the same physical properties as the initial glass.

Reformulation constrained by glass composition may not make possible the same reductions in furnace temperature that could be provided by deliberately changing the glass composition. However, energy savings do arise from a decrease in melting energy associated with the reformulated batch composition, as highlighted by the industrial trials described in Fig. 7.17. In addition to those benefits of introducing BFS which have been discussed in Section 7.5.7.1, further benefits are provided

Table 7.18 Compositionally constrained reformulation of a hypothetical colorless SLS container glass batch

Batch components	Original batch Wt/kg	New batch Wt/kg
Sand	571.41	580.00
Feldspar	18.43	0
Nepheline syenite	31.45	32.00
BFS	0	25.00
Limestone	125.98	115.00
Dolomite	65.15	57.00
Sodium carbonate	179.72	184.00
Sodium sulfate	7.86	7.00
Total batch weight/kg	1000.00	1000.00
Theoretical glass weight/kg	836.88	842.68
Theoretical yield/%	83.69	84.27
Theoretical batch CO_2/kg	161.11	154.36
Glass oxides/wt%		
SiO_2	71.30	71.30
Al_2O_3	1.41	1.41
Na_2O	13.45	13.45
K_2O	0.67	0.64
MgO	1.70	1.70
CaO	10.90	10.92
SO_3	0.53	0.53
Fe_2O_3	0.03	0.03
TiO_2	0.02	0.02

Data adapted from [16, 21, 133, 138, 140, 157]

by an increase, in this example, of ~0.8% in final glass yield. As shown in Table 7.18, nearly 7 kg more glass is produced by 1000 kg of the reformulated batch composition containing BFS. This is because 7 kg less batch CO_2 is evolved and emitted during melting, which itself carries an energy saving before any of the other factors are considered. The energy required to evolve CO_2 is shown in Fig. 7.14. Further examples of compositionally constrained reformulation include

- Changing cullet levels
- Introducing new batch materials. Such materials can include mined or synthetic silicates and aluminosilicates or new silica sources such as glassy sand
- Using new sources or supplies of existing raw materials
- Altering the batch water or carbon content
- Introducing different batch sulfates, nitrates, or hydroxides and mixtures thereof
- Altering additions of decolorizer following changes in cullet levels, origin, or quality

7.6.1.2 Reformulation Constrained by Batch

Batch-constrained reformulation involves modification of the glass composition using only those raw materials already in use. Such a situation would arise if a manufacturer intends to implement a reformulated glass but does not have sufficient batch-handling capacity available for new raw materials or if installation of further capacity is not feasible. As discussed in Section 7.6.2, considering the benefits of reformulation only in terms of batch costs demonstrates an overly simplistic approach; however, such a view is understandable and batch-constrained reformulation can be a practical option under some circumstances.

Some examples of batch-constrained reformulation have been discussed previously (see Section 7.5). A simple yet effective approach is to maximize the levels of batch cullet, as discussed in Section 7.5.2, while ensuring that this does not have a deleterious effect on the final glass product or the smooth operation of melting and forming processes. It is assumed that the substantial majority of glass manufacturers have already optimized their batch cullet levels so this means of reformulation, while being a vital component in the toolbox of the glass technologist, is already widely used and a worked example here would be somewhat redundant.

Here we will consider two worked examples. The first is a float glass composition in which the CaO:MgO ratio is increased by reformulation, resulting in decreased melt viscosity and thereby allowing furnace temperatures to be decreased. Pecoraro et al. [223] reformulated the following US float glass composition (wt%): 72.53 SiO_2, 13.79 Na_2O, 0.02 K_2O, 0.03 Al_2O_3, 0.2 SO_3, and 0.5 Fe_2O_3, for which $(SiO_2 + Al_2O_3) = 72.56$; $(Na_2O + K_2O) = 13.81$; and $(CaO + MgO) = 12.85$. By modifying the CaO:MgO ratio (which is most readily achieved by altering the ratio of limestone:dolomite in the batch), they were able to manipulate the viscosity and crystallization behavior as summarized in Table 7.19. The replacement of 1.3 wt% MgO by CaO provides a decrease of 20°C in melting temperature, T_M, i.e., the temperature at which $\eta = 10^2$ dPa s. Only slight modifications to the forming and annealing characteristics of the reformulated glass were necessary, and these could be easily compensated by altering processing parameters. The value of ΔT_{FL} decreased from 34 to 26°C, however, as noted by Jones and Ohlberg [61], typical values of $\Delta T_{FL} \geq 10°C$ are usually sufficient to prevent problems associated with crystallization occurring during production.

In addition to decreased melting energy and emissions, decreasing the MgO:CaO ratio can provide a number of other benefits:

- As discussed in Section 7.5.5.1 and shown in Fig. 7.15, it has been demonstrated on a laboratory scale that a decrease in MgO:CaO ratio from ~0.3–0.2 can in some cases decrease batch-free times, thereby aiding the melting and refining processes and potentially allowing increased furnace pull rates. A minimum in T_{Liq} occurs at ~2.5 wt% MgO in the tridymite primary phase field (see Section 7.5.6.2), suggesting this MgO content as one potential target.

Table 7.19 Selected modeled effects of altering the CaO:MgO ratio in a US float glass composition

CaO/wt%	MgO/wt%	$T_M/°C$	$T_{Form}/°C$	$T_{Bend}/°C$	$T_{Anneal}/°C$	$\Delta T_{FL}/°C$
9.20	3.65	1423	1020	728	550	34
9.40	3.45	1421	1019	729	550	34
9.50	3.35	1419	1018	729	551	33
9.59	3.25	1418	1018	729	551	33
9.69	3.15	1418	1018	729	551	33
9.79	3.05	1416	1017	729	551	33
9.99	2.85	1414	1016	729	552	31
10.20	2.64	1412	1015	729	552	29
10.30	2.54	1411	1015	729	553	28
10.40	2.44	1409	1014	729	553	27
10.50	2.34	1403	1014	729	553	26

Data adapted from [223]

- Consider the relationship linking temperature with refractory corrosion shown in Fig. 7.9 and described in Section 7.4.4.2. The decrease of 20°C in furnace temperature that this reformulation makes possible is estimated to decrease refractory corrosion by ~17.5% if furnace temperatures are decreased from 1500 to 1480°C. This estimation is within the limits of the test used by Kato and Araki [113], and actual refractory corrosion in service may differ from this illustrative figure.
- Any replacement of dolomite ($MgCO_3 \cdot CaCO_3$) by limestone ($CaCO_3$) results in a small decrease in batch CO_2 emissions. For the above example, replacing 1.3 wt% MgO by CaO so that CaO content increases from 9.20 to 10.50 wt% will reduce batch CO_2 emissions by an estimated ~2%. Using the example discussed in Section 7.3.3.2, a decrease in batch CO_2 by 2% from a 250 tpd flat glass furnace equates to a saving of 1.4 tpd in evolved batch CO_2. The associated reductions in emissions are in addition to those that will be made possible by decreasing the furnace temperature

The second worked example describes reformulation of boron-free A-glass compositions. The forming temperature varies with application, as defined in Section 7.4.1. For fiber manufacture the forming temperature is the temperature at which $\eta = 10^3$ dPa s. This contrasts with the working point temperature at which $\eta = 10^4$ dPa s. As a result T_{Liq} for fiberglass compositions is greater than for container or float glass. Wallenberger et al. [60] replaced SiO_2 by CaO in a stepwise manner and observed the effects on the fiber-forming temperature, T_{Liq} and ΔT_{FL}. As shown in Table 7.20, the reformulated glasses are high in RO. Viscosities have been modeled here using Lakatos [29]. Glass 3 shows a reduction in modeled melting temperature of 68°C while retaining an acceptable value of ΔT_{FL}

Modifications of the CaO/MgO and CaO/SiO$_2$ ratios are methodologies that can be applied to all SLS glass compositions and applications. However, the scale of modification and of benefits will also vary with glass type and application. For

Table 7.20 Reformulated boron-free A-glasses

	Glass 1	Glass 2	Glass 3	Glass 4	Glass 5
SiO_2/wt%	71.80	69.90	68.75	66.60	63.30
Al_2O_3/wt%	1.00	1.00	1.00	1.00	1.00
MgO/wt%	3.80	3.80	3.80	3.80	3.80
CaO/wt%	8.80	10.70	11.85	14.00	17.30
Na_2O/wt%	13.60	13.60	13.60	13.60	13.60
K_2O/wt%	0.60	0.60	0.60	0.60	0.60
Fe_2O_3/wt%	0.40	0.40	0.40	0.40	0.40
Measured T [log $(\eta/dPa\ s) = 3$]/°C	1199	1169	1140	1122	1082
Measured T_{Liq}/°C	1018	1063	1075	1123	1173
Measured ΔT_{FL}/°C	181	106	65	−1	−91
Modeled T [log $(\eta/dPa\ s) = 2$]/°C	1437	1395	1369	1321	1248
Modeled T [log $(\eta/dPa\ s) = 3$]/°C	1184	1157	1140	1109	1062
Modeled T [log $(\eta/dPa\ s) = 4$]/°C	1032	1013	1001	979	945

All measured data adapted from [60]. Viscosities modeled here for comparative purposes using [29]. Note that T_F for fiber manufacture corresponds to log $(\eta/dPa\ s) = 3$

example, when applying these methodologies (which are clearly applicable to fiber and flat glass manufacture) to container glasses such as those discussed by Bingham [151] and Bingham and Marshall [21] differences arise. The value of $\Delta T_{FL} = +6$°C, which is arguably typical for container glass, presents less scope for modification of CaO/MgO or CaO/SiO_2 ratios since any decrease in ΔT_{FL} would make its value negative, resulting in a glass that is more prone to crystallization than its predecessor. Replacement of 1.0 wt% MgO by CaO in an SLS container glass resulted in a 12°C (modeled) reduction in T_M, the temperature at which $\eta = 10^2$ dPa s [151]. However, the associated ΔT_{FL} decreased from +6 to −20°C. This is illustrative of the fact that the reformulation methodology used must consider, in addition to physical properties, the constraints resulting from the manufacturing method in question.

7.6.1.3 Unconstrained Reformulation

The term "unconstrained" is, perhaps, slightly inaccurate since there will always be many constraints placed upon the chemical reformulation of SLS glass. However, the term "unconstrained" is valid for the purpose of distinguishing this, the most adventurous method of reformulation, from the preceding two methods which are arguably more conservative. The chief constraints upon this type of reformulation are economic ones, although some scientific and technological questions also remain.

Bingham [151, 215, 224] and Bingham and Marshall [21] discussed unconstrained reformulation of SLS container glass compositions. A few selected compositions and modeled and measured physical properties are shown here in Table 7.21.

Table 7.21 Reformulated SLS container glass compositions, batches, and modeled and measured physical properties

Nominal glass wt%	Glass B	Glass C16B	Glass C44	Glass C60A	Glass C60B
SiO$_2$	72.0	70.0	71.75	70.0	69.0
Al$_2$O$_3$	1.4	1.4	1.4	1.4	1.5
B$_2$O$_3$	0	1.7	1.7	1.0	1.0
Li$_2$O	0	0	0.25	0	0.4
Na$_2$O	13.4	13.4	13.4	13.4	14.0
K$_2$O	0.4	2.4	0.4	2.4	2.5
MgO	1.7	0	1.7	0.7	0.5
CaO	10.9	10.9	10.9	10.9	10.9
SO$_3$ + Fe$_2$O$_3$ + TiO$_2$	0.2	0.2	0.2	0.2	0.2
Measured T [log (η/dPa s) = 2]/°C (modeled)	1458 (1446)	1405 (1397)	1430 (1430)	1408 (1405)	1376 (1365)
Measured T [log (η/dPa s) = 3]/°C (modeled)	1188 (1178)	1134 (1126)	1175 (1163)	1155 (1137)	1129 (1099)
Measured T [log (η/dPa s) – 4]/°C (modeled)	1021 (1027)	980 (986)	1010 (1013)	995 (996)	970 (962)
Measured T_{Liq}/°C (modeled)	1015 (1070)	942 (983)	1032 (n/a)	955 (1001)	964 (n/a)
ΔT_{FL}/°C	+6	+38	–22	+41	+6
Density/g cm^{-3}	2.507	2.493	2.508	2.518	2.561
Durability (USP 25 acid)/ml	6.69	n/m	n/m	7.58	6.44
Refractory corrosion/%	17.3	n/m	n/m	19.0	n/m

Adapted from [21, 151, 215, 224]. Modeled data calculated using [29, 47] within the compositional range of each model

While it is shown that modeled viscosities generally agree well with measured viscosities it is clear that the liquidus model employed [47] does not adequately predict liquidus temperature of SLS glasses with the degree of accuracy required for reformulation. More accurate liquidus models in the compositional regions of interest, and which contain a larger number of possible components, are identified as one immediate research requirement for industrial glassmaking (see Section 7.4.2.2).

The compositions shown in Table 7.21 were developed during a recent successful UK project to reformulate SLS container glass in order to reduce melting temperatures (see Section 7.6.1.3). The research began with thorough laboratory-based investigations [21, 151, 215, 224] and continued with a series of five full-scale industrial trials carried out at UK container glass manufacturers [138]. The initial aim of the project was to establish the technological feasibility of substantially decreasing furnace temperatures, and thereby reduce the environmental impact of container glass manufacture while retaining all of the desirable properties of the glass. Several formulations were developed that provide a range of melting temperature reductions while maintaining the required working properties, ΔT_{FL}, and chemical durability.

Three glasses underwent full-scale industrial trials resulting in a total of 20,000 tonnes of glass being melted. Target compositions, batches taken from Noble and Marshall [138] and predicted properties are shown in Table 7.22. The "Standard" glass was a representative benchmark composition. Glass 1 contained an addition of 0.12 wt% Li_2O which was achieved by replacing batch nepheline syenite with spodumene. This glass was produced at full scale over a total of 10 weeks with full Li_2O addition for 8 of the 10 weeks. A total of 12,000 tonnes of glass was melted using this reformulated composition and no production problems were reported. This small batch change allowed furnace crown temperatures to be reduced by 30°C and the measured direct reduction in melting energy was 1.21%. Associated emissions were not measured; however, substantial reductions in thermal NO_x and particulate generation should have been expected (see Sections 7.3.3.3 and 7.3.3.5).

Glass 2 represented a more complex reformulation with a combination of Al_2O_3, Li_2O, Na_2O, K_2O, and B_2O_3 partially replacing SiO_2 and MgO. Four new raw materials were used: spodumene, lithium carbonate, potassium carbonate, and colemanite. This presented logistical and economic difficulties owing to the large number of raw materials and the limited availability of batch feed infrastructure at the glass plant, highlighting the difficulties that can arise if a number of components change. Glass 2 provided a melting temperature reduction of 81°C and energy saving of 4.8%. All three of the glass compositions melted at the expected temperatures producing well-refined glasses. Each was successfully formed into bottles.

Table 7.22 shows the published batch compositions, estimated target glass compositions based on these batches, and modeled viscosity and liquidus temperatures using [29] and [47], respectively. Note that the model for T_{Liq} [47] overestimates its value by several tens of °C when compared to measured values of T_{Liq} for SLS container glasses, as illustrated in Table 7.21 and discussed in Section 7.4.2.2. As a result of the inaccuracy of the T_{Liq} models available, estimation of ΔT_{FL} is not possible.

7.6.1.4 Other Industrial Trials and Implementation

Raw materials cost optimization has been documented in articles by Bingham [225], Westerlund et al. [218], and Hatakka and Karlsson [219]. Bingham's [225] study described the incorporation of a new source of dolomite into UK container glass batches. This involved partial replacement of 1.5 wt% Na_2O by MgO, with small changes in SiO_2 and CaO contents. The result was an increase of 22°C in the temperature at which $\log (\eta/dPa\ s) = 3$ and a similar increase in melting temperature may be assumed. However, T_{Liq} increased by 50°C and therefore evidence of devitrification during production of the new glass was carefully sought. Monitored parameters included density and homogeneity, daily seed and stone counts, and daily fuel usage. Incident-free production continued for 6 weeks, so the new composition remained permanently in production. No increase in fuel requirement for melting

Table 7.22 Batch recipes and modelled properties of full-scale trial UK container glasses

Batch components	Standard glass	Glass 1	Glass 2
Sand/kg	580	581	550
Nepheline syenite/kg	32	0	0
BFS/kg	25	26	25
Salt cake/kg	7	6	7
Limestone/kg	115	117	137
Dolomite/kg	57	53	10
Sodium carbonate/kg	184	191	202
Potassium carbonate/kg	0	0	31
Colemanite/kg	0	0	19
Spodumene/kg	0	31	51
Lithium carbonate/kg	0	0	4
Total/kg	1000	1005	1036
Theoretical glass weight/kg	842.68	846.06	867.62
Theoretical glass yield/%	84.27	84.19	83.75
Theoretical batch CO_2/kg	154.36	156.24	158.94
Theoretical composition			
SiO_2/wt%	71.30	71.81	68.22
Al_2O_3/wt%	1.41	1.19	1.57
CaO/wt%	10.92	10.87	10.93
MgO/wt%	1.70	1.60	0.49
Li_2O/wt%	0	0.18	0.48
Na_2O/wt%	13.45	13.54	14.00
K_2O/wt%	0.64	0.31	2.73
B_2O_3/wt%	0	0	1.01
SO_3/wt%	0.53	0.46	0.51
TiO_2/wt%	0.02	0.02	0.02
Fe_2O_3/wt%	0.03	0.03	0.04
Modeled T [log $(\eta/dPa\ s) = 2$]/°C [29]	1423	1417	1346
Modeled T [log $(\eta/dPa\ s) = 3$]/°C [29]	1168	1160	1085
Modeled T [log $(\eta/dPa\ s) = 4$]/°C [29]	1019	1012	951
Modeled T_{Liq}/°C [47]	1073	1071	985

Data adapted from [16, 21, 133, 138, 140, 157]

was observed, although an increase of 0.25% in forehearth fuel was required. It is likely that furnace fuel usage and furnace temperatures had not been optimized, as evidenced by the lack of significant change in melting fuel requirement despite an increase in melting temperature by > 20°C. Furnace and control technology has improved markedly in sophistication and accuracy since 1980. If the same trials were carried out today with the aim of optimizing energy usage in an economically viable manner, the "allowable" increase of 50°C in T_{Liq} available from the original glass formulation could be utilized, through careful reformulation, by increasing CaO content. Limestone is among the lowest cost raw materials (see Fig. 7.11) and optimization of CaO content would decrease viscosity at melting, allowing furnace temperatures to decrease while maintaining ΔT_{FL} at an acceptable value, thereby having no negative effect on devitrification behavior and having little impact on batch costs. However, the increase in CaO could also affect the working range of the glass, and such a change must also be taken into consideration. These works demonstrate the importance of, and benefits that can accrue from, optimization of the RO content of SLS glass to minimize high temperature viscosity while maintaining acceptable forming characteristics and ΔT_{FL}, and having minimal impact on batch costs.

Industrial trials using 0.8 wt% B_2O_3 in green SLS glasses were carried out in France [226] using anhydrous borax batch additions. Results were not fully conclusive owing to the large number of unforeseen variables that occurred during the trials. However, the addition of B_2O_3 enabled the tonnage of glass melted per day to be increased from 61 to 65 with a fuel saving of 4.5%, and the observed benefits were sufficient to persuade the glass manufacturer to implement 0.8 wt% B_2O_3 as a permanent batch constituent.

7.6.2 Reformulation Benefits and Pitfalls

In a hypothetical situation a glass manufacturing company, in the light of rising energy costs and increasing environmental constraints, considers modification of its glass batch formula to reduce melting energy, thereby saving fuel and reducing environmental emissions. Such a decision is not made lightly and the outcome carries an element of risk. The manufacturer will rightly demand strong evidence from its technical advisers that any batch changes will not negatively impact upon production volume, glass quality, or costs. Estimating the costs associated with glass reformulation is not straightforward, owing to the number of factors involved. The danger exists that the cost factors considered before making a decision will be oversimplified, for example, as follows:

Question (1)   What is the raw material cost increase associated with batch reformulation?

Question (2)   Is this cost increase fully offset by fuel savings?

If the answer to Question 2 is "no" then it is possible that this will lead management to giving no further consideration to reformulation. The danger of applying such a simplistic view of the process is that it fails to take account of many factors contributing to costs. Unfortunately it can be difficult to accurately estimate a cash value for every one of these factors. Nevertheless it is becoming essential that they should be taken into account. An inexhaustive list of pertinent factors impacting upon costs is as follows:

- Energy prices will probably continue their general upward trend in the future making the answer to Question 2 more likely to become "yes." Unless manufacturers have reformulated glass strategies available to meet these new challenges, they will begin to lose money relative to competitors that do have a strategy ready to implement.
- Reformulation can increase the volume of glass produced per unit mass of raw materials used, for example, by increasing the cullet ratio, replacing dolomite by limestone and replacing feldspar by BFS, and thereby decreasing the unit cost per article produced.
- Reformulation may allow higher manufacturing volumes, for example, by increasing relative machine speed (RMS) of container glass or providing more rapid refining and conditioning.
- Reformulation may reduce the percentage of rejected ware, for example, as demonstrated by the implementation of BFS in a container glass furnace (see Section 7.5.7.1).
- Lower annealing temperatures may accompany reformulation (as suggested, for example, by T_d data from [21]). Reductions in the temperature of the annealing lehr will thereby reduce lehr (annealing furnace) energy consumption.
- Lower emissions resulting from reformulation may allow the manufacturer to avoid legislative requirements to install and run expensive abatement technologies; less costly technologies may instead be feasible. The energy requirements of abatement technologies, if already installed, may be reduced due to lower loads.
- Re-introduction into the batch of filtrates or particulates resulting from abatement will reduce batch costs.
- The potential public relations value of being "greener" than one's competitors can impact on sales, generate new business, and/or strengthen ties with existing customers. It is difficult to estimate the value of this factor.
- Extended furnace, regenerator, and refractory lifetimes. As discussed in Section 7.4.4.2, there is substantial evidence to indicate that lower furnace temperatures will increase refractory lifetimes, thus making substantial savings given the high cost of furnace rebuild.
- Carbon trading (as is now in place throughout the EU) allows any carbon surplus that may arise from lower emissions to be sold on the market, generating revenue for the company. It is possible that similar schemes will soon be in place throughout North America and the rest of the developed world.
- As the volume of any raw material used by industry increases its price falls as supply chain industries compete to meet the new demand. For example, soda ash is a man-made raw material but due to its high production volume it is inexpensive relative to other man-made raw materials that have not yet been employed in bulk for glassmaking. As the uptake by the glass industry of new raw materials becomes more widespread its production volume will increase and hence its price may fall.

The above list is a small selection of the reasons why the decision to reformulate glasses and glass batches for energy efficiency and environmental benefit is a difficult one. Manufacturers should attempt to evaluate all pertinent factors before making a decision regarding reformulation, avoiding overly simplistic appraisals of the pros and cons. Furthermore, even if a decision is made not to proceed with reformulation *at a given time*, it is important that reformulation strategies are discussed and appropriate measures are considered, ready for use in the future as required.

7.6.3 Research Requirements and Closing Remarks

The following is a brief list of what are, in the author's view, some of the current and future research needs that will facilitate more energy-efficient, environmentally friendly industrial SLS glass development and production.

- Development of more accurate models for prediction of T_{Liq} and crystallization behavior within relevant compositional regions.
- A fundamental understanding of the long-term in-service effects of glasses upon refractories as a function of chemical composition and raw material selection. This will allow construction of accurate models to predict the effects of glass formulation routes during early-stage development.
- Greater assessment of the potential viability of sustainable raw materials including wastes arising from other processes.
- Investigation of methods for enabling batch cullet levels to be further increased without having a deleterious effect on glass quality.

Undoubtedly the need to counteract global warming will result in dramatic changes to be made to high-energy industrial processes including glassmaking. If the glass industry, particularly in the Western world, fails to adapt accordingly it will die out. This is a process that has arguably already begun. Possible solutions specific to SLS glass, but which are outside the sole control of manufacturers, include standardization of container dimensions and increasing the use of the return and reuse system for container glass. The use of cullet in glassmaking will continue to increase as greater quantities of high-quality recycled glass become available, and the technologies to sort and remove unwanted contaminants become more advanced. Maximizing batch cullet levels must be pursued in the drive toward lower fuel consumption, lower emissions, and reduced dependency upon virgin raw materials.

Improvements in melting technology are continually being made, with reduction of SEC, and emissions being the key goals. Newer melting technologies such as all-electric melting, microwave melting, and modified conventional gas- and oil-fired furnaces, should see increased utilization in the coming years. Whichever technological improvements and changes do take place, it is essential that industrial SLS glass batches also evolve by reformulation. Continued reliance on existing compositions will no longer represent the best policy from economic or environmental perspectives. In this section it has been shown that many opportunities exist for decreasing SEC and the emission of harmful NO_x, SO_x, particulates, and heavy metals arising from the manufacture of SLS glass (and indeed across the entire glass industry) through primary means. Opportunities even exist for the most conservative glassmaker to minimize batch costs and modify existing compositions for reduced melting temperatures and melting energies. Moreover, furnace lifetimes could be increased as a result of any reformulation, a factor which has largely been overlooked and yet which, potentially, represents a great cost saving. What will be required from all interested parties is the determination to make it happen. Closer, focussed collaboration between industry and academia and the availability of appropriate government support, grants, and financial incentives will all aid this process.

List of Abbreviations/Glossary of Terms

Batch	Glassmaking raw materials as supplied to the furnace
Batch-free	Molten glass containing no residual batch or relics
BAT	Best Available Technology
BFS	Blast furnace slag
D	Devitrification index
EGA	Evolved gas analysis
EP	Electrostatic precipitator
EU	European Union
η	Viscosity
IR	Infrared
Lehr	Annealing furnace
LOI	Loss on ignition
NBO	Non-bridging oxygen
Nm^{-3}	Normal m^{-3}; measured at standard temperature and pressure.
NO_x	Oxides of nitrogen
NWF	Network former
NWM	Network modifier
PPM	Parts per million
RMS	Relative machine speed
SEC	Specific energy consumption
SHC	Specific heat capacity
SLS	Soda–lime–silica
SO_x	Oxides of sulfur
T	Temperature
T_A	Annealing temperature
T_{CMax}	Temperature corresponding to maximum crystallization rate
T_d	Dilatometric softening point
T_F	Forming temperature
ΔT_{FL}	Temperature difference between forming and liquidus

T_{Liq}	Liquidus temperature
T_M	Melting temperature
T_S	Littleton softening point
T_{Str}	Strain point
T_W	Working point
TPD	Tonnes per day
TPY	Tonnes per year
USP	United States Pharmacopoeia
UV	Ultraviolet
VFT	Vogel–Fulcher–Tamann
WRI	Working range index

References

1. A. K. Lyle and F. V. Tooley, Glass composition design and development, in *Handbook of glass manufacture*, Edition 3, Section 1, F. V. Tooley, ed., Ashlee Publishing Co. Inc., New York, pp. 1–17 (1984).
2. The Carbon Trust, Good Practice Guide 127, Energy efficient environmental control in the glass industry, London, UK (2000).
3. The Carbon Trust, Energy Consumption Guide ECG027, Energy use in the container glass industry, London, UK (2005).
4. C. Madivate, F. Müller and W. Wilsmann, Glastech. Ber. Glass Sci. Technol., 69, 167–178 (1996).
5. C. M. de O. Madivate, J. Am. Ceram. Soc., 81, 3300–3306 (1998).
6. C. P. Ross and G. L. Tincher, *Glass melting technology: a technical and economic assessment*, M. Rasmussen, ed., Glass Manufacturing Industry Council, Westerville (2004).
7. Comité Permanent des Industries du Verre Européenes (CPIV), Glass Industry, the emission trading directive and competition: presentation of the glass industry, Brussels, Belgium (2000).
8. R. G. C. Beerkens and H. P. H. Muysenberg, Glastech. Ber., 65, 217–224 (1992).
9. R. G. C. Beerkens, H. A. C. Van Limpt and G. Jacobs, Glass Sci. Technol., 77, 47–57 (2004).
10. G. Enninga, K. Dytrych and H. Barklage-Hilgefort, Glastech. Ber., 65, 186–191 (1992).
11. W. C. Bauer and F. V. Tooley, Advanced methods of batch preparation, in *Handbook of glass manufacture*, Edition 3, Vol. II, Section 23, F. V. Tooley, ed., Ashlee Publishing Co. Inc., New York, pp. 1139–1157 (1984).
12. Integrated Pollution Prevention and Control (IPPC), Sector Guidance Note IPPC SG2, Secretary of State's Guidance for A2 Activities in the Glassmaking Sector, Department for the Environment, Food and Rural Affairs, London, UK (2006).
13. R. A. Drake, Environmental control in the glass industry, in *Handbook of glass manufacture*, Edition 3, Vol. II, Section 21, F. V. Tooley, ed., Ashlee Publishing Co. Inc., New York, pp. 1053–1109 (1984).
14. H. Barklage-Hilgefort, Glastech. Ber., 62, 113–120 (1989).
15. H. Barklage-Hilgefort, Glastech. Ber., 63, 101–110 (1990).
16. P. J. Doyle, *Glass making today*, Portcullis Press, Redhill (1979).
17. R. S. Arrandale, Furnaces, furnace design and related topics, in *Handbook of glass manufacture*, Edition 3, Vol. I, Section 5, F. V. Tooley, ed., Ashlee Publishing Co. Inc., New York, 249–386 (1984).
18. The Carbon Trust, Energy and carbon conversions fact sheet, London, UK (2006).
19. British Glass Manufacturer's Confederation (BGMC), UK glass manufacture: A mass balance study, Sheffield, UK (2003).

20. F. W. Krämer, Solubility of gases in glass melts, in *Properties of glass-forming melts*, L. D. Pye, A. Montenero and I. Joseph, eds., Taylor and Francis Publications, Boca Raton, pp. 405–482 (2005).
21. P. A. Bingham and M. Marshall, Glass Technol., 46, 11–19 (2005).
22. H. Barklage-Hilgefort and W. Sieger, Glastech. Ber., 62, 151–157 (1989).
23. I. Shulver, Glastech. Ber. Glass Sci. Technol., 67, 318–321 (1994).
24. A. Kasper, E. Carduck, M. Manges, H. Stadelmann and J. Klinkers, Ceram. Eng. Sci. Proc., 27, 203–214 (2006).
25. R. G. C. Beerkens and J. Van Limpt, Glastech. Ber. Glass Sci. Technol., 74, 245–257 (2001).
26. M. B. Volf, *Chemical approach to glass: Glass science and technology*, Vol. 7, Elsevier, Amsterdam (1984).
27. M. Cable and M. A. Chaudhry, Glass Technol., 16, 125–134 (1975).
28. R. J. Ryder, E. C. Taylor and K. B. Tanner, Glass Technol., 21, 199–205 (1980).
29. T. Lakatos, Glasteknisk. Tidskr., 31, 51–54 (1976).
30. U. Kircher, Glastech. Ber., 66, 279–283 (1993).
31. U. Kircher, Glastech. Ber., 70, 52–57 (1997).
32. H. A. Schaeffer, Glastech. Ber. Glass Sci. Technol., 69, 101–106 (1996).
33. A. Smrček, Glass, 83, 28–30 (2006).
34. H. Van Limpt and R. G. C. Beerkens, Glass Technol. Eur. J. Glass Sci. Technol. A, 48, 113–118 (2007).
35. A. Fluegel (2007), http://www.glassproperties.com.
36. A. K. Varshneya, *Fundamentals of inorganic glasses*, Edition 2, Society of Glass Technology, Sheffield (2006).
37. A. K. Lyle, Glass Ind., 48, 252–258 (1967).
38. N. F. Zhernovaya, V. I. Onishchuk, V. A. Kurnikov and F. E. Zhernovoi, Glass Ceram., 58, 329–331 (2001).
39. M. B. Volf, *Mathematical approach to glass: Glass science and technology*, Vol. 9, Elsevier, Amsterdam (1988).
40. Y. Bottinga and D. Weill, Am. J. Sci., 272, 438–475 (1972).
41. H. R. Shaw, Amer. J. Sci., 272, 870–893 (1972).
42. K. C. Lyon, J. Res. Nat. Bur. Std., 78A, 497–504 (1974).
43. T. Lakatos, L. G. Johansson and B. Simmingsköld, Glasteknisk. Tidskr., 27, 25–28 (1972).
44. T. Lakatos, L. G. Johansson and B. Simmingsköld, Glass Technol., 13, 88–95 (1972).
45. T. Lakatos, L. G. Johansson and B. Simmingsköld, Glasteknisk. Tidskr., 30, 7–8 (1975).
46. T. Lakatos and L. G. Johansson, Glasteknisk. Tidskr., 31, 31–35 (1977).
47. R. R. Cuartas, Bol. Soc. Esp. Ceram. Vidr., 23, 105–111 (1984).
48. A. I. Priven, Glass Phys. Chem., 23, 333–343 (1997).
49. A. Fluegel, Statistical analysis of viscosity, electrical resistivity, and further glass melt properties, in *High temperature glass melt property database for process modelling*, T. P. Seward, III and T. Vascott, eds., The American Ceramic Society, Westerville, pp. 187–256 (2005).
50. A. Fluegel, Glass Technol. Eur. J. Glass Sci. Technol. A, 48, 13–30 (2007).
51. G. Bonetti, Riv. Staz. Sper. Vetro, 28, 175–190 (1998).
52. P. Hrma, D. E. Smith, J. Matyáš, J. D. Yeager, J. V. Jones and E. N. Boulos, Glass Technol. Eur. J. Glass Sci. Technol. A, 47, 78–90 (2006).
53. G. W. Morey, J. Am. Ceram. Soc., 13, 683–713 (1930).
54. G. W. Morey, *The properties of glass*, American Chemical Society Monograph, Reinhold Publishing Corporation, New York (1938).
55. Owens-Illinois Glass Company General Research Laboratory, J. Am. Ceram. Soc., 27, 221–224 (1944).
56. F. T. Wallenberger, R. J. Hicks and A. T. Bierhals, J. Non-Cryst. Solids, 349, 377–387 (2004).
57. F. T. Wallenberger, R. J. Hicks and A. T. Bierhals, Ceram. Trans., 170, 181–199 (2004).
58. F. T. Wallenberger, R. J. Hicks and A. T. Bierhals, Glastech. Ber. Glass Sci. Technol., 77C, 170–183 (2004).

59. F. T. Wallenberger and R. J. Hicks, Glass Technol. Eur. J. Glass Sci. Technol. A., 47, 148–152 (2006).
60. F. T. Wallenberger, R. J. Hicks, P. N. Simcic and A. T. Bierhals, Glass Technol. Eur. J. Glass Sci. Technol. A., 48, 305–315 (2007).
61. J. V. Jones and S. M. Ohlberg, US Patent 5,071,796 (1991).
62. C. Dreyfus and G. Dreyfus, J. Non-Cryst. Solids, 318, 63–78 (2003).
63. G. Bonetti, Riv. Staz. Sper. Vetro, 24, 155–169 (1994).
64. G. Bonetti, Riv. Staz. Sper. Vetro, 29, 245–259 (1999).
65. L. Šašek, M. Bartuška and V. Van Thong, Silikaty, 17, 207–217 (1973).
66. C. L. Babcock, *Silicate glass technology methods*, Wiley and Sons, New York, pp. 222–236 (1977).
67. R. Backman, K. H. Karlsson, M. Cable and N. P. Pennington, Phys. Chem. Glasses, 38, 103–109 (1997).
68. K. H. Karlsson, R. Backman, M. Cable, J. Peelen and J. Hermans, Glastech. Ber. Glass Sci. Technol., 74, 187–191 (2001).
69. D. E. Sharp and L. B. Ginther, J. Am. Ceram. Soc., 34, 260–271 (1951).
70. J. Moore and D. E. Sharp, J. Am. Ceram. Soc., 41, 461–463 (1958).
71. O. A. Prokhorenko, Radiative thermal conductivity of melts, in *High temperature glass melt property database for process modelling*, Seward, T. P. III and Vascott, T. eds., The American Ceramic Society, Westerville, pp. 95–118 (2005).
72. N. P. Bansal and R. H. Doremus, *Handbook of glass properties*, Academic Press, Orlando (1986).
73. M. K. Choudhary and R. M. Potter, Heat transfer in glass-forming melts, in *Properties of glass-forming melts*, L. D. Pye, A. Montenero and I. Joseph, eds., Taylor and Francis Publications, Boca Raton, pp. 249–294 (2005).
74. T. P. Seward and T. Vascott, Introduction, in *High temperature glass melt property database for process modelling*, T. P. Seward, III and T. Vascott, eds., The American Ceramic Society, Westerville, pp. 1–26 (2005).
75. J. Endrys, F. Geotti-Bianchini and L. De Riu, Glastech. Ber. Glass Sci. Technol., 70, 126–136 (1997).
76. R. W. Douglas and M. S. Zaman, Phys. Chem. Glasses, 10, 125–132 (1969).
77. T. Bates, Ligand field theory and absorption spectra of transition metal ions in glass, in *Modern aspects of the vitreous state*, J. D. Mackenzie, ed., Butterworths Publications, London, Vol. 2, pp. 195–254 (1962).
78. W. A. Weyl, *Coloured glasses*, Society of Glass Technology, Sheffield (1951).
79. G. H. Sigel, Optical absorption of glasses, in *Treatise on materials science and technology*, Vol. 26, M. Tomozawa and R. H. Doremus, eds., Academic Press, New York, pp. 5–89 (1977).
80. Waste Resource Action Programme (WRAP), Container lite final report – Demonstrating glass lightweighting, London, UK (2007).
81. P. K. Gupta, Glass fibers for composite materials, in *Composite materials series 2: Fibre reinforcements for composite materials*, A. R. Bunsell, ed., Elsevier, Amsterdam (1988).
82. K. L. Loewenstein, *The manufacturing technology of continuous glass fibres*, Edition 3, Elsevier, Amsterdam (1993).
83. G. Bonetti, Riv. Staz. Sper. Vetro, 27, 289–301 (1997).
84. L. Penberthy, Glass composition design and development, in *Handbook of glass manufacture*, Edition 3, Section 1, F. V. Tooley, ed., Ashlee Publishing Co. Inc., New York, pp. 387–400 (1984).
85. F. Scarfe, Glass Technol., 21, 37–50 (1980).
86. J. Stanek, *Electric melting of glass, glass science and technology 1*, Elsevier, Amsterdam, Holland (1977).
87. W. Trier, *Glass furnaces: Design construction and operation*, Translation by K. L. Loewenstein, Society of Glass Technology, Sheffield (1987).

88. A. K. Varshneya, T. Vascott, R. Karuppanan and J. M. Jones, Electrical resistivity, in *High temperature glass melt property database for process modelling*, T. P. Seward, III and T. Vascott, eds., The American Ceramic Society, Westerville, pp. 173–186 (2005).

89. O. V. Mazurin and O. A. Prokhorenko, Electrical conductivity of glass melts, in *Properties of glass-forming melts*, L. D. Pye, A. Montenero and I. Joseph, eds., Taylor and Francis Publications, Boca Raton, pp. 295–338 (2005).

90. J. E. Shelby, *Introduction to glass science and technology*, Society of Chemistry, Letchworth (1997).

91. H. Müller-Simon and K. Gitzhofer, Glass Technol. Eur. J. Glass Sci. Technol. A, 49, 83–90 (2008).

92. D. M. Rue, J. Servaites and W. Wolf, *Industrial glass bandwidth analysis*, Gas Technology Institute, Des Plaines (2006).

93. R. L. Schwenninger, D. A. Hanekamp and H. R. Foster, US Patent 4,849,004 (1989).

94. K. Natterman, US Patent 6,401,492 (2002).

95. K. Natterman, US Patent 6,588,233 (2003).

96. F. Karetta, J. Witte, K. D. Duch, D. Gobike, W. Muench, A. Jakway, R. Eichholz and F. T. Lentes, US Patent 7,231,788 (2007).

97. G. A. Pecoraro, L. J. Shelestak and J. E. Cooper, US Patent 4,919,700 (1990).

98. H. Kobayashi, S. E. Jaynes and R. G. C. Beerkens, World Patent WO/2004/092086 (2004).

99. H. Kobayashi and R. G. C. Beerkens, US Patent 5,922,097 (1999).

100. S. Rudolph, H. Förster and F. Gebhardt, Glastech. Ber., 63, 198–203 (1990).

101. G. Roeth, T. Pfeiffer and K. D. Duch, US Patent 6,769,272 (2004).

102. M. P. Knox and G. J. Copley, Glass Technol., 38, 91–96 (1997).

103. Carbon Trust, Future Practice Report 68, Enhanced energy efficiency through the ultrasonic refining of glass, London, UK (1998).

104. A. M. Huber and B. D. Tinianov, US Patent Application US 2005/0115276 A1 (2005).

105. V. Tonarová and L. Nmec, Sklar. Keram., 56, 77–80 (2006).

106. E. R. Begley, Refractories, in *Handbook of glass manufacture*, Edition 3, Section 7. F. V. Tooley, ed., Ashlee Publishing Co. Inc., New York, pp. 401–454 (1984).

107. A. D. Davis, Refractories (Update), in *Handbook of glass manufacture*, Edition 3, Section 7. F. V. Tooley, ed., Ashlee Publishing Co. Inc., New York, pp. 454-1 to 454-24 (1984).

108. M. Cable, in *Corrosion of advanced ceramics*, K. G. Nickeled., ed., Kluwer Academic Publ., Dordrecht, pp. 285–296 (1994).

109. M. Dunkl, Glastech. Ber. Glass Sci. Technol., 67, 325–334 (1994).

110. V. K. Pavlovskii and Y. S. Sobolev, Glass Ceram., 48, 558–561 (1991).

111. F. Day and J. P. Ambrosone, J. Am. Ceram. Soc., 34, 163–164 (1951).

112. Technical University of Berlin (TUB), Glass, 69, 58–60 (1992).

113. K. Kato and N. Araki, J. Non-Cryst. Solids, 80, 681–687 (1986).

114. A. G. Clare, A. Kucuk, D. R. Wing and L. E. Jones, The measurement of the density and surface tension of glass melts using the sessile drop method, in *High temperature glass melt property database for process modelling*, T. P. Seward, III and T. Vascott, eds., The American Ceramic Society, Westerville, pp. 119–130 (2005).

115. D. A. Weirauch, The surface tension of glass-forming melts, in *Properties of glass-forming melts*, L. D. Pye, and A. Montenero, I. Joseph, eds., Taylor and Francis Publications, Boca Raton, pp. 143–192 (2005).

116. A. Kucuk, A. G. Clare and L. Jones, Glass Technol., 40, 149–153 (1999).

117. R. G. Newton, Glass Technol., 26, 21–35 (1985).

118. R. H. Doremus, in *Treatise on materials science and technology*, Vol. 17 M. Tomozawa and R. H. Doremus, eds., Academic Press, New York, pp. 41–69 (1979).

119. A. Paul, J. Mater. Sci., 12, 2246–2268 (1977).

120. T. Geisler, A. Janssen, J. Berndt and A. Putnis, A new model for nuclear waste borosilicate glass alteration, Presented at the MRS Fall Meeting, Boston, USA, December (2008).

121. J.–M. Combes and D. Sachot, US Patent 5,656,559 (1997).
122. B. Yale, A. F. Mason, P. Shorrock and N. A. Edwards, European Patent EP0516354A1 (1992).
123. C. M. Jantzen, J. B. Pickett, K. G. Brown and T. B. Edwards, US Patent 5,846,278 (1998).
124. G. Bonetti, Riv. Staz. Sper. Vetro, 26, 5–15 (1996).
125. C. W. Sinton and W. C. LaCourse, Mater. Res. Bull., 36, 2471–2479 (2001).
126. T. Lakatos and B. Simmingsköld, Glasteknisk. Tidskr., 27, 77–80 (1972).
127. G. Bonetti, Riv. Staz. Sper. Vetro, 27, 111–125 (1997).
128. M. L. Huggins and K. H. Sun, J. Am. Ceram. Soc., 26, 4–11 (1943).
129. K. S. Hong, S. W. Lee and R. E. Speyer, J. Am. Ceram. Soc., 76, 605–608 (1993).
130. T. D. Taylor and K. C. Rowan, J. Am. Ceram. Soc., 66, C227–C228 (1983).
131. M. L. Froberg, US Patent 4,358,304 (1982).
132. J. C. Alexander, US Patent 5,578,102 (1996).
133. F. G. West-Oram, Glass Technol., 20, 222–245 (1979).
134. M. Cable and M. Q. Siddiqui, Glass Technol., 21, 193–198 (1980).
135. W. H. Manring and A. R. Conroy, Glass Ind., 49, 199–203 (1968).
136. UK Environment Agency (UKEA), Specification for flat glass cullet used in flat glass manufacture, London, UK (2008).
137. K. K. Éidukyavichus, V. R. Matseikene, V. V. Balkyavichus, A. A. Shpokauskas, A. A. Laukaitis and L. Y. Kunskaite, Glass Ceram., 61, 77–80 (2004).
138. B. Noble and M. Marshall, Glass, 84, 16–18 (2007).
139. P. Bertuzzi, P. Ercole, C. Ferrero and G. Balan, Glass Mach. Plants Acc., 6, 158–161 (2006).
140. British Standards Institute (BSI) Standard 2975:1988, Methods for sampling and analysis of glass-making sands, London, (1988).
141. V. V. Mirkovich, J. Canad. Ceram. Soc., 44, 43–47 (1975).
142. W. Ohnemuller and A. Solf, US Patent 3,967,974 (1976).
143. H. ZurStrassen and E. Rauschenfels, US Patent 4,047,968 (1977).
144. G. H. Fairchild and J. A. Hockman, US Patent 6,287,997 (2001).
145. G. P. Tomaino and J. A. Hockman, US Patent 6,420,289 (2002).
146. J. Hockman, Bull. Am. Ceram. Soc., 84, 9101–9103 (2005).
147. J. Place, Glasteknisk. Tidskr., 31, 69–70 (1976).
148. M. M. Bhuiyan and M. Cable, Glass Technol., 6, 206–212 (1965).
149. F. V. Tooley, Raw materials, in Handbook of glass manufacture, Edition 3, Section 2, F. V. Tooley, ed., Ashlee Publishing Co. Inc., New York, (1984), pp. 19 to 56–11.
150. E. Kazazoglu, G. Albayrak and A. Yaraman, Glastech. Ber., 56 K, 19–24 (1983).
151. P. A. Bingham, Reformulation of container glass for energy efficiency, Ref. 2002–6–32–1–3, Carbon Trust, London, UK, (2003).
152. K. A. Shahid and F. P. Glasser, Phys. Chem. Glasses, 13, 27–42 (1972).
153. G. W. Morey, J. Am. Ceram. Soc., 13, 714–717 (1930).
154. H. R. Swift, J. Am. Ceram. Soc., 30, 170–174 (1947).
155. R. Jordán-Hernández, N. Vega-Sánchez, M. E. Zayas-Saucedo, H. Arizpe-Chávez, C. Diaz and J. M. Rincón, Bull. Am. Ceram. Soc., 84, 9301–9304 (2005).
156. W. Simpson, Glass Technol., 17, 35–40 (1976).
157. N. Marriott, M. Orhon, E. Akmoran, T. Gun, C. Cabuk, S. Jones and S. Materowski, Glass Technol. Eur. J. Glass Sci. Technol. A., 48, 280–296 (2007).
158. P. Frolow and G. H. Frischat, Glastech. Ber., 66, 143–150 (1993).
159. R. K. McBerthy, Glass Ind., 9, 215 (1928).
160. G. W. Morey, J. Am. Ceram. Soc., 13, 718–724 (1930).
161. W. B. Silverman, J. Am. Ceram. Soc., 22, 378–384 (1939).
162. W. B. Silverman, J. Am. Ceram. Soc., 23, 274–281 (1940).
163. A. K. Lyle, W. Horak and D. E. Sharp, J. Am. Ceram. Soc., 19, 142–147 (1936).
164. Owens-Illinois Glass Company General Research Laboratory, J. Am. Ceram. Soc., 33, 181–186 (1950).

165. S. Sen and F. V. Tooley, J. Am. Ceram. Soc., 38, 175–177 (1955).
166. Y. Tang and G. H. Frischat, Glastech. Ber. Glass Sci. Technol., 68, 213–221 (1995).
167. B. R. Franklin and L. C. Klein, Bull. Am. Ceram. Soc., 62, 209–210 (1983).
168. P. Stevenson, Glass, 68, 500–501 (1991).
169. G. W. Morey, J. Am. Ceram. Soc., 15, 457–475 (1932).
170. G. E. Walker, J. Soc. Glass Technol., 29, 38T–47T (1945).
171. R. S. Allison and W. E. S. Turner, J. Soc. Glass Technol., 38, 297T–365T (1954).
172. Owens-Illinois Glass Company General Research Laboratory, J. Am. Ceram. Soc., 31, 8–14 (1948).
173. A. R. Conroy, W. H. Manring and W. C. Bauer, Glass Ind., 47, 84–89 (1966).
174. W. H. Manring, D. D. Billings, A. R. Conroy and W. C. Bauer, Glass Ind., 48, 374–380 (1967).
175. J. Kloužek, M. Arkosiová, L. Nmec and P. Cinibusová, Glass Technol. Eur. J. Glass Sci. Technol. A., 48, 176–182 (2008).
176. A. Smrček, Glass Sci. Technol., 78, 173–184 (2005).
177. A. Smrček, Glass Sci. Technol., 78, 230–244 (2005).
178. R. E. Loesell and W. R. Lester, Glass Ind., 42, 623–629 (1961).
179. L. E. Stadler and D. Cronin, Glass Ind., 58, 10–33 (1977).
180. L. E. Stadler and D. Cronin, Glass Ind., 59, 10–13 (1978).
181. K. A. Landa, L. Landa, A. V. Longobardo and S. V. Thomsen, US Patent 6,672,108 (2004).
182. V. Gottardi, B. Locardi and G. Paoletti, Riv. Staz. Sper. Vetro, 4, 9–14 (1974).
183. R. Guy, Bull. Cent. Glass Ceram. Res. Int., 8, 79–80 (1961).
184. H. Römer, W. Kiefer, D. Köpsel, P. Nass, E. Rodek, U. Kolberg and T. Pfeiffer, US Patent 6,698,244 (2004).
185. K. Papadopoulos, Phys. Chem. Glasses, 14, 60–65 (1973).
186. M. Ooura and T. Hanada, Glass Technol., 39, 68–73 (1998).
187. P. A. Bingham and R. J. Hand, Mater. Res. Bull., 43, 1679–1693 (2008).
188. R. L. Lehman and W. H. Manring, Glass Ind., 58, 16–18 (1977).
189. P. Del Gaudio, H. Behrens and J. Deubener, J. Non-Cryst. Solids, 353, 223–236 (2007).
190. R. F. Bartholomew, Water in glass, in *Treatise on materials science and technology*, Vol. 22, Glass III, M. Tomozawa and R. H. Doremus, eds., Academic Press, New York, (1982).
191. F. Geotti-Bianchini and L. De Riu, Glastech. Ber. Glass Sci. Technol., 71, 230–242 (1998).
192. F. Geotti-Bianchini, J. T. Brown, A. J. Faber, H. Hessenkemper, S. Kobayashi and I. H. Smith, Glastech. Ber. Glass Sci. Technol., 72, 145–152 (1999).
193. J. Deubener, R. Müller, H. Behrens and G. Heide, J. Non-Cryst. Solids, 330, 268–273 (2003).
194. F. Geotti-Bianchini and L. De Riu, Glastech. Ber. Glass Sci. Technol., 68, 228–240 (1995).
195. R. G. C. Beerkens, Glastech. Ber. Glass Sci. Technol., 68, 369–380 (1995).
196. K. E. Spear and M. D. Allendorf, J. Electrochem. Soc., 149, B551–B559 (2002).
197. Owens-Illinois Glass Company General Research Laboratory, J. Am. Ceram. Soc., 27, 369–372 (1944).
198. A. Smrček, Glass Sci. Technol., 78, 287–294 (2005).
199. Owens-Illinois Glass Company General Research Laboratory, J. Am. Ceram. Soc., 25, 61–69 (1942).
200. K. D. Kim, Glass Technol., 41, 161–164 (2000).
201. Owens-Illinois Glass Company General Research Laboratory, J. Am. Ceram. Soc., 31, 1–8 (1948).
202. L. G. Johansson, Glasteknisk. Tidskr., 31, 37–42 (1976).
203. W. Simpson, Glass, 66, 475–476 (1989).
204. D. E. King, F. J. Pern, J. R. Pitts, C. E. Bingham and A. W. Czandema, *Record IEEE Photovoltaic Specialists Conf.*, pp. 1117–1120, (1997).
205. D. Chia, B. Caudle, G. R. Atkins and M. P. Brungs, Glass Technol., 41, 165–168 (2000).
206. P. M. DiBello, Glass Technol., 30, 160–165 (1989).
207. R. G. C. Beerkens, J. Am. Ceram. Soc., 86, 1893–1899 (2003).

208. H. Müller-Simon, J. Bauer and P. Baumann, Glastech. Ber. Glass Sci. Technol., 74, 283–291 (2001).
209. E. Guadagnino and O. Corumluoğlu, Glastech. Ber. Glass Sci. Technol., 73, 18–27 (2000).
210. L. J. Shelestak and M. Arbab, Glass Sci. Technol., 78, 255–260 (2005).
211. J. V. Jones and E. N. Boulos, US Patent 5,521,128 (1996).
212. E. N. Boulos and J. V. Jones, US Patent 6,408,650 (2002).
213. O. Knapp, Keram. Rundschau, 39, 601–603 (1931).
214. Owens-Illinois Glass Company General Research Laboratory, J. Am. Ceram. Soc., 25, 401–408 (1942).
215. P. A. Bingham, Glass Technol., 45, 255–258 (2004).
216. M. Krauss, A. Lenhart, H. Ratka and U. Kircher, Glastech. Ber. Glass Sci. Technol., 68, 278–284 (1995).
217. A. G. Amrheim, J. J. Hammel and L. J. Shelestak, US Patent 4,270,945 (1981).
218. T. Westerlund, L. Hatakka and K. H. Karlsson, J. Am. Ceram. Soc., 66, 574–579 (1983).
219. L. Hatakka and K. H. Karlsson, Glass Technol., 27, 17–20 (1986).
220. Y. B. Peng, X. Lei and D. E. Day, Glass Technol., 32, 123–130 (1991).
221. M. M. Khaimovich and K. Y. Subbotin, Glass Ceram., 62, 109–112 (2005).
222. M. M. Khaimovich, Glass Ceram., 62, 381–382 (2005).
223. G. A. Pecoraro, L. J. Shelestak and R. Markovic, US Patent 6,878,652 (2005).
224. P. A. Bingham, Glass, 80, 336 (2003).
225. K. Bingham, Ceram. Sci. Eng. Proc., 3, 186–191 (1982).
226. M. C. Patrick and G. H. Bowden, Glass Technol., 3, 52–58 (1962).

Part III
Glass Melting Technology

Chapter 8
Basics of Melting and Glass Formation

Hans-Jürgen Hoffmann

Abstract The energy/enthalpy functions of solids and melts are investigated as a function of temperature. Several thermal effects can be understood on an atomic scale surprisingly well by energy levels and wave functions of the bonding electrons and their interaction with the oscillating atoms. Among these effects are the melting transition, the glass transformation, the thermal expansion, structural phase transitions, and relaxation effects occurring near the glass transition temperature, Tg. Glass formation is favored if sufficient strong directed bonds are present between the constituents and the melting entropy per particle is sufficiently small.

Keywords Melting · Glass formation · Enthalpy function · Entropy function · Thermal properties of solids and melts · Thermal expansion coefficient

8.1 Motivation

Glass science and technology, in particular applications and products made of glasses, must consider effects on different length scales starting with the diameter of atoms and ending with the diameter of monolithic mirror blanks for astrophysics or – even much larger – with the length of fibers for telecommunication.

The development, production, and use of large systems require an adapted language describing macroscopic properties like viscosity and elasticity or strength, which are clearly different from properties on an atomic scale-like distance between nearest neighbors, bond angles, valence of coloring ions, energy levels, or wave functions of the bonding electrons [1]. Thus, different scales require different descriptions characteristic for the approach and the language must be adapted to the properties under consideration. However, one should not consider the problem-oriented description as a hierarchy of value. Each level has its merits

H.-J. Hoffmann (✉)
Institute of Materials Science and Technology: Vitreous Materials, University of Technology of Berlin, Englische Strasse 20, 10587 Berlin, Germany
e-mail: hoffmann.glas@tu-berlin.de

F.T. Wallenberger, P.A. Bingham (eds.), *Fiberglass and Glass Technology*,
DOI 10.1007/978-1-4419-0736-3_8, © Springer Science+Business Media, LLC 2010

and deficiencies. Important is to acknowledge how these levels of description are related to each other and how to transfer the vocabulary to the neighboring levels. Parameters and properties of one level have to be explained by the properties of the more elementary level. Thus, for a full understanding, macroscopic properties must finally be explained by effects and quantities on an atomic scale.

In glass science and technology the description is very often restricted to macroscopic effects alone. Empirical knowledge of parameters necessary for the production and for a given application is considered to be sufficient. However, a more detailed analysis on an atomic scale is required in order to understand where the limitations are and – in particular – how they can be overcome or how the composition or the process can be improved or simply how the properties of glasses can be understood from the composition. Furthermore, one might get lost in trial and error or mysteries or alchemy if one is not able to explain effects properly on a scientific basis by the properties of the constituents.

Examples of the last kind are melting, glass formation, and glass transition, which are dealt with in the present chapter based on [1] and summarizing recent publications. These effects or processes can now be understood qualitatively quite well at least for the inorganic glasses, which contrasts "the mysterious glass transition" coined by Langer in a puzzling "reference frame" of Physics Today [2]. In the present article, many of these mysteries – at least those of chemically bonded glasses – are solved by considering wave functions of the bonding electrons. In particular, the so-called Kauzmann paradox turned out to be based on an incomplete description of the entropy of melts, glasses, and crystals. For many years glass scientists considered the entropy of these systems, only, and neglected their energy. This led to an ever repeated confusion about an "entropy catastrophe" in the past, which seems to be inerasable in the literature since conferences still have been organized to deal with that surmised paradox. In order to avoid useless repetition I refer the reader to the literature [3], which points at several misunderstandings in this field.

The present approach to explain several mysteries of glasses is based on the interpretation of their macroscopic thermodynamic quantities on an atomic scale. Analyzing thermodynamic data of many one-component systems on an atomic scale one can conclude what makes solids melt [4, 5]. From here the glass transformation can be understood, some further basic understanding will be shown as well as some general directions of present research and possible development of future quantitative understanding.

8.2 Former Melting Criteria

Melting is very often explained in the literature by the so-called Lindemann's criterion disregarding its original intention. Lindemann wanted to estimate the frequency of the oscillations considered by Einstein to explain the decrease of the specific heat capacity of solids approaching 0 K [6]. He assumed that the amplitude just below the melting temperature was about the same in units of the average distance between the atoms of a solid (in the order of 5–10%). Since he succeeded in estimating the order of magnitude of the frequencies, other authors have reversed his consideration aiming to predict the melting temperatures. However, from a careful analysis one must conclude that the predictive power is not sufficient, even if one considers melting of the chemical elements alone [7, 8]. According to Lindemann's treatment the melting temperature, T_m, depends on properties of the material in the solid phase alone, whereas from simple thermodynamic considerations T_m as well as the molar melting enthalpy, ΔH_m, and entropy, ΔS_m, depend on the difference of the molar enthalpies, H, and entropies, S, in the solid (index: s) and the molten (index: l) state. This can clearly be seen from the relation

$$\Delta H_m = H_l(T_m) - H_s(T_m) = T_m\big(S_l(T_m) - S_s(T_m)\big) \qquad (8.1)$$

As a consequence of (8.1), T_m is defined by the ratio $\Delta H_m/\Delta S_m$, which depends on the entropies and enthalpies in both the solid and molten state.

The molar enthalpy function, $H(T)$, of Al_2O_3 in the solid and molten state is represented as a function of the temperature, T, in Fig. 8.1 as an example of a one-component system. Below and above T_m the molar enthalpy increases monotonously with T, the slopes $dH(T)/dT$ being the molar specific heat capacities at constant pressure in the solid and molten state, C_{ps} and C_{pl}, whereas $H(T)$ increases discontinuously at T_m with the step height given by (8.1). As ΔH_m depends on properties of both the solid and the liquid state, melting cannot be explained by properties of the solid state alone as in Lindemann's estimation [6] and subsequent adoptions.

Born stated in [11]: "the difference between a solid and a liquid is that the solid has resistance against shearing stress while the liquid has not." Any liquid with shearing viscosity other than zero obviously has resistance to shearing stress.

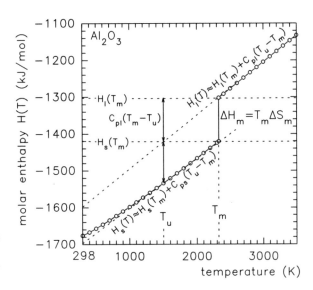

Fig. 8.1 Molar enthalpy of Al_2O_3 as a function of the temperature below and above the melting temperature, T_m. For explanation of the details see text. Data from [9, 10]

Considering the concept of the Deborah number in visco-elasticity, a liquid behaves even elastic if the interaction time is sufficiently short compared to the relaxation time of the liquid. If the resistance to shearing stress would be absent, the constituents of a melt would not attract but would move freely with the consequence that at least three degrees of freedom of the constituents (corresponding to the three directions of the restoring forces in space) would be missing. Then, the specific heat capacity is expected to be smaller by $3R/2$ per mole of particles in the molten state compared to the solid state near the melting temperature.

A histogram of the difference $(C_{pl}-C_{ps})/N$ of about 450 one-component systems is shown in Fig. 8.2. (N is the number of atoms in the formula unit of each one-component system.) In most cases this difference is slightly positive meaning that the molar specific heat capacity is larger above T_m than below. This is partly due to a larger expansion with temperature in the molten state as compared to the solid state, assuming the other contributions being approximately the same. Consequently, there are at least necessarily as many degrees of freedom available to store energy in the solid and in the melt near T_m. Therefore, one can conclude that most elastic interaction forces between the constituents of the melts are still active and only a minor fraction of bonds are open. However, one must not conclude that the vibration spectrum is necessarily the same in the molten and in the solid

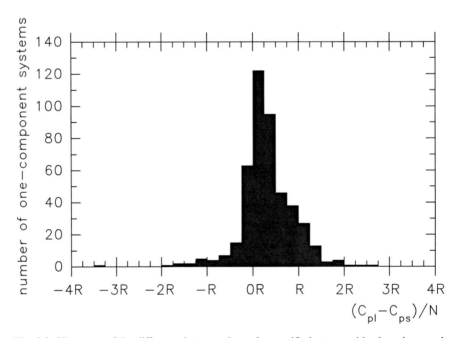

Fig. 8.2 Histogram of the difference between the molar specific heat capacities just above and below the melting temperature divided by the number of atoms of the formula unit, $(C_{pl}-C_{ps})/N$. The axis is divided in units of the molar gas constant, R, with a width of the bucket of $R/4$. Data from [9, 10]

state. This cannot be true since the arrangement of the constituents in the melt differs from that in the crystalline state; in particular, the periodicity is missing in the melt.

Cahn critically discussed in [12] that melting might be induced by the spontaneous creation of intrinsic defects, such as vacancies and intrinsic arrays of dislocations. But what is the mechanism to create these defects? And why are such defects created in Hg at temperatures as low as 234.32 K, whereas in crystals of the neighboring element Au such defects are created at a much higher temperature, namely 1337.33 K [13]? What about the creation of defects for substances that shrink upon melting like Si, Ge, Ce, and Bi? For short, the problem to explain melting would be shifted to explain a mechanism to create such defects.

To terminate the collection of melting relations, it should be mentioned that E. Grüneisen established a correlation between the linear thermal expansion coefficient α and the melting temperature T_m [14] expressed as

$$\alpha \cdot T_m = \text{const.} \tag{8.2}$$

Considering the data of the chemical elements in Fig. 8.3 such a correlation seems to exist. The predictive power, however, is neither sufficient nor convincing, since α varies over half an order of magnitude for elements with similar T_m. Such a correlation fails completely for the rare earth elements (triangles) as can be seen in Fig. 8.3. Particularly in this case one would naively expect a good correlation due to their chemical similarity. However, the relation (8.2) is not confirmed by the experimental data even if some textbooks may make believe it.

8.3 Analysis of the Enthalpy Functions of One-Component Systems

8.3.1 Theoretical Preliminaries

In Section 8.2, a mechanism for the melting transition will be described based on the interpretation of thermodynamic data, in particular the molar enthalpy function, in the pre-melting range below T_m.

The enthalpy, H, is an energetic potential function or Gibbs function of the variables entropy S, pressure p, and particle number N_A, given in its complete differential form (see textbooks on thermodynamics, [16], for example):

$$dH(S, p, N_A) = TdS + Vdp + \mu dN_A \tag{8.3}$$

with the partial derivatives

$$\text{temperature } T = \frac{\partial H}{\partial S}, \tag{8.4}$$

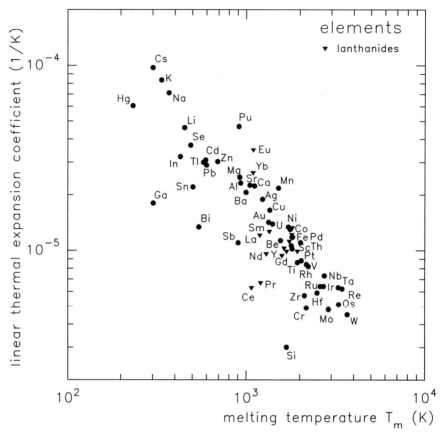

Fig. 8.3 The linear thermal expansion coefficient, α, of the chemical elements at room temperature as a function of the melting temperature, T_m. Data of the rare earth elements are marked by the *triangles*. Data from [13, 15]

$$\text{volume } V = \frac{\partial H}{\partial p}, \tag{8.5}$$

and

$$\text{chemical potential } \mu = \frac{\partial H}{\partial N_A}. \tag{8.6}$$

Considering processes at atmospheric pressure, $p_0 = 101325 \; Pa = $ constant, and one mole of particles, $N_{A0} = 1$ mole= constant (8.3), is reduced to

$$dH(S, p_0, N_{A0}) = TdS. \tag{8.7}$$

Then, entropy can be considered as the carrier of the energy or enthalpy. Entropy dS flowing into the system increases the enthalpy (or energy) with T being the factor of proportionality or coupling constant between energy and entropy. To increase the sensitivity for the following discussion, we rather consider the slope of the molar enthalpy as a function of the temperature, T, which corresponds to the molar specific heat capacity at constant pressure

$$C_p(T) = \frac{dH(S(T), p_0, N_{A0})}{dT} = \frac{dH(S(T), p_0, N_{A0})}{dS}\frac{dS(T)}{dT} = T\frac{dS(T)}{dT} \qquad (8.8)$$

either in the solid or in the liquid state. From (8.8) the molar specific entropy capacity at constant pressure is given by

$$\frac{dS(T)}{dT} = \frac{C_p(T)}{T}, \qquad (8.9)$$

which can also be considered if necessary.

8.3.2 Pre-melting Range and the Contribution to the Molar Specific Heat Capacity by Electrons

If the enthalpy is stored in the lattice vibrations alone, one expects in the case of mono-atomic solids above the Debye temperature for C_{ps} a constant value of approximately $3R$ (R is the universal molar gas constant). The excess of C_{ps} with respect to $3R$ is shown in Fig. 8.4 as a function of the element number together with the sequence of the outermost orbitals of the free atoms filled with electrons. One might argue whether it is allowed to compare properties of solids with properties of isolated atoms. However, the wave functions of the bonding states can be expanded into linear combinations of atomic wave functions of which the outer orbitals certainly play the dominant role. Therefore, the relation of C_{ps} with the outermost atomic orbitals in Fig. 8.4 and of other quantities in several subsequent figures seems to be justified. In fact, this simplification seems to be confirmed by the good correlations with the quantum number representing the angular momentum of the atomic wave functions.

$3R$ is the contribution of one mole of oscillating atoms to the specific heat capacity, C_{Vs}, under constant volume condition and in the limit of high temperatures, if all vibration modes of the solid can be excited such as near the melting temperature. Thus, C_{ps} should have been recalculated into $C_{Vs} = C_{ps} - \alpha_V^2 T_m V_{mol}/\kappa$ for a correct comparison with $3R$ (α_V is the expansion coefficient of the volume, T_m the melting temperature, V_{mol} is the molar volume and κ the isothermal compressibility; see textbooks on thermodynamics [16], for example). This correction has been omitted, since that shift is comparatively small. Figure 8.4 shows that C_{ps} is always larger than $3R$ with the largest excess if d- and f-sub-shells are filled by about $1/4$, $1/2$, or $3/4$. From the good correlation with the filling of the sub-shells one concludes that this excess contribution to C_{ps} is due to electronic transitions. In general, an electronic contribution is seen in C_{ps} of metals at low temperatures in the range of few Kelvin,

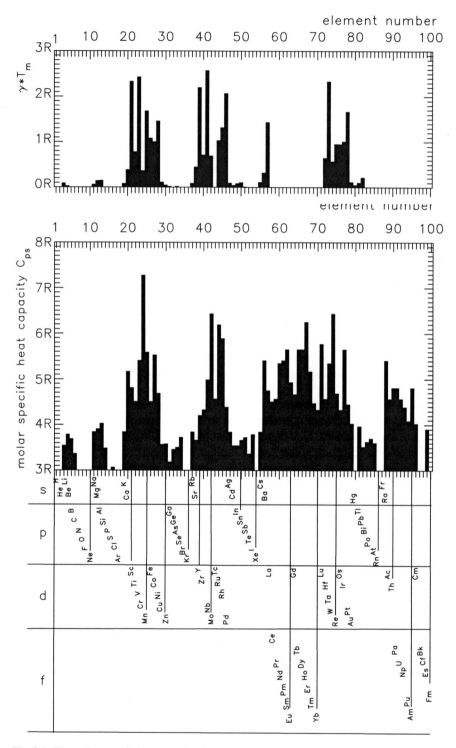

Fig. 8.4 The molar specific heat capacity C_{ps} just below the melting temperature as a function

since the part of the specific heat capacity due to lattice oscillations decreases as βT^3 with decreasing T, whereas the electronic contribution in metals decreases as γT (β and γ are specific constants of the respective metal). The electronic part is clearly detected in measurements of C_{ps} at low temperatures. Extrapolating γT to the melting temperature yields the results shown in the upper part of Fig. 8.4. This extrapolation is based on the assumption of a constant density of electronic states as a function of the energy around the Fermi-level, which does not need to be fulfilled at high temperatures necessarily.

The electronic transitions near the melting temperature have important consequences for the melting [18]. Due to such transitions the probability density functions of the bonding electrons can be considerably changed with the result that the core ions are forced to move to new places determined by the wave functions of the excited states. If these forces are strong enough and the relaxation of the core ions to the new places is fast enough (occurring within the lifetime of the respective excited electronic state) then we have the situation of a melt. The occupation of the excited states changes in a random sequence with time forcing the core ions to accept continuously new places. However, only a minority of bonds are open at the same time, most bonds are closed.

Since the core ions relax according to the random charge distribution of the electrons, the regular order of the solid is broken. The relaxing core ions modify the potential for the bonding electrons and – as a consequence – cause a shift of the ground states of the bonding electrons. Furthermore, the disorder and the electronic transitions shift the frequencies of the vibrations, since the restoring forces are weakened or changed at least. Thus, ΔS_m and $\Delta H_m(T_m)$ are stored essentially in both the phonon and electron systems which are coupled. Considering (8.1) the melting temperature adjusts the entropy and enthalpy needed to fill the respective storage capabilities modified by the new configuration upon melting. The higher the melting temperature, the larger the enthalpy (energy) needed to break the bonds and to accept the new electronic ground state in the melt.

With decreasing temperature both (the phonon and electron) systems decouple near and below the glass transformation temperature, T_g. The melting entropy is certainly large if the average phonon frequency is shifted considerably to lower values due to the melting. Both melting temperature and melting entropy of the chemical elements are shown in Fig. 8.5 as a function of the element number and the highest or last occupied electron state of the free atom for comparison.

Fig. 8.4 (continued) of the element number together with the last orbital of the free atoms occupied by an electron; indicated is the quantum number of the angular momentum $\ell = 0,1,2,3$ corresponding to the symbols s, p, d, and f. The part on *top* shows the molar specific heat capacity of the electrons at the melting temperatures γT_m as extrapolated from low temperatures where it can be determined from the linear part of the molar specific heat capacity given by $C_{ps}(T) = \gamma T + \beta T^3$. Data from [13, 15, 17]

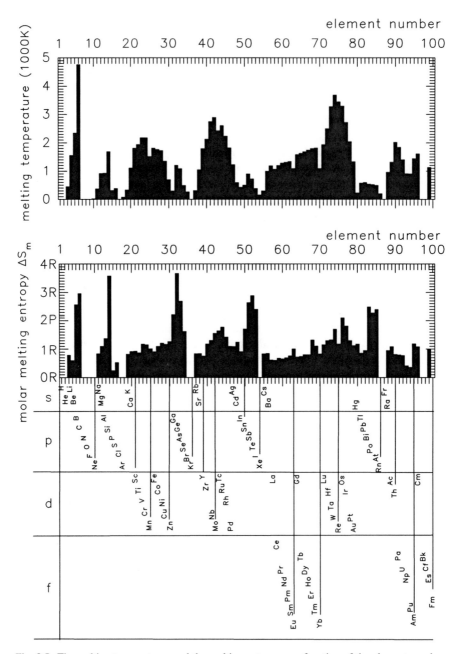

Fig. 8.5 The melting temperatures and the melting entropy as a function of the element number together with the last orbital of the free atoms occupied by an electron; indicated is the quantum number of the angular momentum $\ell = 0, 1, 2, 3$ corresponding to the symbols s, p, d, and f. Data from [13, 15]

Not all electronic transitions to excited states favor necessarily the melting. Important are transitions into those states with a strong deviation of the probability distribution as compared to their ground states. The new probability distribution hasto overcome the forces resulting from the still closed bonds.

The d-shells are known to possess a large density of electron states concentrated in a small energy range. If the energy of those states with a sufficient change of the charge probability distribution is comparatively far above the ground states, these elements will have a high melting temperature, as is well confirmed by the experimental results. On the other hand, s- and p-levels possess a comparatively small density of states and their energy ranges are broad and generally overlap each other or levels of other shells, in particular for those elements with a large element number. Then, only a minor amount of energy is needed to cause a sufficient change of the spatial charge distribution and to induce melting. In the case of Hg this occurs even below 0°C.

In contrast to this, some elements with a small element number favor hybrids of s- and p-orbitals, which causes a considerable decrease of the energy of the corresponding levels, resulting in strong bonds and a large energy difference to the nearest excited levels with a sufficiently strong change of the charge distribution probability. This explains their high melting temperatures and their large melting entropies.

With respect to the melting entropy, one must consider that a large amount of entropy can be stored in the lattice vibrations depending on the shift of the Debye temperature to lower values. In fact, as is seen from Fig. 8.5, the melting entropy is very large for the elements that have a very strong change in the coordination number upon melting like the solid systems with sp^3 or similar hybrids. Generally, in particular for those elements with a small number of electrons in p-shells, the phonon frequencies and their density distribution seem to change strongly.

In order to compare the molar specific heat capacity of compound one-component systems near the melting temperature with those of the elements, C_{ps} has to be normalized with respect to the number of atoms, N, of the compound [19]. In Fig. 8.6 the results are shown as a function of the corresponding T_m. One can clearly see that C_{ps}/N also exceeds the value of about $3R$ for the lattice vibrations considerably. This means that energy and entropy is stored in the electronic system of compounds, too. Then, the same arguments apply to explain melting by suitable electronic transitions as above for the pure chemical elements. A similar distribution is seen for the specific heat capacity per number of atoms of the compound just above the melting temperature, C_{pl}/N [19].

8.4 Melting and the Glass Transformation

Melting and glass formation of chemically bonded inorganic solids is described qualitatively by the scheme of Fig. 8.7 showing occupied electronic levels in the valence band and empty levels in the conduction band. For a metal the band

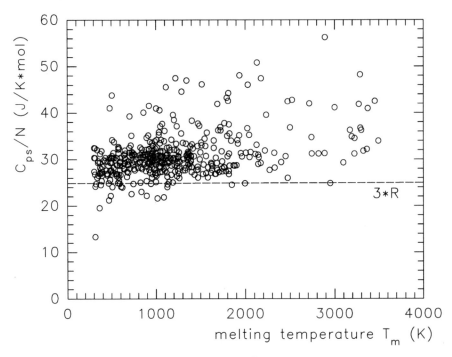

Fig. 8.6 Molar specific heat capacity per number of atoms of one-component systems near T_m, C_{ps}/N as a function of melting temperature. The data include also chemical elements. Data from [9, 10]

gap is zero and the levels of the conduction band may overlap with those of the valence band. Then, electrons shift to empty levels at lower energy. Transitions of electrons from low to high energy states depend on the temperature. As long as such transitions are rare the bonding between neighboring atoms holds and the solid is stable. With increasing temperature, in particular just below the melting temperature, more and more excited levels become accessible. Then, the charge probability distribution of the electrons differs strongly enough from that in the former low-energy states forcing the core ions to accept new places. If the core ions relax fast enough to these new places the solid will melt. Since the electrons change their levels due to thermally induced transitions continuously, the charge distribution is not stable and the core ions have to relax again to new places. This changing arrangement of wave functions and relaxing core ions represents a melt. Thus, melting is basically induced by electronic transitions and the changing probability distribution of the excited electrons, not by vibrations, which is suggested by simulations of the molecular dynamics using classical Newtonian dynamics [20–22]. Such calculations cannot describe high-temperature processes correctly if the electrons and their transitions are neglected. The authors of such calculations ignore that chemical bonding is an electronic effect.

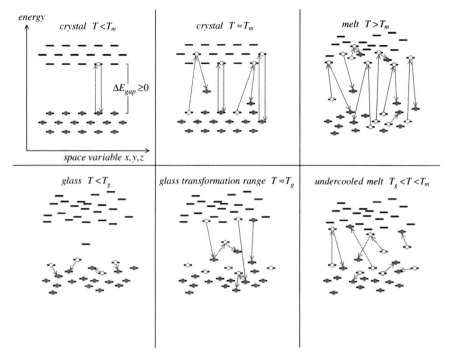

Fig. 8.7 Scheme of the electronic levels with their occupation with electrons and transitions at different temperatures of solids and melts. For explanation see text

In the molten state, the electron levels are not fixed in energy and in space. One level may be shifted to lower energy due to the relaxing atomic environment, whereas other levels are shifted to higher energy. Thus, microscopic reversibility can be preserved without additional assumptions, which is required from basic physics. Unfortunately, such a dynamic behavior cannot be represented in a static drawing like Fig. 8.7.

For increasing temperatures above T_m the exchange will be faster and more electrons will stay in higher energy levels forcing the melt to become more fluid. At very high temperatures sufficient electrons may reach energy levels to decouple atoms from their neighbors completely even in the interior of the melt. Then, boiling starts [23].

With decreasing temperature, the distribution of the outer electrons over excited and ground levels shifts in favor of the low-energy states. Exchange to other excited levels becomes less probable with the result that the viscosity increases. If nearly all bonding electrons occupy low-energy levels in the disordered state the melt may transform into a glass (Fig. 8.7). The transition into a crystal will occur if the symmetry of the wave functions enforces a crystalline or at least regular ordering (for pseudo-crystals) of atoms. This is favored further, if the atoms can be shifted one against the other easily even in the low-energy states, such as for atoms or ions with spherical symmetry of the wave functions of the bonding electrons or if directed bonds are sufficiently weak. Then, bonds need not be reopened for a rearrangement of the neighboring atoms.

Strong directed bonds prohibit such a rearrangement at lower temperatures once a bond is closed. In this case, however, rearrangement is possible as long as the temperature will increase locally if a sufficiently large amount of melting enthalpy is set free by a bond accidentally closing (an electron making a transition from an excited state to the ground level). Then, the temperature in the neighborhood will increase and neighboring bonds can be reopened, which allows rearrangement and ordering of the atoms. Such thermal spiking around a closing bond may induce nucleation and crystal growth [3, 4]. This shows clearly that the classical theories on nucleation and crystal growth require revision since they presume isothermal conditions and neglect thermal spiking and the increase of entropy during crystallization.

In Section 9, the temperature interval in which such a reheating (recalescence) can occur will be estimated and compared with data of known one-component glass systems. This temperature interval has to be passed sufficiently fast in order to avoid nucleation and crystal growth [24]. Then, the disordered state can be frozen in. Thus, the glass transformation is explained by freezing-in electronic transitions or electronic degrees of freedom. In fact, such an effect can be seen by investigations of the specific heat capacity of glasses, C_p, near T_g. Normally, C_p is larger in and above the glass transformation range than at lower temperatures. Since the freezing-in does not occur at a sharp temperature, there is a range of transformation and a standard transformation temperature, T_g, is defined by experimental data of the thermal expansion applying a specified rate of the cooling or heating.

8.5 Effects Occurring in the Glass Transformation Range

The slowing-down of electronic transitions with decreasing temperature near T_g has some consequences. Since the electrons occupy their lowest energy states, their contribution to the specific heat capacity also freezes in, which has already been mentioned. Furthermore, electrons preferentially induce the rearrangement of the atoms via excited states, which is necessary for crystallization of substances with directed bonds. Thus, crystallization during cooling of glass melts is also expected to slow down considerably. In fact, crystallization occurs predominantly above T_g.

ESR and NMR signals are modified by the large fluctuating magnetic moment of electrons making transitions to excited states [25, 26]. All these effects can be used to verify a glass transformation temperature. Also, the thermal expansion changes around T_g, which is the basis for dilatometry as the most prominent standard technique to determine the glass transition temperature. The mechanism of thermal expansion as described in some textbooks is somewhat confusing. Therefore, the deficiencies of such textbook explanations are questioned and a new mechanism is described qualitatively in the following section, which will also fit to the present considerations on melting.

The structure of the solids depends on the rearrangement of the constituents, which is modified via excited states of the electrons (corresponding to broken bonds) and their corresponding wave functions. Since such transitions are hindered and slowing down near T_g, the solid needs some time to accept its new configuration upon changing the temperature. Thus, those quantities depending on the structure or arrangement of the atoms depend also on time if the temperature is changed sufficiently fast to allow for a deviation from the equilibrium state. With decreasing temperature around T_g the time constants become very large until a rearrangement of the constituents is no longer possible in practice. Then, the electron system decouples from the "freezing-in" mechanical order of the atoms representing the "phonon system". Thus, T_g characterizes roughly the temperature range in which one observes this decoupling of the electron system from the phonon system. Among the quantities becoming a function of time near T_g due to the slowing relaxation of the deviation from equilibrium are mechanical parameters (such as the mass density, visco-elastic properties, and stresses), thermodynamic properties (like thermal expansion, specific heat capacities, enthalpy, and entropy), electrical conductivity and dielectric constant, optical parameters (such as the refractive index, absorption constants, photo-elastic coefficients), or chemical parameters (diffusion constants, valence states, reactivity, and deviations from the chemical equilibrium). Abundant literature is available dealing with such relaxation effects because of their relevance for technical applications of glasses. From a theoretical point of view such decoupling is expected to start already on an atomic scale more or less below the melting temperature of one-component systems. There, it may be observed by very fast measuring techniques.

8.6 What Makes Solids and Melts Expand?

Usually, the thermal expansion is explained by asymmetric interaction potentials between the constituents of a solid (see [27], for example). Figure 8.8 shows an asymmetric Lennard-Jones potential in arbitrary units as an example in comparison with the harmonic approximation of the potential energy represented by the dashed curve. The full curve represents the pair potential energy of an oscillating core ion as a function of the distance from another core ion at the zero of the coordinate system. In fact, this potential has been first applied for the interaction of atoms of solid or liquid noble gases. However, potentials with a similar characteristic shape have been used also for other solids, such as the alkali halides.

At low temperatures the oscillating core ion is near the bottom of the potential energy curve. With increasing temperature the ion oscillates with a larger amplitude (or more and more local phonons are excited) and the energy of the vibrating ion is shifted to higher energies. The vibration takes place at constant energy between the inner (left) and the outer (right) point on the potential curve as the turning points of the oscillation. Thus, it seems obvious that the center of the vibrating ion shifts away from the neighboring atom (with its position at the zero $r = 0$). However, such asymmetric potentials do not exist for instance in cubic crystals with their mirror symmetry (which regardless show thermal expansion). The potential curve to the

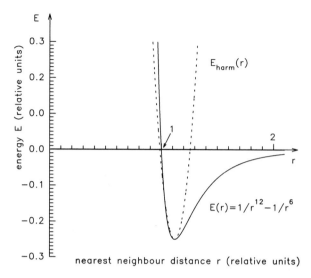

Fig. 8.8 Asymmetric Lennard–Jones potential energy of an atom with respect to another atom at the zero of the *r*-axis

right is the same as to the left and of course also to the front, back, bottom, or top. As a consequence, the center of the vibration is expected not to change.

In addition, sometimes the crystals do not expand but shrink with increasing temperature in contrast to the suggestive potential curve in Fig. 8.8. Then, the explanation of expansion and shrinkage in different temperature ranges becomes difficult. How must the potential curve be modified to explain such effects in the same systems?

Further problems arise from the value of the thermal expansion at low and high temperatures. At low temperature as long as phonons in the potential curve are not excited, the solid should not expand at all. Numerous experimental observations contradict: even at low temperatures expansion or contraction can be observed.

On the other side, the expansion coefficient should increase with temperature above the Debye temperature, since all phonons can be excited and the center of the oscillation between the turning points is shifted the further to the right the higher the energy. In fact, the experiments show very often a constant or a very modestly increasing expansion coefficient at high temperatures.

An alternate explanation, which avoids these difficulties, is the following: The wave functions of the bonding electrons either in a crystal or in a glass are determined by the potentials of the core ions (and other electrons in inner closed shells). Since the atoms oscillate around their equilibrium positions depending on the temperature, the potentials of the bonding electrons are modified by the vibrations and consequently the wave functions, which fit to the respective potential. Thus, the probability distributions of the electrons change with the temperature according to the oscillations. On the other hand, in the changed probability distribution of the electrons the oscillating core ions are forced to relax to new places with lower energy. Both effects, – the atomic rearrangement and the oscillation amplitudes changing with increasing temperature modify the wave functions and, vice versa, the modified wave functions force the atoms to relax to new places – cause expansion and sometimes even contraction depending on the interaction between the oscillating core ions and the wave functions. The same mechanism may even be responsible to induce structural phase transitions (change of the arrangement of the atoms) of crystals and pseudo-crystals depending on the amplitude and kind of the vibrations and the symmetry of the wave functions.

In the melt, the thermal expansion is caused not only by the interaction of the wave functions of the ground states with the oscillating core ions but also by the transitions of electrons into excited states. In the excited states the charge probability distribution is changed stronger causing an additional contribution to the expansion of the melt and in some cases even a contraction at the transition from the solid to the molten state. The low-temperature expansion mechanism via the interaction between the oscillations and the wave functions is supplemented by a – usually stronger – expansion due to electronic transitions into excited states and their interaction with the core ions. In a glass such transitions start to increase considerably near T_g already.

To substantiate this description the thermal expansion coefficients at room temperature are shown in Fig. 8.9 as a function of the element number together with the highest occupied atomic orbital. One can clearly see that the systems with s-orbitals as the last occupied levels show very large expansion coefficients whereas the systems with d-electrons have very low expansion coefficients. This can easily be understood: s-electrons have their highest probability distribution at the center of the atoms; if an atom is oscillating the s-orbitals will stay fixed around the core ion and interact strongly with the neighbors, whereas d-orbitals have a large distribution far away from the center of the core ions. Then, an oscillating core ion is coupled to a minor degree to the surrounding d-electrons, only, resulting in a weaker interaction with the neighboring core ions. For the rare earth elements with their f-shell being filled up, this effect is expected to be even more pronounced. However, the rare earth elements have a mixture of s-, d-, and f-orbitals as the outermost wave functions. Thus, these elements do not represent the ideal situation of f-orbitals. Even in this case, however, one can see a shift to larger values of α, if the f-shell is occupied by one-half or completely. Then, the f-orbitals are no longer effective ("switched off") and the behavior resembles rather that of s-systems.

As a consequence for generalization one cannot predict the tendency of the thermal expansion coefficient of compounds from the properties of the pure elements alone, since the wave functions and the probability distribution of the bonding electrons must be known. The deviation of the constituents from their equilibrium position changes the potential for the bonding electrons and thus effects their energetic position and the probability distribution of the electron density in space. From perturbation theory one expects that the average charge distribution of the localized bonding electrons is modified in space with the mean square vibrational amplitude of the atoms or core ions around the equilibrium location, $(\Delta r)^2$. Hence, one expects

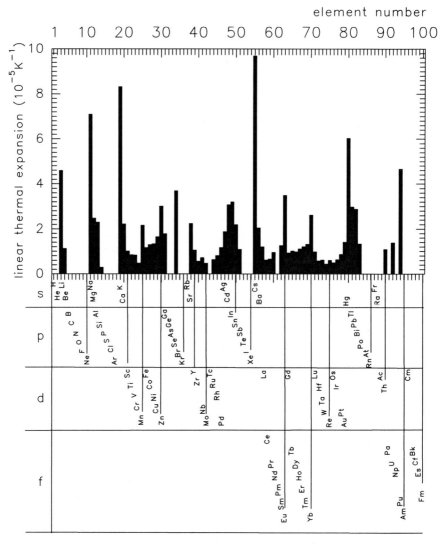

Fig. 8.9 The linear thermal expansion coefficients, α, at room temperature as a function of the element number together with the last orbital of the free atoms occupied by an electron; indicated is the quantum number of the angular momentum $\ell = 0, 1, 2, 3$ corresponding to the symbols s, p, d, and f. Data from [15]

that the total volume expansion is proportional to (or scales with) $(\Delta r)^2$, since the changed charge distribution shifts the core ions in order to minimize the electron interaction (however, with a factor of proportionality characterizing the strength of the coupling between the atomic oscillations and the wave functions). This causes the thermal expansion as has been described above.

Assuming the thermal energy per atom $\frac{1}{N_A} \int_0^T C_p(T)dT$ scales also with $(\Delta r)^2$ (N_A is the number of particles of one mole of the substance), one obtains for the total thermal volume expansion:

$$\frac{\Delta V(T)}{V_{mol}} = \frac{3\Delta\ell(T)}{\ell_0} \sim \frac{1}{N_A} \int_0^T C_p(T)dT \qquad (8.10)$$

(V_{mol} is the molar volume) or for the linear thermal expansion coefficient:

$$\alpha(T) = \frac{1}{\ell_0}\frac{d\Delta\ell(T)}{dT} \sim C_p(T). \qquad (8.11)$$

In fact, a correlation of $\alpha(T)$ with $C_p(T)$ is already known for a long time from theoretical thermodynamic considerations [14]. However, the proportionality constants are expected to be different depending on the materials and the respective wave functions. In a simple Debye model one obtains [28]:

$$\alpha(T) = C_p(T)\frac{1}{V_{mol}(p)}\frac{d\ell n(T_D)}{dp} \qquad (8.12)$$

where in p is the pressure, T_D is the Debye temperature, and $V_{mol}(p)$ is the molar volume. In this model, the constant of proportionality between $\alpha(T)$ and $C_p(T)$, which characterizes the coupling between the oscillating atoms and the bonding or valence electrons, is represented by $\frac{1}{V_{mol}(p)}\frac{d\ell n(T_D)}{dp}$.

If the molar specific heat capacity is constant above the Debye temperature, one also expects an approximately constant expansion coefficient in that temperature range. This is confirmed by many investigations of the thermal expansion of solids (electronic transitions may induce additional contributions to thermal expansion).

E. Grüneisen derived from basic thermodynamic considerations the relation:

$$\alpha = \gamma_G \frac{C_V}{3KV_{mol}} \qquad (8.13)$$

wherein γ_G is the Grüneisen constant, C_V is the molar specific heat capacity at constant volume, K is the isothermal compression modulus, and V_{mol} is the molar volume [14, 27]. However, the interpretation of the different effects was based purely on oscillations of the atoms at that time. Electronic effects have not been taken into account. The ratio $\alpha = \gamma_G/3KV_{mol}$ in (8.13) can be considered as a different, general expression of the coupling between the vibrations and the shift of the core ions via the modified wave functions. Comparing (8.12) with (8.13) yields

$$\frac{1}{V_{mol}(p)}\frac{d\ell n(T_D)}{dp} = \frac{\gamma_G}{3KV_{mol}} \quad \text{or} \quad \frac{d\ell n(T_D)}{dp} = \frac{\gamma_G}{3K} \qquad (8.14)$$

if p is the standard pressure.

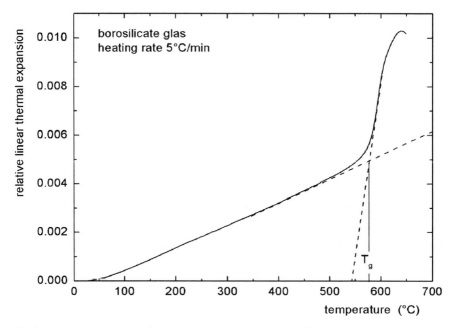

Fig. 8.10 Dilatometer curve of a borosilicate glass as an example for the expansion of a glass and the tangents (*dashed lines*) to determine the glass transformation temperature, T_g

Figure 8.10 shows the dilatometer curve of a borosilicate glass as an example of the expansion of a glass as a function of temperature. The sample expands from room temperature to the glass transition temperature, T_g, at a rather modest rate, which increases by a factor between about 2 and 5 within a small temperature interval near and above T_g.

At higher temperature the length of the sample seems to shrink. This, however, is a result of the push rods pressed against the end faces of the visco-elastic sample with a force sufficiently strong that the sample is slightly compressed and deformed. The glass transformation temperature, T_g, is determined by the intersection of the tangents to the dilatometer curve in the low- and the high-temperature ranges as shown in the figure. (Standards require certain dimensions of the sample and a heating rate of 5 K/min.)

The reason why the expansion becomes so much larger near and above T_g are easily understood by the fact that near T_g the rate of electronic transitions from low to excited states increases exponentially with the reciprocal temperature and becomes sufficiently effective to induce additional expansion. In the excited states the spatial probability distribution of the electrons is different compared to the ground states and forces the core ions to shift to new places. Since the average charge distribution is changed much stronger in the excited states, the forces on the core ions are also stronger causing a larger expansion of the glass near and above T_g. The low-temperature expansion mechanism via the oscillations or phonons is supplemented by a stronger expansion due to electronic transitions into excited states. Since such transitions start to increase considerably near T_g, one can understand why the expansion curve yields about the same values for T_g as the other methods.

8.7 Modulus of Compression of the Chemical Elements

Since the atoms in chemically bonded solids are stuck together by electrons, one expects that the modulus of compression, defined as $K=V_0 dp/dV$, depends also on the wave functions of the bonding electrons. This is verified by Fig. 8.11 showing the modulus of compression as a function of the element number of the periodic system together with the series of the last orbital filled with an electron. One can clearly see that for electrons in s-orbitals K is extraordinary small. Thus, a small pressure is needed to compress a sample. On the other hand, K increases for the p-orbitals and particularly for d-orbitals. The largest values are seen for d-shells filled by about one-half and for boron. One expects that K would be even larger for f-orbitals. However, the rare earth elements are no pure f-electron systems but possess a mixture of s-, d-, and f-orbitals, which renders K smaller. Nevertheless, one can clearly see that the influence of the f-electrons to increase K disappears if the f- shell is occupied by one-half or completely. In both cases, the influence of the f-shell is switched off. The interpretation of these findings is obvious: with increasing quantum number of the angular momentum $\ell = 0,1,2,3,\cdots$ (corresponding to the symbols for the orbitals s, p, d, and f, where the f-shell behaves exceptionally for the reasons already mentioned) the squares of the absolute wave functions possess larger values further away from the core ions corresponding to an increasing probability to find an electron. Upon compression the wave functions for large angular momentum must be modified stronger and a larger amount of energy is required for the admixture of unoccupied s- and p-wave functions. This is confirmed by the fact that d-systems have the smallest molar volumes which require large energies if compressed (see Fig. 8.11, top).

For isotropic materials the compression modulus, K, is related to the elastic or Young's modulus, E, via $K = \frac{E}{3(1-2\mu)}$, where in μ is Poisson's ratio. Thus, neglecting the influence of μ, the statements on K apply mutatis mutandis also on E.

8.8 Necessary Criteria for Glass Formation

Cooling melts of many one-component systems, one rather obtains polycrystalline material than glasses. Therefore, we have to deal with the conditions why a melt may become a structural glass. The disordered structure must be possible and stable over quite a large temperature range. Such a non-crystalline structure is stabilized in solids with directed bonds. This is already a result of the p-electron criterion on the formation of glass due to Winter [29]. All one-component glasses known hitherto

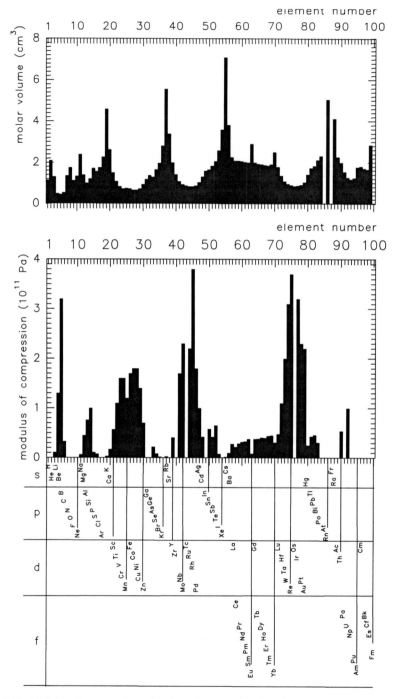

Fig. 8.11 Modulus of compression and molar volume as a function of the element number together with the last orbital of the free atoms occupied by an electron. Data from [15]

possess constituents providing strong directed bonds, in particular via p-orbitals and hybrids. Even H_2O and BeF_2 can be interpreted in this way, since the bonding is due to H- and F-bridges between O and Be at specified locations defined by p-orbitals or hybrids on the surface of these atoms.

However, the presence of directed bonds is only a necessary conditions, since Si (with strong directed bonds due to sp^3-hybridization) will not turn into a glass upon cooling from the melt but crystallize even for very fast cooling rates. A second necessary condition is based on thermodynamic considerations represented in Fig. 8.1 above. Cooling a melt below T_m it is expected to crystallize. However, crystallization necessarily has to start with a nucleus or a crystallite. Since such small crystals are not stable in the vicinity of T_m, crystallization will start at temperatures sufficiently lower than T_m. Thus, the melt will cool down on the enthalpy curve, $H_l(T)$, extrapolated from the molten state. Once it starts to crystallize the system has to undergo a transition from $H_l(T)$ to the enthalpy curve in the solid state, $H_s(T)$. Due to this transition at temperature T_u, e.g., the large amount of melting enthalpy, $H_l(T)$–$H_s(T)$, is set free which drives the temperature of the environment where the crystallization starts back to T_m. This causes bonds already closed at random to reopen and readjust which is necessary for crystal growth. This, however, is possible only as long as the difference between both enthalpy curves obeys

$$H_l(T_u) - H_s(T_u) \geq H_l(T_m) - H_l(T_u) \tag{8.15}$$

The sign of equality defines the minimum temperature difference of undercooling $\Delta T_{min} = (T_m - T_{min})$, for which local remelting is just not expected to occur any more upon binding of particles to crystallites. From this equation one can calculate T_{min}. To simplify the evaluation, the data of the enthalpy functions near the melting temperature are linearly extrapolated to lower temperatures with the slopes C_{pl} and C_{ps} near T_m. This seems to be justified, since the formation of glass usually occurs in an interval not to far below the melting temperature, where this approximation is sufficiently precise. Using this extrapolation of the molar enthalpy functions, $H_l(T)$ and $H_s(T)$, this statement is equivalent to the condition:

$$\begin{aligned} H_l(T_m) - \left\{ H_l(T_m) - C_{pl}(T_m - T_{min}) \right\} = \\ = H_l(T_m) - C_{pl}(T_m - T_{min}) - \left\{ H_s(T_m) - C_{ps}(T_m - T_{min}) \right\} \end{aligned} \tag{8.16}$$

which yields

$$(T_m - T_{min}) = \Delta T_{min} = \Delta H_m / (2C_{pl} - C_{ps}) \tag{8.17}$$

or the minimum temperature difference of undercooling relative to T_m

$$\Delta T_{min} / T_m = \Delta H_m / [T_m (2C_{pl} - C_{ps})] = \Delta S_m / (2C_{pl} - C_{ps}). \tag{8.18}$$

The larger $(T_m - T_{min})$ or $\Delta T_{min} / T_m$, the more difficult it is to avoid crystallization during cooling. Using the linear extrapolation one estimates from Fig. 8.1 that

Al_2O_3 has to be undercooled *theoretically* by about 800 K below T_m in order to yield a stable glass, which is a rather large interval of undercooling. To avoid crystallization, one needs very fast cooling rates, since the crystallization can proceed in a self-catalyzing process due to the large energy released once it started during the cooling. This explains easily the many unsuccessful attempts to transform Al_2O_3 into a glass by cooling melts. The same is true for many semiconductors marked in Fig. 8.12 by diamonds. For elemental and III–V semiconductors, the values of $\Delta T_{min}/T_m$ are very large, which shows that these systems crystallize readily providing good crystals with very high electronic mobility. On the other hand, one must notice that amorphous Si deposited by plasma techniques seems to be very unstable with respect to crystallization even at temperatures far below T_m.

Summarizing, glass formation is favored if

(A) the constituents are connected by directed, sufficiently strong bonds to a sufficient degree (which is necessary to establish a random network) and
(B) the relative temperature interval below the melting temperature T_m, within which bonds can be broken to rearrange the constituents, as estimated by (8.18), is small.

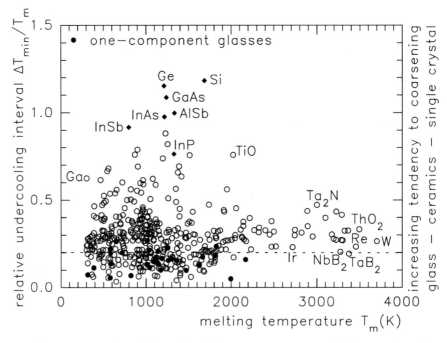

Fig. 8.12 Application of Eq. (8.18) to about 450 one-component systems. Minimum undercooling interval ΔT_{min} relative to T_m, within which bonds can be broken by local reheating to rearrange the constituents, as a function of the melting temperature. Known glasses are characterized by •. Data from [9, 10, 30]

Both conditions (A) and (B) are not sufficient, but necessary. The crystallization has to be overcome by a sufficiently fast cooling rate from the melt. This cooling rate is a function of conditions (A) and (B). The known systems which can be transformed into a glass are marked by the full circles in Fig. 8.12. In fact, all known examples show small relative undercooling intervals $\Delta T_{min}/T_m$ and directed bonds.

The data of the respective one-component glass systems are given in Table 8.1. One must point at the extraordinarily small relative undercooling interval of silica given in the data. Thus, systems based on silica with a major component are very often production glasses, whereas Al_2O_3 in contrast to B_2O_3 does not form a glass, which is quite surprising. It is interesting, however, to note that adding to the conditional glass former Al_2O_3 the network modifier CaO by about the same molar quantity yields a glass. Thus, the traditional differentiation of oxides into network formers and network modifiers seems to have no sense, at least in this case.

Pure B_2O_3 is a one-component system which forms readily a glass. Therefore, the data of $\Delta T_{min}/T_m$ for B_2O_3 seems to be unexpectedly large, which would be worth while re-examining critically.

The relation (8.18) has been discussed in detail in [31]. Both criteria (A) and (B) seem to help avoid unsuccessful attempts to test the glass formation ability of new systems. Since the systems of oxides, chalcogenides, and fluorides have been investigated in the past very thoroughly, the criteria (A) and (B) could be helpful in particular for developments of metallic glasses. Metals are difficult to transform into glasses, since very often the bonds are not directed. Only in the case of sufficiently strong directed bonds and a small $\Delta T_{min}/T_m$ one can expect glass formation. In fact, most metallic glasses contain such elements with strong directed bonds.

Several other criteria on glass formation are known from the literature. They have been critically reviewed in [24, 31]. These criteria are not as general as (A) and (B). Some of them cannot be generalized since they apply to special kinds of glasses or to special view points. The aim is to predict the necessary cooling rate at least for a given one-component system. Some publications already deal with such kinetic criteria of glass formation. However, these papers suffer from the deficiency that they consider nucleation and crystallization as isothermal effects. Since one can see from Fig. 8.1 that crystallization enthalpy is liberated, which causes local heating near the nucleation site, such theories require revision.

The relation (8.18) occurs also in the literature on hypercooling of metal melts [32]. Cooling melts of metals below $(T_m - \Delta T_{min})$ is known in the literature as hypercooling, where the melt is claimed to solidify under non-equilibrium conditions. The hypercooling temperature interval $\Delta T_{hyp} = \Delta H_m/C_{pl} \approx \Delta T_{min}$ corresponds to (8.18), since $C_{pl} \approx C_{ps}$. However, it has not yet been considered before to characterize glass formation, but to distinguish between equilibrium and non-equilibrium conditions of crystallization.

Table 8.1 Melting entropy ΔS_m, melting temperature T_m, and relative minimum undercooling $\Delta T_{min}/T_m$ of one-component systems forming glasses by cooling from the melt. Data from [9, 10, 30]

One-component system	ΔS_m (J/(K·mol))	T_m (K)	$\Delta T_{min}/T_m$	One-component system	ΔS_m (J/(K·mol))	T_m (K)	$\Delta T_{min}/T_m$
P	2.07	317	0.0727	$K_2O \cdot 4B_2O_3$	110.80	1130	0.146
S	4.42	388	0.112	$Rb_2O \cdot SiO_2$	36.64	1143	0.207
As_2S_2	10.47	580	0.0556	$Na_2O \cdot 2SiO_2$	31.03	1147	0.135
As_2O_3	31.66	582	0.221	$Rb_2O \cdot 4SiO_2$	39.27	1173	0.0994
As_2S_3	49.03	585	0.197	$Li_2O \cdot 2B_2O_3$	101.35	1190	0.166
$ZnCl_2$	17.36	591	0.136	$K_2O \cdot SiO_2$	38.56	1249	0.227*
As_2Se_3	62.81	650	0.262*	$CaO \cdot 2B_2O_3$	89.85	1263	0.155
B_2O_3	30.47	723	0.198	$Li_2O \cdot 2SiO_2$	41.20	1307	0.163
BeF_2	5.77	825	0.0683	$Rb_2O \cdot 2SiO_2$	39.94	1363	0.154
$Na_2O \cdot 2B_2O_3$	79.96	1016	0.131	$K_2O \cdot Al_2O_3 \cdot 6SiO_2$	83.58	1473	0.0978
$2PbO \cdot SiO_2$	50.28	1016	0.266	$CaO \cdot MgO \cdot 2SiO_2$	77.21	1665	0.174*
$PbO \cdot SiO_2$	25.12	1037	0.213	$CaO \cdot SiO_2$	30.88	1817	0.184
$K_2O \cdot 4SiO_2$	46.97	1043	0.114	$CaO \cdot Al_2O_3 \cdot 2SiO_2$	91.50	1826	0.235*
$K_2O \cdot 2B_2O_3$	95.84	1088	0.148	SiO_2	4.80	1996	0.0494
$LiBO_2$	30.34	1117	0.173*	$2MgO \cdot SiO_2$	32.79	2171	0.159**

* Poor glass or borderline; ** glass by splat cooling

8.9 Possible Extension to Multi-Component Systems

In practice, one-component systems are used very rarely as glasses but rather multi-component systems. Condition (8.18) applies strictly speaking not only to one-component systems but also to eutectic points, since also in this case all quantities entering that relation are defined. Thus, if the composition of a multi-component system is close to a eutectic point (or sometimes curve), glass formation is also favored if $\Delta T_{min}/T_m$ is small and sufficient strong directed bonds are present (neglecting that a sharp melting temperature is missing).

The separation into different crystallizing components near a eutectic requires some time for diffusion and the different components hinder each other to separate, which favors glass formation in multi-component systems ("confusion principle"). While these kinetic effects certainly are effective, it should be pointed also at the balance of enthalpy and entropy of the transition near the eutectic, which yields that $\Delta T_{min}/T_m$ as defined by relation (8.18) should be sufficiently small for easy glass formation.

If the multi-component system is not close to a eutectic point, its melt will hit the hyper liquidus surface in the space of the composition variables upon cooling. Then, at least one of the components spanning that liquidus surface is expected to crystallize. Whether this occurs depends on the temperature. If it occurs at a temperature lower than $(T_m - \Delta T_{min})$ of the corresponding single component, it probably will not crystallize upon cooling.

A second effect may support or even may dominate this inhibition of crystallization: the sign of the solution enthalpy of the respective crystallizing component, corresponding to the enthalpy flow in or out of the system. If enthalpy and entropy have to be conducted out of the system in order to keep the temperature constant while dissolving the respective component, then – in reverse – the temperature decreases necessarily when that component starts to crystallize. This effect will accelerate the cooling rate and consequently favors glass formation. If, on the other hand, solution enthalpy and entropy have to be brought into the system during the solution of that component, then they will be released upon crystallization, which causes an increase of the temperature of the system and slows down the cooling rate. Then, crystallization seems to be favored.

At the present time, these mechanisms need further elaboration and, in particular, data on dissolution enthalpies, and entropies of suitable glass forming systems.

8.10 Discussion

In the preceding sections melting has been explained by the interaction of electronic wave functions and the probability distribution of electrons with core ions in the solids. Such an interaction has not been taken into account in former theories on melting, in which the presence of electrons is widely ignored. Even today in molecular dynamics calculations using classical potentials the bonding electrons are neglected without justification. This is surprising since chemical bonding is obviously due to electrons and their interactions with the core ions.

The static of solids is generally described using wave functions and the corresponding electron distribution in space and energy is calculated for many systems. Therefore it is hard to understand, why such an interaction has been ignored to explain dynamic effects occurring in solids. In the present article, the interaction between electrons and core ions has been considered to explain qualitatively several dynamic effects. In fact, melting of chemically bonded solids is clearly not a static effect of the lattice but a dynamic effect due to the electron system interacting with the core ions. For that reason, it is essentially a quantum mechanical or wave mechanical effect of the electrons. The same interaction causes thermal expansion in the solid and molten state.

In addition, the glass forming ability of one-component systems can be understood as well as the glass transition temperature on the same basis: in the range around T_g the electronic transitions between excited and low energy states freeze in and the electron system decouples from the core ions and their vibrations. Due to this decoupling the relaxation times of the glass network increase considerably, which explains the numerous relaxation effects and the time dependence of many physical quantities of the glasses near T_g. Thus, the present summary of recent publications seems to clarify and replace the former rather diffuse explanations of T_g and the accompanying relaxation effects.

References

1. H.-J. Hoffmann: Invited lecture presented at JINR Dubna, Russia, on April 13, 2008 during the 12th Workshop on Nucleation Theory and Applications. Printed version: Thermodynamics of melting and the formation of glasses and crystals, in *Nucleation Theory and Applications*, J. W. P. Schmelzer, G. Röpke, and V. B. Priezzhev, eds., ISBN 978-5-9530-0199-1, Dubna, JINR, pp. 235–255 (2008).
2. J. Langer: The mysterious glass transition (Reference frame), pp. 8–9 in Physics Today, February 2007; comments, pp. 15 and 72–75 in Physics Today, January 2008.
3. H.-J. Hoffmann: Energy and entropy of solids, glasses and melts, Glass Sci. Technol. **78** (5), 218–229 (2005).
4. H.-J. Hoffmann: New ideas about melting and the glass transition, Phys. Chem. Glasses **45** (4), 227–237 (2004).
5. H.-J. Hoffmann: Thermodynamic aspects of melting and glass formation, Phys. Chem. Glasses **48** (1), 23–32 (2007).
6. F. A. Lindemann: Über die Berechnung molekularer Eigenfrequenzen, Physik. Z. **11**, 609–612 (1910).
7. A. R. Ubbelohde: *Melting and crystal structure*, Clarendon Press, Oxford (1965).
8. H.-J. Hoffmann: On Lindemann's melting criterion, Mat.-wiss. u. Werkstofftech. (Mater. Sci. Eng. Technol.) **35** (2), 79–81 (2003).
9. I. Barin and O. Knacke: *Thermochemical properties of inorganic substances*, Springer, Berlin (1973).
10. I. Barin, O. Knacke, and O. Kubaschewski: *Thermochemical properties of inorganic substances. Supplement*, Springer, Berlin (1977).
11. M. Born: Thermodynamics of crystals and melting, J. Chem. Phys. **7**, 591–603 (1939).
12. R. W. Cahn: Crystal defects and melting, Nature **273**, 491–492 (1978).

13. Landolt-Börnstein: Numerical Data and Functional Relationships in Science and Technology, *New Series* / Editor in Chief: W. Martienssen, Group IV: Physical Chemistry, Volume 19 Thermodynamic Properties of Inorganic Materials compiled by SGTE, Subvolume A Part 1, ISBN 3-540-64734-1, Springer, Berlin (1999).

14. E. Grüneisen: Zustand des festen Körpers, Handbuch der Physik **10**, 1–59 (1926).

15. M. J. Winter: http://www.webelements.com/

16. H. B. Callen: *Thermodynamics*, John Wiley & Sons, Inc., New York (1960).

17. C. Kittel and H. Kroemer: *Thermal physics*, ISBN 0-7167-1088-9, W. H. Freeman and Co., San Francisco (1980).

18. H.-J. Hoffmann: Reasons for melting of chemical elements and some consequences, Mat.-wiss. u. Werkstofftech. (Mater. Sci. Eng. Technol.) **34** (6), 571–582 (2003).

19. H.-J. Hoffmann: New Concepts to Explain the Melting Transition and the Glass Transformation, Proceedings of the XX International Congress on Glass, Kyoto, September **26** – October 1, (2004).

20. K. Binder, J. Horbach, W. Kob, W. Paul, and F. Varnik: Molecular dynamics simulations, J. Phys. Condens. Matter., **16**, S429–S453 (2004).

21. K. Vollmayr, W. Kob, and K. Binder: Cooling-rate effects in amorphous silica: a computer simulation study, Phys. Rev. **B54**, 15808–15827 (1996).

22. C. A. Angell, J. H. R. Clarke, and L. V. Woodcock: Interaction potentials and glass formation: A survey of computer experiments, Adv. Chem. Phys. **48**, 397–454 (1981).

23. H.-J. Hoffmann: Boiling of the chemical elements, Mat.-wiss. u. Werkstofftech. (Mater. Sci. Eng. Technol.) **35** (9), 562–568 (2004).

24. H.-J. Hoffmann: Conditions for the formation of glasses by cooling melts of one-component systems, Glastechn. Ber. Glass Sci. Technol. 74 (11/12), 324–332 (2001).

25. F. W. Krämer: Magnetische Kernresonanzuntersuchungen über Bewegungsvorgänge in Gläsern. PhD Thesis, Johannes Gutenberg University, Mainz (1970).

26. F. W. Krämer: W. Müller-Warmuth, and H. Dutz, Magnetische Kernresonanzuntersuchungen an Gläsern im Transformationsbereich, Glastechn. Ber. **46** (10), 191–195 (1973).

27. C. Kittel: *Introduction to solid state physics*, third edition, John Wiley, New York (1966).

28. L. D. Landau, E. M. Lifshits, and L. P. Pitaevskii ed.: *Statistical physics*, ISBN 0-080-230733, Pergamon Press, Oxford (1980).

29. A. Winter: Glass formation, J. Am. Ceram. Soc. **40**, 54–58 (1957).

30. O. V. Mazurin, M. V. Streltsina, and T. P. Shvaiko-Shvaikovskaya: *Handbook of glass data*, ISBN 0-444-41689-7, Elsevier, Amsterdam, (1983, 1985, 1987, 1991, 1993).

31. H.-J. Hoffmann: Application of a new necessary criterion for glass formation of one-component systems, Phys. Chem. Glasses **46** (6), 570–578 (2005).

32. P. R. Sahm, I. Egry, and T. Volkmann: *Schmelze, Erstarrung, Grenzflächen – Eine Einführung in die Physik und Technologie flüssiger und fester Metalle*, **ISBN 3-528-06979-1**, Vieweg, Wiesbaden (1999).

Chapter 9
Thermodynamics of Glass Melting

Reinhard Conradt

Abstract First, a model based on linear algebra is described by which the thermodynamic properties of industrial multi-component glasses and glass melts can be accurately predicted from their chemical composition. The model is applied to calculate the heat content of glass melts at high temperatures, the standard heat of formation of glasses from the elements, and the vapor pressures of individual oxides above the melt. An E-fiber glass composition is depicted as an example. Second, the role of individual raw materials in the melting process of E-glass is addressed, with a special focus on the decomposition kinetics and energetic situation of alkaline earth carriers. Finally, the heat of the batch-to-melt conversion is calculated. A simplified reaction path model comprising heat turnover, content of residual solid matter, and an approach to batch viscosity is outlined.

Keywords Heat of formation of glasses · Heat content of glass melts · Thermodynamics of multi-component systems · Dolomite · Limestone · Batch-to-melt conversion · Heat demand of melting

9.1 Approach to the Thermodynamics of Glasses and Glass Melts

It is the purpose of this chapter to develop and outline a description frame for the thermodynamic properties of multi-component glasses and glass melts in general. This comprises, in specific, the standard heat of formation of a glass of arbitrary composition, the heat content of a glass melt at a given temperature, and the chemical potentials of individual oxides of the melt. These quantities are of high practical importance. They form the basis for a quantitative description

R. Conradt (✉)
Department of Glass and Ceramic Composites, RWTH Aachen University, Institute of Mineral Engineering, 52064, Aachen, Germany
e-mail: conradt@ghi.rwth-aachen.de

F.T. Wallenberger, P.A. Bingham (eds.), *Fiberglass and Glass Technology*,
DOI 10.1007/978-1-4419-0736-3_9, © Springer Science+Business Media, LLC 2010

– of the batch-to-melt conversion and the heat balance of a glass furnace (see Section 9.3.2);
– of corrosion and evaporation processes involving a glass melt (see Section 9.2.3);
– even of the relation between viscosity and chemical composition (see Section 9.1.4).

It is true that there are many excellent data collections (e.g., [1], see also the compilation in [2]) relating the properties of glasses and glass melts to their chemical composition. These data collections are usually derived from a broad experimental database which is submitted to a thorough statistical evaluation. The result is typically presented in the form of oxide-specific factors b_j. Individual properties B are calculated in a straightforward way by (typically: linear) interpolation

$$B = \sum_j p_j \cdot b_j$$

where p_j denotes the content of oxide j on a molar or mass-related basis. The famous Appen factors [3] for the calculation of the thermal expansion coefficient may serve as an example. As meaningful as such systems may be, they do not offer any insight into the causes by which a specific oxide modifies the glass properties. Beyond this, they are usually focused on the so-called physical properties of glasses and melts, such as density, viscosity, thermal expansion coefficient and not on the quantities required to quantify thermochemical reactions. The latter quantities may be approached by thermodynamic calculations in a direct way.

9.1.1 Description Frame for the Thermodynamic Properties of Industrial Glass-Forming Systems

The thermodynamic properties of multi-component systems have been approached by different elaborate models, among which are the (modified) quasi-chemical model [4], the cell model [5], the model of ideal mixing of complex components [6, 7]. But even for elaborate computer codes and databases used in computational thermochemistry [8, 9], the generation of reliable data for multi-component systems is still a major problem. The author's own approach [10–12] outlined below is especially well suited for the multi-component systems typical of industrial glasses.

The rigid glass and the glass melt are described by their energetic and entropic difference to a normative state of mineral phases k which would form and coexist at the glass transition temperature T_g under equilibrium conditions. This state has been termed "crystalline reference system" (c.r.s.). In the temperature interval from absolute zero to T_g, the rigid glass differs from the c.r.s. by an enthalpy and entropy of vitrification, H^{vit} and S^{vit}, respectively. In the same way, the melt at liquidus temperature T_{liq} differs by an enthalpy and entropy of fusion: H^{fus} and S^{fus}. The glass and the melt are regarded as a mixture of glassy and melted compounds k, respectively. Heats (enthalpies) and entropies of mixing, which usually make very large contributions in silicate systems if referred to the oxide components j, become negligibly small if referred to the c.r.s. compounds k. The crucial step is the identification of the appropriate set of compounds k. This is described in detail in [11]. According to Gibbs phase rule, the number of oxides j in a glass composition is identical to the number of compounds k in the corresponding c.r.s.; the molar amounts n or masses m of the j and k (given in kmol or kg, respectively, per 100 kg of glass) are thus related by a linear equation system

$$\vec{n}_j = (\nu_{jk}) \cdot \vec{n}_k \Rightarrow \vec{n}_k = (B_{kj}) \cdot \vec{n}_j \qquad (9.1a)$$

$$\vec{m}_j = (\mu_{jk}) \cdot \vec{m}_k \Rightarrow \vec{m}_k = (A_{kj}) \cdot \vec{m}_j \qquad (9.1b)$$

$$(A_{kj}) = (\nu_{jk})^{-1}, \quad (B_{kj}) = (\mu_{jk})^{-1} \qquad (9.1c)$$

Here, ν_{jk} is the matrix element representing how many moles of oxide j are found in compound k; μ_{jk} represents how many kilograms of oxide j are contained in 1 kg of compound k. A_{kj} and B_{kj} are the elements of the inverted matrices (ν_{jk}) and (μ_{jk}), respectively. Table 9.1 presents the main oxides j and compound k, and the matrix elements B_{kj} used for E-glass compositions. The composition of an E-glass depicted as example from [13] is shown in Table 9.2 in terms of both oxides j and compounds k.

Table 9.1 Matrix (μ_{jk}) for the calculation of the normative compounds k of E-glasses from their oxide composition given by the amounts m_j of oxides j in kg/kg glass; the calculation proceeds like $m(k = SiO_2) = 1.000 \cdot m(SiO_2) + 0.752\, m(TiO)_2 - 0.589 \cdot m(Al_2O_3) - 1.491 \cdot m(MgO) - 1.071 \cdot m(CaO) - 4.847 \cdot m(Na_2O) - 3.189 \cdot m(K_2O)$; $m(k = CaO \cdot TiO_2) = 1.702 \cdot m(TiO_2)$

Compound $k =$	Oxide $j =$								
	SiO$_2$	TiO$_2$	Al$_2$O$_3$	B$_2$O$_3$	Fe$_2$O$_3$	MgO	CaO	Na$_2$O	K$_2$O
SiO$_2$	1.000	0.752	–0.589	–	–	–1.491	–1.071	–4.847	–3.189
CaO·TiO$_2$	–	1.702	–	–	–	–	–	–	–
CaO·Al$_2$O$_3$·2SiO$_2$	–	–	2.729	–	–	–	–	–4.489	–2.953
B$_2$O$_3$	–	–	–	1.000	–	–	–	–	–
FeO·Fe$_2$O$_3$	–	–	–	–	1.000	–	–	–	–
CaO·MgO·2SiO$_2$	–	–	–	–	–	5.372	–	–	–
CaO·SiO$_2$	–	–1.454	–1.139	–	–	–2.882	2.071	1.874	1.233
Na$_2$O·Al$_2$O$_3$·6SiO$_2$	–	–	–	–	–	–	–	8.462	–
K$_2$O Al$_2$O$_3$·6SiO$_2$	–	–	–	–	–	–	–	–	5.909

The thermodynamic quantities of a glass or its melt are obtained by the following set of equations:

$$H^{\circ}_{glass} = \sum_k n_k \cdot (H^{\circ}_k + H^{vit}_k) \qquad (9.2a)$$

Table 9.2 Composition on a reference E-glass [13] given in terms of both oxides j and normative compounds k; M = molar mass in kg/kmol; m = mass in kg per 100 kg glass; n = molar amount in kmol per 100 kg glass; a = thermodynamic activity at 1400°C

Oxide j	M_j	m_j	n_j	$\log a_j$	Compound k	M_k	m_k	n_k
SiO_2	60.084	55.15	0.9179	−0.37	SiO_2	60.084	18.75	0.3120
TiO_2	79.898	0.57	0.0071		$CaO \cdot TiO_2$	135.977	0.96	0.0071
Al_2O_3	101.961	14.42	0.1414		$CaO \cdot Al_2O_3 \cdot 2SiO_2$	278.208	36.59	0.1315
B_2O_3	69.619	6.86	0.0985	−0.79	B_2O_3	69.619	6.86	0.0985
Fe_2O_3	159.691	0.44	0.0055		$FeO \cdot Fe_2O_3$	231.537	0.34	0.0015
FeO	71.846		0.0055		$FeO \cdot SiO_2$	131.930	0.14	0.0011
MgO	40.311	4.22	0.1047		$CaO \cdot MgO \cdot 2SiO_2$	216.558	22.67	0.1047
CaO	56.079	17.73	0.3162	−3.33	$CaO \cdot SiO_2$	116.163	8.48	0.0730
Na_2O	61.979	0.61	0.0099	−10.50	$Na_2O \cdot Al_2O_3 \cdot 6SiO_2$	524.444	5.19	0.0099
Sum		100.00					99.98	

$$H^\circ_{1673,\text{liq}} = \sum_k n_k \cdot H^\circ_{1673,\text{liq},k} \tag{9.2b}$$

$$S^\circ_{\text{glass}} = \sum_k n_k \cdot (S^\circ_k + S^{\text{vit}}_k) \tag{9.2c}$$

$$S^\circ_{1673,\text{liq}} = \sum_k n_k \cdot S^\circ_{1673,\text{liq},k} \tag{9.2d}$$

$$c_{P,\text{liq}} = \sum_k n_k \cdot c_{P,\text{liq},k} \tag{9.2e}$$

$$H_{T,\text{liq}} = H^\circ_{1673,\text{liq}} + c_{P,\text{liq}} \cdot (T - 1673) \tag{9.2f}$$

$$S_{T,\text{liq}} = S^\circ_{1673,\text{liq}} + c_{P,\text{liq}} \cdot \ln (T/1673) \tag{9.2g}$$

H_{glass} is the standard enthalpy (heat) of the rigid glass (at 25°C, 1 bar); $H_{1673,\text{liq}}$ is the heat of the melt at 1400°C (= 1673.15 K); $H_{T,\text{liq}}$ is the heat of the melt at arbitrary temperature T; entropies S have the analogous meaning; $c_{P,\text{liq}}$ is the heat capacity of the melt above T_{liq}. The quantities of the individual compounds k used in (9.2a–g) are compiled in Table 9.3. This table does allow to calculate the properties not only of E-glasses but also of A-fiber, C-fiber, stone and slag wool, crystal, low-expansion, container, and float glasses. For an appropriate determination of the c.r.s, see [11].

9.1.2 Heat Content of Glass Melts

For our reference E-glass (see Table 9.2), the following results are obtained [12]:

$$H = -15,111 \text{ kJ/kg} = -4,197.5 \text{ kW h/t}$$

Table 9.3 Thermodynamic data of compounds k employed to represent the crystalline reference systems (c.r.s.) of industrial glasses; enthalpies H in kJ/mol, entropies S and heat capacities c_P in J/(mol K); superscripts: $°$ = standard state at 298.15 K, 1 bar; vit = vitrification; subscripts: liq = liquid state; 1673 = 1673.15 K

k	$-H°$	$S°$	H^{vit}	S^{vit}	$-H_{1673,liq}$	$S_{1673,liq}$	$c_{P,liq}$
$P_2O_5 \cdot 3CaO$	4117.1	236.0	135.1	51.5	3417.1	898.7	324.3
P_2O_5	1504.9	114.4	18.2	9.5	1151.5	586.6	181.6
Fe_2O_3	823.4	87.4	45.2	17.2	550.2	370.3	142.3
$FeO \cdot Fe_2O_3$	1108.8	151.0	82.8	31.4	677.8	579.9	213.4
$FeO \cdot SiO_2$	1196.2	92.8	36.7	13.8	962.3	342.7	139.7
$2FeO \cdot SiO_2$	1471.1	145.2	55.2	20.5	1118.8	512.1	240.6
$MnO \cdot SiO_2$	1320.9	102.5	40.2	15.1	1085.3	345.2	151.5
$2ZnO \cdot SiO_2$	1643.1	131.4	82.4	31.4	1261.1	494.5	174.5
$ZrO_2 \cdot SiO_2$	2034.7	84.5	86.6	32.6	1686.2	381.2	149.4
$CaO \cdot TiO_2$	1660.6	93.7	67.4	25.5	1365.7	360.2	124.7
$BaO \cdot Al_2O_3 \cdot 2SiO_2$	4222.1	236.8	130.5	95.4	3454.3	1198.3	473.2
$BaO \cdot 2SiO_2$	2553.1	154.0	81.6	26.8	2171.1	533.5	241.4
$BaO \cdot SiO_2$	1618.0	104.6	56.5	41.0	1349.8	361.1	146.4
$Li_2O \cdot Al_2O_3 \cdot 4SiO_2$	6036.7	308.8	184.1	12.1	5235.4	1173.2	498.7
$Li_2O \cdot SiO_2$	1648.5	79.9	16.7	6.3	1416.7	339.7	167.4
$K_2O \cdot Al_2O_3 \cdot 6SiO_2$	7914.0	439.3	106.3	29.3	6924.9	1559.4	765.7
$K_2O \cdot Al_2O_3 \cdot 2SiO_2$	4217.1	266.1	80.4	22.1	3903.7	666.5	517.6
$K_2O \cdot 4SiO_2$	4315.8	265.7	26.4	21.3	3697.8	983.7	410.0
$K_2O \cdot 2SiO_2$	2508.7	190.6	12.6	23.9	2153.1	595.4	275.3
$Na_2O \cdot Al_2O_3 \cdot 6SiO_2$	7841.2	420.1	125.0	28.4	6870.1	1512.5	648.1
$Na_2O \cdot Al_2O_3 \cdot 2SiO_2$	4163.5	248.5	92.0	27.9	3614.1	856.9	423.8
B_2O_3	1273.5	54.0	18.2	11.3	1088.7	271.1	129.7
$Na_2O \cdot B_2O_3 \cdot 4SiO_2$	5710.9	270.0	42.7	21.1	4988.0	1090.2	637.6
$Na_2O \cdot 4B_2O_3$	5902.8	276.1	58.3	40.1	4986.7	1275.5	704.2
$Na_2O \cdot 2B_2O_3$	3284.9	189.5	48.8	26.6	2735.9	780.3	444.8
$Na_2O \cdot B_2O_3$	1958.1	147.1	43.6	19.5	1585.7	538.7	292.9
$2MgO \cdot 2Al_2O_3 \cdot 5SiO_2$	9113.2	407.1	135.8	41.4	7994.8	1606.2	1031.8
$MgO \cdot SiO_2$	1548.5	67.8	46.6	13.6	1318.0	296.2	146.4
$2MgO \cdot SiO_2$	2176.9	95.4	61.4	11.0	1876.1	402.9	205.0
$CaO \cdot MgO \cdot 2SiO_2$	3202.4	143.1	92.3	25.7	2733.4	621.7	355.6
$2CaO \cdot MgO \cdot 2SiO_2$	3876.9	209.2	106.7	32.0	3319.2	775.3	426.8
$CaO \cdot Al_2O_3 \cdot 2SiO_2$	4223.7	202.5	103.0	37.7	3628.8	791.2	380.7
$2CaO \cdot Al_2O_3 \cdot SiO_2$	3989.4	198.3	129.9	49.4	3374.0	787.8	299.2
$CaO \cdot SiO_2$	1635.1	83.1	49.8	18.8	1382.0	329.7	146.4
$2CaO \cdot SiO_2$	2328.4	120.5	101.3	38.5	1868.2	509.2	174.5
$Na_2O \cdot 2SiO_2$	2473.6	164.4	29.3	13.2	2102.5	588.7	261.1
$Na_2O \cdot SiO_2$	1563.1	113.8	37.7	9.8	1288.3	415.1	179.1
$Na_2O \cdot 3CaO \cdot 6SiO_2$	8363.8	461.9	77.3	20.5	7372.6	1555.6	786.6
$Na_2O \cdot 2CaO \cdot 3SiO_2$	4883.6	277.8	57.7	13.4	4240.9	990.4	470.3
$2Na_2O \cdot CaO \cdot 3SiO_2$	4763.0	309.6	87.0	22.6	4029.6	1107.9	501.2
SiO_2	908.3	43.5	6.9	4.0	809.6	157.3	86.2

$$H^{\text{vit}} = 328 \text{ kJ/kg} = 91.0 \text{ kW h/t}$$

$$H_{\text{glass}} = -14{,}783 \text{ kJ/kg} = -4{,}106.5 \text{ kW h/t}$$

$$H_{1673,\text{liq}} = -13{,}035 \text{ kJ kg} = -3{,}621.0 \text{ kW h/t}$$

$$c_{P,\text{liq}} = 1{,}456 \text{ J/(kg K)} = 404.4 \text{ W h/(t K)}$$

$$S^{\text{vit}} = 120 \text{ J/(kg K)} = 33.3 \text{ W h/(t K)}$$

All quantities are given in SI units J, kg, K. In order to allow an easy comparison to electrical energy, the quantities are also given in kW h/t and W h/(t K) for heats and entropies, respectively; 1 t = 1,000 kg. From the above data, a number of data with high practical importance are derived. As an immediate example, the heat content of a given glass melt (relative to 25°C) at arbitrary temperature T is given by

$$\Delta H_{T,\text{liq}} = H_{T,\text{liq}} - H^{\circ}_{\text{glass}} \tag{9.3}$$

This is the amount of heat taken from a furnace by the glass pull p. With $\Delta H_{T,\text{liq}}$ given in kW h/t and p in t/h, the quantity

$$Q' = \Delta H_{T,\text{liq}} \cdot p \tag{9.4}$$

Q' in kW yields a lower threshold (disregarding heat losses) of the power required for glass melting. $\Delta H_{T,\text{liq}}$ and Q' are important contributions to the heat and power balance of a glass furnace. For our reference E-glass, the following results are obtained:

$$\Delta H_{T,\text{liq}} = 1{,}748 \text{ kJ/kg} = 485.5 \text{ kW h/t for } 25 - 1400°C$$
$$= 1{,}602 \text{ kJ/kg} = 445.1 \text{ kW h/t for } 25 - 1300°C$$

It is true that the heat content of a melt may also be estimated from existing oxide increment systems [14–16]. As shown in Table 9.4 for the example of a mineral fiber glass, however, the direct thermodynamic approach is more accurate. Since the increment systems are based on a quite restricted composition range only, the thermodynamic approach is also more versatile compositionally.

Table 9.4 Heat content $\Delta H_{T,\text{liq}}$ in kW h/t of a mineral fiber glass melt with a composition of 58.2 SiO_2, 1.1 Al_2O_3, 3.4 Fe_2O_3, 9.0 MgO, 23.5 CaO, 4.6 Na_2O, 0.2 K_2O (wt%) at different temperatures; calculated and experimental values (inverse drop calorimetry)

T, °C	1408	1360	1352
After Schwiete and Ziegler [14]	448	429	426
After Moore and Sharp [15]	441	424	421
After Gudovich and Primenko [16]	391	374	371
Own model	465	447	444
Experimental value ($\pm$ 21)	472	445	440

9.1.3 Chemical Potentials and Vapor Pressures of Individual Oxides

How does an individual oxide influence the chemical reactivity of a glass melt? For this question, frequently asked by the technologist, no simple or straightforward answer is available. The reactivity of an individual oxide does not only depend on the properties of the oxide itself but also on its environment, i.e., on the composition of the melt. Scientifically speaking, the question aims at the chemical potential $\mu_{j,T}$ of an oxide j at a given temperature T and $P = 1$ bar. The chemical potential is related to the Gibbs energy G_T of the pure oxide j and the so-called thermodynamic activity $a_{j,T}$ by

$$\mu_{j,T} = G_T + R \cdot T \cdot \ln a_{j,T} \tag{9.5}$$

$R = 8.314$ J/(mol K) $=$ gas constant. As shown before [11], activities a_j are swiftly calculated from the c.r.s. by

$$\ln a_{j,T} = \sum_k A_{jk} \cdot \left(\frac{G_{k,T}}{R \cdot T} + \ln x_k \right) - \frac{G_{j,T}}{R \cdot T} \tag{9.6}$$

The $G_{k,T}$ and the $G_{j,T}$ are taken from Tables 9.2 and 9.5, respectively; x_k denotes the molar fraction of compound k given by $n_k/\Sigma n_k$. For the reference E-glass melt, activity data at $T = 1400$°C are given in Table 9.2.

An immediate example of application of activity data is the calculation of evaporation losses from glass melts. The evaporation rate J_i of a vapor species i (typically given in units of kg/m^2 and h) reads

$$J_i = \beta_i \cdot (M_i/RT) \cdot P_i \tag{9.7}$$

Table 9.5 Gibbs energies G in kJ/mol of oxides j in equilibrium at absolute temperatures as indicated by the subscripts

Oxide j	$-G_{j,1473}$	$-G_{j,1673}$	$-G_{j,1873}$	Oxide j	$-G_{j,1473}$	$-G_{j,1673}$	$-G_{j,1873}$
SiO_2	1043.8	1073.8	1105.6	MgO	695.3	716.7	739.3
TiO_2	1098.1	1131.5	1166.9	CaO	748.9	773.4	799.3
ZrO_2	1255.0	1289.3	1325.4	BaO	716.1	748.6	782.6
Al_2O_3	1876.7	1924.6	1975.7	MnO	531.4	560.5	591.0
B_2O_3	1489.9	1542.5	1598.2	ZnO	471.2	496.5	523.2
Fe_2O_3	1109.8	1173.9	1241.6	Li_2O	738.1	771.4	810.6
FeO	425.6	456.8	492.2	Na_2O	628.3	681.5	737.1
P_2O_5	1932.3	2120.0	2312.0	K_2O	616.6	670.4	727.9

where β_i is the mass transfer coefficient, M_i the molar mass, and P_i the equilibrium vapor pressure of species i. With reservation to more sophisticated approaches [17, 18], the mass transfer coefficient β_i may be estimated by

$$\beta_i \approx \frac{2}{3} \cdot v^{1/2} \cdot \mu^{-1/6} \cdot L^{-1/2} \cdot D_i^{2/3} \tag{9.8}$$

where v and μ denote the mean flow velocity and the cinematic viscosity of the atmosphere above the melt, L the width of the melt surface in flow direction, and D_i the diffusion coefficient of the evaporating species i. The remaining problem is the determination of the equilibrium pressure P_i. In Table 9.6, evaporation reactions and formulae for P_i are compiled for 17 vapor species typical of industrial glass melts. In Table 9.7, data required to calculate the equilibrium constants K_i of the individual evaporation reactions $i = 1$–17 and the corresponding diffusion coefficients D_i are given. With the activities a_j of the oxides involved known, the P_i are readily calculated. For our reference E-glass melt at 1400°C, the equilibrium vapor pressures of species HBO_2, H_3BO_3, NaOH, $NaBO_2$ at 1400°C amount to 15.8, 0.26, <0.01, and 12.2 mbar, respectively.

Table 9.6 Evaporation reactions of alkali, boron, lead, and halogenide species i from glass melts, given in terms of an overall reaction equation and a resulting partial pressure P_i of species i

Species no.	Educts	$\leftrightarrow$	Vapor species i	Partial pressure P_i
1, 2	$R_2O(l) + H_2O$	$\leftrightarrow$	2 ROH, R = Na, K	$(K_{1,2} \cdot a_{R2O} \cdot P_{H2O})^{1/2}$
3, 4	$R_2SO_4(l)$	$\leftrightarrow$	$2\ R_2SO_4$, R = Na, K	$K_{3,4} \cdot a_{R2SO4}$
5, 6	$R_2O(l) + B_2O_3(l)$	$\leftrightarrow$	$2\ RBO_2$, R = Na, K	$(K_{5,6} \cdot a_{R2O} \cdot a_{B2O3})^{1/2}$
7	$B_2O_3(l) + H_2O$	$\leftrightarrow$	$2\ HBO_2$	$(K_7 \cdot a_{B2O3} \cdot P_{H2O})^{1/2}$
8	$B_2O_3(l) + 3\ H_2O$	$\leftrightarrow$	$2\ H_3BO_3$	$(K_8 \cdot a_{B2O3} \cdot P_{H2O}{}^3)^{1/2}$
9–12	PbO(l)	$\leftrightarrow$	$1/\underline{n}\ Pb_nO_n$, $n = 1$–4	$K_{9-12} \cdot a_{PbO}{}^n$
13	$PbO(l) + H_2O$	$\leftrightarrow$	$Pb(OH)_2$	$K_{13} \cdot a_{PbO} \cdot P_{H2O}$
14, 16	NaX(l)	$\leftrightarrow$	NaX, X = F, Cl	$K_{14,16} \cdot a_{NaX}$
15, 17	$NaX(l) + H_2O$	$\leftrightarrow$	$HX + \frac{1}{2}\ Na_2O(l)$, X = F, Cl	$K_{15,17} \cdot a_{NaX} \cdot (P_{H2O}/a_{Na2O})^{1/2}$

Table 9.7 Molar masses M_i, factors a_0, a_1, a_2 for the calculation of equilibrium constants K_i and $D_{0,i}$ for the calculation of diffusion coefficients of vapor species $i = 1$ to 17 (see Table 9.6) as a function of temperature; the calculation is performed as $\log K_i = a_0 + \frac{a_1}{T} + \frac{a_2}{T^2}$, $D_i = D_{0,i} \cdot T^{1.7}$, T in 1000 K

No.	M_i(g/mol)	a_0	a_1	a_2	$D_{0,i}$ (cm²/s)	No.	M_i (g/mol)	a_0	a_1	$D_{0,i}$ (cm²/s)
1	40.0	6.034	-9.087		1.36	9	223.2	6.847	-13.176	0.97
2	56.1	2.896	4.217	-4.946	1.12	10	446.4	3.393	-6.562	0.74
3	142.0	4.88	-14.443		0.48	11	669.6	1.919	-4.197	0.59
4	174.2	4.777	-13.987		0.43	12	892.8	0.92	-2.321	0.54
5	65.8	14.794	-12.27		0.97	13	241.2	8.385	-14.518	0.85
6	81.9	10.425	4.666	-8.453	0.88	14	42.0	5.673	-11.749	1.34
7	43.8	8.306	-17.355		1.17	15	20.0	3.099	-10.891	1.86
8	61.8	-6.324	4.935	-2.185	0.92	16	58.4	4.995	-8.951	0.93
						17	36.5	2.486	-11.856	1.14

9.1.4 Entropy and Viscosity

The theory by Adam and Gibbs [19] predicts a linear relationship between the decadal logarithm $\log \eta$ of the viscosity η and the reciprocal value of the product of the so-called configurational entropy $S_c(T)$ and absolute temperature:

$$\log \eta = A + D \cdot \frac{T}{T \cdot S_c(T)} \tag{9.9}$$

The value of $S_c(T)$ is assessed by

$$S_c(T) \approx S^{vit} - \Delta c_P \cdot \ln \frac{T_g}{T} \tag{9.10}$$

where Δc_P is the difference between $c_{P,\text{liq}}$ and $c_{P,Tg}$, given by

$$\Delta c_P = c_{P,\text{liq}} - c_{P,Tg} \tag{9.11a}$$

$$c_{P,Tg} = 3 \cdot R/M_{el} \tag{9.11b}$$

M_{el} is the average molar mass of the elements in the glass ($M_{el} = 8.046$ cm^3/mol for the reference E-glass). This yields

$$\log \eta = A + D \cdot \frac{T_g}{T \cdot S^{vit}} \cdot \frac{1}{1 - \dfrac{\Delta c_P}{S^{vit}} \cdot \ln \dfrac{T_g}{T}} \tag{9.12}$$

The relation has been verified by experiment for glasses of predominantly scientific interest [20] and has been extended with success to industrial multi-component glasses [21]. As accurate viscosity data are usually available for industrial glasses, we may employ (9.12) to determine Δc_P and S^{vit} from viscosity rather than predicting viscosity from thermodynamic data. Figure 9.1 shows an Adam–Gibbs plot for the reference E-glass. The data derived from this plot are $\Delta c_P = 237$ J/(kg K) and $S^{vit} = 181$ J/(kg K) as compared to 244 and 120 J/(kg K), respectively, derived from Table 9.3.

Another interesting detail is presented in Fig. 9.1:

Experimental data of the electrical conductivity κ [13] are also plotted as $-\log \kappa$ against the Adam–Gibbs abscissa. In contrast to a conventional Arrhenius plot $\log \kappa$ vs. $1/T$ yielding a systematic bias from linearity for high temperatures, the Adam–Gibbs plot provides a perfectly straight line for $-\log \kappa$. This confirms the relation

$$-\log \frac{\kappa}{\kappa_\infty} = b \cdot \log \frac{\eta}{\eta_\infty} \tag{9.13}$$

already proposed by Babcock [22]. For the reference E-glass, the high-T limits are

$$\kappa_\infty = 10^{1.95}/\Omega/\text{cm}$$

and

$$\eta_\infty = 10^{-3.32} \text{dPa s}$$
$$b = 0.7$$

Fig. 9.1 Adam–Gibbs plot for the reference E-glass comprising both viscosity η and electrical conductivity κ, r^2 = squared regression coefficient; the intercepts for $T \to \infty$ are $\log \eta_\infty = -3.32$ and $-\log \kappa = -1.95$, respectively; T_g, $T(\log \eta = 7.6)$, $T(\log \eta = 2)$: 663, 845, 1380°C; $-\log \kappa = 2.6, 2.1, 1.7$: 1200, 1300, 1400°C

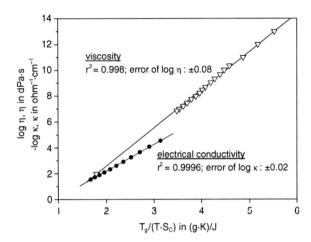

9.2 The Role of Individual Raw Materials

9.2.1 Sand

As a source of silica, natural sand is among the most important raw materials for industrial glass melting. Unfortunately, no matter how much it would be desirable from the point of view of chemical purity, a natural sand never consists of 100% quartz. Table 9.8 gives an overview of the impurity levels of different sand qualities from Europe and Southeast Asia. Sand from Belgium is among the world's top qualities, only surpassed by some sites in the eastern part of the USA and Canada. As the SEM image in Fig. 9.2 demonstrates, silica sand is a complex mineral system comprising, besides quartz, a number of side minerals. Among these, the types $M^{II}R^{III}_2O_4$ with M^{II} = Fe, Mg, Mn, R^{III} = Fe, Al, Cr are especially feared because of their high liquidus temperatures and their influence on the spectral transmission of the glass melt. Figure 9.3 shows the tremendous influence of impurities on the entire melting process for the example of iron impurities in a sand used in a small production (15 t/day) of a C-fiber glass. Due to the local supply of sand, the melt had a level of 0.07 wt% total Fe_2O_3. The pull of primary filaments prior to flame jet fiberization was extremely unstable due to the low IR transmission of the melt. By a quite radical change of the fining strategy, the Fe^{2+}/total Fe ratio was shifted from 0.15 to 0.06, which did not only stabilize the pull of primary filaments in the fiberization process but also had a remarkable impact on energy consumption.

Table 9.8 Impurity levels (wt%) in some selected sand qualities from Europe and Southeast Asia

Origin	TiO_2	Al_2O_3	Fe_2O_3	Origin	TiO_2	Al_2O_3	Fe_2O_3
Germany				Poland I	0.150	0.350	0.035
Frechen	0.059	0.151	0.027	Poland II	0.080	0.800	0.030
Haltern	0.036	0.122	0.034	Thailand			
Hohenbocka	0.037	0.141	0.028	Rayong I	0.020	0.040	0.040
Welferdingen	0.017	0.069	0.028	Rayong II	0.100	0.140	0.090
Amberg	0.030	1.100	0.015	Indonesia	0.050	n.d.	0.030
Belgium	0.016	0.040	0.009	Kambodia	0.050	n.d.	0.040

During glass melting, the dissolution of the sand grains determines the required stay time of the melt in the furnace to a large extent. For this reason, several experimental studies have been performed on the kinetics of sand dissolution as a function of grain size distribution, temperature, and glass composition [23–25]. Most of these examinations are, however, devoted to mass glass compositions using soda

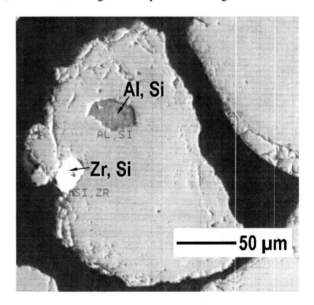

Fig. 9.2 Individual sand grain from quarry Frechen, Germany, with side phases embedded in the quartz matrix

ash in the batch. As concluded from the left graph in Fig. 9.4 and verified by experiments from multiple sources, most of the sand in such batches is attacked and consumed by the batch melting reactions already, the major part of which takes place within a narrow temperature interval between 860 and 1100°C. The onset temperature (see Fig. 9.4) sensibly depends on the minor additions in the batch. It is only the constitutional silica which has to be dissolved by diffusion into the rough melt [28]. In an E-glass batch, the situation is quite different. The boron carrier – if at all – is the only available flux. Let us consider anhydrous boric acid as boron

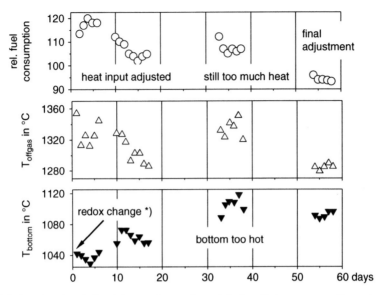

Fig. 9.3 Results of an industrial campaign performed during 60 days on a small gas-fired furnace used to melt a C-fiber glass; a sand with 0.07 wt% Fe_2O_3 was used in the batch; during the campaign, the ratio Fe^{2+}/Fe_{total} was shifted from 0.15 to 0.06; *) redox change accomplished by replacing Na_2SO_4 by $Sb_2O_3 + NaNO_3$

carrier. Figure 9.4, right graph, shows the seemingly favorable position of the B_2O_3–SiO_2 liquidus line. However, due to the presence of alkaline earth oxide carriers in the batch, the early melting B_2O_3 is readily consumed to form alkaline earth borates (see Fig. 9.5). The effect is even enhanced by the fact that the open porosity is quite low for a sand grain, but high for CaO or MgO formed by the decomposition of an alkaline earth carbonate. Thus, there is no essential advantage of B_2O_3 over a natural calcium borate like colemanite as boron carrier.

9.2.2 Boron Carriers

Industrial-grade boron oxide raw materials available on the market are boric acid H_3BO_3, anhydrous boric acid B_2O_3, borax $Na_2O \cdot 2B_2O_3 \cdot 10H_2O$, as well as the pentahydrate or anhydrous forms of borax. These boron carriers offer the advantage of a high chemical purity. Alternatively, there is the option to use a number of naturally occurring boron minerals (with natural impurity levels), among which are natural borax, colemanite $2CaO \cdot 3B_2O_3 \cdot 5H_2O$, ulexite $Na_2O \cdot 2CaO \cdot 5B_2O_3 \cdot 16H_2O$, hydroboranite $CaO \cdot MgO \cdot 3B_2O_3 \cdot 6H_2O$. Of course, the use of Na_2O-containing raw materials in E-glass batches is quite limited. The actual choice will depend on the price per kg B_2O_3 equivalent, on availability, and on energetic considerations: 1 mol of hydrate water attached to a boron mineral requires an additional amount of 3215 kJ/kg H_2O (0.9 kW h/kg) for the promotion of the chemical reaction plus

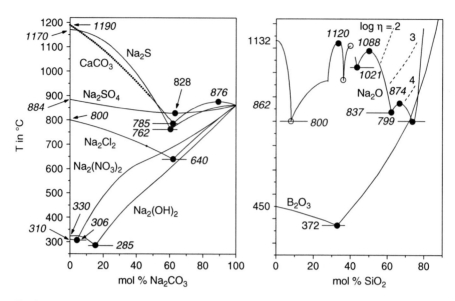

Fig. 9.4 Phase diagrams presenting the liquidus lines of binary systems after data from [26]; *left* diagram: Na_2CO_3 plus a second salt-like component; *right* diagram: $Na_2O–SiO_2$ and $B_2O_3–SiO_2$; in the silica-rich part of the system $Na_2O–SiO_2$, *dashed lines* present iso-viscosity lines (log η, η in dPa s) [27]

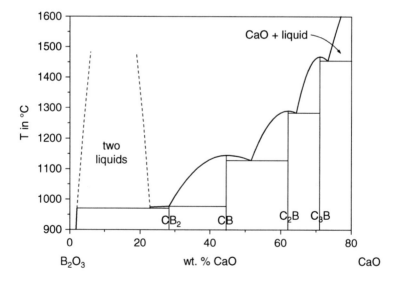

Fig. 9.5 Phase diagram of the binary system $CaO–B_2O_3$, after data from [26]

another 3100 or 736 kJ/kg for heating up to 1400 or 400°C (conventional or cold top furnace), respectively. The standard heats of formation of the boron carriers in kJ/mol are −1094.1 (H_3BO_3), −1273.5 (B_2O_3), −6288.6 (borax decahydrate), −4789.9 (pentaborate), −3291.1 (anhydrous borax), −6939.6 (colemanite), −13520.0 (ulexite), −9294.3 (inyoite), and −4490.2 (kernite). The data for colemanite and inyoite were determined on natural minerals [29]. By calculation, −7029.9 and −9330.0 kJ/mol are obtained for the stoichiometric phases $C_2B_3H_5$ and $C_2B_3H_{13}$, respectively. For a precise compilation of thermodynamic data of boron minerals, see [30]. Lately, the heat of formation of ulexite has been determined as −13524.5 kJ/mol [31]. A recent development may shift the interest of European fiber producers from synthetic toward natural boron carriers: The European Parliament recently enacted a stringent regulation REACH (registration, evaluation, and authorization of chemicals) which imposes severe restrictions on the use of synthetic, however, not on natural boron minerals. As a matter of fact, the compositions of natural minerals may deviate considerably from the corresponding stoichiometric phases. In Table 9.9, the properties of some natural boron carriers from Turkish origin are contrasted to synthetic boron carriers.

Table 9.9 Compilation of phase content, grain size distribution, and boron oxide content of some natural boron carriers from Turkish origin; for comparison, the boron oxide content of synthetic boron carriers is also given

Product and phase content		Grain size distribution		B_2O_3, wt%
Boric acid, H_3BO_3				79.44
Stoichiometric borax pentahydrate				47.80
Stoichiometric borax decahydrate				36.51
Anhydrous borax $Na_2B_4O_7$				69.20
Natural tincal		<60 μm	4%	29–32
Borax decahydrate	80–90%	<1000 μm	99%	
Calcite + dolomite	2–6%			
Humidity	5–10%			
Stoichiometric ulexite				42.95
Ground natural ulexite		<45 μm	5%	37
$C_2B_6H_6$	3–10%	<75 μm	87%	
NCB_5H_8	72–90%	<250 μm	99.5%	
Calcite + dolomite	5–16%	<600 μm	99.98%	
Humidity	5–8%			
Stoichiometric colemanite				50.81
Ground natural colemanite		No data available		36–49
Colemanite	33–48%			
Ulexite	<2.5%			
Calcite + dolomite	7–20%			
Humidity	<1%			

The phase relations of Na–Ca–B–O–H minerals are plotted in Fig. 9.6 in a semiquantitative way.

Fig. 9.6 Phase stability of boron-containing minerals, qualitatively presented as a function of water activity (*abscissa*) and of the ion product $[Ca^{2+}]^2 \cdot [Na^+]^{-3} \cdot [H^+]^{-1}$ (*ordinate*), after [32]; the system is considered to be closed with respect to total boron

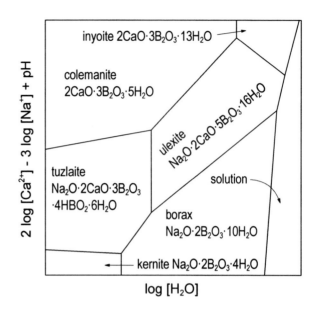

ordinate axis: $2 \log [Ca^{2+}] - 3 \log [Na^+] + pH$

inyoite $2CaO \cdot 3B_2O_3 \cdot 13H_2O$

colemanite $2CaO \cdot 3B_2O_3 \cdot 5H_2O$

ulexite $Na_2O \cdot 2CaO \cdot 5B_2O_3 \cdot 16H_2O$

tuzlaite $Na_2O \cdot 2CaO \cdot 3B_2O_3 \cdot 4HBO_2 \cdot 6H_2O$

solution

borax $Na_2O \cdot 2B_2O_3 \cdot 10H_2O$

kernite $Na_2O \cdot 2B_2O_3 \cdot 4H_2O$

$\log [H_2O]$

9.2.3 Dolomite and Limestone

Dolomite and limestone are the minerals typically used as carriers of MgO and CaO. They are added to the batch as carbonates, or alternatively, in their partially or fully calcined form as "dolime" $MgO + CaCO_3$ or as burnt dolomite $MgO + CaO$ and burnt lime CaO, respectively. The decomposition of alkaline earth carbonates makes the largest contribution to the chemical heat demand of the batch-to-melt conversion. Thus, from the point of view of on-site production efficiency, it may be advisable to use partially or fully calcined products in the batch. This proposition remains valid even if little or no advantage is expected with respect to global ecology in terms of overall CO_2 release and overall energy consumption. The decision will depend on the balance of raw material vs. energy costs for an individual production site. In order to assess such a balance, reliable data on the energetic situation of the different raw material options are required. When, however, inspecting literature data, there is a striking uncertainty, especially with respect to the heats of formation of dolomite. This issue deserves a closer look.

In Table 9.10, heats of formation of dolomite from the elements as taken from several renowned databases are contrasted. The uncertainty amounts to 30 kJ/mol, which is equivalent to 163 kJ/kg dolomite or 311 kJ/kg MgO + CaO equivalent.

From the point of view of mineralogy, natural dolomite and limestone are no pure phases, but rather minerals from the system $CaCO_3$–$MgCO_3$–$FeCO_3$–$MnCO_3$, accompanied by minor amounts of quartz, olivine, and feldspatic minerals. They display a most complex polycrystalline microstructure of coexisting carbonates, even in the individual grains, ranging from coarse to fine and crypto-crystalline phases. Figure 9.7 illustrates the phase relations in the ternary sub-system

$CaCO_3$–$CaMg(CO_3)_2$–$CaFe(CO_3)_2$, redesigned after data from [26]. According to Fig. 9.7, calcite may dissolve considerable amounts of Mg. By contrast, dolomite may dissolve much Fe (such a dolomite would not be used in glass industry), however, hardly any excess Ca. Thus, a natural glass-grade dolomite always contains at least two kinds of phases, i.e., Mg-saturated limestone and Ca-saturated dolomite.

Table 9.10 Compilation of literature data on the standard heat of formation $H°$ of dolomite from the elements; n.n. = unspecified

$H°$ in kJ/mol $CaMg(CO_3)_2$	Kind	Source	
−2315.0 ± 5.0	n.n.	Kubaschweski et al.	[33]
−2329.9	n.n.	Philpotts	[34]
−2331.7	n.n.	Mchedlov-Petrossyan	[35]
−2324.5	n.n.	Robie et al.	[36]
−2314.2 ± 0.5	Disordered	Navrotsky et al.	[37–39]
−2300.2 ± 0.6	Ordered	Navrotsky et al.	[37–39]
−2325.7	n.n.	Saxena et al.	[40]
		FACT-SAGE 5.2	[9]
−2326.3	$CaCO_3 \cdot MgCO_3$	HSC Chemistry	[41]
−2317.6	Disordered	HSC Chemistry	[41]
−2329.9	Ordered	HSC Chemistry	[41]

The mismatch of the thermal expansion coefficients of these phases (5.5 vs. 9.3 × 10^{-6} K^{-1}) must be considered as one of the reasons for a most undesirable phenomenon termed decripitation. Decripitation denotes an explosive diminuition of the dolomite grains upon heating, yielding a considerable dust carryover from the batch to the atmosphere. The disposition to decripitation may be judged by the Ca excess in a dolomite. Thus, it is especially the "good," i.e., Fe-poor, dolomites which are prone to decripitation. By the way, grain sizes of <100 and >500 μm hardly make any contribution to the phenomenon [42].

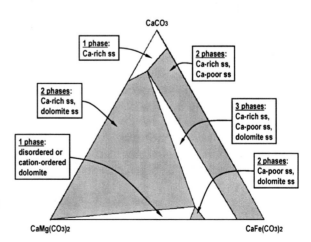

Fig. 9.7 Phase diagram of the system $CaCO_3$–$CaMg(CO_3)_2$–$CaFe(CO_3)_2$ showing the stability fields of one-, two-, and three-phase equilibria

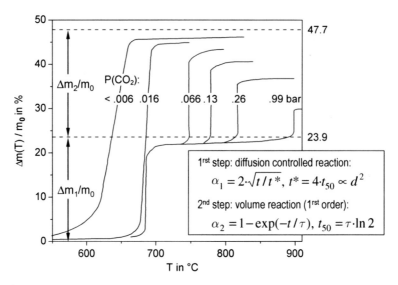

Fig. 9.8 Mass loss $\Delta m(T)$ of dolomite (referred to its initial mass m_0) as a function of temperature for different CO_2 partial pressures, after [43]; *dashed lines* mark the losses Δm_1 and Δm_2 of complete turnover $CaMg(CO_3)_2 \rightarrow MgO + CaCO_3 + CO_2$ and $CaCO_3 \rightarrow CaO + CO_2$, respectively; kinetic equations for the relative turnover $\alpha_1 = \Delta m(T)/\Delta m_1$ and $\alpha_2 = \Delta m(T)/\Delta m_2$; t^*, τ, t_{50}: time demand to reach $\alpha = 1.00, 0.63, 0.50$, respectively

Figure 9.8 shows the two-step decomposition behavior of dolomite as determined by experiment [43]. The first decomposition step is diffusion controlled, hence, dependent on the grain size, while the second step is a grain-size-independent first-order volume reaction [44] similar to the decomposition of limestone.

Let us turn back to the energetic situation of dolomite: A very careful study [37–39] may help to resolve the discrepancies found in Table 9.10. In Fig. 9.9, the experimentally determined heats of formation from the pure carbonates for minerals from the binary systems $CaMg(CO_3)_2$–$CaCO_3$ and $–CaFe(CO_3)_2$ are shown. The composition ranges of the left and right graphs are identical to the baseline and the left side, respectively, of the triangle in Fig. 9.7. Even small amounts of Ca excess in dolomite considerably shift the resulting heat of formation of the mineral. The Fe vs. Mg substitution yields less strong effects. The smallest shift is observed for Mg excess in pure limestone. Let us consider the heat of formation of $Ca_{1+x}(Fe_{(1-x)\cdot y}Mg_{(1-x)\cdot(1-y)})(CO_3)_2$. For $x = 1$, the formula denotes pure limestone, given as $Ca_2(CO_3)_2$. For $x = 0$, it is $Ca(Fe_yMg_{1-y})(CO_3)_2$ with pure dolomite and pure ankerite as end members ($y = 0$ and 1, respectively). Then the standard enthalpies of formation, given in units of kilojoule per mole of formula unit, are calculated from the chemical composition as

$$H^\circ_{dolo} = -2314.2 + 129.4 \cdot x + 74.0 \cdot y \qquad (9.14)$$

for the one-phase dolomite solid solution and

$$H^\circ_{\text{lime}} = -2413.8 - 9.97 \cdot (1 - x) \tag{9.15}$$

for the one-phase limestone solid solution. The stoichiometric coefficients x and y are derived from the analytical mass ratios $u = \text{MgO/CaO}$ and $v = \text{FeO/MgO}$ as

$$x = (0.7188 - u)/(0.7188 + u) \tag{9.16a}$$

$$y = v/(1.7832 + v) \tag{9.16b}$$

with the figures 0.7188 and 1.7832 representing the molar mass ratios of MgO/CaO and FeO/MgO, respectively. Thus, the standard enthalpy of a natural dolomite and limestone can be swiftly calculated from analytical data.

Figure 9.10 (left graph) presents the results already shown in Fig. 9.9 (left graph), however, translated to heats of formation from the elements and complemented by values for the two-phase region. These data are converted to mass-related heats of formation (Fig. 9.10, right graph). This is the information relevant for the technologist. The graph may be used to directly read the heats of formation of natural dolomites and limestones as a function of chemical composition.

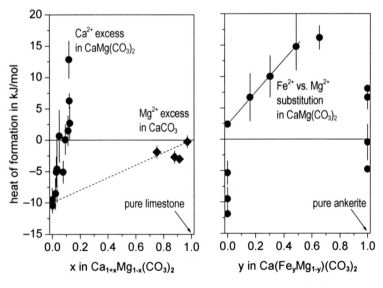

Fig. 9.9 Heats of formation from the pure carbonates for minerals from the systems $CaMg(CO_3)_2$–$Ca_2(CO_3)_2$ (*left* graph) and $CaMg(CO_3)_2$–$CaFe(CO_3)_2$ (*right* graph); data from [37–39]

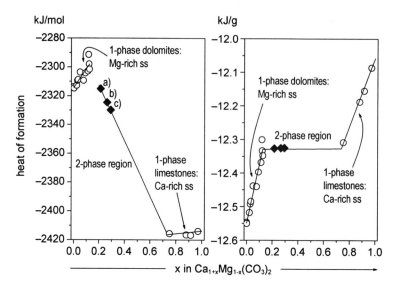

Fig. 9.10 Heat of formation from the elements for minerals from the binary cut $CaMg(CO_3)_2$–$Ca_2(CO_3)_2$; related to molar amount (*left*) and mass (*right*); the deviation of literature data might stem from experiments on two-phase dolomites; compare a) 2315, b) 2324.5, and c) 2329.9 kJ/mol in Table 9.10; if related to mass, the differences vanish

9.3 The Batch-to-Melt Conversion

Batch melting as the first high-T step of glass melting is crucial for the success of all consecutive production steps. In terms of mass balance, the rate of batch melting has to match with the pull rate. Otherwise, quality defects emerge. In terms of heat balance, the batch-to-melt conversion draws an amount of heat approx. equal to the exploited heat of the entire melting process. As this happens in a time interval of 40–60 min only, which is much shorter than the mean residence time of melt in the basin, the batch is an area of extremely high power demand. In terms of chemistry, it is during the short period of batch melting that the redox equilibria in the melt are essentially established.

9.3.1 Stages of Batch Melting

In order to gain a complete picture, batch melting has to be discussed on at least four levels of observation. These are the following:

– the local thermochemical reactions,
– the level of micro-kinetics, comprising grain-to-grain effects, grain wetting by liquid phases,
– the level of meso-kinetics, i.e., of transport phenomena of matter and heat within the batch blanket and across its boundaries,
– the overall mass, heat, power, and entropy balance of the furnace.

The different stages of batch melting may differ from glass type to glass type. There is, however, an overarching pattern by which batch melting may be presented in a unified way [28, 45, 46]. The following stages are observed:

  I. *Open-pore stage; warming-up stage:*

- physical warming of a granular bulk solid,
- loss of physical batch water,
- decomposition of hydrates,
- decomposition of dolomite,
- burning-off of organics,
- binary solid state reactions with a very low turnover rate,
- reactive conversion of low-T fluxes ($NaOH$, $NaNO_3$, B_2O_3) to new solid phases,
- high emission peaks of volatiles.

 II. *Closed-pore stage; reaction foam stage:*

- onset of melting by the formation of a widespread (mostly salt-like) low-viscosity primary melt,
- silicate formation at a high turnover rate (10^3–10^4 times faster than during stage I), reactive decomposition of carbonates,
- gas percolation through the batch, bloating and foaming,
- establishment of the intrinsic redox state between pore volume gases and melt,
- emission of volatiles calms down.

III. *Volume void filling; establishment of a rough melt:*

- silicate formation reactions and CO_2 generation calm down,
- release of percolating gases, volume shrinkage,
- high liquidus solids remain floating in the batch,
- thermal conductivity and optical transmission increase.

As already stated in Section 9.2.1, batches without alkali carbonates may differ significantly for the behavior of typical mass glass batches. Nevertheless, they pass through the very same stages. It is only the relative proportion of the stages which makes the difference.

9.3.2 Heat Demand of the Batch-to-Melt Conversion

Earlier work [47] on the calculation of the heat demand of batch melting yielded considerable success for batches with a small number of chemically pure raw materials. The former calculation strategy was based on the formulation of a gapless(!) sequence of chemical and physical reactions linking the stage of batch at 25°C to the stage of glass melt at a given temperature. For a realistic industrial batch, this is virtually impossible to accomplish. Beyond this, the strategy gives up a noble principle of thermodynamics, i.e., the path independence of the properties of thermodynamic states. With the successful thermodynamic quantification of the states of industrial glasses and glass melts (Sections 9.1.1 and 9.1.2), we may fully exploit the principle of path independence and present the batch-to-melt conversion by the following hypothetical reaction:

$$\text{batch } (25°C) \rightarrow \text{glass } (25°C) + \text{batch gases } (25°C) \qquad (9.17)$$

The energy difference between the right- and left-hand sides of Equation (9.17) is the standard heat of formation of glass and batch gases from the raw materials, $\Delta H°_{chem}$, also termed chemical heat demand of batch melting. $\Delta H°_{chem}$ is calculated as

$$\Delta H°_{chem} = H°_{glass} + H°_{gas} - H°_{batch} \qquad (9.18)$$

where $H°_{glass}$ is determined after Equation (9.2a), and $H°_{gas}$, $H°_{batch}$ are the weighted sums of standard heats of the individual batch gases and raw materials, respectively. In Table 9.11, the calculation procedure is demonstrated for two different batches – batch 1 using dolomite and limestone and batch 2 using fully burnt dolomite and lime – yielding a glass identical with our reference E-glass (see Table 9.2). The values of $\Delta H°_{chem}$ for both batches differ considerably.

By the reactions

$$\text{glass } (25°C) \rightarrow \text{glass } (T = T_{ex}) \qquad (9.19a)$$

Table 9.11 Calculation of the chemical heat demand $\Delta H°_{chem}$ of two different batches given in amounts of kg per 1000 kg of glass, both yielding the reference E-glass (see Table 9.2); M = molar mass, $H°$ = standard enthalpy

	M g/mol	$H°$ kW h/kg	Batch 1 Kg	kW h	Batch 2 kg	kW h
Sand	60.084	−4.2112	−562.00	2366.7	−562.00	2366.7
Al$_2$O$_3$	101.961	−4.5652	−144.00	657.4	−144.00	657.4
3H$_2$O·B$_2$O$_3$	123.664	−4.9152	−97.30	478.2	−97.30	478.2
Na$_2$O·2B$_2$O$_3$·5H$_2$O	291.292	−4.5676	−28.90	132.0	−28.90	132.0
Dolomite	184.410	−3.4859	−192.80	672.1	–	0.0
Burnt dolomite	96.390	−3.5634	–	0.0	−100.80	359.2
Limestone	100.089	−3.3495	−211.50	708.4	–	0.0
Burnt lime	56.079	−3.1449	–	0.0	−118.50	372.7
I Sum of batch			−1236.50	5014.8	−1051.50	4366.2
CO$_2$	44.010	−2.4837	185.04	−459.6	0.00	0.0
H$_2$O	18.015	−3.7284	51.46	−191.9	51.50	−192.0
II Sum of gases			236.50	−651.4	51.50	−192.0
III Glass		−4.1065	1000.00	−4106.5	1000.00	−4106.5
$\Delta H°_{chem} = $ I + II + III				256.9		67.7

$$\text{batch gases } (25°C) \rightarrow \text{batch gases } (T = T_{\text{off}}) \qquad (9.19b)$$

the final stage of the batch-to-melt conversion is described. It is assumed that the glass melt is heated to T_{ex}, i.e., the temperature at which the melt is pulled from the basin, and the batch gases are heated to the offgas temperature T_{off}. The amounts of heat involved are $\Delta H_{T\text{ex,liq}}$ and $\Delta H_{T\text{off,gas}}$, respectively; the former quantity is calculated after Equation (9.3), the latter one from tabulated standard data on CO_2 and H_2O. Let $T_{\text{ex}} = 1300°C$, then $\Delta H_{T\text{ex,liq}}$ amounts to 445 kW h/t of produced glass. Let $T_{\text{off}} = 1400°C$ for a conventional glass furnace vs. 400°C for a cold top electrical melter. This yields 127 vs. 30 kW h/t of produced glass for batch 1 and 44 vs. 11 kW h for batch 2. The so-called exploited heat H_{ex} is given by the sum of chemical heat demand and heat content of the melt, while the heat content of the gases is allotted to the offgas losses:

$$H_{\text{ex}} = (1 - y_C) \cdot \Delta H_{\text{chem}}^\circ + \Delta H_{T\text{ex,liq}} \qquad (9.20)$$

where y_C denotes a fraction of cullet (referred to the total amount of glass produced) which does not contribute to the chemical energy demand. H_{ex} is the key figure to judge the energy efficiency of a furnace and to compare the performances of different batches in different furnaces. The overall efficiency η_{ex} is the ratio of H_{ex} and the amount of energy H_{in} actually used. For details, see [48].

9.3.3 Modeling of the Batch-to-Melt Conversion Reaction Path

For the calculation of the heat demand of batch melting, it has been an advantage to do without information on the actual reaction path. Yet, most valuable additional information may be drawn from a reaction path analysis. For the sake of simplicity, this is demonstrated for a TiO_2- and Fe_2O_3-free version of the reference E-glass. The resulting glass has a composition (by wt.) of 55.7 SiO_2, 14.6 Al_2O_3, 6.9 B_2O_3, 4.3 MgO, 17.9 CaO, and 0.6 Na_2O and a glass transition temperature $T_g \approx 750°C$. The sequence and the amounts of solid phases precipitating from the melt upon cooling from 1600°C were calculated by a commercial software [9]. The results are shown in Fig. 9.11. The batch corresponding to the above glass contains (in g per 100 g glass) 55.7 sand, 22.3 gibbsite Al(OH)$_3$, 15.3 colemanite $C_3B_2H_5$, 19.5 dolomite, 9.4 limestone, and 2.9 borax pentahydrate. When heating up this batch, dehydration is observed as the first step: borax pentahydrate 100–163°C, gibbsite 331°C, colemanite 420°C (idealized behavior; for the real dehydration behavior, see [49, 50]). At approx. 700°C, dolomite releases the CO_2 attached to MgO. It is not until 743°C (the melting point of dehydrated borax) that a molten phase is formed. This phase is, however, readily resorbed by the formation of solid phases. Considerable amounts of melt are not formed below 1000°C; until then, the batch appears to be "dry." Melting does not gain momentum until 1100°C is reached. Note that the curves are equilibrium curves. The curve denoted by "melt" represents the maximum amount of melt

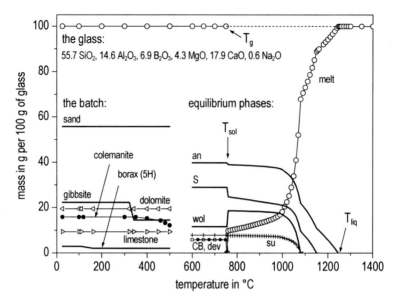

Fig. 9.11 Phase stability diagram for a simplified E-glass composition in terms of phase content as a function of temperature; below 500°C, the amounts of the batch materials sand, colemanite, gibbsite, pentaborate, dolomite, and limestone are shown instead of the equilibrium phases; an = anorthite CAS_2, S = SiO_2, wol = wollastonite CS, CB = calcium borate, devitrite, su = suanite M_2B; the solidus, liquidus, and glass transition temperatures T_{sol}, T_{liq}, T_g, respectively, are marked

attainable at a given temperature – in the absence of any kinetic constraints. The corresponding amounts of solid phases represent what is left in the rough melt. The phase constituting the liquidus temperature is anorthite CAS_2. Upon slow cooling, anorthite would be the first crystalline phase observed. Note that crystalline silica prevails until 1175°C. It is only beyond this temperature that the sand grains can be dissolved completely. In view of this, let us inspect again the kind and the amounts of phases found in the crystalline reference system in Table 9.2. The amount of constitutional silica in the c.r.s. may be adopted as a lower threshold of the solid phase content of the rough melt. For soda lime glass batches, it has been shown [28] that the amount of constitutional silica corresponds to the time actually required to reach the stage of clear melt [51].

The reaction path of batch-to-melt conversion can also be calculated in a most convenient way by using the principles elaborated in Section 9.1.1.. This is demonstrated in Table 9.12 for a simple soda lime silicate glass batch. The path starts from the batch at 25°C and passes through milestone states reached at arbitrarily selected temperatures. These are the physical melting of soda ash (860°C), reactive decomposition of soda ash and formation of a binary $Na_2O–SiO_2$ melt (900°C), reactive decomposition of limestone and formation of a ternary melt still containing solid silica (1000 and 1200°C), formation of a clear melt (1400 and 1200°C), and cooling

Table 9.12 Reaction path of a simple soda lime silicate glass batch yielding a glass of 74 SiO_2, 10 CaO, 16 Na_2O (by wt); reaction path given as a function of temperature in amounts (kg per 1000 kg glass) of solid phases, CO_2 gas, and melt; ΔH = enthalpy difference to the cold batch in kW h per 1000 kg glass
y_{solid} = massfractionsolid/(solid + melt); η = viscosity in dPa s

Temperature, °C	25	860	900	1000	1200	1400	1200	25
Solid phases								
Quartz l/h	740.0	740.0	–	–	–	–	–	–
Limestone	178.5	178.5	178.5	–	–	–	–	–
Soda ash	273.6	–	–	–	–	–	–	–
Cristobalite	–	–	480.8	287.0	287.0	–	–	–
Total solids	1192.1	918.5	659.3	287.0	287.0			
Liquid or glassy components								
Soda ash	–	273.6	–	–	–	–	–	–
$Na_2O \cdot SiO_2$	–	–	103.6	–	–	–	–	–
$Na_2O \cdot 2SiO_2$	–	–	315.6	361.9	361.9	361.9	361.9	361.9
$Na_2O \cdot 3CaO \cdot 6SiO_2$	–	–	–	351.1	351.1	351.1	351.1	351.1
SiO_2	–	–	–	–	–	287.0	287.0	287.0
Total melt	–	273.6	419.2	713.0	713.0	1000.0	1000.0	1000.0
Batch gases								
CO_2	–	–	113.6	192.1	192.1	192.1	192.1	192.1
Heat, content of dispersed solids, viscosity of liquid phase								
ΔH in kW h	0	350	431	498	585	681	589	134
y_{solid}	1.00	0.77	0.66	0.29	0.29	0	0	0
log η (melt)	–	–2.0	1.9	2.7	1.8	2.1	2.8	–

to the glassy state (25°C). The heat balance is given in terms of the enthalpy difference ΔH to the initial state. Thus, the ΔH in the last column is identical with the chemical heat demand $\Delta H°_{chem}$ discussed in Section 9.3.2.. As a special feature, the viscosity of the melt is calculated for every state. It is interesting to note that, in spite of a steady increase of temperature from 860 to 1400°C, the viscosity does not decrease steadily, but rather passes through minima and maxima. Simultaneously, the mass fraction y_{solid} of solid matter in the melt decreases steadily. Using these data to estimate the effective viscosity of the melting batch from the state of granular bulk solid to the state of clear melt, e.g., by employing the Krieger–Dougherty equation [52], would be most rewarding. Unfortunately, for the time being, there is no strategy at hand to estimate the amount of dispersed gas bubbles in the batch.

References

1. C. V. Mazurin, M. V. Streltsina and T. P. Shvaiko-Shvaikovskaya, *Handbook of glass data*, Part A: Silica glass and binary silicate glasses. Part B: Single-component and binary non-silicate oxide glasses. Part C: Ternary silicate glasses. Part D: Ternary non-silicate glasses. Part E: Single-component, binary, and ternary oxide glasses. Elsevier, Amsterdam (1983, 1985, 1987, 1991, 1993).

2. H. Scholze, *Glass – nature, structure, and properties*, Springer Verlag, Berlin (1991).
3. A. A. Appen, Berechnung der optischen Eigenschaften, der Dichte und des Ausdehnungskoeffizienten von Silikatgläsern aus ihrer Zusammensetzung. (Calculation of optical properties, density, and thermal expansion coefficient of silicate glasses from their composition), Rep. Acad. Sci. USSR, 69, 841–844 (1949).
4. A. D. Pelton and M. Blander, Thermodynamic analysis of ordered liquid solutions by a modified quasi-chemical approach – application to silicate slags, Metall. Trans. B, 17B, 805–815 (1986).
5. H. Gaye. *Donneés thermochimiques et cinétiques relatives à certains matériaux sidérurgiques*. in: H. Gay, D. Colombet, eds., Commission de la Communautés Européennes Convention CEEC No. 7210-CF/301 TCM-RE 1064, Bruxelles (1984).
6. D. W. Bonnel and J. W. Hastie, Ideal mixing of complex components, High Temp. Sci., 26, 313–334 (1990).
7. B. A. Shakhmatkin, N. M. Vedishcheva, M. M. Schultz and A. C. Wright, The thermodynamic properties of oxide glasses and glass forming liquids and their chemical structure, J. Non-Cryst. Solids, 177, 249–256 (1994).
8. G. Eriksson and K. Hack, ChemSage – A computer program for the calculation of complex chemical equilibria, Metall. Trans. B, 21B, 1013 (1990).
9. FACTSAGE Software Ver. 5.2; Thermfact Montreal and GTT Technologies Aachen, (2004).
10. R. Conradt, Thermochemistry and structure of oxide glasses. In: *Analysis of the composition and structure of glass and glass ceramics*, H. Bach, D. Krause, eds., Springer Verlag, Berlin, pp. 232–254 (1999).
11. R. Conradt, Chemical structure, medium range order, and crystalline reference state of multicomponent liquids and glasses, J. Non-Cryst. Solids, 345 and 346, 16–23 (2004).
12. R. Conradt, The industrial glass-melting process. In: *The SGTE casebook. Thermodynamics at work*, K. Hack, ed. CRC Press, Boca Raton (2008).
13. T. P. Seward, III, T. Vascott eds., *High temperature glass melt property database for modeling*. The American Ceramic Society, Westerville (2005).
14. H. E. Schwiete and G. Ziegler, Beitrag zur spezifischen Wärme der Gläser. (Contribution to specific heat of glasses), Glastech. Ber., 28, 137–146 (1955).
15. J. Moore and D. E. Sharp, Note on calculation of effect of temperature and composition on specific heat of glass, J. Am. Ceram. Soc., 41, 461–463 (1958).
16. O. D. Gudovich and V. I. Primenko, Calculation of the thermal capacity of silicate glasses and melts, Soc. J. Glass Phys. Chem., 11, 206–211 (1985).
17. *VDI Wärmeatlas*, Springer Verlag, Berlin (1997).
18. R. B. Bird, W. E. Stewart and E. N. Lightfoot, *Transport phenomena*. John Wiley and Sons, New York (1960).
19. G. Adam and J. H. Gibbs, On the temperature dependence of cooperative relaxation properties in glass-forming liquids, J. Chem. Phys., 43, 139–146 (1965).
20. P. Richet and Y. Bottinga, Rheology and configurational entropy of silicate melts, Rev. Mineral., 32, 6593 (1995).
21. R. Conradt, Harmonization of viscosimetric and thermodynamic data for industrial multi-component glasses and glass melts. In: *Melt chemistry, relaxation, and solidification kinetics of glasses*. H. Li, C. S. Ray, D. M. Strachan, R. Weber and Y. Yue, eds., Ceramic transactions Vol. 170, The American Ceramic Society, Westerville (2005).
22. C. L. Babcock, Viscosity and electrical conductivity of molten glasses, J. Am. Ceram. Soc., 17, 329–342 (1934).
23. P. Hrma, Complexities of batch melting. In: *Proceedings of 1st International Conference on Advances in the Fusion of Glass*. Alfred University, New York (1988).
24. M. E. Savard and R. F. Speyer, Effect of particle size on the fusion of soda-lime-silicate glass containing NaCl, J. Am. Ceram. Soc., 76, 671–677 (1993).
25. O. Verheijen, Thermal and chemical behavior of glass forming batches. Thesis, Technical University Eindhoven, The Netherlands, (2003).

26. Phase Equilibria Diagrams. *CD-ROM database, Version 2.1.* The American Ceramic Society, Westerville, (1998).
27. H. F. Mark , D. F. Othmer, , C. G. Overberger , G. T. Seaborg , eds., *Kirk-Othmer encyclopedia of chemical technology*, Vol. 20. Wiley, New York (1982).
28. O. Dubois and R. Conradt, Experimental study on the effect of cullet and batch water on the melting behavior of flint and amber container glass batches, Glass Sci. Technol., 77, 137–148 (2004).
29. V. M. Gurevich and V. A. Sokolov, Enthalpy of formation of inyoite and colemanite. Geokhim. Akad. Nauk. SSSR 3 (1976), pp. 455–457.
30. L. M. Anovitz and B. S. Hemingway, Thermodynamics of boron minerals: Summary of structural, volumetric, and thermochemical data, Rev. Mineral., 33, 181–261 (1996).
31. C. Ruoyu, L. Jun, X. Shuping and G. Shiyang, Thermochemistry of ulexite, Thermochim. Acta, 306, 1–5 (1997).
32. V. Bermanec, K. Furic, M. Rajic and G. Kniewald, Thermal stability and vibrational spectra of the sheet botare tuzlaite, $NaCaB_5O_8(OH)_2 \cdot 3H_2O$, Am. Mineral., 88, 271–276 (2003).
33. O. Kubaschweski, C. B. Alcock and P. J. Spencer *Materials thermochemistry*. Pergamon Press, London (1993).
34. A. R. Philpotts, *Principles of igneous and metamorphic petrology*. Prentice Hall, Englewood Cliffs (1990).
35. V. I. Babushkin, G. M. Matveyev and O. P. Mchedlov-Petrossyan, *Thermodynamics of silicates*. Springer Verlag, Berlin (1985).
36. R. A. Robie, B. S. Hemingway and J. R. Fisher, Thermodynamic properties of minerals and related substances at 298.15 K and 1 bar (10^5 Pascals) pressure and at high temperatures, Geol. Surv. Bull., 1452, 456, US Gov. Printing Office, Washington (1978).
37. A. Navrotsky and C. Capobianco, Enthalpies of formation of dolomite and magnesia calcites, Am. Mineral., 72, 782–787 (1987).
38. L. Chai, A. Navrotsky and R. J. Reeder, Energetics of calcium-rich dolomite, Geochim. Cosmochim. Acta, 59, 939–944 (1995).
39. L. Chai and A. Navrotsky, Synthesis, characterization, and energetics of solid solution along the dolomite-ankerite join, an implication for the stability of ordered $CaFe(CO_3)_2$, Am. Mineral., 81, 1141–1147 (1996).
40. S. K. Saxena, N. Chatterjee, Y. Fei and G. Shen, *Thermodynamic data on oxides and silicates*. Springer Verlag, Berlin (1993).
41. A. Roine; *HSC chemistry software version 3.0*, Outokumpu Research Oy, Pori (2000).
42. M. Santani, D. Dollimore, F. W. Wilburn and K. Alexander, Isolation and idenfication of the intermediate and final products in the thermal decomposition of dolomite in an atmosphere of carbon dioxide, Thermochim. Acta, 367 (8), 285–295 (2001).
43. D. A. Young *Decomposition of solids. The international encyclopedia of physical chemistry and chemical physics, Topic 21. Vol. 1.* E. A.Guggenheim et al. eds. Pergamon Press, London (1966).
44. M. Olszak-Humienik and J. Mozejko, Kinetics of thermal decomposition of dolomite, J. Therm. Anal. Calorim., 56, 829–833 (1999).
45. W. Trier, *Glasschmelzöfen – Konstruktion und Betriebsverhalten. (Glass furnaces – design, construction and operation).* Springer Verlag, Berlin (1984).
46. A. Ungan and R. Viskanta, Melting behavior of continuously charged batch blankets in glass melting furnaces, Glastech. Ber., 59, 279–291 (1986).
47. C. Kröger, Theoretischer Wärmebedarf der Glasschmelzprozesse. (Theoretical heat demand of the glass melting processes), Glastech. Ber., 26, 202–214 (1953).
48. R. Conradt, The glass melting process – treated as a cyclic process on an imperfect heat exchanger, In: *Advances in fusion and processing of glass III*, J. R. Varner, T. P. Seward, and H. A. Schaeffer, eds., Ceramic Transactions Vol. 141, The American Ceramic Society, Westerville, pp. 35–44 (2003).

49. T. Hatakeyama and Z. Liu, *Handbook of thermal analysis*. John Wiley and Sons, Cambridge (1998).
50. I. Waclawska, Thermal behavior of mechanically amorphized colemanite, J. Therm. Anal., 48, 145–154 (1997).
51. P. W. Hodkin, H. W. Howes and W. E. S. Turner, Der Einfluß von Scherben auf die Schmelzgeschwindigkeiten und andere Eigenschaften des Natron-Kalk-Kieselsäure-Glases, Glastechn. Ber. T, 681–692 (1928/1929).
52. I. M. Krieger and T. J. Dougherty, A mechanism for non-Newonian flow in suspensions of rigid spheres, Trans. Soc. Rheol., 3 (1959).

Chapter 10
Glass Melt Stability

Helmut A. Schaeffer and Hayo Müller-Simon

Abstract The employment of sensors during glass melting represents a major prerequisite for an improved process control leading to higher production yields. In situ sensoring techniques can be divided into two groups: on the one hand, techniques which extract information of glass melt properties, e.g., oxidation state and concentrations of relevant polyvalent species (such as iron, sulfur, chromium) and on the other hand, techniques which monitor the furnace atmosphere with respect to toxic emissions (e.g., SO_2, NO_x) and combustion species (e.g., CO, CO_2, H_2O). Nowadays it is feasible not only to install early warning systems indicating deviations from target glass properties, but also to implement process control systems which enforce a stable and reproducible glass melting. Examples are given for the redox control of green glass melting utilizing high portions of recycled cullet and the redox control of amber glass melting.

Keywords Glassmelt · Glassmelt properties · Glass quality · In situ sensors · Redox control · Voltammetry · Recycled glass cullet

10.1 Introduction

The major scientific and technological challenges in the manufacturing of glass are nowadays the improvement of glass quality and thus the improvement of reproducibility of glass product specifications.

Mathematical modeling of the glass melting process and monitoring and controlling of glass melt properties play a decisive role in pursuing these objectives. This holds not only for the flat glass industry where it is mandatory to produce almost optical glass quality in the bulk and highly reproducible glass surfaces for coating purposes, but also for the container glass industry where customer requirements

H.A. Schaeffer (✉)
Formerly Affiliated with Research Association of the German Glass Industry (HVG),
63071 Offenbach, Germany
e-mail: helmut.schaeffer@gmx.net

F.T. Wallenberger, P.A. Bingham (eds.), *Fiberglass and Glass Technology*,
DOI 10.1007/978-1-4419-0736-3_10, © Springer Science+Business Media, LLC 2010

become more stringent with respect to glass homogeneity and reproducibility of color.

The required quality optimizations of the glass melting process have to meet certain energetic, ecological, and economic constraints.

Technological measures in energy saving are determined by the glass furnace design, by the choice of fuel (gas, oil, electricity) and type of combustion (air/fuel, oxygen/fuel), by the use of recycled cullet, by recycling of filter dust, and by the exploitation of waste gas heat.

Ecological constraints result from the fulfillment of legal regulations of waste gas cleaning and thus of reducing total emissions (CO_2, CO, SO_x, NO_x, and particulates).

Finally, the permanent challenge of cost-reducing measures, especially by increasing the productivity and the production yields has accelerated the implementation of process automation, and in particular the employment of in situ sensors.

Therefore, many efforts were undertaken by developing appropriate sensors in recent years [1–4]. The term sensor in connection with glass melting arose mainly when electrochemical sensors for measurements in the glass melt were introduced. However, sensors are defined as devices which transform physical or chemical properties or their alterations into an electric current or voltage. Thus, also a simple thermocouple must be regarded as a sensor. A particular feature of sensors to be used in the glass melting process is their required high-temperature heat resistance.

10.2 Target Properties of Glass Melt and Glass Product

In glass melting two types of information must be distinguished: information which is required for a reliable process management and information about properties of the produced glass. Even though data about glass properties are wanted as early as possible, it is not meaningful to measure these quantities before the melting process is finished, i.e., measurements should be carried out after the refining stage. Thus, sensors which provide this type of data are installed in the feeder channel. Typical glass property related quantities are color, oxygen partial pressure, water content, and viscosity of the melt.

In order to achieve the target properties of the final glass product, it is necessary to maintain a constant chemical composition via batch additions. However, there exist "hidden" parameters which can lead to deviations of the target properties. Such hidden parameters are the oxidation state of the glass melt due to oxidizing and/or reducing agents in the batch and depending also – however, to a lesser degree – on the oxygen partial pressure in the furnace atmosphere and furthermore on the water content of the glass melt, which again depends on the moisture of the raw materials and the humidity of the furnace atmosphere.

The terms "oxidation state" and "redox state" are descriptive; the measurable quantity is the oxygen partial pressure of the glass melt which affects glass properties predominantly via the valence state ratios of polyvalent elements which are added to the glass batch either as coloring (e.g., Fe, Cr, Se, Co, Mn) and refining elements (S, As, Sb) or are introduced unintentionally as impurities (Fe, Ti) together with the raw materials.

In Table 10.1 properties are listed which can be influenced by the oxidation state of the glass melt and which can affect relevant properties during glass melting and glass forming [5]. Variations in water content do possess a further impact on the listed properties [6].

Table 10.1 Impact of oxidation state on properties relevant for melting, forming, and product specifications

Melting process	Forming process	Final product
Solubility and diffusivity of gases (refining, bubble formation)	Viscosity	Optical transmittance
Coloring and decoloring	Visco-elastic behavior	Radiation resistance (solarization)
Volatilization	Surface tension (wetting, adherence)	Semiconduction
Corrosion of refractory and electrode material	Crystallization	Mechanical strength
	Phase separation	
"Melt history"	"Workability"	"Reproducibility"

In other words, variations in the oxidation state and water content have an impact on the melt history, on the "workability" during forming processes, and on the reproducibility of the final glass product.

10.2.1 Batch-Related Fluctuations

In the container glass production the adding of high portions of recycled cullet is common practice. In many glass plants in Western Europe the introduction of cullet amounts to 80–100% for green, 60–80% for amber, and 50–70% for flint glasses. Thus, recycled cullet has become the major component in the batch.

Recycled cullet is characterized by a great variation in composition. Moreover, organic contaminations strongly influence the oxidation state of the glass melt and subsequently the refining behavior and the color of a glass. Also the humidity of the batch and the cullet has an influence on the melting behavior.

In this conjunction it has to be remembered that industrially melted glasses are typically not in equilibrium with the oxygen partial pressure of the furnace atmosphere – due to the low diffusivity of O_2 in the glass melt [7] – but are determined by the oxidizing and predominantly by the reducing additions of the glass batch, the convective flows in the glass tank, and the temperature history. Therefore, the interest is focused to detect reducing contaminations in the batch as early as possible in order to minimize fluctuations of the redox state.

In the past, investigations were carried out to determine these detrimental quantities online immediately before the batch enters the glass furnace. Humidity measurements have been performed online by means of microwave, and organic contaminations can be detected by combining microwave and neutron absorption techniques [8].

10.2.2 Combustion-Related Fluctuations

One pronounced correlation exists between the type of employed fuel for combustion and the resulting content of H_2O in the flue gas. Typical values for oil combustion are about 10%, for natural gas about 18%, and for oxy/fuel firing as high as 60%. On the other hand, basically no water is generated when all-electric melting is employed. These differently water-loaded furnace atmospheres are reflected in different OH contents of the final glass products and thus in different glass properties (e.g., specific density, refractive index, spectral properties).

It has to be noted that gas solubilities in the glass melt are correlated with the respective partial pressures of the gases in the furnace atmosphere, i.e., incorporation and release of gases represent a source of permanent fluctuations of the glass melting process; as a striking example stands the phenomenon of foaming due to a release of SO_2 caused either by an increase of water vapor in the furnace atmosphere or by changing the oxygen partial pressure. Therefore, in situ sensing of the partial pressure of water, oxygen, and other gases is desirable.

Furthermore, the flue gas composition is important not only with respect to environmental issues but also regarding the glass melting process. The type of combustion has an effect on the sulfate decomposition in the batch [9]. Normally combustion is near-stoichiometric in order to minimize NO_x emissions. However, this is a boundary condition which induces sulfate decomposition and leads to an increase of SO_x emissions. Since the sulfate decomposition is based on the reaction with CO which in turn is not in equilibrium with O_2 during combustion, a direct monitoring of CO is of interest, too.

10.2.3 Process-Related Fluctuations

Temperature is a basic quantity of the glass melting process. A sufficient high temperature in the furnace chamber ensures that the required heat is provided. The temperature distribution along the furnace axis is in part responsible for the formation of a hot spot resulting in a strong vertical glass flow. The temperature at the bottom of the furnace is responsible for a sufficient low viscosity of the glass melt so that an adequate amount of bottom glass participates in the glass convection current. The quality of the final glass melt is determined by its residence time in the furnace which primarily depends "internally" on the glass convective flows, but "externally" on the pull rate.

An exact determination of the glass melt convections by means of temperature measurements is only possible with the help of mathematical modeling. However, this coupling of mathematical models with online temperature measurements is difficult. Most of the thermocouples are located within the refractory material to protect them against corrosive attack. Thus, the measured temperatures deviate from the real temperatures at the surface of the superstructure. On the other hand, a thermocouple which is inserted in the furnace chamber interacts with the gaseous atmosphere due to convective heat transfer and interacts further with the furnace walls and glass melt surface due to radiative heat transfer.

An additional factor which hampers constant glass melting conditions is the wear of the furnaces refractories, especially those in contact with the glass melt; the most critical being refractories in the position of the throat. With increasing furnace life, the throughput to the working end rises due to the larger opening of the throat which subsequently induces a higher backflow linked with lower heat efficiency of the furnace and higher process instabilities.

10.3 In Situ Sensors

As compared to the production of other materials, the process of industrial glass melting is lagging behind in utilizing suitable in situ sensors for the online characterization of glass melt properties. The same holds for devices which monitor the composition of the furnace atmosphere. The employment of sensors represents a significant prerequisite for an improved process control, thus leading to a higher glass melt stability.

Several principles have been exploited in order to achieve information about the glass melting or combustion process. The corresponding sensors are based on mechanical, optical, electrical, and electrochemical measurements, cp. Table 10.2 and Fig. 10.1.

Table 10.2 Measuring principles applied in sensors

Measuring principle	Measuring effect	Determined quantity
Mechanical	Rotating, vibrational viscosimeter, ultrasonic	Viscosity
Electrical	Seebeck effect	Temperature
Electrochemical	Electrochemical cell current/potential curve	Oxygen partial pressure concentration of polyvalent elements
Optical	Emitted radiation	Valence state of polyvalent elements (color) water content
	Laser-induced light emission	Chemical composition
	Laser absorption (IR)	Gas species
Mass	Mass analysis	Gas species

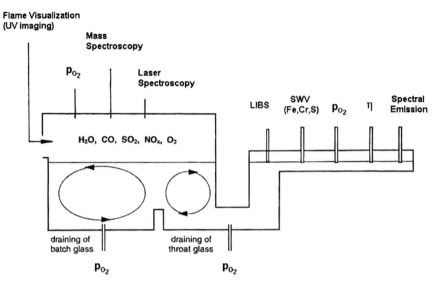

Fig. 10.1 In situ sensors for glass melting

10.3.1 Sensors for Monitoring Glass Melt Properties

10.3.1.1 Viscosity

A constant viscosity is most important for an effective glass forming process.
Therefore, online measurement of viscosity is of great interest. Variations of the
composition of recycled cullet can cause considerable changes not only in the
absolute viscosity but also in the steepness of the viscosity–temperature curve.

Attempts were undertaken to measure glass viscosity online in the feeder
channel, with the help of ultrasonic techniques [1] and by means of a rotation
viscosimeter [10]. However, up to now none of the devices was developed to
commercial availability.

One fundamental problem of viscosity measurement is its strong dependence on
temperature. In the feeder channel, temperature varies with respect to the pull rate.
In order to detect viscosity variations related to alterations in glass composition
(e.g., different OH content), the viscosity must be normalized with respect to temperature, a fact which requires data of the temperature variations in advance. The
"decoupling" of the compositional and thermal impact on viscosity is difficult to
solve.

10.3.1.2 Redox Measurement

Starting in the early 1980s zirconia-based solid electrolyte sensors have been developed for measuring the oxygen partial pressure in glass melts [11–14]. Based
on these investigations sensors for industrial application were designed [15–17].

Electrochemical oxygen sensors represent electrochemical cells which consist of two spaces with different oxygen partial pressure separated by an oxygen ion-conducting material. Today mostly yttria-stabilized zirconia is used as oxygen ion conductor. The zirconia is contacted on both sides with platinum providing the following cell:

$$Pt, pO_2(r) \mid ZrO_2 \mid Pt \mid glass\ melt, pO_2(m)$$

with $[pO_2(r)]$ the reference oxygen partial pressure and $[pO_2(m)]$ the oxygen partial pressure of the glass melt. Such a cell generates an electromotive force (emf) depending on the difference between the oxygen partial pressures on both Pt electrodes. From the emf the oxygen partial pressure of the glass melt can be calculated by means of the Nernstian equation once the reference electrode is exposed to a defined oxygen partial pressure, typically air. The measuring principle does not change if one gas space is replaced by a solution or melt which is in equilibrium with a given oxygen partial pressure.

Under industrial conditions the sensor is exposed to a considerable temperature gradient. In that case a thermally induced voltage superimposes the electrochemical emf. The thermally induced voltage can be considered as a linear function of the temperature difference between the measuring and the reference electrode and can be calculated by means of the standard Seebeck coefficient. In practice the unknown oxygen partial pressure of the glass melt can be determined with sufficient accuracy according to

$$\ln pO_2(m) = \frac{4F}{RT_m}(E - k_s(T_m - T_r)) + \frac{T_r}{T_m} \ln pO_2(r),$$

with E =emf, T_m glass melt temperature, T_r temperature of the reference electrode, k_s the standard Seebeck coefficient, F the Faraday constant, and R the gas constant.

The great advantage of this measuring technique is that such a sensor requires no calibration as long as it works at temperatures above 800°C (the condition of pure oxygen ion conduction in the zirconia electrolyte has to be fulfilled).

Figure 10.2 shows a combined electrochemical sensor developed for the use in industrial glass melts. For some years, experience has been gained in the container glass industry by placing the sensor in contact with the glass melt in the forehearth (1200–1250°C), a location of measurement optimally suited for characterizing the conditioned glass melt prior to feeding into the glass forming machines. The average lifetime of such a sensor is about 12 months being limited by the corrosion of the zirconia.

10.3.1.3 Voltammetric Sensor

Electrochemical sensors have been developed in order to measure the concentrations of sulfur, iron, selenium, and chromium in glass melts. In particular square-wave voltammetry has proved to be a powerful tool in conjunction with oxygen partial

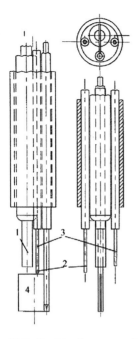
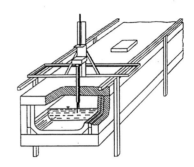

1   reference electrode of yttria-stabilized
    zirconia
2   mesuring electrode (Pt) for oxygen par-
    tial pressure
3   working electrode (Pt) for voltammetry
4   counter electrode (Pt) for voltammetry

Fig. 10.2 Combined oxygen and concentration sensor for glass melts

pressure measurements [18], cp. Fig. 10.2. Therefore, the sensor is able to detect the concentrations of the most important polyvalent elements which determine the color and refining behavior.

The voltammetric sensor provides current/potential curves which show characteristic peaks for polyvalent elements such as iron, chromium, and sulfur, cp. Fig. 10.3.

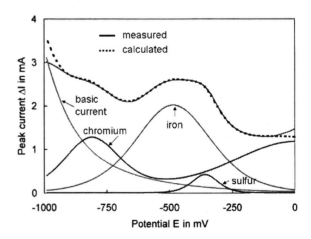

Fig. 10.3 Voltammetric
current/potential curve of a
green container glass

The interpretation of voltammetric curves requires a mathematical deconvolution procedure which separates the various individual peaks. The peak position characterizes the potential of a specific polyvalent element and the peak current is proportional to the concentration of the respective polyvalent element. The peak current is given by

$$I_P = KA \, C_0 \sqrt{D_0} \frac{(zF)^2}{RT} \frac{\Delta E}{\sqrt{\tau}},$$

with electrode area A, diffusion coefficient D_0 and concentration C_0 of the polyvalent element, number of exchanged electrical charges z, temperature T, height ΔE, and width τ of the potential square wave. The factor K considers the deviation of the geometrical shape of the electrodes from the ideal shape, inductive influences due to the extention of the electrical connections, etc. Since K is not known exactly, the sensor has to be calibrated which can be achieved by means of a calibration curve according to Fig. 10.4, where the concentrations are calculated from the peak current densities. Figure 10.4 also demonstrates the excellent agreement between sulfur data obtained by the in situ sensor and by wet chemical analysis and thus proves the advantage of a continuous in situ measurement as compared to the discontinuous time-consuming wet chemical analysis [19].

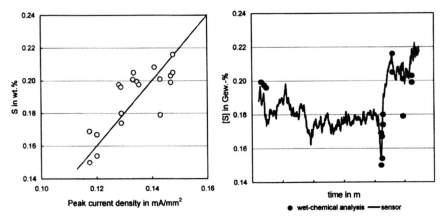

Fig. 10.4 Calibration curve (*left*) and variation of the calibrated signal of a sulfur sensor compared to wet chemical analysis

10.3.1.4 Emission Spectroscopy

Furthermore, light emitted from the glass melt provides information on the composition, especially on the concentration of coloring ions. Hence emission spectroscopy can be a valuable tool for in situ quality control provided that the surface of the glass melt is colder than the bulk glass. In the case of polyvalent elements, data can be obtained not only on their total concentrations but also on their redox ratios [20–22]. Moreover, this technique allows the detection of water in glass utilizing the characteristic spectral emission lines of H_2O and OH in the infrared. It is envisaged that emission spectroscopy can be installed above the feeder channel in order to monitor the surface of the glass melt.

10.3.1.5 Laser-Induced Breakdown Spectroscopy (LIBS)

A further method of high potential for the in situ online measurement of glass components is the laser-induced breakdown spectroscopy. This technique has been successfully demonstrated in the steel industry for online process control. The distance between the laser (Nd:YAG pulsed laser) and the surface of the steel melt can amount to up to 3 m. This technique is based on the generation of a laser-induced microplasma followed by the analysis of the emitted light containing the characteristic spectral lines of the elements present. LIBS can detect most of the elements which are relevant in industrial glass melting with a sensitivity in the order of ppm [23–26]. Additionally, no sample preparation is required and for process control – in most cases – a calibration is not necessary since relative changes of specific concentration ratios are indicative for process deviations. Again the feeder channel can be considered as the most suitable location for an online LIBS detection.

10.3.2 Sensors for Monitoring Species in the Combustion Space

The characterization of the furnace atmosphere and its continuous monitoring and control is motivated by two major aspects. The most obvious one is caused by the environmental regulations limiting the amount of toxic species and particulates in the flue gas. The second aspect is related to energetic considerations, i.e., driven by the interest to fully exploit the combustion potential of a given fuel and to generate a maximum of heat flux from the combustion space into the glass melt. Finally, there is an interest to monitor the long-term detrimental impact of volatile species (e.g., NaOH, boron-containing species) which result in a corrosive attack of the refractories of the superstructure and of the checkers in the regenerators. Of course, in most cases all aspects have to be treated as a joint and correlated endeavor in establishing an environmentally and energetically sound and stable glass melting.

10.3.2.1 Sensors for Environmental Measurements

Mostly the measurements of environmentally relevant species are performed offline, i.e., in the form of a bypass or sampling analysis. Sometimes such measurements are carried out in the flue gas channel, but so far never in the furnace atmosphere.

Extractive or online flue gas analysis is often based on non-dispersive IR or UV absorption spectroscopy for the detection of CO, SO_x, NO, and CO_2 on chemiluminescence for NO_x and on paramagnetic analysis of O_2. Very often conventional wet chemical analysis combined with ion-selective electrodes is employed, e.g., for the analysis of HCl, HF, and NH_3 [27].

10.3.2.2 Sensors for Optimizing Combustion Efficiency

One of the few established in situ techniques represents the flame visualization via a furnace periscope as well as the UV imaging for visualization of non-luminescent flames (e.g., oxy/fuel firing) [28].

The same holds for the continuous in situ measurement of oxygen partial pressure in the combustion space using solid electrolyte sensors. These sensors are used to measure the oxygen concentration directly in the exhaust ports or in the top part of the regenerators and allow the accurate control of the air/fuel or oxygen/fuel ratio.

Mass spectrometry in combination with a heated capillary gas inlet system– as often used in the steel industry – could be a further technique to be employed for online gas analysis in the furnace atmosphere.

Presently a most attractive technique appears to be laser absorption spectroscopy in the near- or mid-infrared. First, near-infrared measurements (0.6–2 μm) for the determination of CO in a glass melting furnace were carried out by Faber et al. [29] utilizing a commercially available tunable single-mode diode laser. However, in the mid-infrared range (2–20 μm) nearly all gaseous combustion species exhibit fundamental absorption lines that are orders of magnitude stronger as in the near infrared.

These lines can be probed with novel mid-infrared diode lasers; they permit highly sensitive and selective measurements even at high temperatures [30]. Therefore, it can be expected that in the near future, combustion reaction products such as not only H_2O, CO, and CO_2 but also SO_2 as well as NO can be continuously monitored in the combustion space by passing the laser light across the width of the glass melting furnace.

10.4 Examples of Glass Melt Stability Control

10.4.1 Redox Control of Glass Melting with High Portions of Recycled Glass

The most pronounced redox problems occur in glass melting with high portions of recycled cullet and recycled filter dust. Variations of redox state can be caused by mixed cullet (flint and amber besides green), organic contaminations, and the addition of filter dust. An uncontrolled variation of the redox state can result in increasing seed count or color problems. Figure 10.5 shows the variation of the oxygen partial pressure of the glass melt together with the seed count and the dominant wavelength over a period of 40 days in a green container glass melt [31].

The redox state was not controlled and thus the oxygen partial pressure varied by more than one order of magnitude. At the highest oxygen partial pressure an

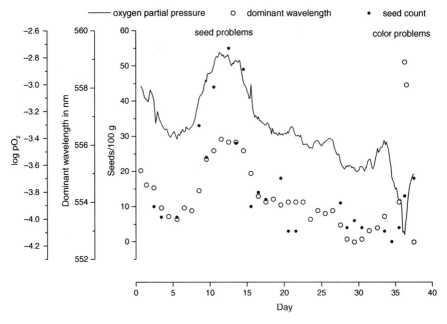

Fig. 10.5 Variation of oxygen partial pressure, seed count, and dominant wavelength in a green container glass melt over a period of 40 days

increase in seed formation was detected. In earlier investigations, an increased seed susceptibility was always found in conjunction with decreasing oxygen partial pressure due to a lowering of sulfur solubility [32]. However, in the present case the increase of the oxygen partial pressure is caused by the increased addition of filter dust which also increases the sulfur concentration in the glass melt. Oxygen balance calculations show that under these circumstances at the given oxygen partial pressure the higher sulfur content leads to a supersaturation of the melt with sulfite (S^{4+}) and subsequently to an increase in seed formation. Obviously, at the given oxygen partial pressure the introduced sulfur partially loses its oxygen, i.e., it cannot be incorporated as sulfate (S^{6+}) and leaves the glass melt as SO_2.

Figure 10.6 displays that the observed increased seed count can be related to an increase of the calculated sulfite concentration.

A further problem which frequently occurs in green glass melting is the over-reduction to amber color. This problem occurred on the 35th day (see Fig. 10.5). Normally the dominant wavelength decreases with decreasing oxygen partial pressure because the absorption of the Fe^{2+} band increases and that of the Fe^{3+} band decreases. However, in a sufficiently reduced green glass melt the sulfur is eventually reduced to sulfide (S^{2-}) which in combination with Fe^{3+} forms the amber chromophore which in turn causes an intense increase of the dominant wavelength due to the lowering of the UV transmission, cp. Fig. 10.5.

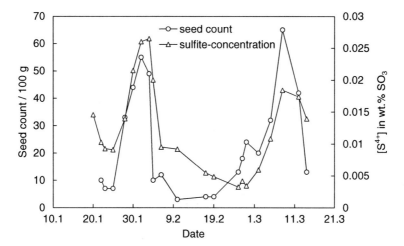

Fig. 10.6 Variation of seed count and calculated sulfite concentration in a green container glass melt at 1200°C

10.4.2 Redox Control of Amber Glass Melting

The typical amber coloring of glass melts is due to a complex electron transfer interaction of two polyvalent species during the cooling process of the glass melt, namely iron and sulfur.

$$S^{4+} + 6\,Fe^{2+} \Leftrightarrow S^{2-} + 6\,Fe^{3+}$$

The intensity of the amber chromophore depends on the product of concentrations of Fe^{3+} and S^{2-} in the final glass product, i.e., $[Fe^{3+}]\,[S^{2-}]$. This product depends not only on the batch additions of iron and sulfur, but also on the redox state during glass melting.

Figure 10.7 shows for given iron and sulfur additions in the batch the calculated intensities of the amber chromophore as a function of oxygen partial pressure for three different temperatures. In the range of oxygen partial pressures between 10^{-6} and 10^{-2} bar the amber chromophore forms during cooling, i.e., at melting temperatures the redox equilibrium between iron and sulfur is shifted to Fe^{2+} and S^{4+}[33].

In Fig. 10.8 the temperature dependence of the intensity and of the absorption coefficient (at 24800 cm^{-1}) of the amber chromophore is depicted [34]. The absorption coefficient decreases linearly with temperature up to 500°C, decreases exponentially for higher temperatures, and coincides with the calculated intensity of the amber chromophore.

Figure 10.7 also displays the ambiguity in amber coloring since there exist two different redox states at about 10^{-5} and 10^{-7} bar resulting in the identical coloring and transmission of the final glass product (cp. 200°C curve). It is interesting to note

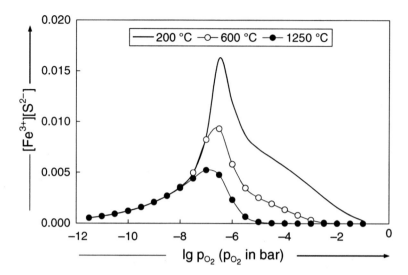

Fig. 10.7 Amber chromophore concentration as a function of oxygen partial pressure for different temperatures

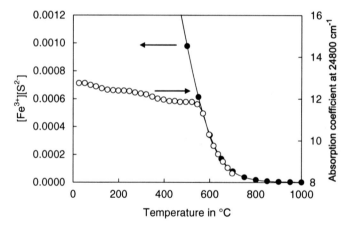

Fig. 10.8 Comparison of calculated amber chromophore concentration and measured absorption coefficient

that the two redox states correspond to different melting conditions: at 10^{-5} bar the glass melt is basically free of the amber chromophore, i.e., highly transparent, also for the heat radiation, whereas at 10^{-7} bar there exists hardly any difference between final product and glass melt with respect to the concentration of chromophore [33].

A further redox influence is given in amber glass melting – in particular when employing mixed cullet or portions of green glass cullet – due to the interaction of chromium originating from the green glass with the amber chromophore. The detrimental effect of chromium consists in a weakening of the chromophore by reducing Fe^{3+} to Fe^{2+} :

$$Fe^{3+} + Cr^{2+} \Leftrightarrow Fe^{2+} + Cr^{3+}.$$

The formation and stability of the amber chromophore serve as a convincing example for the importance of a controlled redox behavior during glass melting since variations in the concentration of the chromophore not only change the intensity of the amber color, but also induce fluctuations in heat transfer with subsequent changes in the temperature distribution and altered convections in the glass furnace.

10.5 Conclusions and Outlook

It was shown that there are several measuring principles and devices which can be applied for in situ sensoring in the glass melting process. Such sensors can be used to monitor and control glass properties as well as process relevant quantities. The use of online sensors lies not only in the possibility of providing prewarning systems. Even if these sensors are positioned at the end of the melting process – for instance in the feeder channel – they have an eminent advantage compared to laboratory measurements, because no sample preparation is needed and the continuous measurement allows an easier detection of trend developments. The latter aspect enables process control on a larger scale, because most variations in the glass tank take place over a period of days.

Glass melting is a very complex process, especially under industrial conditions. The behavior of redox-dependent glass properties can serve as an example. The redox state of a glass melt is primarily determined by the reducing and oxidizing agents in the batch. However, balances and equilibria compete during batch reactions, thus an accurate prediction of glass properties by means of a batch characterization is impossible. In practice a control has to based on measurements of the iron redox ratio or the oxygen partial pressure in the melt. However, the iron redox state and the oxygen partial pressure depend on temperature, i.e., an appropriate model is required which relates the measured quantities to the target properties.

Nowadays the impact of viscosity, foaming, fining, and convective flows on glass melting can be elucidated by mathematical modeling [35–37] which, however, requires the input of in situ data collected by online installed sensors in the glass tank and in the combustion space in order to ensure a more stable glass production.

References

1. M. J. Plodinec and R. D. Costley, Development of instrumentation for in situ measurements of molten glass properties. Proceedings of the XVIII International Congress on Glass, San Francisco, (1998).

2. J. E. Fenstermacher, Sensors or the care and feeding of a successful glass. *GlassResearcher*, 8, 1–3 (1999).

3. A. J. Faber and R. G. C. Beerkens, Sensors for glass melting processes. *Glastech. Ber. Glass Sci. Technol.*, 73(C2), 111–123 (2000).

4. R. Denno, Sensors for glass melting – needs and availability. *Glass Technol. Eur. J. Glass Sci. Technol. A*, 48, 55–58 (2007).

5. H. A. Schaeffer, Scientific and technological challenges of industrial glass melting. *Solid State Ionics*, 105, 265–270 (1998).

6. E. N. Boulos and N. J. Kreidl, Water in glass: A review. *J. Canad. Ceram. Soc.*, 41, 83–90 (1972).

7. H. A. Schaeffer, H. Lachenmayr and L. D. Pye, Oxidation studies of amber glass. *Glastech. Ber.*, 56 K(1), 602–607 (1983).

8. U. Roger and R. Dreier, Erfassung der organischen Bestandteile und der Feuchtigkeit von Recycling-Scherben. *Glastech. Ber. Glass Sci. Technol.*, 68, N156 (1995).

9. H. Müller-Simon et al., Sulfur mass flows in industrial glass melting. Proceedings of the XX. International Congress on Glass, Kyoto, (2004).

10. U. Roger, H. Hessenkemper and P. Roth, Glass conditioning by viscosity control. *Glastech. Ber. Glass Sci. Technol.*, 69, 242–245 (1996).

11. H. A. Schaeffer, T. Frey, I. Löh and F. G. K. Baucke, Oxidation state of equilibrated and non-equilibrated glass melts. *J. Non-Cryst. Solids*, 49, 179–188 (1982).

12. F. G. K. Baucke, Development of electrochemical cells employing oxide ceramics for measuring oxygen partial pressures in laboratory and technical glass melts. *Glastech. Ber.*, 56 K, 307–312 (1983).

13. A. Lenhart and H. A. Schaeffer, Elektrochemische Messung der Sauerstoffaktivität in Glasschmelzen. *Glastech. Ber.*, 58, 139–147 (1985).

14. A. Lenhart and H. A. Schaeffer, The determination of oxidation state and redox behavior of glass melts using electrochemical sensors. XIV. International Congress on Glass, New Delhi, Vol. 1 147–154 (1986).

15. H. Müller-Simon and K. W. Mergler, Electrochemical measurements of oxygen activity of glass melts in glass melting furnaces. *Glastech. Ber.*, 61, 293–299 (1988).

16. F. G. K. Baucke, R. D. Werner, H. Müller-Simon and K. W. Mergler, Application of oxygen sensors in industrial glass melting furnaces. *Glastech. Ber. Glass Sci. Technol.*, 69, 57–63 (1996).

17. P. Laimböck, In-line oxygen sensors for the glass melt and the float bath. *Adv. Mater. Res.*, 39–40, 443–446, (Trans Tech Publications, Switzerland) (2008).

18. C. Montel, C. Rüssel and E. Freude, Square-wave voltammetry as a method for the quantitative in situ determination of polyvalent elements in molten glass. *Glastech. Ber.*, 61, 59–63 (1988).

19. H. Müller-Simon, Control of the glass melting process using online sensors. Proceedings of the XX. International Congress on Glass, Kyoto, (2004).

20. D. Gödecke, M. Müller and C. Rüssel, Thermal radiation of chromium-doped glass melts. *Glastech. Ber. Glass Sci. Technol.*, 74, 277–282 (2001).

21. D. Gödecke, M. Müller and C. Rüssel, Absorption and emission of semi-transparent glass melts. *Phys. Chem. Glasses*, 43C, 232–237 (2002).

22. C. Rüssel, Redox state of glasses. *Glastech. Ber. Glass Sci. Technol.*, 77C, 149–159 (2004).

23. G. Zikratov et al., Laser-induced breakdown spectroscopy of hafnium-doped vitrified glass. *Glass Technol.*, 40, 84–88 (1999).

24. C. Y. Su et al., Glass composition measurement using laser-induced breakdown spectrometry. *Glass Technol.*, 41, 16–21 (2000).

25. J. Yun, R. Klenze and J. Kim, Laser-induced breakdown spectroscopy for the on-line multi-element analysis of highly radioactive glass melt. Part I: Characterization and evaluation of the method. *Appl. Spectrosc.*, 56, 437–448 (2002).

26. K. Loebe, A. Uhl and H. Lucht, Microanalysis of tool steel and glass with laser-induced breakdown spectroscopy. *Appl. Opt.*, 42 (30), 6166–6173 (2003).
27. U. Kircher, Continuous measurements of emissions. in HVG Advanced Course Measuring techniques for the glass melting process (in German) (1996).
28. H.-J. Voss and K. W. Mergler, Visual Monitoring of non-luminescent flames in high temperature glass melt furnaces. (in German). *Glastech. Ber.*, 51, 96–103 (1978).
29. A. J. Faber, R. Koch, H. van Limpt and R. G. C. Beerkens, Co-sensors for hot exhaust gases to control combustion processes. *Glastech. Ber. Glass Sci. Technol.*, 73(C2), 370–378 (2000).
30. L. Wondraczek, G. Heide, G. H. Frischat et al., Mid-infrared laser absorption spectroscopy for process and emission control in the glass melting industry. *Glass Sci. Technol.*, 77, 68–76 (2004).
31. H. Müller-Simon, Online redox control in industrial glass melting. *Soc. Glass Technol. Ind. Sec.*, IX (2–4), (2002).
32. M. Beutinger, Employment of recycled glass in the container glass melt. (in German). *Glastech. Ber. Glass Sci. Technol.*, 68, N51–N58 (1995).
33. H. Müller-Simon, Temperature dependence of the redox state of iron and sulphur in amber glass melts. *Glastech. Ber. Glass Sci. Technol.*, 70, 389–391 (1997).
34. M. Müller, C. Rüssel and O. Claußen, UV-VIS spectroscopic investigations of amber glass at high temperatures. *Glastech. Ber. Glass Sci. Technol.*, 72, 362–366 (1999).
35. R. G. C. Beerkens, Modelling of the melting process in industrial glass furnaces, in *Mathematical simulation in glass technology. schott series on glass and ceramics*, D. Krause, H. Loch, eds., Science, Technology and Applications, Springer Verlag, Berlin, 17–73 (2002)
36. O. S. Verheijen, O. M. G. C. Op den Camp, Advanced operation support system for redox control. *Avances in Fusion and Processing of Glass III*, in J. R. Varner, T. P. Seward, H. A. Schaeffer, ed., Ceramic Transactions, Vol. 4, 421–428 (2004)
37. E. Muijsenberg, Advanced model predictive control for glass production optimisation. Glass Trend Workshop, (2006).

Chapter 11
Plasma Melting Technology and Applications

J. Ronald Gonterman and M.A. Weinstein

Abstract A plasma arc melter is a modular high-intensity skull melter capable of rapidly melting a wide variety of materials, both conductive and nonconductive. Although its commercial use to melt and process metals is well known, the method is less well known as a method of melting glass. Extensive research has been conducted by several organizations into the use of skull melting of glass using plasma arcs. This research has shown plasma melting to be a promising technology that can achieve high efficiencies, high temperatures, extreme flexibility, low capital cost, rapid changeovers of glass formulas, and minimal scrap. Plasma melting lends itself to modular melting in which each step of the glass melting process is partitioned into functional modules, which can greatly improve melting efficiency and throughput. Also, plasma arc melting has been shown to be a promising technology for rapidly and inexpensively producing "synthetic minerals" melted from common commercial oxides.

Keywords Glass melting · Plasma · Plasma glass melting · Plasma torches · Plasma electrodes · High-temperature glass melting · Skull melting · High-intensity glass melting · Synthetic minerals · Glass quality · Glass chemistry · Glass volatilization · Plasmelt · JM · AGY · BTU/ton · Pounds per hour · Torch life · Torch stability · E-glass · AR-glass · S-glass · C-glass · Frit · Energy efficiency · Kilowatts per pound · Plasma torch life · Remotely coupled transferred arc · Wollastonite melting

11.1 Concepts of Modular and Skull Melting

The glass melting process is known to be composed of separate and distinct stages as outlined in a pioneering paper on modular melting [1]. Efficiencies can be improved by partitioning the melting process in functional modules. Any high-intensity melter

J.R. Gonterman (✉)
Plasmelt Glass Technologies, LLC, Boulder, CO 80301, USA
e-mail: ron@plasmelt.com

F.T. Wallenberger, P.A. Bingham (eds.), *Fiberglass and Glass Technology*,
DOI 10.1007/978-1-4419-0736-3_11, © Springer Science+Business Media, LLC 2010

which melts glass more rapidly and with smaller melting volume as compared to traditional glass melters can be characterized as a modular melter. A plasma melting furnace is considered to be a modular melter.

A further subset of modular melting utilizes another concept called skull melting. In this type of melter, the unreacted batch and devitrified glass form a layer between the melter wall and the hot molten glass zone to provide the insulation needed to allow the sidewalls to run at an acceptable temperature. Skull melting uses no refractory liner and, instead, typically utilizes a low-cost metal shell to contain the hot zone. For some applications, the elimination of traditional refractory has distinct advantages: the heatup/cooldown schedule can be much more aggressive since there is no refractory that is susceptible to thermal shock; defects caused by small fragments of abraded refractory are eliminated thus enhancing the quality of the glass; the cost associated with the purchase of refractory is eliminated; complex shapes for the melter sidewalls and bottom are possible and are not limited by the commercially available refractory shapes.

Water cooling, which may be used to impose an even steeper thermal gradient between the hot glass and the cold sidewalls, is optional. The main beneficial effect of this additional cooling zone is the ability to run the hot zone at very high temperatures. Hotter operation may be desirable to enhance the desired glass melting reactions and increased throughputs. Also, elevated temperatures are required for certain glass types. The use of a water cooling jacket also gives one more degree of freedom in controlling the thickness of the batch blanket build-up on the sidewall and bottom.

However, operating without water cooling is always desirable if the melting operation can be achieved without such cooling. The energy penalty is greater if cooling water is routinely used. For example, over 40,000 J of energy are required to raise one liter of water $10°C$. For a skull melter without water cooling, the sidewall temperatures are limited by the ambient cooling that is achieved with air. In operation, as the melter heatup begins, the hot zone increases to glass melting temperatures, the thermal gradient becomes higher, the thickness of the batch blanket at the sidewalls decreases, and the walls heat up. Since the thickness of this insulating zone becomes reduced at higher melting zone temperatures and glass throughputs, the ultimate service temperature of the metal used in the sidewall construction becomes the upper limit on the temperatures that can be used in service.

Skull melting is known to have disadvantages that must be overcome. Upon the initial heating of fresh batch, any raw material that contains gases must release them: the CO_2 from limestone/dolomites and water from clays and borates. These gases must be reacted and allowed to escape in order to achieve a quiescent batch blanket to protect the melter steel parts during the subsequent furnace campaign. This degassing stage is a normal heatup step that must be built into each melter heatup schedule. Subsequently re-heating a skull that was previously used does not require this degassing. If the glass compositions between the old campaign and the new campaign are the same, it is always desirable to re-heat the skull melter and to make use of the glass lining that was previously in service.

An additional disadvantage of skull melting is the possibility of supplying devitrified glass defects from the colder sidewall areas of the melter. In practice, this is not an issue if the melter is operated in a stable mode without disturbing the thermal gradient. Any area of molten glass is susceptible to forming crystals of the primary, secondary, or tertiary phases of the glass if held at sub-liquidus temperatures. The formation of these crystalline areas is not a disadvantage of plasma melting as long as the sub-liquidus glass stays in the stagnant areas where glass flow is too slow to allow any significant movement out of the melter discharge orifice. When the thermal gradient changes – due to an operational change in the power setting or in the throughput of the glass flow – some of this undesirable movement can occur. However, the very high thermal gradient that is employed in the hot zone causes any crystalline material to be quickly resorbed into the glass. Defects from glass devitrification are essentially non-existent in plasma melting if the melter is properly operated.

In summary, a plasma melter is a modular skull melter that can provide rapid batch materials decomposition and rapid glass melting. The downstream processes of fining, mixing, and homogenization must be completed by additional modular processing.

11.2   The Technology of High-Intensity DC-Arc Plasmas

A thermal plasma is a partially ionized gas where a portion of the electrons is free rather than being bound to a particular atom or molecule. The ability of electric charges to move through the plasma makes it electrically conductive. Plasmas respond to electromagnetic field and thermal effects. Plasmas have properties different to solids, liquids, or gases and are considered to be a distinct state of matter.

Plasmas have been used extensively in many industries for melting, reacting, and processing materials. Examples include tong dish heating, metal refining, semiconductor manufacturing, and hazardous waste refining. In most of these cases, the plasma is coupled with a conductive substrate or being used in a low-energy requirement application. Traditional plasma technology has not been used in high-throughput glass melting applications because most glass melting requires high energy for melting and many glasses are not electrically conductive until molten.

Thermal plasmas are generally divided into three categories: direct current (DC), alternating current (AC), and radio frequency. Due to the amount of power required for most industrial glass melting processes (> 500 kW), radio frequency and AC plasmas are not used. With DC-generated plasmas, both non-transferred- and transferred-arc modes of operation are commonly used (see Fig. 11.1). Non-transferred arcs are typically used in gas heating and thermal spray applications. Where high energy flux is required, the transferred arc is most commonly used due to its very high transfer efficiency.

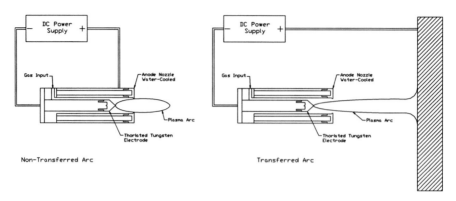

Fig. 11.1 Non-transferred-arc plasmas

Conventional transferred-arc plasma melting of glass materials requires pre-melting of the material to create an electrically conductive molten phase. This initial processing step adds to the start-up time. Also, either the molten bath is contained in an anodic crucible or anodes are submerged into the bath to provide a return electrical conduction path. The direct current that flows through the system can cause undesirable electrolytic reactions at the anode surface. The electrolytic reactions can lead to anode corrosion and contamination of the molten material.

To eliminate these fundamental shortcomings, Plasmelt Glass Technologies, LLC [2], is using a transferred-arc system that uses two torches of opposite polarity. The cathode torch consists of a water-cooled, thoriated-tungsten electrode. The anode torch consists of a high-purity, water-cooled copper surface. Both electrodes are surrounded by a water-cooled nozzle. The nozzle focuses the gas around the electrode and provides arc stabilization. Figure 11.2 illustrates the process.

The relative positions of the torches determine the impedance and melting efficiency. Steeper angles and longer horizontal distance will increase the impedance.

In typical glass melting, the transferred-arc plasma provides heat to the glass material with three different heat transfer mechanisms. These are

- Conductive
- Radiant
- Joule heating (in certain plasma configurations)

11.2.1 Conductive

Conduction is the transfer of heat by *direct contact* of particles of matter. Because of the extremely high temperature, the conductive heat transfer mechanism provides rapid glass melting. Depending on the configuration, arc temperatures in excess of 20,000 K are common. The physical size of the arc dictates the actual heat transfer rate to the glass or batch being processed. With the conductive heat mechanism,

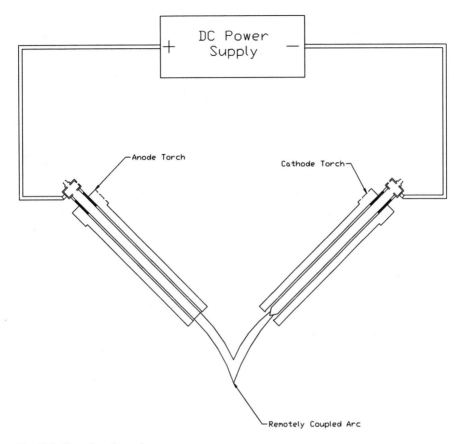

Fig. 11.2 Transferred-arc plasmas

the energy transfer rate is simply the ratio of the material temperature to the arc temperature multiplied by a constant. The high temperature, high energy transfer rate is useful for certain glass types, but not others. For some glass compositions, extreme volatilization of certain components is realized.

11.2.2 Radiant

Radiant energy from plasma is probably the most novel feature in glass melting. Radiation is the transfer of heat energy through empty space. It is the complete process in which energy is emitted by one body, transmitted through an intervening medium or space, and absorbed by another body. Thermal radiation, even at a single temperature, occurs at a wide range of frequencies. The total amount of radiation, of all frequencies, increases rapidly as the temperature rises. (It grows as T^4, where T is the absolute temperature of the body.) For example, an object at the temperature

Fig. 11.3 Photo of plasma torches melting glass batch

of a kitchen oven (about twice room temperature in absolute terms – 600 vs. 300 K) radiates 16 times as much power per unit area. An object at the temperature of the filament in an incandescent bulb (roughly 3000 K or 10 times room temperature) radiates 10,000 times as much per unit area.

The efficiency of the radiant energy from the plasma arc is related to the glass absorption characteristics and the melter configuration.

11.2.3 Joule Heating

Joule heating, also known as ohmic heating and resistive heating, is the process by which the passage of an electric current through a conductor releases heat. It was first studied by James Prescott Joule in 1841. Joule immersed a length of wire in a fixed mass of water and measured the temperature rise due to a known current flowing through the wire for a 30-min period. By varying the current and the length of the wire, he deduced that the heat produced was proportional to the electrical resistance of the wire multiplied by the square of the current. This relationship is known as Joule's first law. The SI unit of energy was subsequently named the joule and given the symbol J. The commonly known unit of power, the watt, is equivalent to one joule per second. Joule heating is referred to as ohmic heating or resistive heating because of its relationship to Ohm's law.

Joule heating is currently used in many glass melting applications.

Typical 3-phase electric melters are common in the industry. In these melters, molybdenum electrodes are inserted and electric current is passed through the melt. This can be a very effective means of heating. The heat transfer to the melt is I^2R. With the effective resistance of the glass being very high (0.1–5 Ω in. or 0.25–12.7 Ω cm), a small amount of current will provide a great deal of heat transfer directly to the glass. The negatives to the molybdenum electrodes in the melt pool are often adverse anodic or cathodic reactions with the glass, which contaminates the glass with small levels of molybdenum. The remotely coupled transferred-arc plasma configuration eliminates the adverse reactions.

For certain plasma configurations, the Joule heating portion of the energy transfer provides a 100% efficient heat transfer mechanism. Joule heating, combined with the radiant and conductive heat transfer (HT), can often provide the optimal configuration of heat transfer means. In applications described later in this chapter, the conductive HT is used for the preliminary batch melting. After the material is partially liquid, the radiant HT causes continued reaction because the material can better absorb the radiant energy. After the material is at the molten state, Joule heating is the predominant HT that provides the final glass reactions and conditioning.

11.3 Brief History of Plasma Melting of Glass

For over 25 years, ongoing attempts have been made to perfect the melting of glass using plasma arcs. The prospect of improved energy efficiency, very high power density, small footprint, low capital cost of the melter, high temperature capability, rapid heatup and cooldown, and rapid changeover from one composition to another all continue to intrigue glass scientists. Below is a brief list of organizations that have been active in plasma melting research for glasses.

11.3.1 Johns-Manville

In the mid-1980s, Johns-Manville combined forces with Tetronics to conduct research into plasma melting as applied to a rotating base metal melter, which utilized remotely coupled transferred plasma arcs. This collaboration yielded one patent [3] – Method of Melting Materials and Apparatus Thereof, US Patent No. 5,028,248, July 2, 1991. A second patent [4] was issued to Johns-Manville (Schuler International) as a result of continued work by JM – Method for the Melting, Combustion or Incineration of Materials and Apparatus Thereof, US Patent No. 5,548,611, August 20, 1996.

This body of work concentrated on the melting operations involving fiberglass compositions and quoted promising research melter results up to glass throughputs of 1200 lb/h. E-glass, calcium silicate, fiberglass scrap, and insulation compositions were all used for these melting experiments. The melter was small (approximately 1 m in diameter) and was powered by 1 MW power supply yielding a much higher than normal power density. The temperature (and glass flow rate) was controlled by manipulating the power and positions of the two plasma torches. There was only minor emphasis on the quality of glass that resulted from the melting operations and no fiberization experiments were conducted.

For business and technical reasons, JM made a decision not to pursue this method of melting for their production operations.

11.3.2 British Glass Institute

A project funded by the Research Group of British Glass and the British Department of Trade and Industry was conducted in the early 1990s. The stated objective of the project was to compare the efficiencies of plasma-arc melting with those of submerged electrode melting and to demonstrate the energy savings from the inherently high melting efficiency of plasmas. The results of this two-phase project were published in a final report in 1994.

The furnace built by the project team had a small footprint of 0.6 m^2 and a depth of 400 mm. A refiner section was also constructed using both sonic horns in combination with electrodes capable of boosting the temperatures in the refiner to allow the glass to degas. The plasma torches were large and capable of imparting 250 kWh/h into the melter. Furnace campaigns were conducted in which the small melter was fired exclusively using the submerged electrodes and compared to other campaigns in which the melter was fired with the plasma arcs.

Results of the study were published in a final report [5] which concluded that for this traditional box design of a refractory lined non-rotating melter, the plasma-arc operation had higher energy losses due to the requirement for water cooling and the higher radiation losses. They further concluded that the higher energy densities were possible with plasma-arc melting (250 kVA) compared to submerged electrodes (100 kVA) and this characteristic could best be applied to small tonnage, specialized applications that require intermittent production or rapid product changes. However, the project team made no recommendations for the optimal furnace design to use plasma arcs.

11.3.3 Plasmelt Glass Technologies, LLC

The United States Department of Energy (DOE) became active in the funding of plasma-based glass melting with the sponsorship of a 2003 to 2006 melting project at Plasmelt Glass Technologies, LLC, in Boulder, Colorado. This work was aimed at building on the already completed Manville research work, which used a rotating furnace, to demonstrate that a high-efficiency plasma process that could yield acceptable glass quality using fiberglass compositions. Although the melting platform used for this work was similar to that used by JM, the work was extended beyond the JM work mainly in the area of plasma torch design and 3-axis torch positioning using state of the art robotics. Detailed glass quality assessments of the plasma-melted glasses were also conducted by the project team.

The Plasmelt research was focused on developing a process aimed at high-quality glass applications that have commercial applications. During the project, a plasma melting process was successfully demonstrated that was capable of producing glass patties and nuggets with sufficiently high quality to allow the glass to be fiberized into a small (10–13 μm) diameter fibers. This marble re-melt fiberizing process is a rigorous test that cannot be passed by glass of poor homogeneity/quality. Fibers as small as 4.5 μm in diameter were also fiberized, but had poorer performance due to higher break rates. The source of the fiber breaks was not due to any discrete stones or other crystalline defects. The seed level of this glass was high due to incomplete fining in the very short residence time that is available in the plasma melter. These seeds were the main cause of the fiberizing breaks in the smaller diameter fibers. An add-on refiner stage is necessary to lower the seed levels to acceptable levels that do not impact fiberization.

Other than Plasmelt, no other known company in the world has successfully made continuous glass fibers from plasma-melted glass. The absence of batch stones, molybdenum metal, devitrification, or any other defects in this plasma-melted glass is very encouraging. Finally, the tensile testing results of glass fibers, comparing the plasma-melted glass and the glass melted in standard production operations, show that the average tensile strength of the two different glasses is not statistically different. This successful fiberizing work not only is of major significance to the fiberglass process but also suggests that plasma-melted glass has a quality level that may be acceptable in other non-fiberglass segments of the glass industry.

The plasma melter's operational parameters must be tailored to accommodate the different compositions with their different melting characteristics. Each glass has its own requirements, and therefore the initial melting of a new glass composition must be used to determine the optimal process setup. These trials may require several iterations.

At the completion of the DOE project, Plasmelt aggressively began to market the plasma technology platform for implementation into specialized niches within the glass industry.

11.3.4 Japanese Consortium Project

Work, published in 2008, was conducted by a Japanese consortium [6] and was sponsored by the Strategic Development of Energy Conservation Technology Project of NEDO (New Energy and Industry Technology Development Organization, Japan). The project team performed research comparing three types of melting and the effects on the vitrification and decomposition of raw materials used in soda lime glass as well as an alkali-free glass composition.

The melting devices used for this research included a comparison of three types:

a. an RF plasma apparatus
b. a 12-phase AC arc
c. oxygen burner apparatus

Each of these devices allowed the injection of batch particles and mixtures down a hollow feeder tube with various carrier gases in order to provide some pre-melting and pre-decomposition. The powders were quenched and examined by various analytical techniques – SEM, XRD, TG, and ICP.

Using this methodology, the team investigated the relationship between the highest temperatures of RF plasmas, the intermediate temperatures of AC-arc plasmas, and the lowest temperatures of the oxygen burner. The degree of vitrification as determined by the formation of rounded spheres of glass, surface roughness, chemical analyses, the thermal decomposition curves, and x-ray diffraction was used to judge the completion of the decomposition and melting processes. The more complete the vitrification process, the lower the content of bubbles and the lower the quantity of incompletely reacted batch ingredients, which leads to a reduction in the formation of bubbles in molten glass, and effectively shortens the fining time.

Conclusions from the study showed that the RF plasmas led to the greatest degree of decomposition and vitrification, whereas the oxygen burner led to the least. Treatment by AC plasmas was intermediate. The study aimed to show the beneficial effects of releasing the decomposed gases from the raw material during powder injection. The resulting improvement of melting efficiency and the reduced use of fossil fuels were claimed to decrease the emissions of greenhouse gases.

11.4 DOE Research Project – 2003–2006

11.4.1 Acknowledgments

This work described in this chapter was conducted by Plasmelt Glass Technologies, LLC. The majority of the funding was provided by the US Department of Energy [7] with in-kind support provided by AGY Holding Corporation and Johns-Manville. The final report on this project is available online [7].

11.4.2 Experimental Setup of the Plasmelt Melting System

To demonstrate plasma melting using a remotely coupled transferred-arc plasma, the following system illustrated in Fig. 11.4a,b was designed and fabricated.

A water-cooled volumetric screw feeder conveys the batch into the reactor shell. The rotating reactor shell conveys the material from the entry point through the plasma arcs. The rotation speed and feed rate determine the batch thickness before presentation to the plasma arcs.

The anode torch is connected to the positive side of a DC power supply and the cathode is connected to the negative side of the supply. A computer-controlled positioning system moves the torches during melter operation. The torch position is extremely important for the melting process and final glass temperature. Often, one of the torches is placed near the exit orifice. The torch position is then used to control the orifice temperature and flow rate.

In the center of the rotating shell is a molybdenum orifice. The orifice size and configuration is changed based on glass characteristics. Extensive development resources have been used on this portion of the melter design to optimize flow rate, glass residence time, glass contamination while ensuring that the self-skulling features of the plasma melter are maintained.

The hood assembly is water-cooled in the Plasmelt melter. In addition to shielding the environment from the plasmas, the water-cooled hood holds the torches and positioning mechanism. The exhaust blower is controlled by a variable frequency drive. The exhaust rate controls the reactor temperature and pulls any condensate from the reactor. This condensate can be collected in a baghouse or other collection device as required by the specific site location.

Because the plasma melting process is very rapid, extremely tight control of all parameters is critical. The Plasmelt system uses a closed-loop PLC system for integrating all adjustments based on the measurable outputs. The measurable outputs are stream temperature and exhaust temperature.

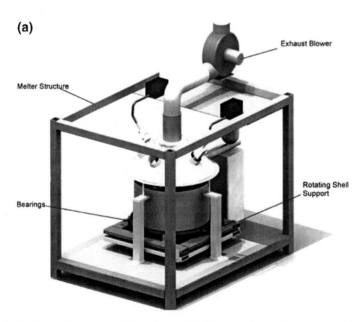

Fig. 11.4 (a) Plasmelt plasma melting system. (b) Cross section – plasmelt remotely coupled transferred-arc melter (protected by US Patent 5,548,611)

(b)

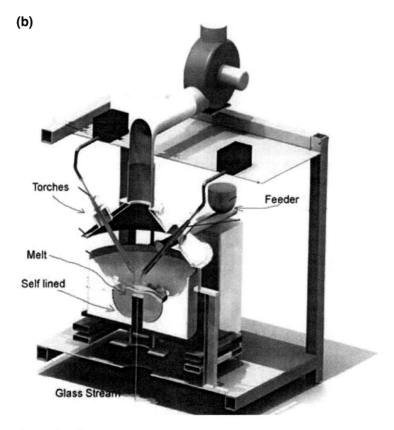

Fig. 11.4 (continued)

11.4.3 Technical Challenges of Plasma Glass Melting

Several technical challenges have prevented plasma melting technology from routinely being adopted for production of glassmaking operations. These include

- Torch life
- High cost of purge gases (argon, nitrogen, hydrogen)
- Melter instability and inconsistent glass quality
- Limited throughput of the plasma-based melting system
- Overall energy efficiency

Torch life and stability In the early work on plasma melting, torches were short-lived and unreliable. Torch lives are influenced by both design and operation. Proprietary designs have evolved by several companies – Westinghouse, Tetronics, Plasmelt, and PEC. Those proprietary designs that are deemed most successful were developed to yield high energy input, good articulation in the X–Y–Z planes, low maintenance costs, and the capability for on-the-fly rapid replacement. One such design has been developed by Plasmelt and has many advantages/features not heretofore seen in torch designs. A successful torch design is one in which it will operate for up to 1 week in a production operation without failure. Replacement times of 15 min and rebuild costs of less than $20 per torch are also deemed to be necessary criteria.

Purge gas costs The second most expensive direct cost – after electricity – is for the purge gases that are required by most plasma technologies. The inert gases provide a consistent electron path and are used to prevent oxidation of the orifice.

Plasmelt has demonstrated innovative torch designs as well as certain proprietary operational modifications, which can lower the volume (and cost) of purge gases required per hour.

Melter stability Modular plasma melters have low volume of glass/batch holdup as compared with more traditional melters. For example, it is common for a 1-m diameter plasma melter to contain glass at a depth of 0.5 m. Glass/batch holdup is approximately 150–200 lb. Assuming plug flow, at pull rates of 200 lb/h, complete turnover of glass in the melter can occur in 1 h. This feature can be used to advantage when the melter is being changed over from one composition to another. However, the quality/consistency of the input batch and the resultant melter operation must be very stable to yield an operation in which consistent temperature glass exits the melter. Extremely fast reactive control of all parameters is of paramount importance to the stability of the system. Consistent feed rates are best achieved with loss-in-weight feeders. Loss-in-weight feeders use load cells that adjust the volumetric feed rate based on actual weight delivered. Also, load cells can be installed on the melter shell to determine if weight is being gained or lost. This data is then used to adjust the torch positions and power levels to stabilize and control melter operation.

Melter throughput From proprietary research reports, it is known that some plasma operations have demonstrated 100–1200 lb/h in a 1-m plasma melter [8]. The broad range is in recognition that some compositions are more easily melted than others. Those compositions least susceptible to melt segregation are generally the most amenable to melting in a modular plasma melter. Those compositions that are prone to this melt segregation can yield two immiscible melts – one is silica rich, one is silica poor. Once silica defects, often appearing as cristobalite, are formed from this melt segregation problem, no amount of subsequent mixing and conditioning in the operation can correct this problem. The key to overcoming the tendency to melt segregate is to slow down the melting rates by controlling the input power, temperature distribution, and batch particle size distribution.

Energy efficiency Melter efficiency is largely influenced by the glass compositions melted and the melter operational setup. Literature citations for plasma melting show 5–6 million BTU/ton [9]. While these numbers are not as low as those cited for the most efficient electrical melters [10], they do exceed many of the efficiencies achieved by many of the traditional gas-fired melters.

11.4.4 Glasses Melted: Results and Broad Implications

11.4.4.1 Glass Melting Trials

During the 3-year Plasmelt research program, the majority of the melting process and plasma torch development work was conducted on E-glass similar in composition to the "Standard E-glass" column of Table 11.1. Extensive glass melting trials on other compositions have also been conducted in plasma melters by the other organizations listed above in Section 11.3.

Table 11.1 Chemistries of plasmelt-melted E-glasses vs standard E-glass

Oxides	Plasma-melted E-glass produced in Boulder, CO Lab					"Standard" E-glass
	Hour 1	Hour 2	Hour 3	Hour 4	Hour 5	
SiO_2	54.39	–	54.32	–	54.27	53.55
Fe_2O_3	0.279	–	0.279	–	0.278	0.26
FeO	0.133	–	0.133	–	0.127	0.09
TiO_2	0.57	–	0.57	–	0.57	0.58
Al_2O_3	15.06	–	15.01	–	15	14.78
Cr_2O_3	0.01		0.01	–	0.01	
CaO	23.05	–	22.91	–	23.07	22.35
SrO	0.167	–	0.166	–	0.165	
MgO	0.52	–	0.52	–	0.51	0.52
Na_2O	0.33	–	0.37	–	0.37	0.64
K_2O	0.03	–	0.04	–	0.03	
Fluorine	0.21	0.25	0.27	0.29	0.3	0.45
B_2O_3	5.42	5.56	5.57	5.66	5.67	6.94
MoO_3	0.0054	0.0079	0.0049	0.0043	0.0082	
CuO	0.004	0.005	0.005	0.004	0.004	
SO_3	< 0.01	–	< 0.01	–	< 0.01	

All values are expressed as wt%.

The E-glass melting work was done with commercially available glass batch ingredients (silica flour, limestone, clay, colemanite, and fluorspar) mixed in an AGY fiberglass production batch mixer. These batches were weighed, mixed, placed in drums, and truck-shipped to the Boulder Lab, where the melting trials were conducted.

To demonstrate the broader applicability of the plasma melting process, several other non-E-glass compositions were also melted during exploratory trials. These trials were conducted with commercially available raw materials that are commonly used by the fiberglass industry. All batch ingredients were weighed, then mixed in an Eirich Intensive Mixer, which was leased for these trials from the Eirich Machines, Inc. of Gurnee, Illinois. The complete list includes

- E-glass – An electrically resistant glass fiber composition used as a reinforcement for many hundreds of fiber-reinforced plastic applications and products (for the composition used, see Table 11.1, Column 7 entitled "Standard E-glass")

- S–R–T glasses – The S-glass family includes S-2, R-glass, and T-glass. All of these fiberglass compositions are commercially produced by various companies. All of these glasses belong to a compositional family that demonstrates higher tensile strength properties than traditional E-glass. S-2 was the only member of this family that was melted during the DOE work and contains approximately 65% SiO_2, 25% Al_2O_3, and 10% MgO.
- C-glass – A fiberglass composition known to have good chemical resistance to mineral acids (for composition, refer to Table 11.2)
- Low DK-glass – A composition which is known to have low dielectric constant and low dissipation factors that is useful to the electronics industry (composition used was published by a Japanese commercial supplier and contains ~22% B_2O_3)
- AR-glass – A fiberglass composition known to have good resistance to highly alkaline environments such as concrete reinforcement (for composition, refer to Table 11.2)
- Low flux E-glass without boron and fluorine – an E-type fiberglass composition that is being used in the fiberglass industry to lower emissions and batch costs (for composition, refer to Table 11.2)
- B-glass – A composition used for fine fiber that is known to possess good properties for micro-filtration products (specific composition used was proprietary, but generally contained ~13% alkali, ~6% alumina, ~11% boron oxide, ~5% barium oxide, ~4% zinc oxide, ~2% calcia with the balance made up of silica)
- Soda–lime–silica – a common glass type used in containers and window panes
- Scrap fiberglass – both continuous and discontinuous (such as insulation fibers)

Most of these glasses melted well and drained from the various plasma melters. The quality of the glasses ranged from high quality (on-target chemical composition, no visible batch stones, and minor-to-zero cord) to low quality (high levels of unmelted batch and/or melt segregation with evidence of abundant silica scum formation and cord). High defect levels are usually indicative that the melter parameters have not been optimized for the glass melting behaviors. From the extensive experience of the principal investigators of various plasma projects, it is known that a plasma melter's operational parameters must be tailored to accommodate the different compositions with their different melting characteristics. Each glass has unique plasma process requirements, and therefore each time a new composition is melted for the first time, the optimal process setup on the melter must be determined through a series of process definition trials. These trials may require several iterations.

Plasma melting, due to its high temperature environment, imposes its fingerprint on all glasses melted. Through the use of the inert gases to purge the plasma arcs, oxygen is excluded and many glasses show a slightly reduced chemical redox state as compared with gas-fired or electrical melters. Below is a table of E-glass chemistries melted by Plasmelt within 1 day melting run. To determine the stability of the plasma glass melting process, samples were collected each hour then analyzed chemically by a glass laboratory skilled in routine E-glass chemical analyses.

Table 11.2 Chemical analyses (performed by JM) of three plasma-melted glasses showing starting, finished, and calculated retention values

Oxide	C-glass			Low flux E-glass			AR-glass		
	Final glass chemistry	Starting glass chemistry	Calculated retentions (%)	Final glass chemistry	Starting glass chemistry	Calculated retentions (%)	Final glass chemistry	Starting glass chemistry	Calculated retentions (%)
SiO_2	66.5	66.39	100.20	58.7	59.49	98.70	60.8	59.37	102.40
Al_2O_3	6.52	6.44	101.30	13.6	13.23	102.80	0.23	0.19	123.00
Fe_2O_3	0.37	0.39	94.50	0.425	0.39	108.90	0.02	0.03	59.30
TiO_2	0.01			0.07			0.044	0.04	99.60
B_2O_3	0.01			0			0		
Na_2O	13.2	13.21	99.90	1.35	1.12	120.80	18.4	19.34	95.10
K_2O	0.009			0.036			0		
Li_2O	0			0.001			0		
CaO	9.26	9.41	98.40	22.6	22.66	99.70	0.07		
MgO	4.13	4.16	99.20	3.19	3.11	102.40	0.02		
SrO	0.004			0			0.001		
BaO	0.002			0.003			0.001		
ZnO	0.003			0.009			0.002		
SO_3	0.008			0			0.029		
ZrO_2	0.003			0.007			20.3	21.03	96.50
MoO_3	0.004						0.005		
CuO	0.0003			0.0012					

In addition, E-glass patties were examined microscopically to inspect for the presence of any crystalline defects. None were found. All patties contained numerous seeds.

Chemical analyses from Table 11.1 suggest

- Plasma-melted glasses were reasonably consistent from hour to hour.
- Plasma glass was slightly more reduced (as shown by the higher FeO values) than standard gas-fired glasses.
- Volatile constituents (boron, fluorine, alkali) had slightly lower retentions in the plasma-melted glasses.
- The copper and molybdenum content in the glass, contributed from wear of the copper electrodes and the molybdenum orifice, was low.

These chemical analyses suggest that plasma-melted E-glasses can have a consistent chemistry and quality over a several hour glass melting trial run.

Table 11.2 contains other chemical analyses run on other glass compositions melted by Plasmelt – C-glass, low flux E-glass, and AR-glass. The analyses were performed on glass patties collected during plasma melting runs done in the Boulder Lab. Each set of patties was inspected microscopically for the presence of stones and seeds. C-glass quality was high with no visible crystalline defects and very minor cordiness. The melting trials showed excellent meltability with no performance issues. Alkali retention was excellent. C-glass required no optimization of the melting process and demonstrated good compatibility with the plasma melting process.

Both the low flux E-glass and AR-glass were harder melting, more prone to melt segregation, and demonstrated more cordiness with occasional stones of incompletely reacted batch ingredients, often manifest as cristobalite. Both of these glasses are promising candidates for plasma melting but will require melting process optimization of the same sort that has been completed on E-glass before these glasses could be used in a production operation.

Of all glasses melted, low DK-glass was by far the most difficult composition. The very high (greater than 20%) boron level was prone to volatilization in the high-temperature plasma environment and promoted extreme cordiness in the glass patties. This glass has a reputation of being difficult to melt in any type of melter so plasma melting joins a long list of problem melting processes for this family of glasses.

11.4.5 Synthetic Minerals Processing Implications

Plasma melters are well suited for the melting and production of synthetic minerals. The major attributes of plasma melting that are sought out by these synthetic minerals producers include the high temperature capabilities; rapid turnover from light-off to shutdown; ability to shutdown the melter full of molten material; very short start-up time required (15–30 min); low capital costs; and high glass throughput capability of a melter with a small footprint.

Several investigators including Plasmelt Glass Technologies, Synsil, NASA, USGS, and Zybek Advanced Products have experimented with the plasma melting of minerals or simulated chemistries of synthetic minerals, including

  Wollastonite – Calcium silicate
  Albite feldspar – Sodium aluminum silicate
  Anorthite – Calcium aluminum silicate
  Kaolinite – Aluminosilicate
  Zircon – Zirconium silicate
  Enstatite – Calcium magnesium silicate
  Cordierite – Magnesium aluminum silicate
  Spinel – Calcium aluminum silicate
  Pigeonite – Ferromagnesian silicate
  Augite – Calcium ferromagnesian silicate
  Fayalite – Iron silicate

Each of the batches for these minerals was mixed from commercially available raw material oxides and brought to molten temperatures. Cooling rates and collection methodologies determined the final form of the material (i.e., glass or crystal).

For example, calcium silicate melts were made to produce a final product of a wollastonite composition – 48.3% CaO and 51.7% SiO_2. The input batch was calculated using commercially available silica flour and limestone mixed in the proper proportion to yield the final wollastonite composition. The mixed batch was charged with the volumetric feeder and melted in the plasma melter. A dynamic molten stream was then established to produce wollastonite "cullet" (i.e., quenched glass). Various controlled cooling rates were able to produce varying degrees of crystallinity from 100% (or less) in the final wollastonite product. This plasma melting method of batching with oxides and producing naturally occurring minerals is termed "synthetic minerals processing."

11.4.6 Energy Efficiency vs. Throughput

11.4.6.1 Energy Efficiency

The energy efficiency of plasma melting is directly related to the throughput. Plasma melting has a significant energy overhead. Figure 11.5 illustrates the measured energy efficiency for actual runs made with E-glass at Plasmelt Glass Technologies during a development program. Energy consumption is commonly quoted within the US glass industry as million BTU's per ton of glass melted, which is designated in Figure 11.5 as "mm btu/ton."

As the graph illustrates, the plasma melting process may be able to reach 5 mm btu/ton with increased production throughput 450–550 lbs/hr glass throughput

mm btu/ton

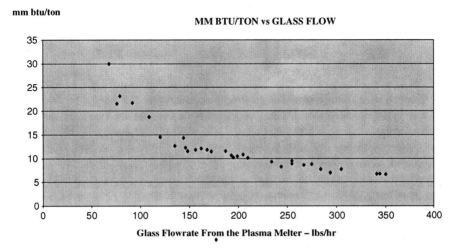

Fig. 11.5 Energy consumption vs. Glass throughput

range. Most of the glass melting runs made during this DOE program were made at the lower throughput range (75–350 lbs/h) to emphasize glass quality and process stability.

The glass composition also plays a large role in the efficiency of the process. The majority of the work for this program was directed toward E-glass. Some glasses have proven to be harder to melt than E-glass, whereas others are much easier.

11.4.6.2 Energy Balance

Based on the work presented in the previous section, it is known that the increased power efficiencies will be achieved at higher throughputs. The increased efficiencies at the higher throughputs are primarily due to the ohmic heat transfer mechanism that is realized by the amperage and torch position. It is only possible to operate in these positions at the higher feed rates.

The highest efficiency datapoint in Fig. 11.5 was at 350 lbs/hr. The average operational conditions at this level were

- 415 V
- 712 A
- 295 kW

To calculate the energy used per pound of glass:

$$\text{Power} = \text{Volts} \times \text{Amps} = 415\ \text{V} \times 712\ \text{A} = 295\ \text{kW}$$

$$\text{kW/Pound of Glass} = \frac{295\ \text{kW}}{300\ \text{lbs/h}} = 0.98\,\frac{\text{kW}}{\text{lbs/h}}$$

$$1 \text{ kW} = 3412.4 \text{ BTU/h}$$

$$0.98 \frac{\text{kW}}{\text{lbs/h}} \times 3412.1 \frac{\text{BTU}}{\text{h}} = 3343.9 \frac{\text{BTU}}{\text{lbs}} \Rightarrow 6.69 \text{ MBTU/ton}$$

Compared to standard melting means, this efficiency is still very poor. The losses of energy at various points in the process include the torches, hood, exhaust, and miscellaneous losses, which total about 78% of the input energy. This extremely poor energy balance is due to the glass throughput still being too low. Since these losses are more or less constant, a significantly higher glass throughput will not cause significantly more losses, and therefore more energy efficiency is delivered to the glass itself. Table 11.3 below summarizes these losses.

Table 11.3 Energy losses at various points in plasma melting

DC power	300 Flow rate	kW	Delta T	Energy (kw)	(%)
Torch/power supply water loss	55	GPM	4	30.7	10
Hood water losses	10	GPM	50	69.7	23
Exhaust losses	1,200	ACFM	175	62.6	21
Misc losses				71.0	24
Glass energy	350	PPH	2600	65.8	22
				299.8	

11.5 Future Applications for Plasma Melting

Best-Fit Industrial Applications for future commercial plasma melting applications include

- Fiberglass (both continuous and insulation)
- Specialty glasses, e.g., S–R–T glasses, frit glasses
- New, test market products – New glass-product development projects often require less-than production volumes.
- Minerals melting – Certain specialty minerals cannot be melted with conventional glass furnace processes due to their lower temperature restrictions. Plasma-fired skull melters have no such limit.
- Low volume operations – Some routine glass production operations only require a few hundred pounds per hour per glass product.
- Unpredictable production operational schedules – Glass products having erratic production runs cannot be made in large furnaces that require several days of heatup or glass changeover.
- Multiple glass compositions in one factory – Factories sometimes require multiple glasses be campaigned back to back. Scrap generated during these conversions can be unusable.

- Melter boost for commercial melters – Using a plasma torch system as a booster on commercial glass furnaces has never been demonstrated, but remains a promising field for future research.
- Scrap re-melt – The high temperature of plasmas can burn off organics typically found on some scrap glass re-melt operations.
- Low capital cost operations – The capital cost of a plasma melter is less than a conventional melter, making it attractive for installation in low capital facilities.

11.6 Summary and Conclusions

Plasma melting of glass is a technology that has been investigated by several organizations over a period of 25 years. During that time, many innovations have been developed that has moved the technology closer to the reliability that is required for a production operation. Each investigator has concluded that plasma melting technology does fulfill a niche within the overall glassmaking requirements. The very high power density in combination with the rapid start-up/shutdown and the rapid turnover rates are features not routinely enjoyed (or needed) in more traditional operations. Very high temperature capability – up to 1750°C – makes this technology the natural choice for materials operations that require such high temperatures.

References

1. Beerkens, R. Prof., Modular melting, Am. Ceram. Soc. Bull., 83 (4), 28–32, April, 2004.
2. Gonterman, J. R. and Weinstein, M. A., Plasma high-intensity glassmelting, Am. Ceram. Soc. Bull., 83 (10), 5–7, October, 2004.
3. Williams, J. K., et al., Method of melting materials and apparatus thereof, US Patent No. 5,028,248 dated July 2, 1991.
4. Cusick, M. J., et al., Method for the melting, combustion or incineration of materials and apparatus thereof, US Patent No. 5,548,611 dated August 20, 1996.
5. Dalton, D. A., Plasma and electrical systems in glass manufacturing, IEE Colloquium [Digest], 229, pp. 3/1–3/2 (1994).
6. Yaochun Yao, et al., An innovative energy-saving in-flight melting technology and its application to glass production, Sci. Technol. Adv. Mater., 9, 1–8 (2008).
7. Gonterman, J. R. and Weinstein, M. A., High Intensity Plasma Glass Melter Project Final DOE Report, 10-27-2006. Available at www1.eere.energy.gov/industry/glass/pdfs/894643_plasmelt.pdf
8. Johns-Manville Internal Reports – Proprietary.
9. Gonterman, J. R. and Weinstein, M. A., "Glass-on-Demand," an invited paper presented by Gonterman at the Advances in the Fusion of Glass Conference, Dresden, Germany in June, 2006.
10. Ross, P., and Tincher, G.L., Glass melting technology: A technical and economic assessment, GMIC, October, 2004.

Index

Note: The letters '*f*' and '*t*' following locators denote figures and tables respectively.

A

AC, *see* Alternating current (AC)
Acid-leached E- and A-glass fabrics, 21–22
Acid leaching, 15, 19, 22, 45*t*, 48, 58
Acid-resistant glass fibers, 46*t*, 48
Aerospace–rotors
    application, 218
    critical fitness for use properties, 218–219
    market trends and future needs, 219
A-glass, 21–22, 24, 45–46, 48, 92, 117–121,
    199*t*, 201*t*, 269, 289, 336–337
A-glass-reinforced composites, 119
A-glass variants, 120
AGY Holding Corporation, 440
Alkali-resistant glass fibers, 45–48
    examples
      ArcoteX™, 45
      CemFil™, 45
    reinforcement of cement composites, use,
      46–47
Alternating current (AC), 180, 433
Alumina, Al_2O_3, 310–313
Alumina–borosilicate glass, *see* E-glass fibers
Aluminate glass fibers, 11, 13, 45, 60–77
Aluminate glass fibers from fragile melts,
    60–66
    downdrawn from supercooled melts
      single/bicomponent fluoride fibers, 61
      single/double crucible process, 60–61
    updrawn from supercooled melts, 63*f*
      aluminate glass fibers, 62
      hybrid fiber-forming processes, 65–66
      quaternary calcium aluminate fibers,
        64–65
      tellurite glass fibers, 62
Aluminate glass fibers from inviscid melts,
    66–77
    fiber formation from inviscid jets, 68
      CLH process, 68

IMS process, 68
    RJS process, 68
    jet formation from inviscid melts, 66–68
      straight fiber and frozen Rayleigh
        waves, 67*f*
Aluminosilicate glass fibers, 14*t*, 99–115, 198,
    320, 328
Amber chromophore, 288, 326, 424–427
    concentration/oxygen partial pressure at
      different temperatures, 426*f*
    formation and stability of, 427
    temperature dependence of inten-
      sity/absorption coefficient, 426*f*
Amber glass, 231, 246–247, 249, 288, 310,
    326, 425–427
Amber glass melting, 425–427
    amber coloring of glass melts, cause, 425
    final glass product, dependent factors, 425
    redox influence in, 426–427
    See also Amber chromophore
Amorphous alumina *vs.* single-crystal sapphire
    fibers, 82–83
Amorphous resins
    PC, 166
    PPO, 166
    PSU/PESU, 166
Amorphous YAG *vs.* single-crystal YAG
    fibers, 83
Annealing point, 23, 201*t*, 206, 279–280
Antistatic agents, 130
Aramids, 31, 38, 40, 126, 167, 176, 194, 205,
    211–215, 217
    aromatic polyamide, derived from, 212
    drawbacks, 213
    families of
      meta-aramid (*m*-aramid), 212
      para-aramid (*p*-aramid), 212
      polypara-phenylene terephthalamide,
        chemical name, 212

F.T. Wallenberger, P.A. Bingham (eds.), *Fiberglass and Glass Technology*,
DOI 10.1007/978-1-4419-0736-3, © Springer Science+Business Media, LLC 2010

ArcoteX™, 45, 47
Armor, 38, 214, 216–218
ASTM E-glass standard
    general reinforcement applications, 93
    PCB applications, 93
    B₂O₃ and fluorine fluxes, use in, 99
    in US, role, 93
Atmospheric emissions, limits, 271–272
Automotive–belts, hoses, and mufflers
    application, 220
    Chevrolet Corvette body, 220
    critical fitness for use properties, 220–221
    market trends/future needs, 221

B

Babcock's model, *see* Liquidus models
Backscattered electron images (BEI), 58–59
Ballistics, 36, 38, 208, 216–217
Baria, BaO, 323
Basalt glass, 35, 37, 45–48, 206–210, 233
Basalt Fiber & Composite Materials
        Technology Development
        (BFCMTD), 210
BAT, *see* Best available technologies (BAT)
Batch-free times, 308
Batch materials, consolidation of, 300–302
Batch melting, stages/levels of
    level of meso-kinetics, 404
    level of micro-kinetics, 404
    local thermochemical reactions, 404
    overall mass/heat/power/entropy balance of
        furnace, 404
    unified classification
        closed-pore stage; reaction foam stage,
            405
        open-pore stage; warming-up stage, 405
        volume void filling, 405
Batch-related fluctuations in glass melting,
        415–416
    microwave and neutron absorption
        techniques, detection by, 416
Batch-to melt conversion, 386, 400, 404–409
BEI, *see* Backscattered electron images (BEI)
Beryllia (BeO), 39, 40, 203, 204
Best available technologies (BAT), 271,
        274–275, 277–278, 344
BFCMTD, *see* Basalt Fiber & Composite
        Materials Technology Development
        (BFCMTD)
BFS, *see* Blast furnace slag (BFS)
Bicomponent silicate glass fibers
    hollow porous sheath/core, 58
    hollow sheath/core

aircraft design and construction, use
        in, 57
    S-glass fibers *vs.* E-glass fibers, 57
    sheath/core and side-by-side, 56
    solid side-by-side, 58–59
    BEI of bicomponent glass fiber, 59*f*
    SEI profiles of calcium and magnesium,
        59*f*
Bingham, P.A., 232, 295, 307, 309, 316,
        337, 339
Birefringence, 14
Blast furnace slag (BFS), 234, 236, 298,
        308–309, 318–319, 333
BMC, *see* Bulk Molding Compounds (BMC)
Bone bioactive glass fibers, 53
Boric oxide, B₂O₃, 316–318
Boron carriers, role in glass melts, 397–399
Boron- and fluorine-free E-glass fibers, 29–30
Borosilicate glass, 6, 12–16, 21, 23, 25–26,
        28–33, 45*t*, 48–49, 57–58, 60,
        92–93, 95, 99, 109, 111–113, 120,
        127, 185, 188, 190, 201, 250, 253,
        255–256, 262, 276–277, 294, 297,
        318, 374*f*
Borosilicate E-glass fibers
    boron- and fluorine-free E-glass fibers, 30
    commercial, 29–30
    with energy-friendly compositions, design
        criteria, 31
    See also Energy-friendly glass fibers,
        design of
    industrial specifications, 29
    ASTM E-glass specification, 29*t*
British Department of Trade and Industry, 438
British Glass Institute
    project, objective of, 438
    results of study, 439
BS standards, UK, 93
Bulk molding compounds (BMC), 139*t*,
        141–142, 148–149, 156*t*
    batch/continuous process, 148
Buried passives technology, 193
"Bushing," 6, 9–12, 15, 23–28, 41–42, 48–49,
        54, 57–58, 66, 68, 82, 95, 98, 100,
        103, 117, 128–129, 203, 205, 439

C

Cahn's mechanism, 359
    creation of intrinsic defects, 359
Calcia, CaO, 307–309
Calcium aluminate glass fiber, quaternary, 62,
        64–65, 73
    potential applications, 65

spectral transmission of, 65*f*
structural and optical fiber properties, 64
updrawn from fragile melts, 64*t*
Carbon fibers, 17, 36–38, 40, 43, 53, 65, 83, 169, 210, 212–213
  characteristics/properties, 212
  PAN-based carbon fibers, 212
  pitch-based fibers, 212
  process of manufacture, 212
The cell model, 386
CemFil™, 45–47
Centrifugal molding, 142, 144–145
C-glass variant, 120
Chalcogenide glass fibers, 66
Chemical durability of glass
  definition, 297
  modeling of SLS glass, 298
  SiO_2 substitution on hydrolytic durability of SLS glass, effects of, 298*f*
Chemical durability, soda-lime-silica glasses, 231, 297–299, 308, 313–314
Chinese C- (or CC-) glass, 24, 26*t*, 34, 46*t*, 48, 60, 117–118, 120–121
Chlorides and fluorides, 322–323
Chopped strand mat (CSM), 34, 138–139, 143, 145, 150*t*, 169
Chopped strands, 33, 139–140, 148, 158, 198
  DMC reinforcement, use in, 140
Classical Newtonian dynamics, 366
CLH process, *see* Containerless laser-heated (CLH) fiber-forming process
Closed-pore stage, 405
Coefficient of thermal expansion (CTE), 16, 127*t*, 182–183, 193–194, 257, 324
Colored glasses, 261, 288–289
Combustion-related fluctuations, glass melting
  flue gas composition, importance, 416
  gas solubilities/their partial pressures, correlation, 416
Commercial borosilicate E-glass fibers, 29
Commercial E-glass products and applications, 33–34
Commercial/experimental glass fibers
  aluminate glass fibers
    from fragile melts, *see* Aluminate glass fibers from fragile melts
    from inviscid melts, *see* Aluminate glass fibers from inviscid melts
  glass fiber formation, principles of
    fiber-forming processes, generic, 9–10
    fibers from fragile/inviscid melts, 4*t*, 11
    fibers from strong melts/solutions, 10–11

glass melt formation, principles of
    fragile viscous melts, behavior of, 8–9
    glass melt properties, 4–8
    inviscid glass melts, behavior of, 9
    strong viscous melts, behavior of, 8
  silica fibers, sliver and fabrics
    pure, *see* Silica, sliver and fabrics, pure
    tensile strength of high/ultrahigh-temperature glass, 22*t*
    ultrapure, *see* Silica fibers, ultrapure
  silicate glass fibers
    general-purpose, *see* Silicate glass fibers, general-purpose
    non-round, bicomponent and hollow silicate fibers, 54–59
    special-purpose, *see* Silicate glass fibers, special-purpose
    from strong viscous melts, *see* Silicate glass fibers from strong viscous melts
  single-crystal alumina fibers
    alumina and aluminate fibers, future of, 82–83
    from inviscid melts, 77–82
  structure of melts and fibers
    fiber structure *vs.* modulus, 12–14
    fiber structure *vs.* strength, 14–15
    from glass melts to fibers, 11–12
    melt structure *vs.* liquidus, 12
Composite design and engineering
  composite mechanical properties
    bidirectional (orthotropic) reinforcement, 133–134
    levels of study, 131
    short fibers, 134–137
    test methods, 137–138
    unidirectional continuous fibers, 131–132
  composites for wind turbines
    blade design methodologies, 170–172
    blade-manufacturing techniques, 169–170
    raw materials, 169
  continuous fibers for reinforcement, 125–126
  E-glass fibers, 127
  fiberglass manufacturing, 128–129
  fiberglass size, 129–130
  products
    chopped strands, 140
    fabrics woven from rovings, 141
    glass mats, 138–140
    glass yarn, 141

Composite design and engineering (*cont.*)
   milled fibers, 140
   non-woven fabrics, 141
   rovings, 140
   reinforced thermoplastic materials, *see*
      Reinforced thermoplastic materials
   thermoset composite material
      applications, 142*f*
      fabrication process, parameters, 142
      fillers, 154–155
      liquid resin processing techniques, *see*
         Techniques, liquid resin processing
      release agents, 155–156
      thermosetting matrix resins, *see* Resins,
         thermosetting
Composites, 17, 22, 31, 34, 37–38, 40, 43–44,
      47–49, 51, 54–55, 66, 78, 82, 114,
      117, 119, 125–126, 129–131, 133,
      138–142, 146, 152–154, 156, 164,
      168, 177
Composites for wind turbines
   blade design methodologies, 170–172
      ASTM D3479/D3039, test standard
         used, 171
      blade design, example of, 170*f*
      fatigue mechanism, 171
      fatigue test data on epoxy matrix/glass
         fabric specimen, 172*f*
      log–log model, 171
      S–N regression parameter estimates,
         172*t*
   blade-manufacturing techniques,
      169–170
      blade components, design, 170
      RTM, 170
      VARTM, 170
   composite technology, advantages/benefits,
      168–169
   raw materials, 169
   wind energy park, 168*f*
   wind, renewable energy source, 168
Compositional design principles, 91–99
Compositional reformulation for reduced
      energy use and cost, 92, 98–99,
      105–107, 112, 117, 119
Compounding process/compound, defined,
      125–126
Compressive strength, 138, 150*t*, 213–215,
      217–218
Conductive DC-arc plasma(s), 433
Configurational entropy $S_c(T)$, 394
Conradt, R., 385

Container glass, 48, 117, 119, 230–232,
      234–237, 245–251, 254, 258,
      261–262, 273, 277, 280, 282–284,
      288–289, 291, 295, 297–298,
      303–304, 307, 309, 311, 315–316,
      318–319, 321, 325, 327–328, 334*t*,
      337, 342–343, 413, 415, 419, 423
   green container glass, development of, 248*f*
   white container glass, development of, 247*f*
Containerless laser-heated (CLH) fiber-forming
      process
   mullite composition glass fibers, 70
   process concept, 68–69
   YAG glasses and glass fibers, 69–70
Continuous filament-forming process, 128*f*
Continuous glass fibers, 9, 56, 66, 144–145,
      160, 202, 215, 439
Continuous laminating
   continuous liquid resin processing
      technique, 145–146
   non-continuous liquid resin processing
      techniques, 145*t*
Continuous updrawing process, 63*f*
Corrosion, refractory
   corrosion rate as function of Arrhenius
      function, 293
   corrosion tests, importance, 293
   downward drilling, 293
   by molten glass, key factors, 293
   ZrO_2/zircon refractories, 294*t*
      corrosion loss as function of
         temperature and glass composition,
         295*f*
   Noyes–Nernst equation, 294
Coupling agents, 130, 155
Cracking, 47, 134–136, 171
Crimp effect, 141
Crystalline reference system (c.r.s.), 386,
      389*t*, 408
Crystal, 7, 11, 13–14, 65–66, 68, 70, 77–83,
      98, 119, 229, 230, 233, 235, 237,
      242, 251, 259–261, 281–286, 313,
      317, 320, 323, 367*f*, 368, 370, 377,
      388, 448
Crystallization/devitrification, 7, 12
   EDG process, 14
CSM, *see* Chopped strand mat (CSM)
CTE, *see* Coefficient of thermal expansion
      (CTE)
Cullet, 302–304
   effect, 236–237
   foreign cullet, 236
   in-house cullet, 236, 256

optimization, tool for minimizing
    SEC, 332
recycling, ecological advantages,
    236, 415
Curing agents, categories
    catalytic, 150
    coreactive, 150

D

Davy process, 62, 64
DC, *see* Direct current (DC)
De-bonding, 135, 136
Debye model, 373
Debye temperature, 361, 365, 370, 373
Defense–hard composite armor
    application, 216–217
    critical fitness for use properties, 217
    iron triangle for armor systems, 217*f*
    market trends and needs, 218
Deformation ratio, 55
Delta temperature, 4, 8–9, 12, 24, 30*t*, 36,
    49, 68, 98, 100–109, 111–114,
    116–117, 119–121
Density, 231, 299
Design of energy-friendly glass fibers, 91
    environmental regulations and emission
        control, 92
    industry standards and specifications,
        92–93
        ASTM E-glass standard D-570–00, 93*t*
Design requirements, soda-lime-silica glass,
    268–269
Devitrification, 7, 12, 23, 61, 233, 265,
    281–286, 309, 327–328, 339, 341,
    433, 439
D-glass, 49–50, 60, 92, 186*t*, 188–191,
    199–201
    compositional improvements,
        challenges/problems
        boron volatility, 188
        glass melting, 188
        hollow filaments, 189
        homogenization, 188
        limited manufacturing options, 189
        PCB-related difficulties, 189
        poorer forming behavior, 189
    limitations, 190–191
Dielectric constant (Dk), 3, 16–17, 33–34,
    49–53, 57, 112, 179–180, 186–189,
    369, 445
    performance of PCB, vital factor, 180
Dielectric dissipation factor, 181
Dielectric loss (Df), 49, 180–181, 190

DIN standards, Germany, 93
Diode lasers, mid-infrared, 423
Direct current (DC), 26–27, 180, 433–434,
    441, 450*t*
DMC, *see* Dough molding compound (DMC)
DOE, *see* US Department of Energy (DOE)
DOE research project (2003–2006), 440–450
    energy efficiency *vs.* throughput
        energy balance, 449–450
        energy efficiency, 448
    glasses melting, results/implications, 444
        trials, 444
    plasma glass melting, technical challenges
        of, 442–443
    plasmelt melting system, experimental
        setup of, 440–441
        energy efficiency, 443
        melter stability, 443
        melter throughput, 443
        purge gas costs, 443
        torch life and stability, 442–443
    synthetic minerals processing implications,
        447–448
        'cullet' (quenched glass), 448
        synthetic minerals processing(batching
            with oxides), 448
Dolomite, 29, 156*t*, 235, 239*t*, 245–246,
    251, 262–265, 300, 304, 307–310,
    334–336, 339–340, 342, 399*t*,
    400–408, 432
Double-crucible melts spinning process, 61*f*
Dough molding compound (DMC), 139*t*, 140
    pressure molding application, 140
Downward drilling, 293
Dry spinning process, 9, 11, 15, 18–22
    acrylic fibers, fabrication of, 19

E

Economics, 230, 341–343
ECR-glass, 24, 26*t*, 45, 60, 92, 99, 114–116,
    120–121
ECR-glass variants, energy- and eco friendly,
    114–116
    corrosion-resistant ECR-glass, commercial,
        114
EDG, *see* Edge-defined film-fed growth (EFG)
    process
Edge-defined film-fed growth (EFG) process,
    14, 77–79
    process versatility, 78
    sapphire fibers, growth of, 78
EFG process, *see* Edge-defined film-fed growth
    (EFG) process

E-glass fibers, 6–8, 12–14, 16, 19, 20–34,
        36–38, 43, 45–52, 54, 57–58,
        60, 64, 72–73, 75–76, 92–93, 95,
        99–114, 116t, 119, 127, 183–188,
        190–191, 198, 201–203, 205–206,
        208–209, 211t, 213, 215, 220,
        222, 276–277, 294–295, 316, 318,
        387–388, 391–392, 394–395, 397,
        406–407, 437, 444–449
    AR coating, benefits, 47–48
    ASTM E-glass standard D-570-00, 93t
        general reinforcement applications, 93
        PCB applications, 93, 127
    composition, 185
    definition, 127
    products and applications, 33–34
    properties and fiber structures
        electrical bulk properties, 33t
        mechanical properties, 31
        physical properties, 32–33, 127t
    textile industry, categories, 185
    See also Borosilicate E-glass fibers
E-glass-reinforced composites, 119
E-glass variants with 2–10% B_2O_3,
        energy-friendly
    effect of boron at equal delta temperatures,
        113
    quaternary SiO_2-Al_2O_3-CaO-B_2O_3 phase
        diagram, 111
    trend line design of, 111–113
Einstein coefficient (k_E), 75
Eirich Intensive Mixer, 444
Elastic modulus (E), 70, 178, 181–182, 186,
        203, 206, 208, 213–214, 218
    alumina, effect on, 203–204
    BeO additions, effect on, 203
    methods for improving glass fiber modulus,
        203t
    M-glass, high-modulus fiberglass, 204
    rare earth oxides, effect on, 204
    specific modulus (E_{sp}), equation, 204
Electrical properties, 289–291
Electrochemical oxygen sensors, 419
Electronics, 38, 83, 175, 180, 191–192,
        194, 445
Electrostatic precipitators (EPs), 275, 329
Elongation, 31, 35–36, 126–127, 132, 151,
        161t, 201t, 201f, 203, 208–209,
        212–213, 296
Emission control systems, 92
Emission spectroscopy, 421–422
Energy consumption vs. glass throughput, 449f
Energy efficiency vs. throughput, 448–450

energy balance, 449–450
energy efficiency, 448–449
Energy-friendly glass fibers, design of
    aluminosilicate glass fibers
        ECR-glass variants, see ECR-glass
            variants, energy- and eco friendly
        E-glass variants with <2% B2O3,
            99–111
        E-glass variants with 2–10% B_2O_3,
            111–113
    compositional reformulation for reduced
        energy use and cost, 92, 98–99,
        105–107, 112, 117, 119
    designing new compositions, principles,
        91–99
        compositional, energy, and
            environmental issues, see
            Design issues of energy-friendly
            glass fibers
        trend line design, see Fiberglass (new)
            compositions, trend line design of
    design requirements
        commercial glass compositions, 269
        environmental legislation, compliance
            with, 268
        standardization of glass compositions,
            criteria, 268
    environmental issues
        atmospheric emission limits, 269, 272t
        pollution prevention and control,
            see Pollution prevention/control,
            eco-friendly glasses
        SEC, 269–271
    fundamental glass properties
        chemical durability, 297–299
        conductivity and heat transfer, 286–291
        density and thermo-mechanical
            properties, 299
        devitrification and crystal growth,
            281–286
        interfaces, surfaces, and gases, 291–296
        viscosity–temperature relationship,
            279–281
    glass reformulation methodologies
        benefits and pitfalls, 341–343
        research requirements, 343–344
        worked examples and implementation,
            330–341
    SLS glasses, design of
        alumina, Al_2O_3, 309–313
        baria, BaO, 323
        batch processing, preheating, and
            melting, 300–302

boric oxide, B_2O_3 316–318
calcia, CaO, 307–309
chlorides and fluorides, 322–323
cullet, 302–304
economics of batch selection, 300f
energy-saving technologies, 270
lithia, Li_2O, 315–316
magnesia, MgO, 309
multivalent constituents, 324–327
nitrates, 329–330
potassia, K_2O, 313–315
recycled filter dust, 329
silica, SiO_2 304–305
soda, Na_2O, 305–307
strontia, SrO, 324
sulfate, SO_3 318–321
water, H_2O, 321–322
zinc oxide, ZnO, 323–324
soda–lime–silica (S–L–S) glass fibers
A- and C-glass compositions, 117–119
Energy losses in plasma melting, 450t
Enthalpy function, 357, 359–365, 377
Enthalpy of fusion (H^{fus}), 386
Enthalpy of vitrification (H^{vit}), 386
Entropy and viscosity, glasses/glass melts
Adam–Gibbs plot for reference E-glass, 395f
configurational entropy $S_c(T)$, 394
Entropy of fusion (S^{fus}), 386
Entropy of vitrification (S^{vit}), 386
Epoxy (EP) resins, 144, 146, 148, 150–151, 169, 182, 194
characteristics of, 151
hardener, impact on cured resin, 151
EP resins, see Epoxy (EP) resins
EPs, see Electrostatic precipitators (EPs)

F
Fabrics, 15–23, 33–34, 44, 48, 55f, 57, 61, 141, 143–146, 150–153, 169, 170–172, 176–177, 181–182, 184, 188, 190, 192–193, 210, 217–218, 275, 277–278, 329
Fabrics woven from rovings, 141
Faraday constant (F), 419
Fatigue, 57, 140t, 151, 164, 169, 171–172, 218–220
Fiber cross section technology, 54
Fiber-forming melts, crystallization rates of, 8f
Fiber-forming processes
generic, 10f
dry spinning, 9, 11
melt spinning, 9

Fiber-forming temperature (log 3 FT)
in design of new fiberglass compositions, 95–98
MgO and TiO_2, impact on, 103
Fiber-forming viscosity (FV), 4, 6, 18, 52, 68
Fiberglass, 4, 6, 8, 12, 15, 24f, 26–31, 33–34, 44, 48–49, 77, 91, 92, 94–99, 120t, 126, 128–130, 133–134, 139–141, 151–152, 158–159, 161, 163, 167–170, 172t, 175, 176–179, 180–186, 188–189, 191–194, 198, 204, 207, 209–210, 212–213, 215–222, 279, 284, 304, 336, 437, 439, 444–445, 450
manufacturing
continuous filament-forming process, 128f
size, 129–130
functional groupings in, 130f
provides lubrication, 129
thermal degradation prevention, by additives, 130
Fiberglass (new) compositions, trend line design of
commercial and experimental, 120t
compositions, energy use, and emissions
addition/removal of flux, 99
compositional reformulation, 99
glass databases and compositional models, 94
melt properties required
delta temperature, 98
fiber-forming and liquidus temperatures, 95–98
melt viscosity/melt temperature, relationship, 96f
ternary SiO_2-Al_2O_3-CaO system, phase diagram of, 97f
principles/aim, 94–95
Fiberglass yarn, fabric, and laminate boards
process steps of manufacture, 176f
Fiberization, 27, 128, 188, 395, 438–439
Fiber pullout, 135
Fiber-reinforced composites (FRC), 40, 131, 142, 154, 164, 171, 210, 218–219, 289
comparison of properties of variety of high-performance fibers, 211t
Fiber rupture, 135
ruptured specimen, 136f
Fibers/melts, structure of
fiber structure vs. modulus, 12–14
effect of alumina on, 13t

Fibers/melts, structure of (*cont.*)
   fiber structure *vs.* strength, 14–15
     effect of composition on tensile
      strength, 14*t*
     spin orientation, effects, 14
     surface flaws/non-uniformities, effects,
      14
   from glass melts to fibers, 11–12
   melt structure *vs.* liquidus, 12
     energetic/environmentally friendly
      fiberglass, design of, 12
Fiber structure
   vs. modulus, 12–14
   vs. strength, 14–15
Fiber weave effect (FWE), 184
Fictive temperature (T_f), 202
Filament-forming process, continuous, 128*f*
Filament winding, 140, 142, 144, 146
Fillers, 140, 147, 154–155, 156*t*, 183–184
   characteristics of, 156*t*
   final composite, characteristics, 155
   ideal filler, criteria, 154
   inorganic/organic, 154
   magnesium oxide, use in SMC, 155
   matrix resin/filler, reqirements for bonding,
     155
     organo-functional silanes, used
      in, 155
Flame visualization, techniques, 423
Flammability, 150–153, 166, 178–179,
   215, 217
Flat glass, 230–231, 236, 242–245, 250,
   262, 273, 278, 284, 298, 303,
   310, 316–319, 321, 325, 327–328,
   336–337, 413
Flexural modulus, 131, 138, 150*t*
Fluoride fibers, single/bicomponent, 61
Fluorine and boron-free A-glass, 117
Fluorine-free C-glass, 117–119
Fluorophosphate glass fibers, 66
Flux
   definition, 99
   examples, 99, 114
Fragile viscous melts
   behavior of, 8–9
   contents, 3
FRC, *see* Fiber-reinforced composites (FRC)
Fulcher curve, 4, 6*f*
   of borosilicate and boron-free E-glass
     melts, 6*f*
Furnace periscope, 423
FV, *see* Fiber-forming viscosity (FV)
FWE, *see* Fiber weave effect (FWE)

G
Gel coat, curing liquid, 143
   purpose of use in molding, 143
Glass
   composition, importance, 127
   low CTE component, 183
   viscosity–temperature relationship, 127
Glass compositional families
   D-glass, compositional improvements,
     188–191
     challenges/problems, 188–189
     limitations, 190–191
   E-glass, improvements, 184–188
     boron-free E-glass, 186
     challenges/limitations, 187–188
     E-glass specification as per ASTM
      D-578–00, 185*t*
     improving dielectric properties,
      186–187
     The Industry Standard, 185–186
     textile industry, categories, 185
Glass databases and compositional models, 94
Glass (energy-friendly) fibers, design
   requirements
   commercial glass compositions, 269
     SLS glass compositions, 269*t*
   environmental legislation, compliance
     with, 268
   standardization of glass compositions,
     criteria, 268
Glass fiber formation, principles of
   fiber-forming processes, generic, 9–10
   fibers from fragile/inviscid melts, 11
     single-crystal sapphire/YAG, examples,
      11
   fibers from strong melts/solutions, 10–11
     silica glass fibers, example, 11
Glass fibers
   with bone bioactive oxide compositions,
     53–54
   with high chemical stability, 46*t*
     acid-resistant glass fibers, 48
     alkali-resistant glass fibers, 45–48
     chemical resistance of glass fibers,
      44–45
   with high densities/dielectric constants,
     50–51
   with low dielectric constants, 49–50
   with super- and semiconducting properties,
     53
   with very high dielectric constants, 51–53
Glass fibers, commercial/experimental
   aluminate glass fibers

glass fibers from fragile melts, 60–66
glass fibers from inviscid melts, 66–77
glass fiber formation, principles of
   fiber-forming processes, generic, 9–10
   fibers from fragile/inviscid melts, 11
   fibers from strong melts/solutions,
      10–11
glass melt formation, principles of
   fragile viscous melts, behavior of, 8–9
   glass melt properties, 4–8
   inviscid glass melts, behavior of, 9
   strong viscous melts, behavior of, 8
silica fibers, sliver, and fabrics
   pure, 19–23
   ultrapure, 15–19
silicate glass fibers
   formation from strong viscous melts,
      see Silicate glass fibers from strong
      viscous melts
   general-purpose, *see* Silicate glass
      fibers, general-purpose
   non-round, bicomponent and hollow
      silicate fibers, 54–59
   special-purpose, *see* Silicate glass
      fibers, special-purpose
single-crystal alumina fibers
   alumina and aluminate fibers, future of,
      82–83
   from inviscid melts, 77–82
structure of melts and fibers
   fiber structure *vs.* modulus, 12–14
   fiber structure *vs.* strength, 14–15
   from glass melts to fibers, 11–12
   melt structure *vs.* liquidus, 12
Glass fiber strength, 14*t*, 198–203
   Griffith equation for calculation of stress,
      200
   single-fiber tensile strength for differnt
      glass fiber compositions, 201*f*
   specific strength (σ_{sp}), equation for, 202
   theoretical strength, Orowan's expression,
      199
   types, 202
Glass (industrial) making, principles of
   cullet effect, 236–237
   demands on the glass melt, 228–231
      chemical resistivity, higher, 229
      color demands, 228
      iron content, impact on color of glass,
         228
      legislative requirements, 229
      production cost, low, 228
      workability, 229

economics, 228
      melting costs, 228
      raw material costs, 228
   glass refining, 237–240
   meltability, parameters
      calculation of glass properties, factors
         for, 231*t*
      calculation of viscosity factors for, 233*t*
      decreased melting temperature, uses,
         232
      furnace design and charging system,
         232
      glass composition, 232
      particle size of raw materials, 232
      raw material selection, 232
   raw materials, choice of, 235–236
   workability, 233–235
      definition, 233
Glass mats, 138–140
Glass mat thermoplastic (GMT), 157,
      159–160
Glass melt formation, principles of
   fragile viscous melts, behavior of, 8–9
   glass melt properties, 4–8
   inviscid glass melts, behavior of, 9
   strong viscous melts, behavior of, 8
Glass melting and fiber formation, 3–15
   See also Glass fibers,
      commercial/experimental
Glass melting technology
   enthalpy functions of one-component
      systems, analysis of
      pre-melting range/molar specific heat
         capacity, 361–365
      theoretical preliminaries, 359–361
   expansion of solids and melts, cause,
      369–375
      dilatometer curve of a borosilicate
         glass, 374*f*
      Lennard–Jones potential energy of an
         atom with another atom, 370*f*
      lowering temperature, effects, 371
      thermal volume expansion, expression,
         373
   glass formation, criteria for, 375–380
   glass melt/glass product, properties of,
      414–417
      batch-related fluctuations, *see*
         Batch-related fluctuations in glass
         melting
      combustion-related fluctuations, *see*
         Combustion-related fluctuations,
         glass melting

Glass melting technology (*cont.*)
    constant chemical composition, criteria
      to achieve, 414
    oxidation state, impact on, 415*t*
    parameters, 414
    process-related fluctuations, 416–417
    related quantities, 414
  glass transformation range, effects in
    crystallization, slow down of, 368
    electron system decoupling, 369
    ESR/NMR signals, modification, 368
    relaxation effects, parameters, 369
  mathematical modeling/control of
    properties, role in, 413–414
  melting and glass transformation, 365–368
    electronic transitions, effect on,
      366–367
    nucleation and crystal growth, 368
    scheme, 367*f*
  melting criteria
    creation of intrinsic defects, Cahn's
      mechanism, 359
    Lindemann's criterion, 356–357
    melting temperatures, predictions, 357
    one-component system, example, 366*f*
  modulus of compression of chemical
    elements, 375–376
  monitoring properties using in situ sensors,
    418
    emission spectroscopy, 421–422
    LIBS, 422
    redox measurement, 418–419
    viscosity, 418
    voltammetric sensor, *see* Voltammetric
      sensor
  monitoring species in combustion space
    using in situ sensors
    combustion efficiency optimization,
      423
    environmental measurements, 422–423
  motivation, 355–356
    Kauzmann paradox, 356
    "the mysterious glass transition,"
      Langer, 356
  multi-component systems, extension to,
    381
  quality optimization, constraints
    ecological, 414
    economic, 414
    energetic, 414
  stability control, examples
    amber glass melting, redox control of,
      see Amber glass melting

    melting with high portions of recycled
      glass, 423–424
  trials, 444–447
Glass melting, thermodynamics of
  batch-to-melt conversion
    batch melting, stages of, 404–405
    heat demand of, 405–407
    phase stability diagram for simplified
      E-glass composition, 408*f*
    reaction path, modeling of,
      405–407, 409*t*
  glasses/glass melts
    chemical potentials/vapor pressures of
      individual oxides, 391–393
    entropy and viscosity, 394
    heat content of glass melts, 388–390
    industrial glass-forming systems,
      thermodynamic properties, *see*
      Thermodynamic properties,
      industrial glass-forming systems
  individual raw materials, role of
    boron carriers, 397–399
    dolomite and limestone, 400–403
    sand, 396–397
Glass-melting viscosity, 232
Glass melt properties, 4–8, 413, 417–422
Glass melt stability, 413–427
Glass properties, fundamental
  chemical durability, 297–299
  conductivity and heat transfer, 286–291
    electrical properties, 289–291
    specific heat capacity, 286
    thermal conductivity/optical properties,
      287–289
  density and thermo-mechanical properties,
    299
  devitrification and crystal growth,
    281–286
    liquidus models, 284–286
    methods of avoiding devitrification,
      282–284
    ternary SiO_2–CaO–Na_2O system, phase
      diagram for, 283*f*
  interfaces, surfaces, and gases, 291–296
    chemical durability, 297–299
    density and thermo-mechanical
      properties, 299
    refining, 291–292
    refractory corrosion, 293–296
    surface energy, 296
    See also Refining
  viscosity–temperature relationship, 280*f*
    methods of measuring, 279

viscosity models, 281
viscosity set points for SLS glass, 279*t*
Glass refining, 237–240
Glass reformulation methodologies, 330–344
    benefits and pitfalls, 341–343
    examples and implementation
        batch constrained reformulation,
            333–335, 338*t*
        component/parameter checklist for SLS
            glass, 331*t*
        compositionally constrained
            reformulation, 331–332, 334*t*
        other industrial trials and
            implementation, 339–341
        unconstrained reformulation, 337–339
    physical properties/parameters as a
        function of reformulation, 333*t*
    principles, 332
    research requirements, 343–344
Glass reinforced plastics (GRP), 143
Glass softening point, 206
Glass transition temperature (T_g), 4, 16, 23, 34,
    101–102, 165–167, 368, 374, 382,
    386, 407–408
Glass volatilization, 188, 254, 276–278, 281,
    289, 292–293, 301, 318, 326, 329,
    415*t*, 435, 447
Glass yarn, 48, 141, 176, 181, 184, 186,
    189, 220
Global warming, 273, 343
GMT, *see* Glass mat thermoplastic (GMT)
Gonterman, R.J, 431–451
Green glass melting, 424
Griffith equation, 200

H
Hand Lay-Up (HLU), 143
    GRP production, use in, 143
Hausrath, R.L., 197–222
HDI, *see* High density interconnects (HDI)
HDT, *see* Heat deflection temperature (HDT)
Heat content of glass melts, 388–390
Heat deflection temperature (HDT), 166
Heat of formation of glass, 385, 406
Heat demand of melting, 405–407
Heat transfer (HT), 434–435, 437
High density interconnects (HDI), 192, 193
High-intensity DC-arc plasmas, technology of
    conductive
        direct contact of particles of matter, 434
    Joule heating
        Joule's first law, 436
        ohmic heating, 436

Ohm's law, 436
SI unit of energy (J), 436
unit of power (Watt), 436
radiant
    in glass melting, 437
High-modulus–high-temperature glass fibers,
    39–40, 39*t*
High-modulus (HM) glass fibers, 13, 35,
    38–39, 204
Heat-resistant polymers, 166–167
High-strength–high-temperature glass fibers
    process and products, 34–38
    properties and applications, 38
High-strength (HS) glass fibers, 34–36,
    197–222
    characteristics
        compositional ranges (wt%), 198*t*
        elastic modulus, 203–205
        strength, *see* Glass fiber strength
        thermal stability, 205–206
    competitive material landscape
        carbon fibers, 212
        polymer fibers, 212–214
    continuous glass fibers, advantages
        compressive strength, 215
        flammability or oxidation resistance,
            215
        low cost, 215
        strength and modulus, 215
        thermal stability, 215
    glass compositional families
        HiPer-tex™, 209
        K-glass, 209
        properties, 201*t*
        R-glass, 208
        S-glass, 197, 206–208
        S-1 Glass™, 209
    markets and applications
        aerospace – rotors and interiors,
            218–219
        automotive – belts, hoses, and mufflers,
            220–221
        defense – hard composite armor,
            216–218
        high-strength fiber market, overview,
            216*f*
        industrial reinforcements – pressure
            vessels, 221–222
    US Patent 3,402,055 by Owens Corning,
        198
"High-temperature" polymers, definition, 166
    properties of, effect of fiberglass
        reinforcement on, 167*t*

HiPer-tex™, 209, 222
HLU, *see* Hand Lay-Up (HLU)
Hoffmann, H.J, 355–381
Hollow filament, 181, 186, 189
HS4 glass, 207–208
HT, *see* Heat transfer (HT)
Hybrid fabric constructions, 184
Hybrid fiber-forming processes, 65–66

I

Ideal mixing of complex components, model
    of, 386
ILSS, *see* Interlaminar shear strength (ILSS)
Impact strength, 138, 160, 162
IMS process, *see* Inviscid melt spinning (IMS)
    process
Industrial glasses
    compositions, examples
        container glass batch charge, examples
            of, 262–265
        perspectives, 261
    compositions of
        colored glasses, 260
        container glass, 244–249
        flat glass, 242–243
        history, 240–241
        lead crystal, 258–260
        lead-free utility glass, 250–252
        production costs, vital factor, 227
        technical glass, 252–258
        viscosity, effect on, 232
    industrial glass making, principles of
        cullet effect, 236–237
        demands on the glass melt, 228–231
        economics, 228
        glass refining, 237–240
        meltability, 231–233
        raw materials, choice of, 235–236
        workability, 233–235
        See also Glass (industrial) making,
            principles of
Industrial reinforcements – pressure vessels
    application, 221–222
    critical fitness for use properties, 222
    market trends and future needs, 222
Industry E-glass specifications, 29, 185–186
Inhibitors (or retarders), 149
Injection molding technology, 157
In situ sensors
    for glass melting, 418*f*
    for monitoring glass melt properties,
        418–422
        advantage of, 419

for monitoring species in the combustion
    space, 422–423
principles applied in sensors, measuring,
    417*t*
Institute for Printed Circuits (IPC), 192
Interconnect Technology Research Institute
    (ITRI), 191
Interlaminar shear strength (ILSS), 138–139
    short span flexural test, 138
International Organization for Standardization
    (ISO), 137–140, 297
    dynamic testing of fiberglass-reinforced
        composites, standards for, 140*t*
    mechanical testing of fiberglass-reinforced
        composites, standards for, 139*t*
International Technology Roadmap for
    Semiconductors (ITRS), 191
Inviscid glass melts
    behavior of, 9
    contents, 4
Inviscid melt spinning (IMS) process, 68,
    70–76
    mechanism of jet solidification, 73–76
        inviscid calcium aluminate jets,
            chemical stabilization/properties of,
            75–76
    metal fibers formation in a reactive
        environment, 70–72
    oxide glass fiber formation in a reactive
        environment, 72–73
IPC, *see* Institute for Printed Circuits (IPC)
ISO, *see* International Organization for
    Standardization (ISO)
ITRI, *see* Interconnect Technology Research
    Institute (ITRI)
ITRS, *see* International Technology Roadmap
    for Semiconductors (ITRS)

J

Japanese consortium project
    degree of vitrification, 440
    melting devices
        an RF plasma apparatus, 440
        oxygen burner apparatus, 440
        12-phase AC arc, 440
Johns-manville, 437–438, 440
Joule heating, 190, 434, 436–437
    Joule's first law, 436
    Ohm's law, 436
    SI unit of energy (J), 436
    unit of power (Watt), 436
Joule, James Prescott, 436
Joule's first law, 436

K

Kauzmann paradox, 356
Kevlar, 34, 212–213
  para-aramid, 212
K-glass, 207*t*, 209
"Knee," 134
"Knuckles," 141, 184
Krieger–Dougherty equation, 409

L

Lakatos models, 281
Laminate, 32, 33, 57, 132–134*f*, 139*f*, 143,
    150–152, 169, 171, 176–178,
    180–185, 189, 193–194
Laser absorption spectroscopy, 423
Laser-heated float zone (LHFZ) method, 53,
    77–82
Laser-heated pedestal growth (LHPG) process,
    70, 77–82
  high T_c superconducting fibers, 81–82
  single-crystal fibers, growth of, 79–81
Laser-induced breakdown spectroscopy
    (LIBS), 422
Lawton, E.L., 125–172
LCP, *see* Liquid Crystal Polymer (LCP)
Lead crystal, 258–260
Lead-free utility glass, 250–252
Lead glass, 51–52, 55, 232, 257, 259–260
Lewis acids/bases, 150
LFT, *see* Long Fiber Thermoplastic (LFT)
L-glass, 186*t*, 189, 193
LHFZ method, *see* Laser-heated float zone
    (LHFZ) method
LHPG process, *see* Laser-heated pedestal
    growth (LHPG) process
LIBS, *see* Laser-induced breakdown
    spectroscopy (LIBS)
Limestone, 235, 239*t*, 245–246, 251, 253–254,
    263–265, 300–301, 304, 307–308,
    334–336, 340–342, 400–404,
    406–408, 432, 444, 448
Lindemann, 357
Lindemann's criterion for glass melting, 357
Liquid crystal polymer (LCP), 167
  lyotropic/thermotropic, 167
Liquid resin processing techniques, 142–148
Liquidus models, 284–286, 338
Liquidus temperature (L_T)
  crystal formation at, 6
  in design of new fiberglass compositions,
    96
  soda-lime-silica glass, 231, 281, 286, 307,
    309, 313, 317–318

Lithia, Li_2O, 315–316
Lithium ion batteries, 99
Littleton softening point, 264, 279
"Loewenstein, private communication (1997)",
    117
Log (viscosity), *see* Glass-melting viscosity
Long Fiber Thermoplastic (LFT), 157,
    159–160, 216*f*
  direct (D-LFT), 160
  granulated (G-LFT), 160
Longobardo, A.V., 175–194
Low weight-reinforced thermoplastic (LWRT)
  mats, 160
Lyotropic/thermotropic LCPs, 167

M

Magnesia, MgO, 309–310
Mass spectrometry, 73, 423
Matrix failure, 135
Mechanical properties, composites
  bidirectional (orthotropic) reinforcement,
    133–134
  levels of study, 131
  short fibers, 134–137
  test methods
    compressive strength, 138
    flexural strength and modulus, 138
    impact strength, 138
    shear strength, 138
    tensile strength and modulus, 137–138
  unidirectional continuous fibers, 131–132
Meltability, 232–233
Melting temperature (T_m)
  mathematical definition, 357
  properties, 357
Melt spinning process, 11, 15, 34, 66,
    70–71, 76
Melt temperature and linear viscosity,
    relationship, 7*f*
Microwave circuitry
  dielectric dissipation factor of PCB, criteria
    for design of, 181
Microwave/ neutron absorption techniques,
    416
Milled fibers, 33, 140
Mod ratio, 54–55
  nylon ribbons/fibers, effect on, 55
Modular/skull melting, 431–433
  glass melting, 431
  skull melting, 431–432
    disadvantage, 432
Modulus of elasticity *(E)*, 126, 131, 198
Molybdenum, 41, 205, 437, 439, 441, 447

Müller-Simon, H., 292, 413–427
Mullite composition glass fibers, 70
Multi-axial fabrics, 141
Multi-component systems, 381, 386

N

Natural gas (NG), 26, 38, 221, 275, 416
NBO, *see* Non-bridging oxygens (NBO)
NEDO, *see* New Energy and Industry
        Technology Development
        Organization (NEDO)
NE-glass, 186t, 189–191, 193
Nernstian equation, 419
Network forming (NWF) oxides, 305
Network modifier (NWM) cations, 305
New Energy and Industry Technology
        Development Organization
        (NEDO), 439
Newtonian fluid, 75, 128, 281
Nomex®, 212
        meta-aramid, 212
Non-bridging oxygens (NBO), 305, 307
Non-round, bicomponent and hollow silicate
        fibers
    bicomponent silicate glass fibers
        hollow porous sheath/core, 58
        hollow sheath/core, 56–57
        sheath/core and side-by-side, 56
        solid side-by-side, 58–59
    fabrication of, 55f
    glass fibers with non-round cross sections
        processes and structures, 54–55
        products and applications, 55–56
Non-transferred-arc plasmas, 434f
Non-woven fabrics, 141
    bidirectional, 141
    crimp effect, 141
    multi-axial, 141
    unidirectional, 141
Noyes–Nernst equation, 294
Nucleation and crystal growth, 368, 379

O

Offline flue gas analysis, 422
Offline/online flue gas analysis, 422
Ohmic heating, *see* Joule heating
Ohm's law, 436
Olefin copolymers, cyclic, 184
One component systems, 356, 358–365, 369,
        375, 378–382
Online flue gas analysis, 422
    mass spectrometry used in, 423
Open-pore stage, 405
"Oxidation state," 414

Oxide (individual), reactivity of
    evaporation reactions from glass melts,
        392t
    factors, 391
    Gibbs energies G in kJ/mol of oxides in
        equilibrium, 392t
    molar mass factors for calculation of
        equilibrium constants, 393t
Oxides of nitrogen (NO_x), 273–275
Oxides of sulphur, SO_x, 275–276
Oxygen and concentration sensor, combined,
        417–418, 420f
    use in industrial glass melts, 419
Oxygen partial pressure, 414–419, 423–427
    Nernstian equation, emf calculation by, 419
Oxynitride glasses, 41, 43–45, 203t, 205

P

PA, *see* Polyamide (PA)
Particulates, 276–278
PBT, *see* Poly(butylene terephthalate) (PBT)
PC, *see* Polycarbonate (PC)
PCB, *see* Printed circuit boards (PCB)
PCB glass fibers, 93, 120
PCB, glass fibers for
    electrical aspects
        dielectric constant, 179–180
        dielectric loss, 180–181
        hollow filaments, 181
        polarization of a dielectric, 180f
        velocity equation for a PCB laminate,
            180
    fiberglass' role in PCB construction,
        177–179
    fiberglass/resin properties used in FR4
        laminate board, 178t
    PCB, design criteria, 177–178
    glass compositional families
        D-glass, compositional improvements,
            188–191
        E-glass, improvements, 184–188
    PCB market, future needs of
        board and yarn makers, impact on,
            192–194
        electronics manufacturer's roadmap,
            191–192
        environmental regulations, 194
        fiberglass use, importance, 194
        HDI manufacture, design criteria,
            192–193
        lead-free processing, 194
        operating frequency range as a function
            of board layers, 192f
    requirements and implications, 176–177

fiberglass yarn/fabric/laminate boards,
  process of manufacture, 176*f*
structural aspects
  elastic modulus, 182
  mechanical strength, 182
  thermal expansion, 182–183
  upper use temperature, 183–184
  weave and fabric construction, 184
Permittivity, 179–181
PESU, *see* Polyethersulfone (PESU)
PET, *see* Poly(ethylene terephthalate) (PET)
PF resins, *see* Phenolic (PF) resins
3-phase electric melters, typical, 437
Phenolic (PF) resins, 152, 217
  characteristics of, 152
  high flammability resistance/low smoke
    emission, 152
  open mold applications, 152
Phonon system, 369
Plasma melter, 433, 439, 441, 443–445,
  447–448, 451
  See also Modular/skull melting
Plasma melting technology/applications
  DOE research project (2003–2006),
    440–450
    acknowledgments, 440
    energy efficiency *vs.* throughput,
      448–450
    glasses melted: results/implications,
      444–447
    plasma glass melting, technical
      challenges of, 442–443
    plasmelt melting system, experimental
      setup of, 440–442
    synthetic minerals processing
      implications, 447–448
  high-intensity DC-Arc plasmas, technology
    of, 433–437
    conductive, 434–435
    Joule heating, 436–437
    radiant, 435–436
  history of
    British glass institute, 438
    Japanese consortium project, 439
    Johns-manville, 437
    Plasmelt glass technologies, LLC,
      438–439
  industrial applications, best-fit, 450
  modular/skull melting, concepts of,
    431–433
  plasma-melted glasses, chemical analyses
    of, 446*t*
Plasma melt process, experimental, 26–27

application, 27
skull-melting concept, benefits, 27
vs. conventional glass furnace technology,
  26
Plasmelt coupled transferred-arc melter, 441*f*
Plasmelt Glass Technologies, 27
  add-on refiner stage, 439
  United States Department of Energy
    (DOE), 438
Plasmelt-melted E-glasses *vs.* standard E-glass,
  chemistry of, 444*t*
Plasmelt plasma melting system, 441
  anode torch, 441
  molybdenum orifice, 441
  torch position, 441
Poisson's ratio, 127*t*, 182, 375
Pollution prevention/control, eco-friendly
    glasses
  carbon dioxide
    sources of emission, 273
    furnace design, 271–272
    furnace designs, BAT, 272*t*
  oxides of nitrogen (NO_x))
    generation as a function of furnace
      temperature, 274*f*
    generation mechanisms, 273–274
    global warming, cause, 273
    NO_x control, primary/secondary
      techniques, 274–275
  oxides of sulfur (SO_x)
    sources of emissions, 275
    SO_x control/reduction, techniques, 275
  volatilization and particulates
    dust emissions from glass furnaces,
      comparison of, 276*t*
    furnace pull rate/temperature on
      particulate emissions, effects of, 278
    glass composition/temperature/
      pressure, effects on, 276
    particulate formation/emission control,
      BAT, 277
    particulates, sources, 277
    volatilized mass loss as a function of
      temperature, 277*t*
Polyacetal (POM), 165
Polyamide (PA), 140, 157, 159–161, 164,
  166–167, 212
Polybenzobisoxazole (PBO) fiber, 211*t*, 214
  high-performance fiber, 214
  zylon, example, 214
Poly(butylene terephthalate) (PBT),
  160–161, 164
Polycaprolactam (PA 6 or Nylon 6), 161, 164

Polycarbonate (PC), 161, 166
Polycondensation reaction of ultrapure silica
        fibers, 18
Polycrystalline fibers
    fiber FP, 13
    nextel 440, 13
    nextel 480, 13
    safimax, 13
Polyester resin, 141, 143, 146–151, 154, 169
    characteristics of, 150*t*
    properties of polyester laminates, 150*t*
Polyethersulfone (PESU), 166
Poly(ethylene terephthalate) (PET), 145,
        155, 164
Poly(hexamethylene adipamide) (PA 66 or
        Nylon 66), 161, 164
Polymer fibers, 54, 198, 210, 212–215
Poly(phenylene oxide) (PPO), 166
Polypropylene (PP), 140, 159–162, 164–165
Polysulfone (PSU), 53, 166
Polyurethanes (PUR), 129, 144, 146, 148,
        151, 153
    applications, 153
    characteristics of, 153
POM, *see* Polyacetal (POM)
Potassia, K$_2$O, 313–315
Power, 436
PP, *see* Polypropylene (PP)
PPO, *see* Poly(phenylene oxide) (PPO)
Pre-impregnated fabrics (Prepregs), 146, 152,
        170, 177
    components/function of laminate prepreg,
        178*f*
Pressure vessels, 38, 144, 208, 216*f*, 221–222
Printed circuit boards (PCB), 29*t*, 33, 38, 49,
        93*t*, 175–194
Printed wiring boards (PWBs), *see* Printed
        circuit boards (PCB)
Products, glass fiber
    chopped strands, 140
    fabrics woven from rovings, 141
    glass mats, 138–140
        chopped strand mat, 138–139
        continuous strand mat, 138
    glass yarn, 141
    milled fibers, 140
    non-woven fabrics, 141
    rovings, 140
PSU, *see* Polysulfone (PSU)
Pultrusion, 138, 140, 142, 145–146, 149*t*, 222
PUR, *see* Polyurethanes (PUR)
Pure silica sliver and fabrics
    acid-leached E- and A-glass fabrics

        process, 21
        products and properties, 21–22
        value-in-use and applications, 22
    from aqueous silicate solutions
        process, 20, 20*f*
        products and properties, 20–21
        value-in-use and applications, 21

Q
Quartz fibers, 16, 17, 22, 48
Quasi-chemical model, 386
Quaternary SiO$_2$-Al$_2$O$_3$-CaO system
    eco-/energy-friendly E-glass compositions,
        110*f*, 110*t*
    from eutectic to commercial compositions,
        102–103
        first-generation fluorine- and B$_2$O$_3$-free
            E-glass, 102*t*, 104*f*
        fluorine- and boron-free E-glass with <1%
            Li$_2$O, 106–108
        energy-friendly E-glass compositions
            with 0.9% Li$_2$O, 108*t*
        Li$_2$O as replacement for Na$_2$O, effects,
            107*f*
        fluorine- and boron-free E-glass with
            ≤1.5% TiO$_2$, 104–106
        design of, 106*t*
        fluorine-free E-glass with 1.5% B$_2$O$_3$,
            108–109
        energy-friendly E-glass compositions
            with ≤1.3% B$_2$O$_3$, 109*t*
        fluorine-free E-glass with ≤1.5% B$_2$O$_3$ and
            <1% Li$_2$O, 109
    phase diagram of, 102–111
    quaternary eutectic with regard to MgO,
        103–104

R
Rapid jet solidification (RJS) process, 68
    amorphous fiberglass ribbons, 77
    amorphous metal ribbons, 76–77
    products and applications, 77
Rare earth oxides, 204
Raw materials, role in glass melts
    boron carriers, 395–397, 399*t*
        binary system CaO–B$_2$O$_3$, phase
            diagram of, 398
        liquidus lines of binary systems, phase
            diagrams of, 398*f*
        phase relations of Na–Ca–B–O–H
            minerals, 400*f*
    dolomite and limestone, 400–403

$CaCO_3$_ $CaMg(CO_3)_2$_ $CaFe(CO_3)_2$
system, one/two/three-phase
equilibria of, 401*f*
standard heat of formation H^f of
dolomite from elements, 401*t*
two-step decomposition behavior of
dolomite, 402*f*
sand
impurity levels (wt%) in selected sand
qualities, Europe/Asia, 396*t*
kinetics of sand dissolution,
394–395, 397*f*
Rayleigh waves, 66–68, 72, 74
REACH, *see* Registration, evaluation,
and authorization of chemicals
(REACH)
"Redox state," 238, 405, 414–415, 423,
425–426, 445
Refining
alternative methods of, 292
alternative refining agents, 318–320,
324–326, 329–330
behaviour, 237–240, 318–320
removal of bubbles, mechanisms, 291
Stokes' law, buoyancy effects by, 291
Refining, alternative methods
alternative refining agents, 292
alternative refining gases, 292
physical refining methods, 292
Reformulated glass compositions, 98–101,
104–118, 262–265, 330–341
Refractory corrosion, 293–296
Registration, evaluation, and authorization of
chemicals (REACH), 399
Reinforced composites, 40, 44, 114, 119, 125,
131, 133, 140*t*, 142, 154, 164, 210,
213, 219–221
Reinforced plastics, 125, 139–140
Reinforced Reaction Injection Molding
(RRIM), 144
Reinforced Thermoplastic Compounds (RTP),
158–159
fiberglass-reinforced thermoplastic
compounding, extruder for, 158*f*
filament diameter/fiber length, relativity,
159*f*
Reinforced thermoplastic materials
injection molding technology, 157
semifinished materials based on
thermoplastics, 158–167
amorphous resins, 165–166
GMT, 159–160
heat-resistant polymers (HT), 166–167

LCP, 167
LFT, 159–160
mechanical properties, 160–163
RTP, 158–159
semicrystalline resins, 164–165
Reinforcement, bidirectional (orthotropic),
133–134
bidirectional reinforced laminate, 133*f*
induced unidirectional strain in, 134*f*
"knee," 134
Reinforcing fibers, 31–32, 34, 43, 56, 73, 78,
82, 125–126, 131, 142, 144, 148,
159, 164
properties, 126
Relative machine speed (RMS), 279
of Russian SLS container glass, 280*t*
Relative permittivity, *see* Dielectric
constant (Dk)
Release agents, 155–156
Remotely coupled transferred arc, 437,
440–441
Renewable energy, source
wind, 168
Resins
definition, 126
thermosetting
EP resins, 150–151
PF resins, 152
PUR, 153
reinforcement with glass fibers,
property trends for, 154*t*
SI resins, 153–154
techniques, initiators/accelerators used
in, 149*t*
UP resins, 148–149
VE resins, 151
See also individual resins
Resin transfer molding (RTM), 142–144,
149*t*, 170
Resistive heating, *see* Joule heating
Resorcinol formaldehyde latex (RFL), 220
Restriction of hazardous substances (RoHS),
194
impact on PCB fabrication, 194
RFL, *see* Resorcinol formaldehyde latex (RFL)
R-glass, 35, 37, 38, 198–201, 203, 205–209,
222, 445
produced by Vetrotex, 208–209
vs. S-glass, 207–208
RJS process, *see* Rapid jet solidification (RJS)
process
RMS, *see* Relative machine speed (RMS)

RoHS, *see* Restriction of hazardous substances
    (RoHS)
Rovings
    assembled rovings, 140
    direct draw rovings, 140
3R process, 275
    See also Oxides of nitrogen (NO$_x$)
RRIM, *see* Reinforced Reaction Injection
    Molding (RRIM)
RTM, *see* Resin Transfer Molding (RTM)
RTP, *see* Reinforced Thermoplastic
    Compounds (RTP)
Rule of mixtures, 131–132, 182–183

S
Saphikon, 13
Sapphire fibers, growth of, 78
Schaeffer, H.A., 413–427
SCR, *see* Selective catalytic reduction (SCR)
SEC, *see* Specific Energy Consumption (SEC)
Secondary electron image (SEI), 58–59
Seebeck coefficient, 419
SEI, *see* Secondary electron image (SEI)
Selective catalytic reduction (SCR), 275
    See also Oxides of nitrogen (NO$_x$))
Selective non-catalytic reduction (SNCR), 275
    See also Oxides of nitrogen (NO$_x$))
Self-reinforcing polymers, *see* Liquid Crystal
    Polymer (LCP)
Semiconductor industry association (SIA), 191
Semicrystalline resins
    PA, 164
    PET/PBT, 164
    POM, 165
    PP, 164–165
    properties of, 165*t*
Sensors
    definition, 414
    for environmental measurements, 422–423
    high-temperature heat resistance, feature
        of, 414
    for optimizing combustion efficiency, 423
        control of air/fuel or oxygen/fuel ratio,
            423
        flame visualization, techniques, 423
        laser absorption spectroscopy, 423
        mass spectrometry, 423
    See also individual sensors
S-glass, 7, 8*f*, 13, 16, 23–25, 27, 31, 35–37,
        42, 48–50, 57, 60, 64*t*, 92, 101, 198,
        207–209, 212–215, 217, 445
S-1 glass, 207–209, 222
S-2 glass®, 198, 200, 202–208, 212–215

Shear strength, 136, 138–139
Shear stress, 131, 137
Sheath/core *vs.* side-by-side bicomponent
    fibers, 56
Sheet Molding Compounds (SMC), 141–142,
        147–149, 152–153, 155–156, 169
    SMC process, flow of materials in, 147*f*
Sheridanite, 310
Short fibers, random, 134–137
    critical length, equation, 136
    fiber/critical length, relativity, 137
    reinforcing efficiency *vs.* fiber length, 137*f*
    response to strain, 135*f*
    stages of fracture, 135
Short span flexural test, 138
SIA, *see* Semiconductor industry association
    (SIA)
Sialons, 43
Silanes, 130, 155
Silfa yarn, 20
Silica, SiO$_2$, 304–305
Silica fibers, ultrapure
    from sol–gels
        process, 18
        products and properties, 18–19
        value-in-use and applications, 19
    from strong viscous melts
        process, 15–16
        products and properties, 16–17
        value-in-use and applications, 16–17
Silica, sliver and fabrics, pure
    acid-leached E- and A-glass fabrics
        process, 21
        products and properties, 21–22
        value-in-use and applications, 22
    from aqueous silicate solutions
        process, 20
        products and properties, 20–21
        value-in-use and applications, 21
Silicate glass fibers, 8, 11, 16, 23–59, 64–65
Silicate glass fibers from strong viscous melts
    commercial melt process, 23–26
        boron-free CC-glass, use in, 24
        glass fibers formed by, 26*t*
        winders/direct-drawing
            winders/choppers, formation, 25*f*
    experimental plasma melt process, 26–27
        application, 27
        skull-melting concept, benefits, 27
        vs. conventional glass furnace
            technology, 26
    glass fiber drawing, modeling of, 28
    strong viscous melts, critical properties, 23

Silicate glass fibers, general-purpose
    borosilicate E-glass fibers, 28–31
    E-glass products and applications, 33–34
    E-glass properties and fiber structures,
        31–33
Silicate glass fibers, special-purpose
    designations of, 34
    glass fibers with bone bioactive oxide
        compositions, 53–54
    glass fibers with high chemical stability,
        44–48
    glass fibers with high densities and
        dielectric constants, 50–51
    glass fibers with low dielectric constants,
        49–50
    glass fibers with super- and semiconducting
        properties, 53
    glass fibers with very high dielectric
        constants, 51–53
    high-modulus–high-temperature glass
        fibers, 39–40
    high-strength–high-temperature glass
        fibers, 34–38
    ultrahigh-modulus glass ceramic fibers,
        40–44
Silicone (SI) resins, 153–154
    characteristics of, 153
Single-crystal fibers, 70, 77–83
Skew phenomena, 184
Skull melting, 27, 431–433
SLS glass, see Soda-lime-silica (SLS) glass
SMC, see Sheet Molding Compounds (SMC)
Smrček, A, 259
SNCR, see Selective non-catalytic reduction
    (SNCR)
Soda, Na_2O, 305–307
Soda-lime-silica (SLS) glass, 12, 15, 116–119,
        229–351
    A- and C-glass compositions, 118t
        fluorine and boron-free A-glass, 117
        fluorine-free c-glass with 5% B_2O_3,
            117–119
        limitation, 119
    thermal conductivity of, 288f
    viscosity-temperature (η-T) curve for, 280f
Soda-lime-silica (SLS) glass,
        composition/design of
    alumina, Al_2O_3, 309–313
        compositions of aluminous raw
            materials, 311t
        effect of BFS on energy and fuel
            consumption, 312f

        effect on liquidus temperature of SLS
            glass, 313f
        effects on chemical durability of SLS
            glass, 314f
        effects on glass properties, 311–313
        raw materials, 309–311
    baria, BaO, 323
    batch processing, preheating, and melting,
        300–302
        batch consolidation, forms, 301
        SEC based on cullet content, 303f
        stages of melting, 300
    boric oxide, B_2O_{33} effects on glass
        properties, 315–316
        raw materials, 316
    calcia, CaO
        effects on glass properties, 308–309
        raw materials, 307–308
    chlorides and fluorides, 322–323
    cullet, 302–304
    economics of batch selection, 300f
    energy-saving technologies, 270
    lithia, Li_2O
        effects on glass properties, 315–316
        raw materials, compositions
            of, 313, 315t
    magnesia, MgO
        effect on liquidus temperature of SLS
            glass, 308f
        effects on glass properties, 309f
        raw materials, 309
    multivalent constituents
        colorants and refining agents, 324–326
        effects on physical properties, 326–327
    nitrates, 329–330
    potassia, K_2O
        effects on glass properties, 314–315
        raw materials, 313
    recycled filter dust, 329
    silica, SiO_{22} effects on glass properties,
        303
        raw materials, 304–305
    soda, Na_2O
        effects on glass properties, 308–309
        molar enthalpy of decomposition at 296
            K, 306f
        raw materials, 305–307
    strontia, SrO, 324
    sulfate, SO_3, 318–321
    water, H_2O, 321–322
    zinc oxide, ZnO, 323–324
Soda lime silicate glass, reaction path of, 409t
Solid electrolyte sensors, zirconia-based, 418

Solid/liquid, difference between
   by Born, 357
   Deborah number, concept of, 358
   See also Glass melting technology, melting
      criteria
Special-purpose glass fibers, 49–54
Specific Energy Consumption (SEC), 269–271,
      289, 303*f*, 332
   energy efficiency, 269–271
   energy-saving technologies in SLS glass
      furnaces, 268–269, 271*f*
   for SLS glass furnaces, average, 270*f*
Specific heat capacity, 286–287, 356, 358,
      361–366, 368, 373
Specific modulus (E_{sp}), equation, 204
Spray deposition, 140, 143
   liquid resin processing technique, 143
Stages of fracture, thermoplastics
   cracking, 135
   de-bonding, 135
   fiber pullout, 135
   fiber rupture, 135
   matrix failure, 135
Standard (glass) transformation temperature,
      T_g, 368, 374*f*
Stefan–Boltzmann constant (T), 287
Stokes' law, 291
   buoyancy effects described by, 291
Strain (ε), 131
   magnification, 134
Stress, kinds of, 131
Strong melts *vs.* fragile melts, 4
Strong viscous melts, behavior of
   in continuous commercial process, 7
   critical properties of, 23
   in stationary process, 7
Strontia, SrO, 324
Structure of melts and fibers, 11–15
   fiber structure *vs.* modulus, 12–14
   fiber structure *vs.* strength, 14–15
   from glass melts to fibers, 11–12
   melt structure *vs.* liquidus, 12
Styrene–butadiene rubbers, 147
Sulfate, SO₃, 318–320
Superconducting fibers, high T$_c$, 81–82
Synthetic fibers, 126, 212
Synthetic minerals, 447–448
Synthetic minerals processing, 448

T

Technical glass, 229, 231, 233, 236, 242,
      253–258
Techniques, liquid resin processing

centrifugal molding, 144–145
continuous laminating, 145–146
filament winding, 144
HLU, 143
non-continuous liquid resin processing
      techniques, 145*t*
pre-combined materials
   BMC, 148
   pre-impregnated fabrics (Prepregs), 146
   SMC, 147–148
pultrusion, 145
RRIM, 144
RTM, 143–144
spray deposition, 143
Technora®, 212
Tellurite glass fibers, 62
Temperature, upper use, 183–184
Tensile modulus, 17*t*, 127*t*, 131, 138, 150*t*,
      152*t*, 161, 165*t*, 167*t*, 213–214
Tensile stress, 131, 198–200
Ternary SiO₂-Al₂O₃-CaO system
   high-temperature applications, use in, 101
   phase diagram of, 97*f*, 100–102
   ternary compositions around eutectic,
      98*t*, 100*t*
   topography of, 97*f*
Tetraethylorthosilicate (TEOS) sol–gels, 15, 18
T-glass, 207–208, 445
Thermal conductivity, 17*t*, 57, 82, 127*t*, 181,
      202, 287–289, 299, 333*t*, 405
Thermal expansion, 16, 32*t*, 127*t*, 150*t*,
      178–183, 213, 231, 255, 257, 299,
      305, 307–308, 313–314, 316–317,
      323–324, 327, 359–360, 368–373,
      386, 401
   CTE, 182
   and resin content, relativity, 183*f*
Thermal expansion coefficient, 299, 308, 316,
      327, 359, 360*f*, 371, 373, 386, 401
   Appen factors for calculation of, 386
Thermocouples, role/function, 417
Thermodynamic properties, industrial
      glass-forming systems
   multi-component systems, models
      the cell model, 386
      Gibbs phase rule, oxide components,
         385, 387*t*
      model of ideal mixing of complex
         components, 386
      quasi-chemical model, 386
      thermodynamic equations, 387–388
Thermoplastics resins
   stages of fracture

cracking, 135
de-bonding, 135
fiber pullout, 135
fiber rupture, 135
matrix failure, 135
Thermosetting matrix resins, 148–153
Thermosetting *vs.* thermoplastic resins, 141
T_{Liq} models, 285
Torch life/stability, 442–443
Transferred-arc plasmas, 435*f*
Traveling solvent zone melting (TSZM), 82
Trend line design, 11, 94–99, 100–101,
  103–104, 107–109, 111–114,
  117–120
Trilobal glass fibers, 55*f*
TSZM, *see* Traveling solvent zone melting
  (TSZM)
Twaron®, 212

U
U-glass, 207–208
UHMWPE, *see* Ultra high molecular weight
  polyethylene (UHMWPE)
Ultrahigh-modulus (UHM) glass ceramic
  fibers
  examples
    Ca–Mg–Si–Al–O–N fiber, 42
    oxynitride fibers, 43*f*
    Si–Al–O–N glass fibers, 42*t*
    Y–Si–Al–O–N fiber, 42
  process and products
    oxygen formation, 41
    silicon formation, 41
    silicon oxidation, 41
  properties and applications, 43–44
Ultra high molecular weight polyethylene
  (UHMWPE)
  drawbacks
    low upper use temperature, 213–214
    poor compressive strength, 213–214
  Spectra and Dyneema, UHMWPE fibers,
    214
Ultrapure silica fibers
  from sol–gels
    process, 18
    products and properties, 18–19
    value-in-use and applications, 19
  from strong viscous melts
    process, 16–17
    products and properties, 16–17
    value-in-use and applications, 16–17
Unidirectional continuous fibers
  rule of mixtures, 131–132

strain effect in, 132*f*
stress–strain diagram, 133*f*
Unsaturated Polyester (UP) resins, 143–149,
  151, 154, 169
  cross-linking reaction in, 148
    cobalt complexes, accelerators, 149
    organic peroxides, initiators, 148
UP resins, *see* Unsaturated Polyester (UP)
  resins
US Department of Energy (DOE), 438, 440

V
Vacuum infusion resin transfer molding
  (VARTM), 170–171
Van der Woude, J.H.A., 125–172
VARTM, *see* Vacuum infusion resin transfer
  molding (VARTM)
VE resins, *see* Vinyl ester (VE) resins
VFT equation, *see* Vogel-Fulcher-Tammann
  (VFT) equation
Vinyl ester (VE) resins, 144–146, 151
  choice of hardener, criteria for adjusting
    characteristics of composites, 152*t*
  composite systems, characteristics of, 151
Viscosity models, 281
Viscosity of Newtonian fluids
  VFT equation for, 281
Viscosity–temperature (η–T) curve
  for SLS glass, 280*f*
Viscous melts
  contents, 3
  fragile/strong, 8–9
Viscosity models, 233, 281
Vogel-Fulcher-Tammann (VFT) equation, 281
Volatilization, 276
Voltammetric sensor, 419–421
  current/potential curve of green container
    glass, 420*f*
  polyvalent element concentration/peak
    current, proportionality, 419, 421*f*
  polyvalent elements, detection of, 420
  in situ/wet chemical analysis, sulfur sensor
    data by, 421*f*
  square-wave voltammetry
    oxygen partial pressure measurements,
    420
Volume void filling, 405

W
Water, H_2O, 321–322
Wallenberger, F.T., 10, 16, 65, 71, 74, 96–97,
  104–105, 110, 114, 118, 125,
  205, 338

Waste Electrical and Electronic Equipment
    (WEEE), 194
Waste gas treatment plants
    control of SO_x emissions, use in, 275
    made of EPs, 275
Weave and fabric construction, 184
    FWE, 184
WEEE, *see* Waste Electrical and Electronic
    Equipment (WEEE)
Weinstein, M.A., 431–451
Wind turbines, composites for
    blade design methodologies, 170–172
        ASTM D3479/D3039, test standard
            used, 171
        blade design, example of, 170*f*
        fatigue mechanism, 171
        fatigue test data on epoxy matrix/glass
            fabric specimen, 172*f*
        log–log model, 171
        S–N regression parameter estimates,
            172*t*
    blade-manufacturing techniques, 169–170
        blade components, design, 170
        RTM, 170
        VARTM, 170

    composite technology, advantages/benefits,
        168–169
    raw materials, 169
    wind energy park, 168*f*
    wind, renewable energy source, 168
Wollastonite melting, 304, 448
Working range index (WRI), 280, 284
WRI, *see* Working range index (WRI)

Y
YAG glasses and glass fibers, 69–70
Yarn, 16, 17*t*, 20, 21, 24, 33, 44, 48, 54–55, 57,
    98, 136, 141, 150*t*, 176, 181, 184,
    186, 189, 192–194, 198, 210, 220
Young's modulus, 182, 197, 199–203, 205,
    208*t*, 299, 375
    See also Elastic modulus
Yttria-stabilized zirconia, 419
    oxygen ion conductor, 419

Z
Zinc Oxide, ZnO, 323 324
Zybek Advanced Products, 448
Zylon, 211*t*, 214–215

LaVergne, TN USA
17 December 2009
167230LV00003B/101/P